Lecture Notes in Networks and Systems

Volume 641

The series "Lecture Notes in Networks and Systems" publishes the latest developments in Networks and Systems—quickly, informally and with high quality. Original research reported in proceedings and post-proceedings represents the core of LNNS.

Volumes published in LNNS embrace all aspects and subfields of, as well as new challenges in, Networks and Systems.

The series contains proceedings and edited volumes in systems and networks, spanning the areas of Cyber-Physical Systems, Autonomous Systems, Sensor Networks, Control Systems, Energy Systems, Automotive Systems, Biological Systems, Vehicular Networking and Connected Vehicles, Aerospace Systems, Automation, Manufacturing, Smart Grids, Nonlinear Systems, Power Systems, Robotics, Social Systems, Economic Systems and other. Of particular value to both the contributors and the readership are the short publication timeframe and the world-wide distribution and exposure which enable both a wide and rapid dissemination of research output.

The series covers the theory, applications, and perspectives on the state of the art and future developments relevant to systems and networks, decision making, control, complex processes and related areas, as embedded in the fields of interdisciplinary and applied sciences, engineering, computer science, physics, economics, social, and life sciences, as well as the paradigms and methodologies behind them.

Indexed by SCOPUS, INSPEC, WTI Frankfurt eG, zbMATH, SCImago.

All books published in the series are submitted for consideration in Web of Science.

For proposals from Asia please contact Aninda Bose (aninda.bose@springer.com).

Hari Om Bansal · Pawan K. Ajmera ·
Sandeep Joshi · Ramesh C. Bansal ·
Chandra Shekhar
Editors

Next Generation Systems and Networks

Proceedings of BITS EEE CON 2022

Springer

Editors
Hari Om Bansal
Department of Electrical and Electronics Engineering
Birla Institute of Technology and Science (BITS)
Pilani, India

Pawan K. Ajmera
Department of Electrical and Electronics Engineering
Birla Institute of Technology and Science (BITS)
Pilani, India

Sandeep Joshi
Department of Electrical and Electronics Engineering
Birla Institute of Technology and Science (BITS)
Pilani, India

Ramesh C. Bansal
Department of Electrical Engineering
University of Sharjah
Sharjah, United Arab Emirates

Chandra Shekhar
Chancellor, Academy of Scientific and Innovative Research (AcSIR)
Ghaziabad, India

ISSN 2367-3370 ISSN 2367-3389 (electronic)
Lecture Notes in Networks and Systems
ISBN 978-981-99-0482-2 ISBN 978-981-99-0483-9 (eBook)
https://doi.org/10.1007/978-981-99-0483-9

This Springer imprint is published by the registered company Springer Nature Singapore Pte Ltd.
The registered company address is: 152 Beach Road, #21-01/04 Gateway East, Singapore 189721, Singapore

Preface

With a greater push globally toward energy-efficient communication networks and the incentives given toward the design of green networks, the research solutions toward optimizing the design of networks and systems are important. The concept of smart networks which encompasses various intelligent networks which include wireless sensor networks smart grids, and autonomous vehicles, to name a few, have led to various data efficient algorithm designs.

With the advent of 5G and beyond networks, the proliferation and dependence of technology in day-to-day human activities have increased manifolds. The exponential growth in the number of user equipment from smartphones to miniature processing devices capable of providing multimedia services to the users along with a promise of high battery life has motivated the research community to find innovative and efficient solutions toward system design. Another aspect of the efficient design is to have power-efficient data and signal processing techniques which in turn also help in optimizing energy consumption.

The use of artificial intelligence to take the processing from the devices to the central servers along with the cloud-based architecture design has provided many solutions for optimizing the power consumption and made user devices intelligent. Further, the integration of various renewable energy sources is required to address the power demand regularly. This integration will need advanced power electronics converters and energy storage devices. To minimize the carbon footprint and ecological imbalance lot of work is going on in the area of green vehicles (XEVs) and on their charging schemes. To tap the power stored in these EVs the new concept of V2G is also being explored.

This book is a compilation of full-length papers presented in the International Conference on Next Generation System and Networks (BITS EEE CON 2022) held during 4–5 November 2022, at BITS Pilani, Pilani Campus. The aim of the conference was to discuss the research and development works being carried out in research organizations, academic institutes and industries and to chart out a road map for reliable and efficient design solutions for the next generation networks and systems. This conference provided a platform for scientists, engineers and academicians to disseminate their knowledge, learn from other's experience and networking.

The topics covered include Energy, Power & Control, Communication, Signal Processing, Electronics, Nanotechnology, etc.

The conference had 4 keynote speakers, delivering talks on smart grid applications, AI-based systems development for computer vision applications, Internet of Things (IoT) & 5G and modern DRAMs. There were 42 papers presented and deliberated upon in eight technical sessions in the conference. The papers included in this book have been peer-reviewed by minimum of two domain experts to ensure the quality and correctness of the technical contents.

Based on the domains covered in the conference, it is expected that this book will serve as a ready reference to anyone who wishes to get updated on the reliable and efficient design solutions for the next generation networks and systems. The conference was sponsored by Micron Technology, CISCO System and Scientific-Delhi. Springer is publication partner and associated in bringing out this book.

The editors thank the BITS Pilani leadership & staff, members of the advisory committee & organizing committee, speakers, reviewers, sponsors, and Springer authorities for their support and cooperation in organizing the conference and bringing out this book.

Pilani, India — Hari Om Bansal
Pilani, India — Pawan K. Ajmera
Pilani, India — Sandeep Joshi
Sharjah, United Arab Emirates — Ramesh C. Bansal
Ghaziabad, India — Chandra Shekhar

Contents

About the Editors

Hari Om Bansal obtained his bachelor's degree in Electrical Engineering from University of Rajasthan in 1998. He received his post-graduate degree from MREC, Jaipur, in Power Systems and Ph.D. from BITS, Pilani, in Electrical Engineering in 2000 and 2005, respectively. Prof. Bansal has been with BITS, Pilani, Pilani campus, since June 2001, and presently, he is serving as a Professor in the Electrical and Electronics Engineering Department. He has contributed to teaching, research, and institution building for over 20 years. He has supervised 5 doctoral theses and worked on 2 government organization-funded projects. He has published over 80 papers in journals and conferences. His fields of interest include power quality, integration of renewable energy sources, hybrid electric vehicles, battery management systems, and application of artificial intelligence techniques in these areas. He is Fellow of the Institution of Engineers (IE), India, Senior Member of IEEE, and Life Member of Indian Society for Technical Education (ISTE).

Pawan K. Ajmera received the B.E. degree in Industrial Electronics Engineering from Dr. B. A. Marathwada University, Aurangabad, India, in 2001 and M. E. degree in Instrumentation Engineering from S. R. T. Marathwada University, Nanded, India, in 2005. He received the Ph.D. degree in Electronics Engineering from SGGS Institute of Engineering and Technology, Nanded, India, in 2012. Since August 2015, he has been with the Department of Electrical and Electronics Engineering, BITS, Pilani, India, where he is currently Assistant Professor and involved in research in the area of signal processing and multimodal biometrics. He is Member of IEEE and has a mix of academic and research experience. The areas of his research interest are digital signal processing, image processing, and unimodal and multimodal biometrics.

Sandeep Joshi received the B.Tech. degree (Hons.) in electronics and communication engineering from Uttar Pradesh Technical University, India, in 2006, the M.E. degree in communication engineering from Birla Institute of Technology and Science (BITS), Pilani, India, in 2009, and the Ph.D. degree in electrical engineering from Indian Institute of Technology (IIT) Delhi in 2019. Since August 2020, he has been with the Department of Electrical and Electronics Engineering, BITS, Pilani, where he is currently working as an Assistant Professor. Prior to that, he worked as Chief Engineer in the Mobile Communications research group at the Samsung R&D Institute Bangalore, where he was involved in research in the area of beyond 5G/6G communication systems. He is Fellow of the Institution of Electronics and Telecommunication Engineers (IETE), India, is Senior Member of IEEE, Life Member of the Indian Society for Technical Education (ISTE), and Life Member of the Institution of Engineers (IE), India. He is Co-founder of Agrix, an AgriTech startup.

Ramesh C. Bansal has over 25 years of teaching, research, academic leadership, and industrial experience. Currently, he is Professor in EE Department at the University of Sharjah, UAE, and Extraordinary Professor at the University of Pretoria, South Africa. In previous postings, he was Professor and Group Head (Power) at the University of Pretoria and worked with the University of Queensland, Australia; USP, Fiji; BITS, Pilani, India. Prof. Bansal has published over 400 journal articles, conference papers, books/book chapters. He has Google citations of over 16000 and an H-index of 60. He has supervised 25 Ph.D. and 5 post-docs. Prof. Bansal has attracted significant funding from industry and government organizations. He is Editor/AE of reputed journals including IEEE Systems Journal, IET-RPG, TESGSE. He is Fellow and CP Engineer of IET-UK, Fellow of IE (India), and Senior Member of IEEE. He has diversified research interests in the areas of Renewable Energy, Power Systems, and Smart Grid.

Chandra Shekhar completed his M.Sc. in Physics and Ph.D. in Semiconductor Electronics from BITS, Pilani, in the years 1971 and 1975, respectively. He joined Central Electronics Engineering Research Institute (CSIR-CEERI), Pilani, as Scientist in 1977 and served as Director from November 2003 to October 2015. His team at CEERI developed the country's first MOS LSI chip, a full-custom dedicated VLSI processor in 6-micron NMOS technology and CMOS Gate Array-based semi-custom VLSI chip. He also led the design of the country's first general-purpose microprocessor, and his team also designed India's first Application Specific Instruction Set Processor for Hindi text-to-speech conversion. He has been awarded the Young Scientist Award by UNESCO/ ROSTSCA; Distinguished Alumnus Award by BITS, Pilani (in 2014); and Distinguished Service Awards by Institution of Engineers and the Institution of Electronics and Telecommunication Engineers. He is currently Chancellor of the Academy of Scientific and Innovative Research (AcSIR).

Overview: Real-Time Video Monitoring for Suspicious Fall Event Detection

Madhuri Agrawal and Shikha Agrawal

Abstract Escorted by the advancement of technology, monitoring of real-time video data is playing an increasingly important role and massive impact on our lives. Therefore, correctly classifying the suspicious fall event detection during real-time video data monitoring is a highly appealing interest in the research area. In this paper, real-time events are considered for monitoring a person living in an indoor environment and performing their daily activity. Suspicious events are considered as detecting the falling of a person. Up to now, numerous techniques have been developed efficiently to classify suspicious fall event detection. An overview of numerous machine learning and deep learning techniques applied by several researchers for the detection of suspicious fall event detection in the monitoring of real-time video data has been considered. The paper also presents several affairs of machine learning and deep learning techniques that may assist focused researchers in working on monitoring video data.

Keywords Video monitoring · Suspicious event detection · Machine learning techniques · Deep learning techniques · Fall detection

1 Introduction

Monitoring video data is defined as analyzing video sequences to check the behavior, activities, and other certain information in a video sequence. Monitoring real-time video data provides strength to the video surveillance system. Video monitoring results in the successful detection of suspicious events, which is a powerful task for a video surveillance system. A 'suspicious event' is an unpredictable activity that rarely occurs and has not been viewed generally. The exact classification between

M. Agrawal (✉) · S. Agrawal
University Institute of Technology, Rajiv Gandhi Proudyogiki Vishwavidyalaya, Bhopal, India
e-mail: madhuriagrawal2000@gmail.com

S. Agrawal
e-mail: shikha@rgpv.ac.in

H. O. Bansal et al. (eds.), *Next Generation Systems and Networks*, Lecture Notes in Networks and Systems 641, https://doi.org/10.1007/978-981-99-0483-9_1

a suspicious event and a normal event is an extensive challenge in suspicious event detection [1]. In this paper, real-time events are considered for monitoring a person living in an indoor environment and performing their own daily activity in a normal manner. Suspicious events are considered as detecting the falling of a person. As falling of a person is not a normal event, it is a major cause of serious injuries for the elderly population, and sometimes it may be a cause of death. Thus, it is considered a suspicious event in real time. It can be used in the field of public healthcare, such as monitoring elderly people to prevent them from fractures and serious injuries. It can also help during the novel coronavirus pandemic. A person in home isolation can be monitored, and due to stress or any other health issues, if a suspicious event is detected, medical attention can be provided rapidly. It can also be used for the safety of elders and young kids living alone. The major aim of the research work is to propose a general monitoring method that will achieve results comparable to humans on standard datasets. There are two types of monitoring in the video data that is manual monitoring and automatic monitoring. In India, most systems are monitored manually, and the problem is that a person is unable to monitor each and every second of the day [1]. There is a need for research in the domain of automatic video monitoring because of security and privacy issues. In India, in most places, monitoring is not practically implemented due to a lack of manpower. Automatic monitoring overcomes the problems of the manual monitoring method. It avoids the total dependency on monitoring manpower. Automatic monitoring has a major challenge in the correct classification of suspicious events. Accurate classification of suspicious events is very important for sustaining efficient monitoring information and for quality of service.

2 Application of Machine Learning Techniques for Monitoring Suspicious Fall Event Detection

Supervised machine learning techniques which are used in various research papers related to detecting suspicious fall event detection in monitoring real-time video data. In this paper, real-time events are considered for monitoring a person living in an indoor environment and performing their own daily activity in a normal manner. Suspicious events are considered as detecting the falling of a person. As the falling of a person is not a normal event, it is considered a suspicious event in real time.

2.1 Support Vector Machine (SVM)

Support vector machine is a widespread supervised machine learning technique used for the classification of data among two classes. There are three primary steps: training, testing, and performance estimation. If data is linearly separable, then SVM allows classifying the data directly; otherwise, by using the kernel function, SVM classifies non-linear data. SVM takes the training data as input and generates a decision boundary known as hyperplane as output to classify the testing data into a relevant class, as shown in Fig. 1. The computational complexity of SVM is $O(n^3)$. Soni et al. [2] used an automated framework for fall detection. Background subtraction was applied to detect a moving person, and then all associated geometric features were extracted to classify a suspicious event as a fall, and SVM was applied to classify falls from other daily activities of a person. Sensitivity was achieved 98.15% and Specificity was achieved 97.10%. Iazzi et al. [3] proposed a method for fall detection on the basis of vertical and horizontal variations of human silhouette area. The human silhouette is extracted and constructed histogram to detect the postures of humans. A multiclass SVM classifier was used for classifying these postures. Based on the rules-method fall analysis was conducted. Results reported showing 93.7% of accuracy. Harrou et al. [4] used a monitoring scheme called a multivariate exponentially weighted moving average (MEWMA) for fall detection. Due to its sensitivity toward minor changes, it is effective in the correct classification of fall detection. To differentiate real falls from gesture fall, SVM was applied to detected sequences. Due to the high performance and generalizability of SVM, the combination of MEWMA-SVM was motivated for higher accuracy along with less time complexity in comparison to the conventional methods. Due to low computational cost, real-time implementation is a bit easy.96.66% of accuracy was achieved in MEWMA-SVM. Zerrouki et al. [5] summarized a relative study among three machine learning techniques (KNN, Naïve Bayes, and SVM). Classification performance is implicitly dependent on the extracted features from human posture after background subtraction. The experiment metric revealed the superiority of SVM and showed 93.96% accuracy.

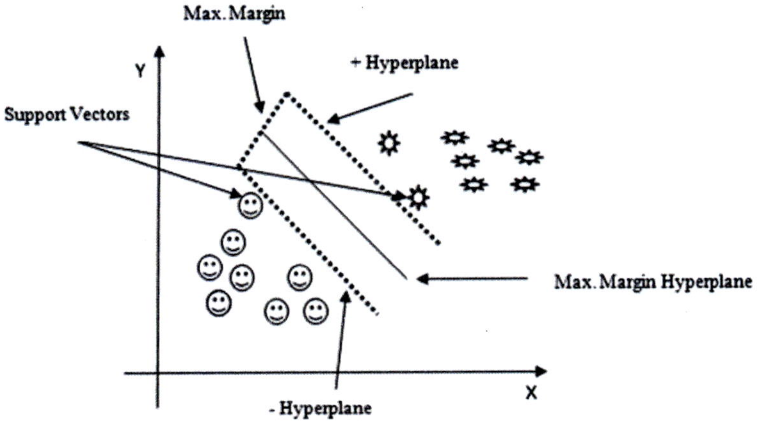

Fig. 1 Support vector machine

2.2 C4.5 Decision Tree

Decision trees are similar to tree-based structure models due to the presence of nodes and branches. Classes are categorized from the input data by using the divide and conquer technique. This tree is used to decide the class label of the testing data set. It explicitly and visually represents decisions and helps in final decision-making. It works on the top-down approach from the root node and uses the best split at each level. A C4.5 decision tree is a famous supervised machine learning technique. It is a good statistical classifier that generates univariate decision trees. An enhancement of the Iterative Dichotomiser3 (ID3) algorithm is the C4.5 decision tree, and it can work with both discrete and continuous attributes. Missing value attributes in training data are also handled carefully by the C4.5 decision tree. Its computational complexity is O(n). Alzahrani et al. [6] worked on Microsoft Kinect v2 with skeleton features. In the framework, after preprocessing, skeleton data is extracted, and features are selected based on velocity and average velocity. Then 'fall' class is predicted by applying the C4.5 decision tree. With all joints' average velocity, it obtained 91.3% accuracy, while normalizing skeleton features and extracting average velocity resulted in an accuracy of 71.8%. Khawandi et al. [7] suggested a multi-sensor monitoring system. It uses a heart rate sensor along with a webcam for classification using the C4.5 decision tree, and if it predicts a 'fall' class, it blows an alarm. It has shown 98% accuracy approx. An error rate is 1.55%. Ramanujam et al. [8] proposed monitoring depends upon posture. It uses infrared cameras attached to digital video recorders. With the help of retro-reflective radium tapes fabricated in a cotton dress, fall detection was detected using the C4.5decision tree. The overall average accuracy achieved is 91.51%.

2.3 K-Nearest Neighbor (KNN)

K-nearest neighbor (KNN) is a simple, non-parametric technique used for the classification and regression in supervised machine learning. KNN assumes that similar things exist in close proximity means they are near to each other. Inputs of KNN are the k closest training data of a given test data in the feature space. It stores all available data and classifies new data based on the similarity measure by a distance function. For the categorical variables, Hamming distance measure is calculated, while for the continuous variables, three different distance measures: Euclidean, Manhattan, and Minkowski are used. When the variables are mixed in that case, the issue of standardization of variables between 0 and 1 is needed to apply. By the maximum of votes of test data neighbors, the output class is assigned to the test data. It uses a powerful technique of assigning weight to the contribution of neighbors to help nearer neighbors. It contributes additionally to the average as compared to the distant neighbors. KNN has high computational complexity as compared to the SVM [5]. Ramanujam et al. [8] proposed monitoring depends upon posture. It uses infrared cameras attached to digital video recorders. With the help of retro-reflective radium tapes fabricated in the cotton dress, fall detection was detected. The overall average accuracy with KNN achieved is 93.1%. Zerrouki et al. [9] worked on an effective human action recognition method. Human body segmentation was performed in input video data. From the background, the body's silhouette and features were extracted. Feature extraction was based on the variations in body shape, which was divided into five partial occupancy areas of the standing body. Each frame area ratio was calculated and used for the action recognition stage as input data. With URFD Dataset, it evaluated 91.09% accuracy, while with UMAFD Dataset, it evaluated 90.40% accuracy.

2.4 *Application of Deep Learning Techniques for Monitoring Suspicious Fall Event Detection*

To provide a major focus on the importance of monitoring video data miscellaneous deep learning techniques have to be appertaining to classify correct suspicious events more accurately and effectively. Deep Learning Technique is an upcoming decade trend. Machine learning is a superset of Deep Learning. Deep Learning is more famous and reliable because of its five major capabilities: automatic feature extraction, end-to-end models, model re-usability, general method, and superior performance. From input data, deep learning performs feature extraction and classification altogether; it requires no preprocessing techniques and gives output. The reason behind the popularity of deep learning is the powerful Graphics Processing Units and the large availability of datasets. The major powers of DL are High-resolution image synthesis, image-to-image translation, image super-resolution, video-to-video translation, and medical image processing. The next subdivision is on the application of some of the Deep Learning techniques for monitoring suspicious fall event detection. The performance of various supervised deep learning techniques is listed in Table 1, and a few popular used datasets are discussed in Table 2.

2.5 *Convolutional Neural Networks (CNN)*

A Convolutional neural network is a popular DL technique that is efficient in extracting meaningful information from video data. It is able to model complex variations and behavior and provide highly accurate predictions. Layers consist of a high number of filters that are trained by the training data. It requires GPUs and labeled Images. Kernels extract relevant features from input by using the convolution operation. Different CNN architecture is used to maximize performance in classification. CNN produces high accuracy as compared to KNN. CNN has high computational complexity. Xiong et al. [10] presented human skeleton sequence representation by estimating pose and extracted features and generating an optimized skeleton sequence which is processed with 3D CLP (consecutive low pooling) neural network and

Table 1 Performance of various supervised deep learning techniques

Paper reference no.	Year	Classification method used	Type of video frame	Performance (%)	Dataset
Xiong et al. [10]	2020	S3D-CNN	Depth video frame	Accuracy: 90.02	UCF101
Kong et al. [11]	2019	CNN	Surveillance Video frame (single RGB)	Sensitivity: 100 Specificity: 99.3	MCF+URF
Zhang et al. [12]	2018	CNN	RGB	Accuracy: 96.04	SDUFall
Marcos et al. [13]	2017	CNN	RGB	Accuracy: 95 Accuracy: 97	URFD FDD
Min et al. [14]	2018	Normalized shape aspect ratio (NSAR)	RGB	Precision: 100 Recall: 93.50	Generated from experiments
Lotfi et al. [15]	2018	Multilayer perceptron neural network (MLP NN)	RGB	Accuracy: 99.24 Precision: 99.60 Sensitivity:99.52 Specificity: 97.38	URFD
Min et al. [16]	2017	RCNN	RGB	Accuracy: 95.50 Precision: 94.44 Recall: 94.95	Generated from experiments
Tran et al. [17]	2015	Convolutional 3D (C3D)	RGB	Accuracy: 85.2	UCF101
Pourazad et al. [18]	2020	CNN+LSTM	RGB	Accuracy: 87	Generated from experiments
Chen et al. [19]	2020	Attention-guided bi-directional LSTM	RGB	Accuracy: 96.7 Recall: 91.8 F-score: 94.8	URFD
Lu et al. [20]	2019	3D-CNN combined with LSTM	RGB	Accuracy: 99.36 Accuracy: 99.27	FDD URFD
Ge et al. [21]	2018	Recurrent convolutional network (RCN)	RGB	Accuracy: 98.96	ACT4^2 dataset
Feng et al. [22]	2018	Attention-guided LSTM	RGB	Avg. Precision: 94.8 Avg. Recall: 91.4 F-score: 93.1	URFD
Hashemi et al. [23]	2018	LSTM	RGB+depth	Precision: 93.23 Recall: 96.12	NTU RGB+D action recognition dataset

(continued)

Table 1 (continued)

Paper reference no.	Year	Classification method used	Type of video frame	Performance (%)	Dataset
Lie et al. [24]	2018	CNN+RNN+LSTM	RGB	Accuracy: 90	Generated from experiments
Hasan et al. [25]	2019	RNN+LSTM	RGB	Sensitivity: 99.0 Specificity: 96.0 Specificity: 97.0	FDD+URFD URFD FDD
Adhikari et al. [26]	2017	CNN	RGB+depth	Accuracy: 74	Generated from experiments

Table 2 Few popular used datasets for suspicious fall event detection

Paper reference no.	Name of dataset	Video type	No. of videos
[27]	Multiple camera fall dataset (Multicam/MCF)	RGB	Total: 24 Fall: 22
[28]	UR fall detection dataset (URFD)	Depth+RGB	Total: 70 Fall: 30 ADL: 40

detects the falling with the 3D-CNN. The network had improved in respect of layer number, single input frame number, and pooling kernel size. It reduces the failure of fall detection, and accuracy was observed 90.02%. Kong et al. [11] presented learning human fall by spatiotemporal representation. To model spatial and temporal representation in videos, background subtraction and rank pooling were employed. After that, it uses view-independent three-stream CNN to classify fall events. First is Silhouettes, second is a silhouette motion history image, and third is a dynamic image for which temporal duration is the same as motion history images. Mainly it uses appearances and motion representations. Due to good generalization capability, more reliable performance is achieved. Results showed sensitivity as the ability to detect falls was 100%, and specificity as the ability to recognize daily life activities was 93.3%. Zhang et al. [12] suggested a trajectory based on weight descriptors in videos to describe the dynamics of human actions effectively and robustly to surrounding environments. From each frame, the CNN feature map was extracted with deep ConvNet and built a trajectory attention map to localize the area optimally. Then CNN feature map was weighted with a trajectory attention map. In the time sequence of a video, to reduce the redundancy of convolutional features cluster pooling method was applied. Lastly, to encode the dynamic of the cluster-pooled sequence, the rank pooling method was applied. Results showed an accuracy of 96.04%. Marcos et al. [13] proposed a scenario-independent video motion model by using input images of optical flow to the convolutional neural networks, followed by the performance of transfer learning to overcome the class imbalance dataset.

With a three-step training phase, it generated a general independent model from environmental features and produced an accuracy of 95% in URFD and 97% in FDD datasets.

2.6 Long Short-Term Memory Network (LSTM)

LSTM is an advanced version of recurrent neural networks and has the capability of learning long-term dependencies. It can remember short-term memories for a very long time. It has a memory cell containing three gates: Input, Forget, and Output. Also, it has a neuron with a self-recurrent. All the above-mentioned gates permit the memory cell to keep and access information over long periods of time. LSTM is designed in such a way that it resolves the problem of vanishing gradient problems without altering the training model. Long-time lags, in a few cases, are bridged using LSTM by handling noise, distributed representations, and continuous values. LSTM provides a huge range of parameters like learning rates and input-output biases and does not require any fine adjustments. The complexity of updating the weight is also reduced. LSTM is composed of three sigmoid gates and one tanh layer to limit the information from passing through the memory cell. LSTM has high computational complexity. LSTM is better than Hidden Markov Model (HMM) because it is not required to keep a finite number of states from beforehand. In order to achieve more efficiency and reduce computational complexity, various variations of LSTMs were designed. Real-time fall detection was presented by Pourazad et al. [18], which is composed of LSTM and CNN modules. A long-term recurrent convolutional network model (LRCN) was selected for video classification, which consists of spatial and temporal classification. It takes the input from RGB video stream, and on the basis of non-intrusive deep learning, it processes in short time windows. Results show real-time fall detection with 87% accuracy. By applying bi-directional LSTM, fall event detection for the complex backgrounds was presented by Chen et al. [19]. Mask-RCNN is used to detect the person in the frame. To extract the features, VGG16 was used, followed by the bi-directional LSTM to provide an attention model for behavior information in forward and backward directions for classification. It results in the improvement of the performance of fall event detection. The result shows 96.7% of accuracy. On video kinematic data, a combination of 3D-CNN with LSTM was used by Lu et al. [20]. A visual attention guide has consolidated a soft attention mechanism for video analysis through LSTM into 3D-CNN. To locate the informative region within each and every frame of the video, 3D-CNN generated feature cube was fed to LSTM. Experiments verified with an accuracy of approx 99%. Ge et al. [21] proposed human fall detection by using co-saliency enhanced deep RCNN. An effective method to enhance dynamic human activities and to put down unwanted backgrounds in videos is a co-saliency detection method. Co-saliency enhanced video frames are given as input to recurrent convolutional network (RCN) in which RNN is realized by LSTM connecting to a set of CNN. A result shows 98.96% test accuracy. Attention-guided LSTM was used by Feng et al. [22] in complex scenes for spatial-temporal fall event detection. YOLO v3 was employed to detect persons in videos. Deep sort tracking was used to track the person, and their trajectories were identified. For each trajectory feature extraction, the VGG16 model was used, then CNN features were extracted, and with attention guided LSTM fall event was detected. The experimental result shows 94.8% as average precision and average recall as 91.4% on the URFD dataset.

3 Conclusions

This overview paper presents recent advancements in monitoring real-time video data for detecting suspicious fall event detection. A suspicious event, such as a fall event, occurs in an extremely tiny period of time, and a fast response is required after a fall event for the survival of an elderly person or the patient. Thus, real-time monitoring in detecting suspicious fall event detection is a very important factor. On comparing sensor-based systems and vision-based systems, earlier researchers concluded that vision-based systems provide better results and it is comfortable for an elderly person as they are not required to wear a sensor. Earlier research work done in this domain reveals the excellence of deep learning techniques over machine learning techniques for detecting suspicious fall events detection. Due to the absence of a simplified standard evaluation metric and procedure, researchers used different evaluation metrics and procedures to evaluate the performance of the suspicious fall event detection. This makes a comparison among suspicious fall event detection systems very difficult. Deep learning overcomes the issues of machine learning techniques and also enhances efficiency in the domain. While many deep learning techniques have been applied till now, there are lots of chances to enhance the processing speed and accuracy along with the increase in the dataset size. Deep learning-based suspicious fall event detection systems do not have a direct connection to the fall rescue operation like emergency response services, and due to specific hardware requirements, it is difficult to deploy in a real-time framework. Thus, in upcoming years research may progress in this direction a lot.

References

1. Agrawal M, Agrawal S (2020) Suspicious event detection in real-time video surveillance system in proc.SCI-2018. Singapore 100:509–516
2. Soni PK, Choudhary A (2019) Automated fall detection from a camera using support vector machine in proc. ICACCP, Gangtok, India, pp 100–106
3. Iazzi A, Rziza M, Thami ROH (2018) Fall detection based on posture analysis and support vector machine in proc. ATSIP, Sousse, Tunisia, pp 98–103
4. Harrou F, Zerrouki N, Sun Y, Houacine A (2017) Vision-based fall detection system for improving safety of elderly people. IEEE Instr Measur Mag 20(6):49–55
5. Zerrouki N, Harrou F, Houacine A, Sun Y (2016) Fall detection using supervised machine learning algorithms: a comparative study in proc. ICMIC, Algiers, Algeria, pp 665–670

6. Alzahrani MS, Jarraya SK, Addallah HB, Ali MS (2019) Comprehensive evaluation of skeleton features-based fall detection from Microsoft Kinect v2 in proc. SIVip, Verlog, London, pp 1431–1439
7. Khawandi S, Ballit A and Daya B (2013)Applying machine learning algorithm in fall detection monitoring system in proc. CICN, Mathura, pp 247–250
8. Ramanujam E, Padmavathi S (2019) A vision-based posture monitoring system for the elderly using intelligent fall detection technique. Comput Commun Netw 1(11):249–269
9. Zerrouki N, Harrou F, Sun Y, Houacine A (2018) Vision-based human action classification using adaptive boosting algorithm. IEEE Sens J 18(12):5115–5121
10. Xiong X, Min W, Zheng WS, Liao P, Yang H et al (2020) S3D-CNN: skeleton-based 3D consecutive-low-pooling neural network for fall detection. Appl Intell 50:3521–3534
11. Kong Y, Huang J, Huang S (2019) Wei Z and Wang S. J Visual Commun Image Represent 59:215–230
12. Zhang Z, Ma X (2019) Wu H and Li Y. IEEE Access 7:4135–4144
13. Marcos AN, Azkune G, carreras IA (2017) Vision-based fall detection with convolutional neural networks. Wirel Commun Mob Comput 2017(9474806):1–16
14. Min W, Zou S, Li j (2018) Human fall detection using normalized shape aspect ratio. Multimedia Tools Appl 78:14331–14353
15. Lotfi A, Albawendi S, Powell H, Appiah K, Langensiepen C (2018) Supporting independent living for older adults; Employing a visual based fall detection through analysing the motion and shape of the human body. IEEE Access 6:70272–70282
16. Min W, Cui H, Rao H (2018) Li Z and Yao L. IEEE Access 6:9324–9335
17. Tran D, Bourdev L, Fergus R, Torresani L, Paluri M et al (2015) Learning spatiotemporal features with 3D convolutional networks in proc. ICCV, Santiago, pp 4489–4497
18. Pourazad MT, Hashemi AS, Nasiopoulos P, Azimi M, Mak M et al (2020) A non-intrusive deep learning based fall detection scheme using video cameras in proc. ICOIN, Barcelona, Spain, pp 443–446
19. Chen Y, Li W, Wang L, Hu J, Ye M (2020) Vision-based fall event detection using attention guided Bi-directional LSTM. IEEE Access 8:161337–161348
20. Lu N, Wu Y, Feng L, Song J (2019) Deep learning for fall detection: 3D-CNN combined with LSTM on video kinematic data. IEEE J Biomed Health Informatics 23(1):314–323
21. Ge C, Gu IYH, Yang J (2018) Co-saliency-enhanced deep recurrent convolutional networks for human fall detection in e-healthcare in proc. EMBC, Honolulu, HI, pp 1572–1575
22. Feng Q, Gao C, Wang L, Zhao Y, Song T et al (2018) Spatio-temporal fall event detection in complex scenes using attention guided LSTM. Pattern Recogn Lett 130:242–249
23. Hashemi AS, Nasiopoulos P, Little JJ, Pourazad MT (2018) Video-based human fall detection in smart homes using deep learning in proc. ISCAS, Florence, pp 100–105
24. Lie WN, Le AT, Lin GH (2018) Human fall-down event detection based on 2D skeletons and deep learning approach in proc. IWAIT, Chiang Mai, pp 100–104
25. Hasan MM, Islam MS and Abdullah S (2019) Robust pose-based human fall detection using recurrent neural network in proc. RAAICON, Dhaka, Bangladesh, pp 48–51
26. Adhikari K, Bouchachia H, Charif HN (2017) Activity recognition for indoor fall detection using convolutional neural network in proc. MVA, Nagoya, pp 81–84
27. MCF dataset. http://www.iro.umontreal.ca/~labimage/Dataset/
28. URFD dataset. http://fenix.univ.rzeszow.pl/mkepski/ds/uf.html

Implementation of Master-Slave Communication Using MQTT Protocol

Darsh Patel, Hitika Dalwadi, Hetvi Patel, Prasham Soni, Yash Battul, and Harsh Kapadia

Abstract With the revolution in industry 4.0, the applications of the Internet of Things (IoT) have wide innovation in the commercial and industrial domains. The significant applications of IoT are deployed in sectors such as embedded systems, sensing technologies, and computer vision. Some of the many aspects to consider while implementing IoT include hardware, connectivity, protocol, server availability, hardware sensors, decision, and analysis. The research paper aims to present and implement the work of master-slave communication using the Message Query Telemetry Transfer (MQTT) protocol, deployed using Node-MCU and Python. The work in discussion focuses on Master-Slave communication between multiple mobile robots and a central system. The application of this project can be branched to domains like defense, household, e-commerce, etc. A local server setup using a router works as a Wi-Fi network or a master control unit. The real-time navigation, monitoring, and control are done with the help of central image processing (master module) which sends and receives data to robots (slave module) using the MQTT protocol. The results show that MQTT is an efficient communication protocol that can be used for master-slave communication because it has efficient synchronization, data loss is minimum, and is error-free.

D. Patel · H. Dalwadi (✉) · H. Patel · P. Soni · Y. Battul · H. Kapadia
Institute of Technology, Nirma University, Ahmedabad, India
e-mail: 19bic021@nirmauni.ac.in

D. Patel
e-mail: 19bic038@nirmauni.ac.in

H. Patel
e-mail: 19bee039@nirmauni.ac.in

P. Soni
e-mail: 19bic048@nirmauni.ac.in

Y. Battul
e-mail: 19bic004@nirmauni.ac.in

H. Kapadia
e-mail: harsh.kapadia@nirmauni.ac.in

H. O. Bansal et al. (eds.), *Next Generation Systems and Networks*, Lecture Notes in Networks and Systems 641, https://doi.org/10.1007/978-981-99-0483-9_2

Keywords MQTT · Master-slave communication · Coordinated control · Autonomous mobile robots · Node-MCU · IoT · Python

1 Introduction

The Internet of Things (IoT) has grown fast during the last two decades around the world. It has greatly empowered businesses and offered them economic value in a variety of industries, including manufacturing, healthcare, automotive, security, transportation, and others.

Internet of Things is a wide domain and has a significant number of protocols that have been developed like MQTT, and Hypertext Transfer Protocol (HTTP) which can be used for implementing master-slave communication. Master-slave communication works on a request-reply mechanism. There is one master device and multiple slave devices. It is an asymmetric type of communication as master-to-slave two communication is possible but slave-to-slave is not possible. For any IOT application there are three major parts, such as hardware, software, and the protocol required. Figure 1 depicts the available hardware, software, and protocols.

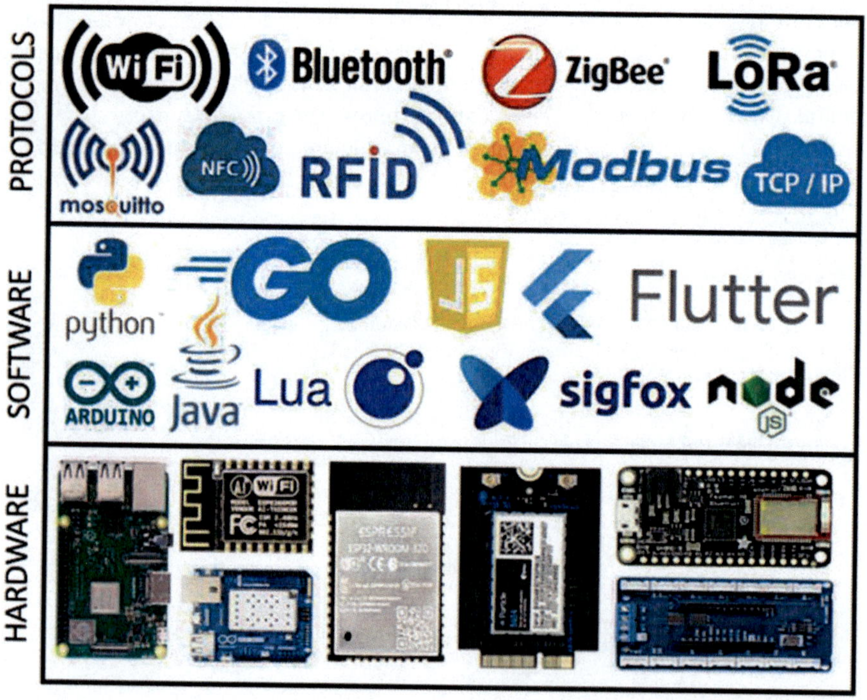

Fig. 1 Relation between various communication protocols, hardware, and software

Various communication protocols can be used for implementing master-slave protocols, for example, Bluetooth Low Energy, HTTP, Zigbee, LoRA (Long Range Radio), Modbus, RFiD (Radio Frequency Identification), and Wi-Fi. Nonetheless, the scope of this paper is limited to the MQTT protocol. MQTT protocol works on the publish-subscribe model rather than the conventionally used client-server model the significant difference between the two is that the publish-subscribe model communicates via a third-party broker, in this case, MQTT Explorer. The broker filters out all the incoming messages and distributes them to the respective subscriber. The filtering method is divided into three parts, subject-based filtering, content-based filtering, and type-based filtering depending upon the application these filtering methods can be altered. For the master-slave application, subject-based filtering is selected which ensures that subscribers get messages from a specific topic and the data transferred is in 'raw' format. To ensure that data is not lost in transmission, a feedback mechanism is established such that once the slave subscribes to the message from the broker it sends a confirmation message to the master module via the broker.

MQTT protocol is implemented for efficient, robust, and coordinated control of autonomous mobile robots. Autonomous mobile robots have prominent applications in major industries such as manufacturing, automation, e-commerce, transportation, surgical, and security. Robotics systems are widely used because of their extended range and resilience. For coordinated control of autonomous mobile robots, a centralized system is required for monitoring and navigation purpose. Computer vision along with Python acting as the master module is employed to achieve a centralized control and navigation system. The data is sent to slave modules, which in our case are the robots, with the help of the MQTT protocol in form of small data packets.

Node-MCU is a micro-controller board, used as a slave module in the proposed system. It establishes communication between the master device and mobile robots. This micro-controller unit contains an ESP8266 Wireless Fidelity (Wi-Fi) module which is mounted System On a Chip (SOC), interfaced with Transmission Control Protocol (TCP)/IP protocol (IP). This provides Wi-Fi network access to the user. With an onboard LX106 Central Processing Unit (CPU), and storage memory of 4 Mega Bytes (Mbytes), this unit can be easily integrated with actuators and drives.

One of the significant aspects of coordinate control of mobile robots is the communication algorithm. The following paper discusses the implementation of a master-slave communication algorithm via MQTT protocol as a mode for wireless data transmission. The master-slave algorithm implemented consists of a single processing unit that is a Python compiler that monitors the live location of mobile robots using machine vision and acts as a master module, and multiple slave modules which are individual robots comprising of Node-MCU that transmit the information received from the master module and controls the robot.

The paper further discusses the master-slave communication algorithm developed for multiple mobile robots' coordinate control and collision avoidance using the MQTT protocol. The work presented in the paper is a case study of master-slave communication in coordinated control of robots where MQTT or the central computer serves as master and robots, as slaves.

2 Literature Review

Researchers around the globe have presented their work around different strategies for various protocols. Applications demanded new and efficient methods for master-slave communications which can be applied in various IoT applications. Executing these applications requires a suitable communications protocol, ensuring optimized results. Some aspects that significantly impact on the choice of a communication protocol include speed of transmission, latency, efficiency, reliability, and range. A lot of the previously mentioned have been studied and tested by the researchers to explore a more optimized communication protocol.

Atmoko et al. in [1] discussed improvisation of the Unmanned Surface Vehicle(USV) scalability, and overcoming persisting local networks by the usage of the MQTT protocol for better efficiency in real-time. Extension of the application incorporating machine learning algorithms for task-solving capability was suggested. The aim was to improve a larger range of control of USV via internet networks. Sasaki et al. [2] presented that HTTP protocol was observed to provide lesser efficiency. The paper reflected upon certain problems concerning the application of HTTP for IoT and suggested a finer alternative, i.e., MQTT, being a promising and more efficient candidate over HTTP. The notion was concluded after conducting a performance comparison between both. It stated the problem associated with data traffic and suggested name/identifier-based communication independent of location, thus ensuring continuous, interruption-free communication. The author suggested a preference for the MQTT protocol over HTTP. Thirupathi et al. [3] explained the design and development of a cloud-based home automation system and the flexibility of enhancement by changing existing features. The explained system consisted of Wi-Fi, cloud MQTT, ESP32, relays, and a power supply unit that helped in controlling home appliances very easily. The performance of the system was evaluated using two different home appliances-a bulb and a fan. It was observed to be a secure, low cost, and flexible home automation system that provided security at the Secure Socket Level (SSL). It is user-friendly and highly efficient.

Kashyap et al. [4] paper tested two service models of communication. First, using serial USB transmission, and second using MQTT protocol deploying Wi-Fi module ESP8266-12. Some common choices of brokers employed include mosquitto, Adafruit, and HiveMQ. The protocol broadly consists of a broker, publisher, client, and subscriber. As a result, it was discovered that since the speed of transfer of data is dependent upon Wi-Fi speed, the speed of data transfer is slower in comparison to serial connection. However, the data in this protocol is never lost while it is in use, it saves the data in a queue. Mesquita et al. [5] discussed Wi-Fi modules with the benefits of low power consumption, external communication operators and gateways avoidance, and reasonably low costs while simultaneously ensuring. Provides low latency, reduces the size of computing hardware, and lower system cost. The authors characterized ESP8266 as ultra-low power, a low-cost module with an undocumented performance of IoT. Factors such as built-in sleep modes, infrastructure parameters, beacon interval, Delivery Traffic Indication Message (DTIM) period, packet delivery ratio, and received signal strength, which is the function of distance and antenna orientation, for assessing area coverage, have been studied and experimented upon.

Atmoko et al. [6] recognized challenges for IoT to bridge physical parameters to databases. It discussed applications such as remote control, radio frequency identification, and wireless sensor network. It presented a study of data acquisition from sensors to the server, allotting sequential ID for each packet of data transmitted from the hardware, employing MYSQL as a database. Preference was given to the ability of MQTT over HTTP for better data transmission. Miorandi et al. [7] presented that Internet of things was referred to as an umbrella term that has been carefully studied for identifying technologies associated, challenges involved, and applications that IoT can branch out to. The common research challenges faced include identification and sensing/actuation. It discussed how security, privacy, data confidentiality, and trust are factored in. The provision of effortlessly integrating the physical environment with a computational/internet database, while employing embedded systems, unlocks new possibilities for research. It paves the way for the development of the Information and Communication Technologies (ICT) sector.

Aroon et al. [8] discussed that the usage of MQTT protocol was proposed over traditional HTTP protocol, assisted with cloud platform benefiting from the lower overall packet size, significantly reducing the communication time. The flexibility and ease of configuration with pre-existing systems of operation along with its working with the cloud for remote control and tracking of robots using GPS module were discussed. Bottone et al. [9] while discussing swarm robots refer to multiple ground robot systems that can be exploited by interfacing it with the exceptional features of cloud computing offers, to make the best use of computational as well as data storage capabilities. Thus, helping one encapsulate remote control, computational effort, and back-end server and external swarm network. The approach of the author was based on the realization of control and coordination between environments of multiple robots using MQTT and stigmergy. The concept allows decentralization of computation, developing computational intelligence from within the environment. Hurdles associated with data traffic have been addressed by these various authors in order to explore better options for a communication protocol that is both efficient and reliable. Comparison between protocols such as HTTP and MQTT has been made, concluding that MQTT is more efficient and deals with the problems such as loss of data packets.

3 Methodology

The implementation of research work is divided into two major parts such as hardware and software. The hardware consists of mobile robots which function as slave modules. Mobile robots publish and subscribe to messages from the master module via Wi-Fi by the means of MQTT protocol using Node-MCU micro-controller which is an open-source Lua-based firmware development board. The firmware runs on ESP8266 Wi-Fi SoC from Espressif Systems, and the hardware is similar to the

ESP-12 module. The mobile robots traverse the arena using the differential drive kinematics in which 2-wheels are mounted on a common axis and each wheel is driven independently for forward or backward motion. The motors used for the application were National Electrical Manufacturers Association 17 (NEMA 17) [10] which is a hybrid stepper motor with a 1.8° step angle and a holding torque of 3.2 kg cm. To control the stepper motor via Node-MCU Moons SR3 Mini stepper motor driver [11] is used because it operates on a 3.3 V pulse generated by Node-MCU micro-controller and is capable of controlling the speed of stepper motor up to 2000, power supply ranges from 12 to 48 V DC supply and the current control mode in which 3 piano switch set the running current with 3 amps as peak maximum.

MQTT protocol is useful for establishing the machine-to-machine type of communication. MQTT can run over TCP/IP and the packet size of the MQTT protocol is small, so the power supply consumption is small enough. MQTT provides publish-subscribe asynchronous communications. In the implementation, publish-subscribe provides better IoT services than the request-response protocol. This improvement is possible because the publish-subscribe protocol client does not require any updates and this operation can be possible with less bandwidth. MQTT also supports requests/responses of a Constrained Application Protocol (COAP). MQTT can transmit data in various forms such as text, binary data, JavaScript Object Notation (JSON), and Extensible Markup Language (XML). MQTT broker acting as server already has stored topics and the client can act as a publisher and send a message to the topic and a subscriber receives that message from the broker of that topic. The drawback of COAP is that it has the problem of packet loss because of the TCPs transmission procedure. MQTT brokers need a username and password for security which is controlled by Transport Layer Security (TLS)—TLS() and SSL().

Figure 2 presents the publish-subscribe method used in communication networks with a common broker (here a router) based on MQTT protocol.

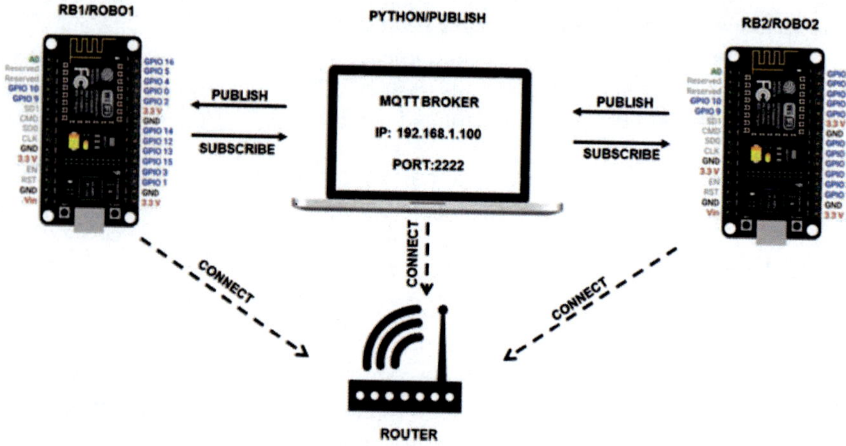

Fig. 2 Block diagram for MQTT communication

The software is further divided into embedded programming of the mobile robots and localization and control algorithms using a central camera. The embedded programming of each robot is done in Arduino integrated development environment (IDE) and the libraries used are:

1. ESP8266WiFi: provides a wide collection of C++ methods and properties to configure and operate an ESP8266 module in the station and/or soft access point mode [12]
2. PubSubClient: provides a wide collection of C++ methods and properties to configure and operate an ESP8266 module in the station and/or soft access point mode. It supports Arduino Ethernet-compatible software [13]
3. AccelStepper: controls various stepper motor and provide coding for various stepper pin configuration [14]

To detect the AruCo marker on each robot machine vision and OpenCV library is utilized. AruCo marker [15] is a synthetic square with an inner binary matrix and each matrix has a specific identification and has its application for pose estimation. detectMarkers() function is used and contains the following parameters:

1. markerCorners
2. markerIds

For communication paho mqtt library is imported into the Python environment; it enables the laptop to connect to the MQTT broker and publish-subscribe to messages. It includes various functions to enable connection with the broker, callback, network looping, and publishing and subscribing. Figure 3 describes the communication sequence of MQTT protocol.

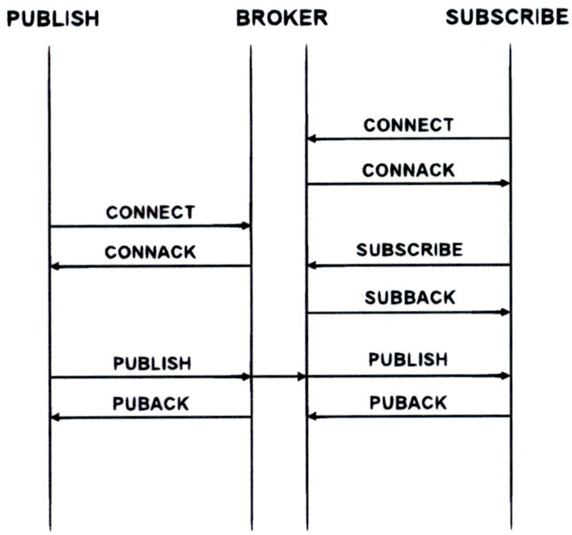

Fig. 3 Communication sequence of MQTT

3.1 Communication Algorithm and Flowchart

Communication and message transmission is divided into two parts master module which is the central computer and slave modules which are mobile robots with Node-MCU as micro-controller.

Master Module: MQTT protocol has been implemented on the master-slave prototype, where the connection is generated via Wi-Fi. To establish the connection with the server Eclipse mosquitto broker was used. The server consists of a static IP address and port address which were provided to slave modules. Broker checks for a slave module with a unique topic that has the same IP address and port address. Once all slave and master modules are connected, the respective modules publish a feedback message to the broker which verifies that the connection is active or lost. In the above-mentioned system, three robots are set to act as a client, which accepts the published command through the master via the broker. The Master publishes the command and the slave acts accordingly. The master here Python monitors the robots using a centralized camera which provides continuous feedback. This feedback helps determine relevant commands which are then communicated to the slave modules.

A dynamic keyword or message for the said client acts as a command to complete the assigned task. The published command which further received by the controller on the robot, which is in turn reciprocated by the execution of the pre-programmed task. Moreover, the algorithm used to complete the task by the bot can be further divided into two parts:

1. All the robots complete the task successfully without any interruption.
2. Occurrence of collision during the task.

During the execution of any task, if the second condition ceases to occur, the robot will continue to execute the pre-defined task. The suggested algorithm for collision avoidance is as follows: if robot 1 (client subscriber 1) is heading toward the subscribed path to reach the destination, robot 1 is at the middle of the subscribed path which checks continuously if any other mobile robot at the same time robot 2 (client subscriber 2) is at the adjoining position which ArUco is scanned by a centralized camera which acts as the master. If the distance between both decreases, the master sends the command to stop the lower priority robot (this command work on the priority basis of robot 1, 2, and 3 accordingly) at that time master sends the "STOP" command, while the higher priority robot continues its path. After the collision has been avoided the master (Python) sends the lower priority robot to resume its task. The process flow sequence of master module is shown in Fig. 4.

Slave Module: Master-Slave Communication operations of the proposed systems are implemented using the Node-MCU micro-controller unit. As a slave robot, this micro-controller unit device primarily connects to Wi-Fi after which, this device establishes publish/subscribe relation with MQTT broker, in this case, Eclipse mosquitto broker. Once the connections had been established between the slave robot and the MQTT broker the slave robot subscribes to the unique topic which was assigned to it. After subscribing to the topic slave robot sends a feedback message to the master device. When the system starts the master device makes decisions

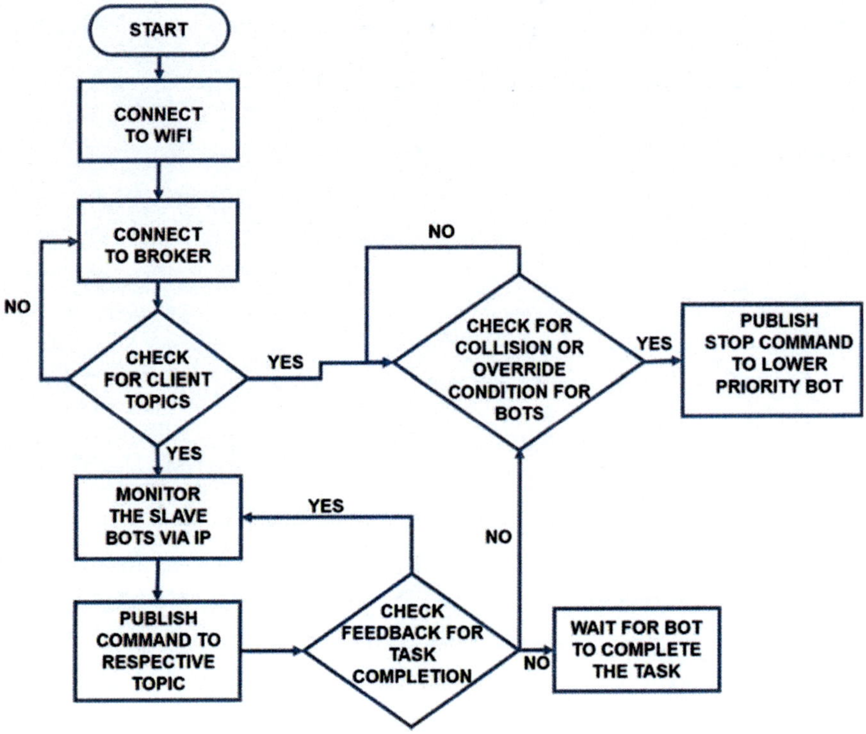

Fig. 4 Flowchart of MQTT master module

according to the feedback from the centralized camera. The master then publishes a message under the corresponding topic to control any particular slave robot. As an advantage of the MQTT protocol, the slave robot continuously checks if the master is publishing any new message to its assigned topic, and if the slave robot detects any new message being published to its assigned topic it overwrites the old message and starts performing the new task. Once the assigned task is completed from the slave's side, it again publishes feedback to the master and waits until it receives any new message. The command flow sequence of the slave module with reference to the master module is shown in Fig. 5.

4 Results

The comparative study of various communication protocols has been determined on the basis of criteria such as modes of message transmission, connection status, and encoding standards [16]. In order to determine the most suitable protocol on the basis of application, the same has been tabulated in Table 1. The total time taken

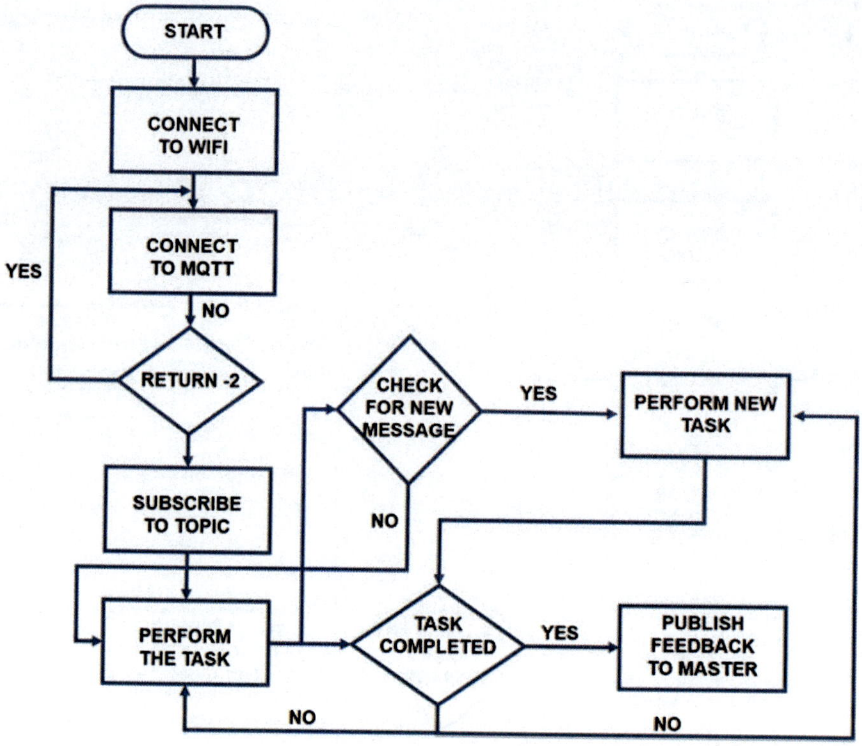

Fig. 5 Flowchart of MQTT slave module

Table 1 Comparision of various communication protocols

Criteria	HTTP	MQTT	ZigBee
Bytes	8	<256	<128
Messaging mode	Synchronous	Asynchronous	Asynchronous
Message distribution	One-One	One-Many	One-Many
Connection status	Unknown	Known	Known
Encoding format	Text	Binary	AES Encryption Standard
Standards	IETF and W3C	OASIS and Eclipse	IEEE 802.11

by the master module to publish the command and the slave module to receive the command is observed during the implementation of master-slave communication. It is of supreme importance while determining the latency of the communication established. This time is recorded and tabulated as follows and is shown in Table 2.

Initially, all the devices send their status to the master module Fig. 6 upon receiving confirmation, the master sends appropriate commands to the respective slave Fig. 7.

Table 2 MQTT data transfer

Sr. No.	Publisher	Subscriber	Data from	Time taken
1	Laptop (Master)	Node-MCU (Slave)	Raw	22–25 ms
2	Node-MCU (Master)	Laptop (Slave)	Raw	22–25 ms

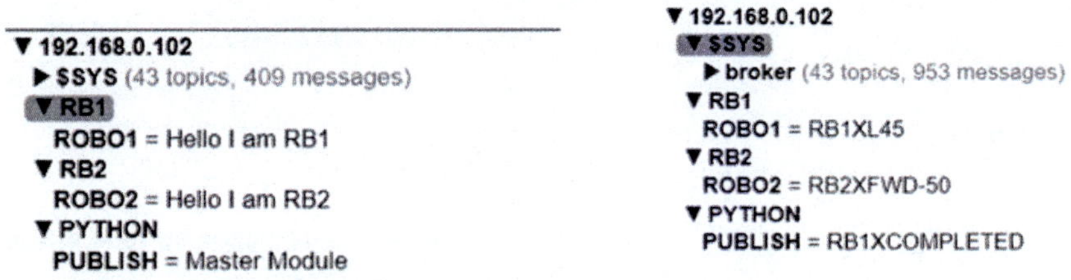

Fig. 6 Connection monitoring on MQTT explorer

```
void MQTTcallback(char* topic, byte* payload, unsigned int length) {
  Serial.print("Message arrived in topic: ");
  Serial.println(topic);
  Serial.print("Message:");
  message="";
  for (int i = 0; i < length; i++) {
    message = message + (char)payload[i];  //Conver *byte to String
  }
  int value = message.toInt();
  Serial.println(value);
  Serial.println(message);
  Serial.println();
  Serial.println("-----------------------");
}
```

Fig. 7 MQTT callback code

Analyzing the factors discussed in Tables 1 and 2, MQTT protocol satisfies the requirement for master-slave communication in coordinate control of mobile robots mainly because its latency is optimal and supports one-many message distribution.

5 Conclusion

This research proposes the development of coordinated control of multiple robots via an Internet network. The algorithm discussed in the research paper allows Python (master module) to publish data to multiple slaves simultaneously. The given algorithm can be replicated for multiple slaves and single master applications in industries such as e-commerce, defense, and manufacturing. MQTT protocol has low latency, hence enabling high-speed communication as compared to HTTP which works on request-response protocol. Moreover, the MQTT broker can connect to a maximum of 15 clients which is widely applicable in master-slave communication for coordinated control of mobile robots. The research can be further developed in the domains of remote monitoring of mobile robots through path planning, monitoring of robots in industrial areas which are hazardous, etc.

References

1. Atmoko RA, Yang D, Alfiani MY, Subiyanto L (2019) Controlling unmanned surface vehicle using MQTT protocol. J Comput Netw, Arch High Perform Comput 1(2):21–28
2. Sasaki Y, Yokotani T (2019) Performance evaluation of MQTT as a communication protocol for IoT and prototyping. Adv Technol Innov 4(1):21–29
3. Thirupathi V, Sagar K (2018) Implementation of home automation system using mqtt protocol and esp32. Int J Eng Adv Technol (IJEAT) ISSN 8 (2018):2249–8958
4. Kashyap M, Sharma V, Gupta N (2018) Taking MQTT and NodeMcu to IOT: communication in Internet of Things. Procedia Comput Sci 132:1611–1618
5. Mesquita J, Guimarães D, Pereira C, Santos F, Almeida L (2018) Assessing the ESP8266 WiFi module for the Internet of Things. In: 2018 IEEE 23rd international conference on emerging technologies and factory automation (ETFA), vol 1. IEEE, pp 784–791
6. Atmoko RA, Riantini R, Hasin MK (2017) IoT real time data acquisition using MQTT protocol. J Phys: Conf Ser 853(1). IOP Publishing
7. Miorandi D, Sicari S, De Pellegrini F, Chlamtac I (2012) Internet of things: vision, applications and research challenges. Ad Hoc Netw 10(7):1497–1516
8. Aroon N (2016) Study of using MQTT cloud platform for remotely control robot and GPS tracking. In: 2016 13th international conference on electrical engineering/electronics, computer, telecommunications and information technology (ECTI-CON). IEEE, pp 1–6
9. Bottone M, Palumbo F, Primiero G, Raimondi F, Stocker R (2016) Implementing virtual pheromones in bdi robots using mqtt and jason (short paper). In: 2016 5th IEEE international conference on cloud networking (Cloudnet). IEEE, pp 196–199
10. Nema 17 Stepper Motor Data sheet. https://components101.com/motors/nema17-stepper-motor
11. SR3 Mini user manual. https://www.symtek.com.tw/wp-content/uploads/2019/05/SR3-Mini-User-Manual.pdf
12. ESP8266WiFi library. https://arduino-esp8266.readthedocs.io.html
13. Arduino client for MQTT. https://pubsubclient.knolleary.net/
14. Accelstepper http://www.airspayce.com/mikem/arduino/AccelStepper/
15. Poroykov A, Kalugin P, Shitov S, Lapitskaya I (2020) Modeling ArUco markers images for accuracy analysis of their 3D pose estimation. In: 30th international conference on computer graphics and machine vision (GraphiCon 2020). Part (Vol. 2)
16. Rani D, Gill NS (2019) Review of various IoT standards and communication protocols. Int J Eng Res Technol 12(5):647–657
17. Yassein MB, Shatnawi MQ, Aljwarneh S, Al-Hatmi R (2017) Internet of Things: survey and open issues of MQTT protocol. In 2017 international conference on engineering & MIS (ICEMIS). IEEE, pp 1–6
18. Liao TL, Lin HR, Wan PY, Yan JJ (2019) Improved attribute-based encryption using chaos synchronization and its application to MQTT security. Appl Sci 9(20):4454

19. Stanford-Clark A, Truong HL (2013) Mqtt for sensor networks (mqtt-sn) protocol specification. Int Bus Mach (IBM) Corp Version 1(2):1–28
20. Nuratch S (2018) Applying the MQTT protocol on embedded system for smart sensors/actuators and IoT applications. In: 2018 15th international conference on electrical engineering/electronics, computer, telecommunications and information technology (ECTI-CON). IEEE, pp 628–631
21. Shakev NG, Topalov AV, Ahmed SA, Shiev KB (2015) Distributed control of robotized nodes in a hybrid wireless sensor network. IFAC-PapersOnLine 48(24):144–149

An Algorithm for Customizing Slicing Floor Plan Design

Pinki Pinki and Krishnendra Shekhawat

Abstract This paper proposes a linear time algorithm for the customization of a slicing floor plan design, which can be done by customizing its modules in the following two ways:

- by modifying the aspect ratio or by modifying either its width or height while retaining its area,
- by modifying its area while keeping either of its aspect ratio or initial width or height.

Both of the aforementioned approaches demonstrate that a slicing floor plan can be generated for any aspect ratio and area while preserving the module adjacencies of the original floor plan. A demonstration has been provided for a devised prototype to validate the viability of the aforementioned approaches.

Keywords Adjacency · Slicing tree · Floor plan · Modules · Slicing floor plan

1 Introduction and Related Work

In literature, there is a lot of work in the field of architectural and VLSI chip design process which revolves around the idea of either maintaining the adjacency of modules or minimizing the area of floor plan. The architectural design process requires both, the adjacency relations and dimensioning, to structure the required floor plan. From a VLSI perspective, the optimization of chip area is the primary objective while structuring a chip design. The increasing complexity of integrated circuits has prompted the development of techniques for VLSI design. There are various optimization techniques that are performed on an initial layout to minimize the cost measures such as chip area and perimeter. In literature, there exist a substantial

P. Pinki (✉) · K. Shekhawat
Department of Mathematics, Birla Institute of Technology and Science, Pilani Campus, 333031 Pilani, India
e-mail: 456pinkiyadav@gmail.com

H. O. Bansal et al. (eds.), *Next Generation Systems and Networks*, Lecture Notes in Networks and Systems 641, https://doi.org/10.1007/978-981-99-0483-9_3

amount of work related to rectangular floor plans with adjacency requirements of its constituent modules which are mentioned as follows.

In 1982, Roth et al. [1] presented the generation of dimensioned rectangular floor plan with specified adjacency requirements. His work involved the division of adjacencies into two parts using a coloring technique and then apply dimensions to them using PERT algorithm. In 1986, Bhasker and Sahni [2] presented an algorithm to generate a rectangular floor plan (if possible) with the same adjacency relations as in the input graph; where the input graph is taken as properly planar triangulated graph. In 2010, Marson and Musse [3] proposed a method for the automated generation of sliceable floor plans using squarified tree maps. Using this approach, the modules with given areas were generated with an aspect ratio close to one. This method does not take into account the adjacency relations of given modules.

Some studies have also been conducted on minimizing the area of rectangular floor plans; a few of them are briefed as follows.

In 1988, Wimer et al. [4] proposed a technique to minimize the area of rectangular floor plan while keeping the area of each building block fixed. The height and width of blocks were allowed to vary. However, the floor plan violated the required adjacency requirements. In 1996, Moh et al. [5] made some amendments to the work done by Wimer [4] by formulating the given problem into a geometric programming problem and solving it using convex optimization techniques while satisfying the adjacency requirements.

Further, a lot of work has also been done on rectangular floor plan optimization. In 1982, Otten [6] presented his work on optimizing a rectangular floor plan where every module has specific dimensions. He chooses the orientations of all the modules to minimize the cost measure. In 1986, Wong et al. [7] worked on area minimization of slicing rectangular floor plans using simulated annealing technique. In 1989, Rosenberg [8] presented a work on minimizing the area in which aspect ratio may be modified in order to achieve minimal area while preserving interconnection requirements. In 2012, Eppstein et al. [9] introduced the idea of area-universal layouts. The layouts (if exist) always met the adjacency requirements and had the magnitude of area of individual block equal to its perimeter. In 2018, Wang et al. [10] presented Graph Approach to Design Generation (GADG) tool. This tool automates the generation of rectangular floor plans and helps in their customization which is done by either adding or removing a module. In 2020, Upasani et al. [11] used linear optimization techniques in order to obtain a feasible dimensioned rectangular floor plan as per user given constraints. In 2021, Shekhawat et al. [12] further improvised the idea by developing a software GPLAN which takes input in the form of dimensions and adjacencies of modules and generates multiple floor plans for the given adjacencies.

From literature, it can be seen that in the existing work, the interaction of user for modifying area and adjacency connections in the corresponding sliceable floor plan is not involved. In contrast to the approaches taken in the aforementioned research, our work is based on re-sizing an existing floor plan rather than creating one to meet the user requirements. In this paper, we have taken a slicing floor plan as input and customized it in two different ways, each one of them preserve the inbuilt adjacency constraints. The main idea and contribution of this paper is to provide a framework for real-time procedural designing and modifying a floor plan as per requirements, where adjacency is preserved. Our approach is simpler and easy to apply.

2 Preliminaries

This section covers terminologies which are used often in the paper and have been previously mentioned in literature [7].

Definition 1 (*FP*) A *floor plan* (FP) is a closed polygon which is divided by either vertical or horizontal line segments into a finite number of non-overlapping polygons. A module in a floor plan is a polygon that is not subdivided.

Definition 2 (*RFP*) A *rectangular floor plan* (RFP) is an enclosing rectangle partitioned into non-overlapping rectangles (by either vertical or horizontal line segment).

Definition 3 (*SFP*) An RFP is slicing if it is a module or there is a segment that slices the rectangle into two pieces that are slicing floor plans themselves.

Definition 4 (*ST*) A *slicing tree* (ST) represents a SFP in the form of a rooted binary tree. In ST, each internal node of the tree is labeled as $*$ and $+$, corresponding to a *vertical division* (V) and a horizontal division (H) respectively. Each leaf of the tree is represented by the rectangles of a floor plan. A slicing tree is an arrangement of two types of divisions (horizontal and vertical) in a SFP. There can be many floor plans represented by the same slicing tree.

For example: In Fig. 1, a SFP and its corresponding slicing tree is given.

Definition 5 (*Block*) A *block* in a floor plan is a rectangle which has either a vertical or a horizontal division that divides the block into two smaller blocks, i.e., child blocks. If the block has a vertical division, then the child block has the same height as that of the parent block. Similarly, if a block has a horizontal division, then the child block has the same width as that of the parent block. A rectangular floor plan is a block in itself.

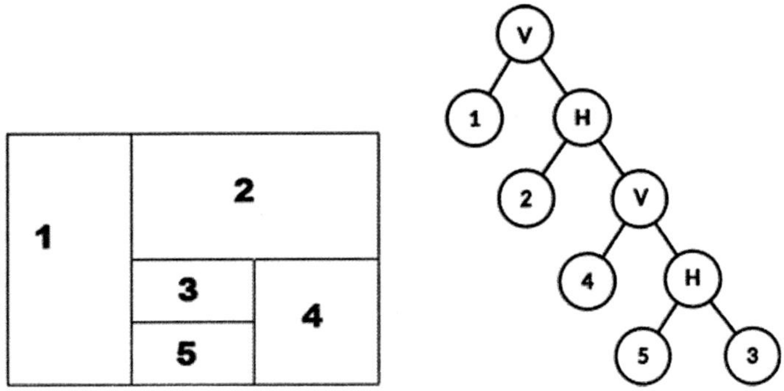

Fig. 1 Slicing rectangular floor plan and its corresponding slicing tree

3 Methodology

This section discusses an algorithm for customizing a slicing floor plan using its slicing tree. The proposed algorithm has the following two parts:

1. **Module Aspect Ratio Modification**: In this part, the aspect ratio of a module is modified by using either changing its height or width, retaining the area of all the modules constant. After modification, the adjacencies among the modules constituting the floor plan remains preserved (Fig. 2).

Algorithm 1[Customizing Algorithm]

(a) A slicing tree of the floor plan F with dimensions
(b) Module Modification
  i. Module which user want to modify (R_C)
  ii. New Dimension (D_n)
(c) Area Modification
  i. Floor plan area to modify

Output: A Modified floor plan
Procedure:

```
if (Module Modification) then
  Search through tree to find R_C
  if (R_C is not found) then
    return doesn't exist
  else if (choose == 1) then                          ▷ For module height change
    R_C.height ← D_n
    R_C.width = R_C.area/R_C.height
  else                                                ▷ For module width change
    R_C.width ← D_n
    R_C.height = R_C.area/R_C.width
  root = R_C.parent
  temp = R_C
  while (root ≠ none) do                              ▷ If root has a vertical division
    if (root == V) then
      root.height = temp.height
      root.width = root.area/root.height
      if (root.left child == temp) then
        Update child(root.right child, temp.height, Invalid)
      else
        Update child(root.left child, temp.height, Invalid)
    else if (root = H) then                           ▷ If root has a Horizontal division
      root.width = temp.width
      root.height = root.area/root.width
    if (root.left child == temp) then
      Update child(root.right child, Invalid, temp.width)
    else
      Update child(root.left child, Invalid, temp.width)
    temp = root
    root = root.parent
else if (Area modification) then
  if (Height constant) then
    Update area height constant(root, new area)
  else if (Width constant) then
    Update area width constant(root, new area)
else if (Aspect ratio constant) then
  Update area aspect ratio constant(root, new area)
exit
```

2. **Area Modification**: In this part, the area of the floor plan is modified by either keeping its height (or width or aspect ratio) fixed. The adjacency between the modules after modification remains undisturbed. Area modification can be performed in either of the following three ways:

   (a) **Modifying the area of a floor plan while keeping height fixed**: In this part, the area of the floor plan is modified by keeping its height constant while its width and aspect ratio are manipulated. The modified floor plan, along-with all of its modules, has the same height as that of the original floor plan (Fig. 3).

```
Procedure: Update area height constant(root, new area)
  old area = root.area
  root.area = new area
  root.width = new area / root.height
  while (root ≠ leaf)
      do
      { Update area height constant(root.left child, root.left child.area * new area / old area)
      Update area height constant(root.right child, root.right child.area * new area / old area)
      root = root.right child }
  exit
```

   (b) **Modifying the area of a floor plan while keeping width fixed**: In this part, a floor plan is re-sized by changing its height and aspect ratio while keeping the width constant, in order to get the desired area. The modified floor plan along-with all of its modules has the same width as that of the original floor plan.

```
Procedure: Update area width constant(root, new area)
  old area = root.area
  root.area = new area
  root.height = new area / root.width
  while (root ≠ leaf)
      do
      { Update area width constant(root.left child, root.left child.area * new area / old area)
      Update area width constant(root.right child, root.right child.area * new area / old area)
      root = root.right child }
  exit
```

   (c) **Modifying the area of a floor plan while keeping aspect ratio fixed**: In this part, a floor plan is re-sized by changing its height and width while keeping the aspect ratio constant, in order to obtain the desired area. The modified floor plan and all of its modules have the same aspect ratio as that of the original floor plan (Fig. 4).

```
Procedure: Update area width constant(root, new area)
  old area = root.area
  root.area = new area
  root.width = √(root.width * root.area/root.height)
  root.height = root.area/root.width
```

```
while (root ≠ leaf)
    do
    { Update area aspect ratio constant(root.left child, root.left child.area * new area / old area)
    Update area aspect ratio constant(root.right child, root.right child.area * new area / old area)
    root = root.right child }
exit
```

Theorem 1 *Customizing Algorithm is well-defined and implemented in $O(n)$ time.*

Proof The input of Algorithm 1 is a slicing tree of a floor plan. The adjacency of modules in the floor plan are the links between the nodes in a slicing tree. In Algorithm 1, we are not modifying the link between nodes of the tree but iterating the nodes over the tree. Hence, the adjacencies in the modified floor plan remain undisturbed. The correctness for Algorithm 1 is as follows:

1. **Precondition**: The floor plan is stored in the form of a binary slicing tree with each node representing a block which is linked to its left child block and right child block. The child blocks are obtained by dividing its parent block either vertically or horizontally. Each node has the dimensional information as well as the information of its division (vertical/ horizontal /if it is leaf of tree) of the block stored.

2. **Postcondition**:

   (a) **Module aspect ratio modification**:
   The nodes in the slicing tree have the dimension information and a node R_c with new dimensions. Each node represents a valid block with the same area with modified dimensions.
   Explanation: In the algorithm, we are changing one of the dimensions of R_c taken as input while retaining its area by setting the other dimension as (area/input dimension). The variable temp, root as R_c, and its parent is initialized. In the loop variant, block stored in temp has been re-sized. After that the parent of temp and its other child is re-sized using the dimension of temp. If a parent block has a vertical (horizontal) division then the height (width) of the parent block and its child block has to be the same. Variable temp is a child block which is already been re-sized. Fix the height (width) of the parent block and the other child using the height of child block stored as temp, respectively. The parent block is re-sized using the height of its child block which is stored as temp and the area of the block. Similarly, the child is updated using procedure Update child(root, height, width).
   The idea behind this procedure is as follows: First we modified the dimension of the block using the input parameter and area. Then, its parent block and its corresponding other child block is updated by verifying the type of division of the block. This is a recursive procedure until the leaf of the tree is reached. In case of leaf of the tree, there is no vertical or horizontal division. Hence, procedure would terminate.

Python 2.7.14 Shell

File Edit Shell Debug Options Window Help

```
Python 2.7.14 (v2.7.14:84471935ed, Sep 16 2017, 20:19:30) [MSC v.1500 32 bit (In
tel)] on win32
Type "copyright", "credits" or "license()" for more information.
>>>
============== RESTART: C:\Users\pc\Downloads\New example 2.py ==============

Enter 1 to run Module modification
Enter 2 to run Area modification
1
Room details before Algorithm is run
id: 1 , height: 4.0 , width: 4.0 , area: 16.0
id: 2 , height: 2.0 , width: 7.0 , area: 14.0
id: 4 , height: 2.0 , width: 7.0 , area: 14.0
id: 3 , height: 4.0 , width: 3.0 , area: 12.0
id: 6 , height: 2.0 , width: 2.0 , area: 4.0
id: 7 , height: 3.0 , width: 2.0 , area: 6.0
id: 8 , height: 5.0 , width: 1.0 , area: 5.0
id: 5 , height: 3.0 , width: 9.0 , area: 27.0
id: 9 , height: 2.0 , width: 7.0 , area: 14.0
id: 11 , height: 2.0 , width: 2.0 , area: 4.0
id: 10 , height: 5.0 , width: 2.0 , area: 10.0
Enter the room id of the room you want to change: 1
Enter 1 to change height, 2 to change width: 2
Enter new width: 6
After updation
id: 1 , height: 2.66666666667 , width: 6.0 , area: 16.0
id: 2 , height: 1.33333333333 , width: 10.5 , area: 14.0
id: 4 , height: 1.33333333333 , width: 10.5 , area: 14.0
id: 3 , height: 2.66666666667 , width: 4.5 , area: 12.0
id: 6 , height: 1.33333333333 , width: 3.0 , area: 4.0
id: 7 , height: 2.0 , width: 3.0 , area: 6.0
id: 8 , height: 3.33333333333 , width: 1.5 , area: 5.0
id: 5 , height: 2.0 , width: 13.5 , area: 27.0
id: 9 , height: 1.33333333333 , width: 10.5 , area: 14.0
id: 11 , height: 1.33333333333 , width: 3.0 , area: 4.0
id: 10 , height: 3.33333333333 , width: 3.0 , area: 10.0
```

(a)

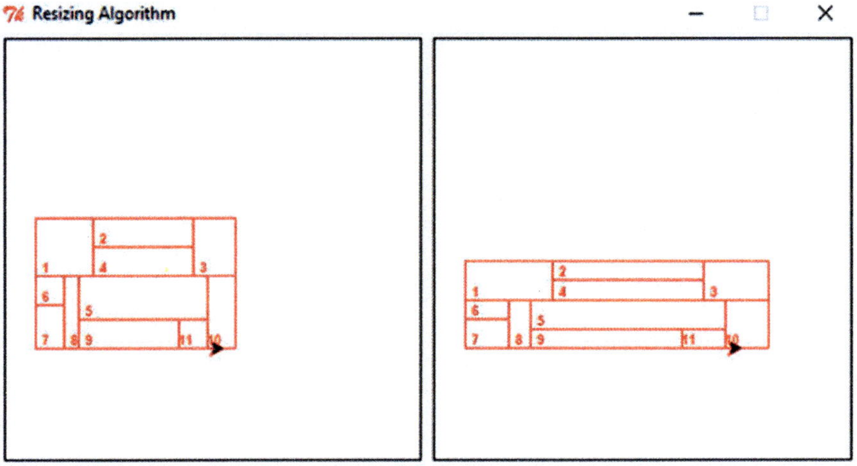

(b)

Fig. 2 Changing width of module "1" from 4 units to 6 units

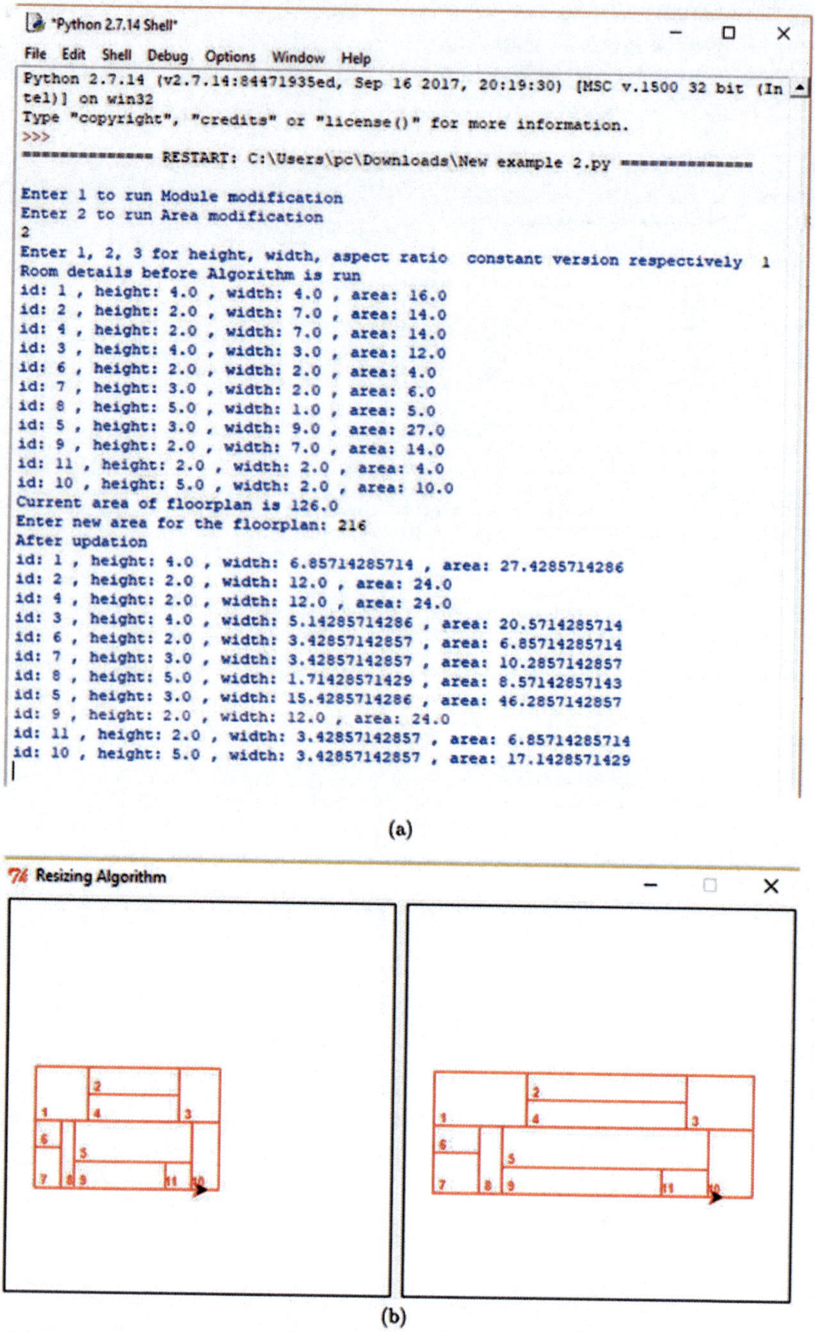

Fig. 3 Modifying the area of floor plan while keeping its height constant

```
*Python 2.7.14 Shell*
File Edit Shell Debug Options Window Help
Python 2.7.14 (v2.7.14:84471935ed, Sep 16 2017, 20:19:30) [MSC v.1500 32 bit (In
tel)] on win32
Type "copyright", "credits" or "license()" for more information.
>>>
============= RESTART: C:\Users\pc\Downloads\New example 2.py =============

Enter 1 to run Module modification
Enter 2 to run Area modification
2
Enter 1, 2, 3 for height, width, aspect ratio  constant version respectively  3
Room details before Algorithm is run
id: 1 , height: 4.0 , width: 4.0 , area: 16.0
id: 2 , height: 2.0 , width: 7.0 , area: 14.0
id: 4 , height: 2.0 , width: 7.0 , area: 14.0
id: 3 , height: 4.0 , width: 3.0 , area: 12.0
id: 6 , height: 2.0 , width: 2.0 , area: 4.0
id: 7 , height: 3.0 , width: 2.0 , area: 6.0
id: 8 , height: 5.0 , width: 1.0 , area: 5.0
id: 5 , height: 3.0 , width: 9.0 , area: 27.0
id: 9 , height: 2.0 , width: 7.0 , area: 14.0
id: 11 , height: 2.0 , width: 2.0 , area: 4.0
id: 10 , height: 5.0 , width: 2.0 , area: 10.0
Current area of floorplan is 126.0
Enter new area for the floorplan: 216
After updation
id: 1 , height: 5.23722936566 , width: 5.23722936566 , area: 27.4285714286
id: 2 , height: 2.61861468283 , width: 9.16515138991 , area: 24.0
id: 4 , height: 2.61861468283 , width: 9.16515138991 , area: 24.0
id: 3 , height: 5.23722936566 , width: 3.92792202425 , area: 20.5714285714
id: 6 , height: 2.61861468283 , width: 2.61861468283 , area: 6.85714285714
id: 7 , height: 3.92792202425 , width: 2.61861468283 , area: 10.2857142857
id: 8 , height: 6.54653670708 , width: 1.30930734142 , area: 8.57142857143
id: 5 , height: 3.92792202425 , width: 11.7837660727 , area: 46.2857142857
id: 9 , height: 2.61861468283 , width: 9.16515138991 , area: 24.0
id: 11 , height: 2.61861468283 , width: 2.61861468283 , area: 6.85714285714
id: 10 , height: 6.54653670708 , width: 2.61861468283 , area: 17.1428571429
```

(a)

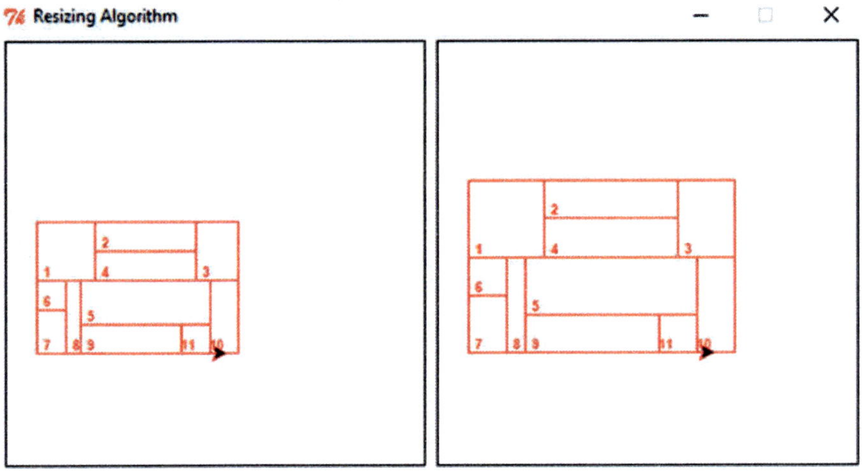

(b)

Fig. 4 Area modification by changing its height and width while keeping its aspect ratio constant

(b) **Area modification**: In this procedure, either of height, width, or aspect ratio of a node is kept unchanged and remaining parameters are calculated with the help of new area. Update function is used with all the three parameters to create three different approaches to update node dimensions recursively. This in turn updates the dimension of whole floor plan.

The worst case time and space complexity of Algorithm 1 is linear in time because in a binary tree with n leaves have $(2 \times (n - 1))$ nodes.

4 Comparison with Existing Work

In literature most of the work is either on generating a floor plan with user-defined constraints [11, 12] or customizing it by addition or deletion of modules [10]. The idea of this paper is simple but it customizes or regenerates any existing rectangular floor plan by either modifying its modules dimensions or its overall dimensions. There are works on customization in the literature (refer to [7, 10], but they did not address the complexity of the method. In contrast to these studies, we proposed a linear time algorithm for designing floor designs based on the idea of slicing trees. Additionally, the algorithm described in this work is put into practice in Python.

5 Conclusion and Future Work

Algorithm 1 illustrates the work presented in this paper and is programmed in python. It provides two different ways of customizing a slicing floor plan, i.e., module aspect ratio modification and module area modification. The work presented in this paper has significant applications in both VLSI and architectural design process. Also, it is quite evident that our approach is practical and feasible. In future, we will try to customize non-slicing floor plans.

References

1. Roth J, Hashimshony R, Wachman A (1982) Turning a graph into a rectangular floor plan. Build Environ 17(3):163–173
2. Bhasker J, Sahni S (1987) A linear time algorithm to check for the existence of a rectangular dual of a planar triangulated graph. Networks 17(3):307–317
3. Marson F, Musse SR (2010) Automatic real-time generation of floor plans based on squarified treemaps algorithm. Int J Comput Games Technol
4. Wimer S, Koren I, Cederbaum I (1988) Floorplans, planar graphs, and layouts. IEEE Trans Circuits Syst 35(3):267–278
5. Moh TS, Chang TS, Hakimi SL (1996) Globally optimal floorplanning for a layout problem. IEEE Trans Circuits Syst I: Fundam Theory Appl 43(9):713–720

6. Otten RH (1982) Automatic floorplan design. In: 19th Design Automation Conference. IEEE, pp 261–267
7. Wong DF, Liu CL (1986) A new algorithm for floorplan design. In: 23rd ACM/IEEE Design Automation Conference. IEEE, pp 101–107
8. Rosenberg E (1989) Optimal module sizing in VLSI floorplanning by nonlinear programming. Zeitschrift für Operations Research 33(2):131–143
9. Eppstein D, Mumford E, Speckmann B, Verbeek K (2012) Area-universal and constrained rectangular layouts. SIAM J Comput 41(3):537–564
10. Wang XY, Yang Y, Zhang K (2018) Customization and generation of floor plans based on graph transformations. Autom Constr 94:405–416
11. Upasani N, Shekhawat K, Sachdeva G (2020) Automated generation of dimensioned rectangular floorplans. Autom Constr 113:103149
12. Shekhawat K, Upasani N, Bisht S, Jain RN (2021) A tool for computer-generated dimensioned floorplans based on given adjacencies. Autom Constr 127:103718

A Prediction Model for Real Estate Price Using Machine Learning

Bhavika Gandhi, Neha Jaitly, Khushi Vats, and Pinki Sagar

Abstract The task of buying or selling a house is troublesome since many factors influence the property valuation such as location, carpet area, and transportation factors. Then there is always a chance of price fluctuation in the future. Thus, one needs to consider both the major determining factors as well as future trends with utmost care. Due to a lack of consistency and fragmentation, real estate and house price prediction is turning out to be an emerging field of research. Since previous years, machine learning has continued to show its impact in various fields like image recognition, speech-to-text translation, fraud detection, etc. Similarly, it can be applied for house price prediction too. The paper is organized into five sections. Section 1 is the introduction and Sect. 2 is concluding the literature. In Sect. 3 existing algorithms are analyzed based on various measures. Section 4 proposes a prediction approach and concludes the paper with future scope.

Keywords House price prediction · Real estate · Machine learning · Prediction · Root mean/square errors · Regression

1 Introduction

Retailing activity is ever-burgeoning and changing in India in the recent past, which has started with economic reforms, liberalization, globalization, and development. One of the biggest decisions of an individual's life is buying or selling an apartment or a house. Every person has his own set of expectations, wants, and needs when it comes to their dream home- buyers look for satisfaction at affordable prices along with what they are looking for. Then there are sellers, who search for a number that can be used as a price tag for the property. This whole process is influenced by a good deal of factors that make property valuation somewhat an inexplicit task.

B. Gandhi · N. Jaitly (✉) · K. Vats · P. Sagar
Department of Computer Science and Engineering, Manav Rachna International Institute of Research and Studies, Faridabad, Haryana, India
e-mail: nehajaitly2@gmail.com

H. O. Bansal et al. (eds.), *Next Generation Systems and Networks*, Lecture Notes in Networks and Systems 641, https://doi.org/10.1007/978-981-99-0483-9_4

It clearly indicates that the real estate price forecast continues to stay as a challenging topic. Since the real estate industry persists as being highly fragmented because of organized and unorganized retailing, house price prediction is turning out to be a prominent research area where due knowledge of machine learning can be applied.

Simply put, Machine Learning is the practice of making computers able to find and learn patterns from the data entered, often for making predictions and subsequent decisions. It is an application of Artificial Intelligence (AI). Among other trending technologies like Internet of Things (IOT), Cloud Computing, Ad-hoc Networks, etc., Machine learning is found to give more cognizance in the widespread area of datasets. Machine learning techniques can therefore be employed to extricate information from the data to get valuable and much more precise results.

The primary aim of this paper is to make use of various machine learning techniques to predict house price. Various machine algorithms are used on the collected datasets and later using different parameters like RMSE (Root Mean Square Error), MSE (Mean Square Error), RMSE (Root Mean Square Error), MAE (Mean Absolute Error), MAPE (Mean Absolute Percentage Errors); the best technique of all aforementioned is determined.

2 Literature Survey

In the literature survey we have done the analysis of half decades by studying the existing algorithms which were executed with different datasets. In Table 1 literature is concluded briefly with the objective, techniques used for the implementation, and key points of the research.

Patil et al. [1] proposed model makes use of CatBoost algorithm along with Robotic Process Automation (RPA). The model has been implemented in stages from Data Extraction, Data Cleaning to applying various algorithms. RPA has been used for Data Scraping purposes. Then, after applying multiple algorithms-XGBoost,

Table 1 Analysis based on RMSLE

Dataset	Algorithms used	Parameters used	
		RMSLE	
		Train set	Test
Housing prices in Beijing	Random forest	0.1298	0.16568
	Extreme gradient boosting	0.16118	0.16603
	Light GBM	0.16687	0.16944
	Hybrid regression	0.14969	0.16372
	Stacked generalization regression	0.16404	0.1635

LightGBM, and CatBoost; it is concluded by careful analysis that CatBoost is a good choice in terms of accuracy.

Wang et al. [2] proposed a model that uses mile-level granularity on publicly released transaction data. Pertinent features are chosen by making use of a feature-selection mechanism. The proposed model Joint Gated Co-attention Based Multi-modal Network (JGC MMN) learns features, records correlations, and later fuses heterogeneous data.

Dabreo et al. [3] proposed model experimented with several Machine Learning Algorithms like XGBoost, Random Forest, Decision Tree, Linear Regression, and Gradient Boosting. Two real datasets are taken from Kaggle from Boston and Melbourne in this research. As per Dabreo et al.'s [3] findings, the best accuracy was provided by XGBoost Regression technique.

Sivasankar et al. [4] proposed a model surveyed to predict housing price using Machine Learning algorithms Lasso Regression, Ridge Regression, Ada-Boost Regression, XGBoost Regression, Decision Tree Regression, and Random Forest Regression. A public output dataset is used and all aforementioned algorithms are used to determine the best among them using parameters like Root Mean Square Error (RMSE). Later, the best three techniques with RMSE value less than 0.12 are integrated to get best output with increased accuracy.

Rawool et al. [5] proposed a model that makes use of various features like ZN, INDUS, CHAS, RAD, etc. Here the raw data is stored in '.csv' file. We are using 2 ML libraries—'pandas' and 'numpy'. Pandas is used to load '.csv' file into Jupiter notebook and also clean the data as well as manipulate the data. For train-test splitting numpy was used. Random forest is predicted better accuracy than other models.

Sakri new et al. [6] proposed a model using particle swarm optimization feature selection to address the dataset dimensionality issues for both classification-based models and regression-based models on Ames housing dataset. Accuracy increased from 81.4% to 84.4% for classification-based models whereas price prediction error decreased from 13.3% to 6.9% for regression-based models.

Zulkifley et al. [7] proposed a model that examined and analyzed the current research on the significant attributes of house price prediction and analyzed the data mining techniques used to predict house price. Technically, houses with a strategic location such as the accessibility to shopping mall or other facilities tend to be more expensive than houses in a rural area with a limited number of facilities. They used SVR, ANN. XG BOOST for the prediction of housing price.

Truong et al. [8] proposed a model that investigates different models for housing price prediction. Three different types of machine learning methods including random forest, XG boost, and light GBM and two machine learning techniques including hybrid regression and stacked generalization. Forest method has the lowest error on training set but is prone to be overfitted as its time complexity is high since the dataset has to be fixed multiple times. The XG boost and light GBM are decent models when comparing accuracy but their time complexities are best with respect to "HOUSING PRICE IN BEIJING" dataset.

Malarvizhi et al. [9] proposed a model that uses the most fundamental machine learning algorithms like decision tree classifier, decision tree regression, and multiple linear regression. Work is implemented using sciki-learn machine learning tool. This helps the users to predict the availability of houses in the city and their prices. Two algorithms, i.e., decision tree regression and multiple linear regression were used in predicting the prices of houses. Comparatively, the performance of multiple linear regression is found to be better than decision tree regression in predicting the house price. In the future the dataset can be prepared with more features and machine learning techniques.

Yu et al. [10] proposed a model that makes use of vector machines, random forests, and XGBoost, later they compared their accuracy using mean square error score. Dataset is divided into two sets in which 80 percent of the data is used for model training and the remaining is used for model testing. At last, various hyperparameters are mixed together to make the lowa house price dataset more accurate.

3 Analysis of Existing Algorithms

Every person desires to buy the house of their dreams. However, the house-hunting process can be confusing and tiring as well as, the person can be caught in the vicious dealings of the sketchy agents. In this paper, we have proposed a model that can help the buyers to get a house as per their expectations at the right price. Even the sellers can make use of this model to find accurate property valuation. Furthermore, this model can be extended to other application areas as well like healthcare, fraud detection, etc. to get higher accuracy rates (Table 2).

Table 2 Analysis based on RMSE mean, RMSE Std Dev. and mean CrossVal

Dataset	Algorithms used	Parameters used		
		RMSE MEAN	Rmse Std. Dev	Mean CrossVal
Boston dataset from Kaggle	XGBoost	3.06	0.75	0.88
	Random forest	3.38	0.76	0.85
	Decision tree	4.189	0.848	0.76
	Linear regression	4.22	0.75	0.69
Melbourne dataset from Kaggle	XGBoost	299,880.4	23,019.5	0.79
	Random forest	312,347.6	19,880.3	0.76
	gradient Boosting	342,911.4	26,544.5	0.73
	Decision tree	421,026.9	15,907.3	0.6
	Linear regression	519,029.1	97,954.6	0.38

4 Proposed Prediction Approach

Step 1: The data can be gathered with the aid of scrapping that collects data from the internet in a format that the machine learning models can use it. Data extraction is another term used to refer to Data Scrapping. Data Scrapping only collects raw data and requires extensive pre-proposing. In some cases, at times, human involvement is also required. Data can also be collected from the local brokers and property dealers.

Step 2: Data pre-processing is done in order to transform the dataset into a clean dataset for better machine learning models. Data pre-proposing techniques are applied to data in raw format which is not feasible for analysis as the data is collected from different sources.

Step 3: House price prediction uses various features and attributes such as location, neighborhood, economy, and structure. So, we will take the basic and most important main features attributes to attain the highest possible accuracy rate.

Step 4: We then use the statistical approach of assumed mean for the regression analysis to find the relationship between variables. For a large dataset, the assumed mean method is preferred over the direct mean method because it simplifies calculating accurate values (Fig. 1).

Next, the mean value is supposed to be predicted from the previous results. Post the prediction of the assumed mean value; we will do the data analysis of the received data from previous steps. Here, we will apply various machines learning algorithm on them and then use different scores such as MSE (Mean Square Error), RMSE (Root Mean Square Error), MAE (Mean Absolute Error), and MAPE (Mean Absolute Percentage Error) to find the accuracy rates. After applying various ML algorithms, we will get their respective accuracy rates and post that we will determine which algorithm is best for housing prices prediction. The purpose behind this model is to convert the numeric data to categorical data since it will help us to attain higher accuracy rates.

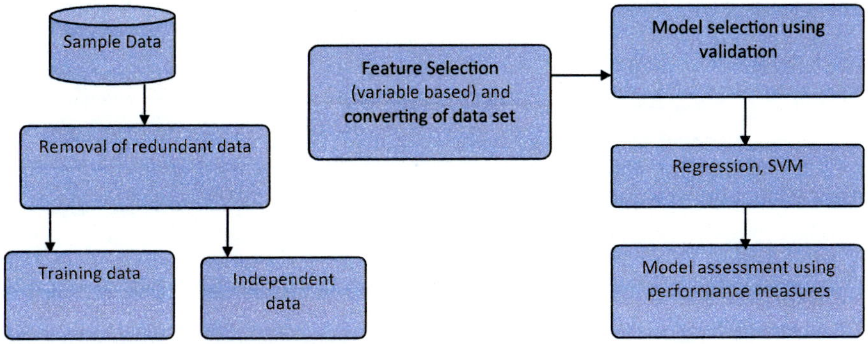

Fig. 1 Housing price prediction for categorical data

Future Scope

The proposed approach is based on prediction of housing rates. Data collected from real estate can be categorical or numerical data and a combination of categorical and numerical data. Our proposed method will convert the categorical data into numerical data so that the accuracy rates can be improved. It can also be very helpful in various application areas like disease prediction, rain forecasting, crop production forecasting, etc. Prediction can be concluded briefly with objectives, techniques used for the implementation, and key points of the research. CatBoost is a good choice in terms of accuracy. XGBoost Regression technique is best as this algorithm has the ability to adapt well with dataset situations and produce accurate and effective results. RMSE values with less than 0.12 are integrated to get the best output with increased accuracy. SVR and ANN can also be used for the prediction of housing price and among these SVR performs better than the rest of the machine learning algorithm. Forest method has the lowest error on the training set but is prone to be overfitted as its time complexity is high since the dataset has to be fixed multiple times. The XG boost and light GBM are decent. Dataset will be prepared with more machine learning techniques.

References

1. Patil P, Shah D, Rajput H, Chheda J (2020) House price prediction using machine learning and RPA. Int Res J Eng Technol (IRJET) 07(03):5560–5563
2. Wang P, Ge C, Zhou Z, Wang X, Li Y, Wang Y (2021) Joint gated co-attention based multi-modal networks for subregion house price prediction. IEEE Trans Knowl Data Eng. https://doi.org/10.1109/TKDE.2021.3093881
3. Dabreo S, Rodrigues S, Rodrigues V, Shah P (2021) Real estate price prediction. Int J Eng Res Technol (IJERT) 10(04):644–649. ISSN: 2278-0181
4. Sivasankar B, Ashok AP, Madhu G, Fousiya S (2020) House price prediction. Int J Comput Sci Eng 08(07):98–102. https://doi.org/10.26438/ijcse/v8i7.98102
5. Rawool AG, Rogye DV, Rane SG, Bharadi VA (2021) House price prediction using machine learning. IRE J 04(11):29–33
6. Sakri SB, Ali Z (2022) Analysis of the dimensionality issues in house price forecasting modeling. In: Fifth international conference of women in data science at Prince Sultan University (WiDS PSU), 28–29 March 2022, Date Added to IEEE Xplore: 03 May 2022, ISBN Information: INSPEC Accession Number: 21776813, IEEE, Riyadh, Saudi Arabia. https://doi.org/10.1109/WiDS-PSU54548.2022.00015
7. Zulkifley NH, Rahman SA, Ubaidullah NH, Ibrahim I (2020) House price prediction using a machine learning model: a survey of literature. Int J Mod Educ Comput Sci (IJMECS) 12(6):46–54. https://doi.org/10.5815/ijmecs.2020.06.04
8. Truong Q, Nguyen M, Dang H, Mei B (2019) Housing price prediction via improved machine learning techniques. In: 2019 International conference on identification, information and knowledge in the internet of things (IIKI2019), pp 433–442. https://doi.org/10.1016/j.procs.2020.06.111
9. Thamarai M, Malarvizhi SP (2020) House price prediction modelling using machine learning. Int J Inf Eng Electron Bus (IJIEEB) 12(2):15–20. https://doi.org/10.5815/ijieeb.com.2020.02.03
10. Yu D. Dalhousie University Halifax, Computer Science, Canada, NS B3H 4R2

Efficient ASIC Implementation of Artificial Neural Network with Posit Representation of Floating-Point Numbers

Abheek Gupta, Anu Gupta, and Rajiv Gupta

Abstract This paper presents a low-power ASIC architecture of a feedforward Artificial Neural Network using Posit representation. The ASIC Posit shows 50% improvement over ASIC using IEEE 754 format in terms of Power and Silicon Area and is also 13% faster while achieving the same accuracy. The same design using the FPGA platform consumes more power than the ASIC design. The designs are done using Cadence RTL Encounter with TSMC 180 nm technology node.

Keywords Artificial neural network · Floating-point representation · Posit · Low power · IEEE 754

1 Introduction

This research paper proposes an Artificial Neural Network (ANN) as Application Specific Integrated Circuit (ASIC) with 50% less power consumption and 13% more speed at comparable accuracy with respect to other published results. Typically, ANN implementations are based on IEEE 754 Floating-Point Representation (IEEE 754). Representing floating-point numbers in digital hardware architectures is a challenge as discussed in Sect. 2. Section 3 describes the benefits of Posit representation which addresses many of the aforementioned issues with IEEE 754 [1]. Section 4 explains the hardware implementation of Posit ANN using ASIC and FPGA. Section 5 discusses the results and their comparisons with published literature results followed by conclusions.

A. Gupta · A. Gupta (✉) · R. Gupta
BITS Pilani, Pilani, India
e-mail: anug@pilani.bits-pilani.ac.in

H. O. Bansal et al. (eds.), *Next Generation Systems and Networks*, Lecture Notes in Networks and Systems 641, https://doi.org/10.1007/978-981-99-0483-9_5

2 IEEE 754

Despite all its benefits, IEEE 754 representation has many limitations. IEEE 754 has a rigid arrangement. The partitions of sign, exponent, and mantissa are fixed in size, and this limits precision leading to rounding errors due to which additional bits have to be reserved. Thus, it is necessary to change to a newer format that requires lesser bit width without affecting the accuracy.

3 Posit Representation

John Gustafson [1] proposed the Posit representation of floating-point numbers. Reciprocals only follow the perfect reflection rule for 0, ±infinity, and integer powers of 2. There are no open intervals. The structure of a Posit is shown in Fig. 1.

- Posits have only two reserved patterns for zero and ±infinity. This reduces the arithmetic inconsistencies that occur due to a large number of reserved patterns in IEEE 754.
- Posits have tapered accuracy for numbers with very large or small exponents. This enables Posits to be able to represent comparable representation accuracy at a lesser number of bits.
- Since the bit widths for Regime, Exponent, and Mantissa are not rigid, many rounding-off errors in computations can be avoided.

As can be observed in Table 1, for a 32-bit number, the dynamic range of Posits is much greater than that of IEEE 754 [1].

These properties of Posits put together make Posits highly suitable for application in devices that need to be portable and power-efficient. Thus, for our application of a portable, in situ water quality monitoring device, ASIC implementation of ANN using Posit representation is highly desirable.

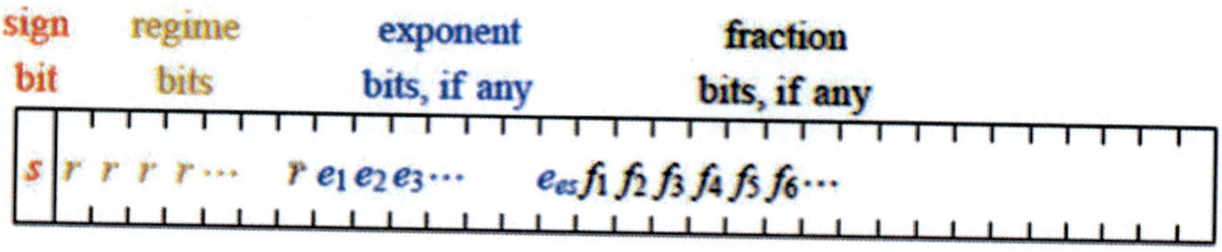

Fig. 1 Format of posit representation [1]

Table 1 Dynamic range comparison between IEEE 754 and Posit

Size, bits	IEEE float exp. size	Approx. IEEE float dynamic range	Posit "es" value	Approx. Posit dynamic range
16	5	6×10^{-8} to 7×10^{4}	1	4×10^{-9} to 3×10^{8}
32	8	1×10^{-45} to 3×10^{38}	3	6×10^{-73} to 2×10^{72}
64	11	5×10^{-324} to 2×10^{308}	4	2×10^{-299} to 4×10^{298}

4 Hardware Implementation

4.1 Structure of ANN

General ANN Architecture—Artificial Neural Networks (ANNs) are typically implemented using software approaches or embedded systems owing to computational accuracy and simplicity. A general structure of an ANN is given in Fig. 2.

An ANN consists of neuron units inspired by the neural structure of the human brain.

Neuron Structure—There are 3 types of neurons in an ANN—the Input neurons, the Hidden neurons, and the Output neurons. Input neurons consist of an activation function. Hidden neurons consist of three arithmetic units—multipliers, adders, and an activation function unit. Output neurons have adders, but no activation function.

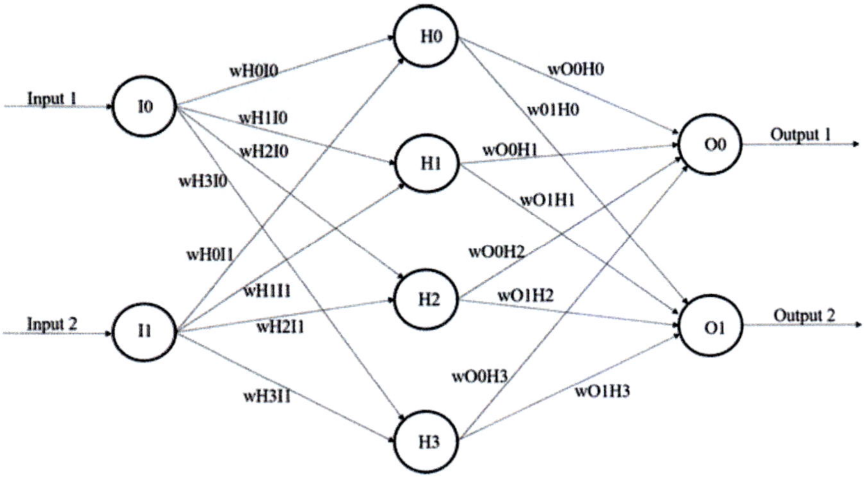

Fig. 2 ANN structure

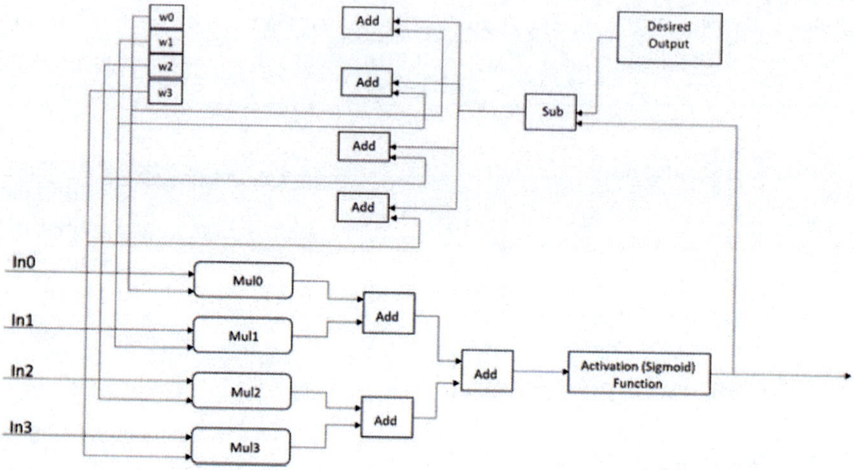

Fig. 3 Structure of a neuron

For the application of water quality monitoring, our neurons are designed to have a sigmoid function as the activation function. The sigmoid function used in this study is the logistic function given by Eq. (1):

$$f(x) = \frac{1}{1 + e^{-x}} \tag{1}$$

Figure 3 shows the internal components of a neuron implemented on Xilinx Zynq 7000 FPGA in the proposed work.

4.2 Design Decisions Made for Proposed Implementation

Many choices exist in the selection of activation function and Posit implementation. This necessitated us to make some design decisions based on our goal to reduce bit size and complexity and increase accuracy as discussed below.

- Decision 1—Choice of Sigmoid as Activation function Sigmoid is suitable for probability prediction with an output range 0 to 1.
- Decision 2—Sigmoid Implementation using Posit—We implemented sigmoid using IEEE 754 and Posit representations. *Sigmoid function with IEEE 754 (using Padé and Non-linear approximation)*—This implementation requires mathematical functions such as exponential and division. Traditionally, Multiply-Accumulate and Division blocks are used which are very intensive operations for floating-point numbers [2]. Amongst these two, Non-linear approximation is more resource-efficient [3]. Thus, we used the Non-Linear approximation-based

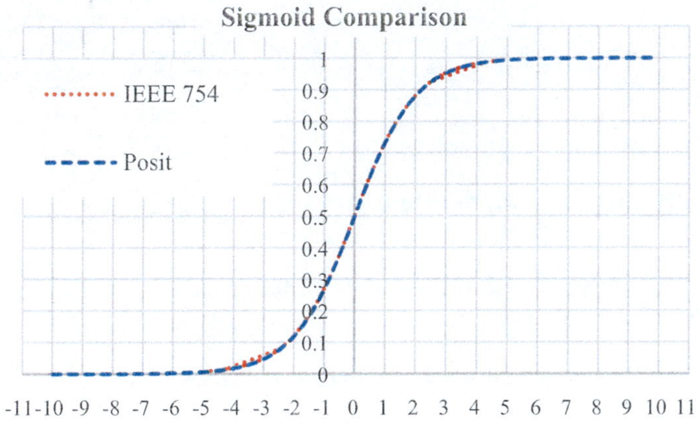

Fig. 4 SF calculation comparison between IEEE 754 and Posit representation

method as the IEEE 754 implementation of an ANN. *Comparison of Representations*—The results thus obtained, for sigmoid calculation in both, have been plotted on the same axes, which display close matching as shown in Fig. 4.

In Fig. 4, SF is applied to real numbers ranging from -10 to $+10$ in both IEEE 754 and Posit. For IEEE 754, an arithmetic unit is designed to calculate the sigmoid value of the numbers. For Posit representation, the first bit is inverted and then shifted right by 2 bits, and 0's are appended to the left to calculate the sigmoid value [1].

- Decision 3—We have bounded the Posit parameters on hardware to get flexibility. This bounded representation is referred to as Parameterised Posit.

From above discussion, it is observed that Posit implementation with fixed point and limited range leads to more data conversion operations, hence increased silicon area, and less usability in case of variable data input conditions, as is typically found in Water quality Indexing.

Parameterised Posit ANN (PPANN)—Posit representation is a flexible representation of floating-point numbers. However, for hardware implementation, we have imposed some limits on all the Posit parameters to achieve the desired accuracy at a much lower bit width. The proposed work tries to address the above challenges by developing full hardware implementation, removing the limitations on the range of input numbers, and not using the fixed point intervention.

Using the above 3 decisions, the complete proposed architecture is described in Sect. 5.

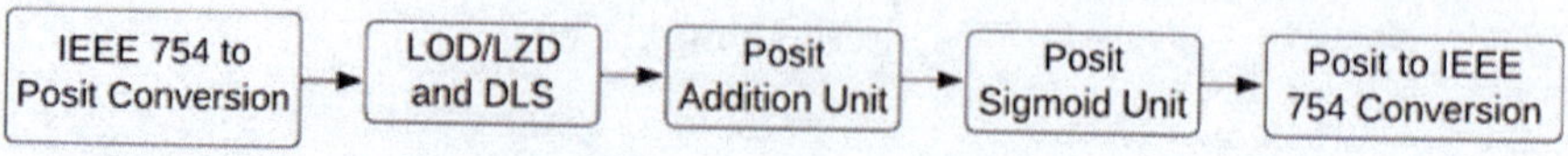

Fig. 5 Design steps for ANN implementation using Posit

5 Parameterised Posit ANN (PPANN) Implementation

The PPANN Implementation consists of 5 design steps as shown in Fig. 5. The operation of each step is shown using flowcharts as given in Figs. 6, 7, and 8:

- Step-1: Flow diagram of IEEE 754 to Posit Conversion.
- Step-2: Leading One/Zero Detector.
- Step-3: Parameterised Posit to IEEE 754 Converter.

5.1 Design Step 1 and Design Step 2—IEEE to Posit Converter and LOD/LZD and DLS

See Figs. 6 and 7.

5.2 Posit to IEEE 754 Converter

See Fig. 8.

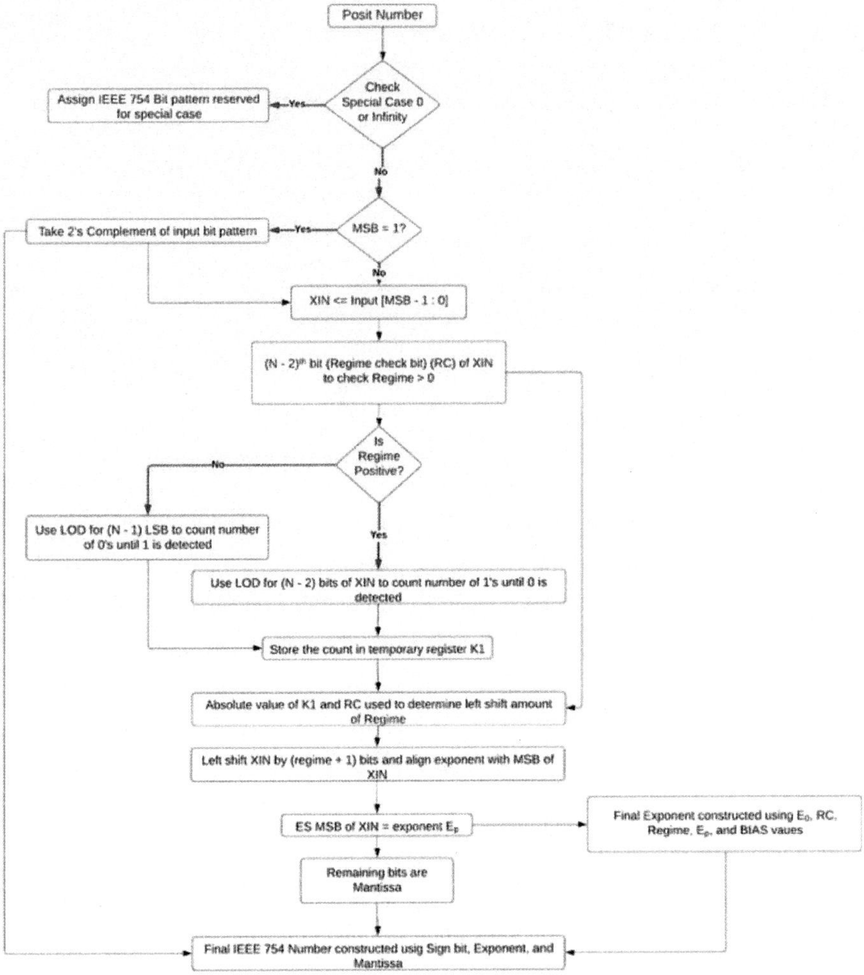

Fig. 6 IEEE 754 to Posit conversion

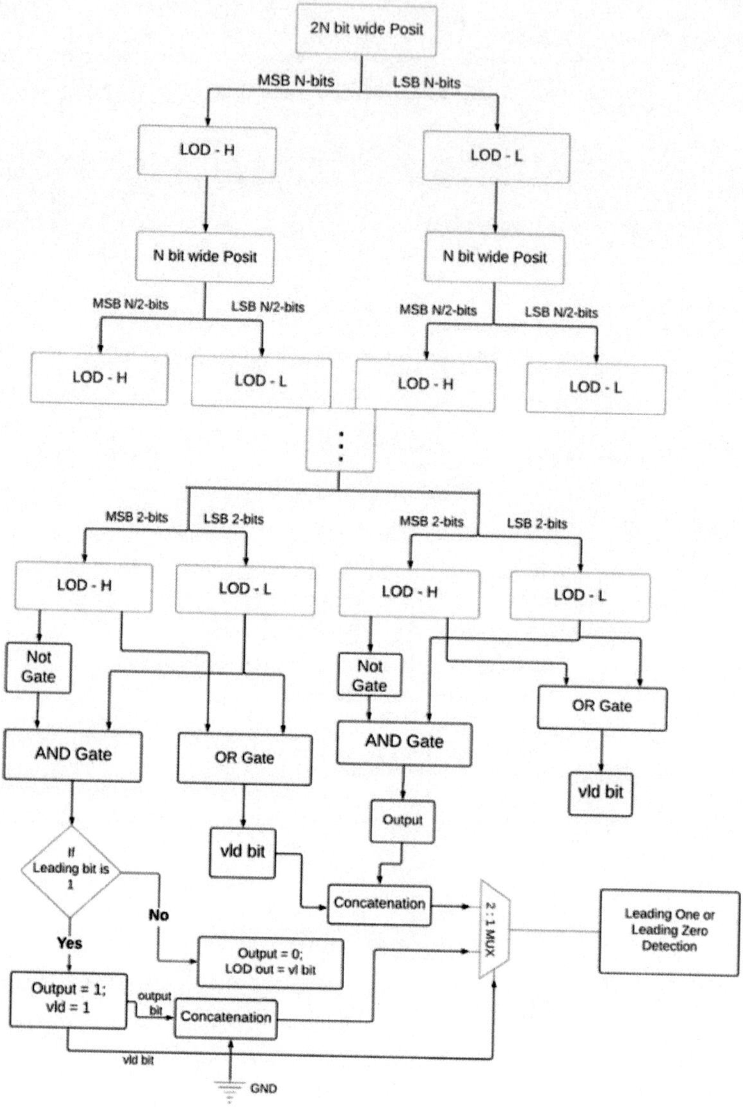

Fig. 7 Leading one/zero detector and dynamic left shifter

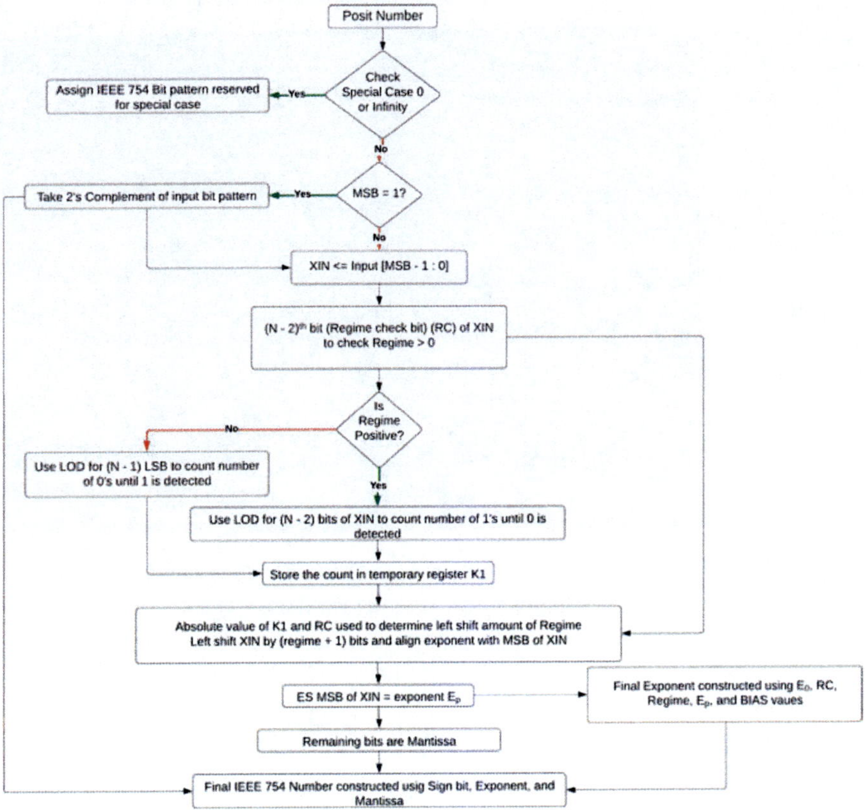

Fig. 8 Posit to IEEE 754 converter

6 Results and Observations of Proposed Smart Water Quality Indexing Device (WQI-Device)

6.1 Snapshot of PPANN

The snapshot of implemented single neuron circuit synthesised using TSMC 180 nm technology node on Cadence RTL Compiler is shown in Fig. 9.

6.2 Comparison of the Results of Proposed ASIC and FPGA Implementation of ANN in IEEE 754 and Posit

The Designed ANNs were trained using 100 samples for Water Quality Indexing. The comparison of results is presented in Figs. 10, 11 and Table 2.

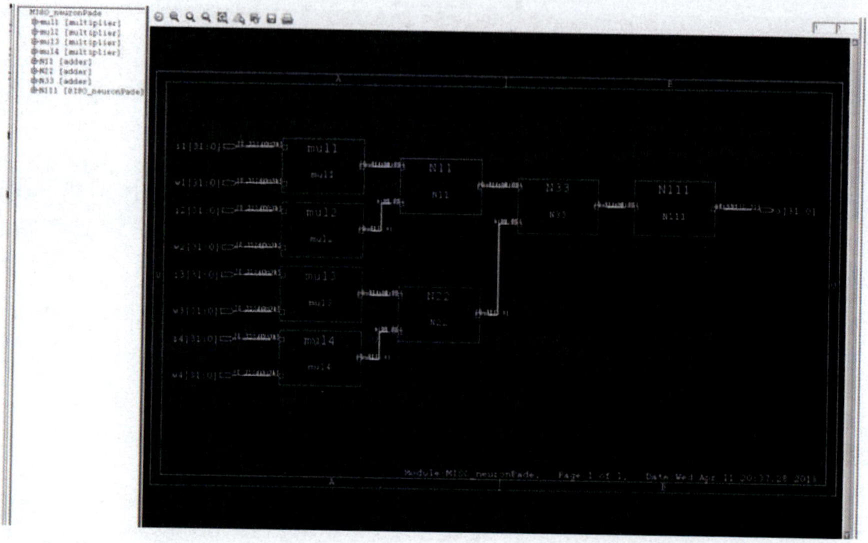

Fig. 9 ASIC implementation of a neuron on Cadence RTL Encounter using TSMC 180 nm standard cell library

Fig. 10 Comparison of proposed ASIC implementation of ANN using IEEE 754 and Posit

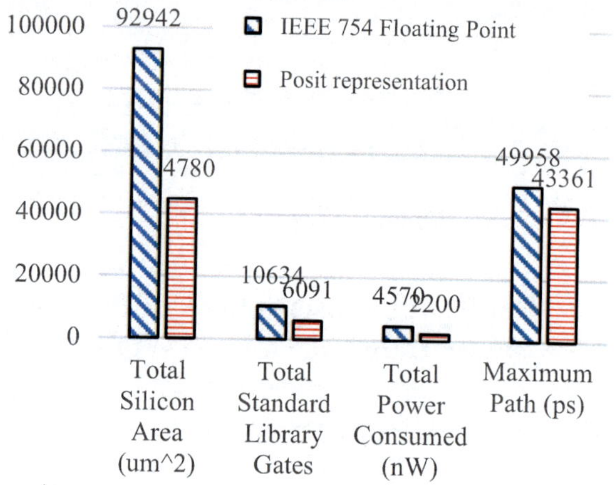

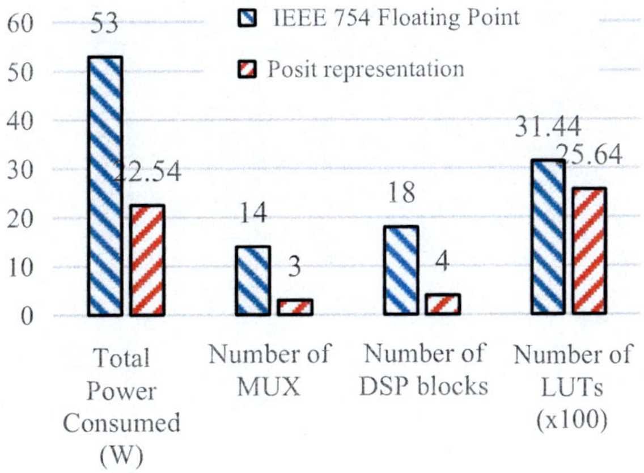

Fig. 11 Comparison of proposed FPGA Implementation of ANN using IEEE 754 and Posit

Observations from Figs. 10 and 11:

- It is observed that Posit implementation consumes 50% less silicon area and standard cells, both.
- Posit has 50% less power consumption IEEE 754 floating-point representation.
- Posit implementation has an advantage of 13.2% over IEEE 754 in Critical Path Delay.
- Posit has comparable accuracy to that of IEEE 754.

Table 2 presents a comparison of ANN-Posit with ANN-IEEE 754.

Observation from Table 2: In comparison to ANN-IEEE 754

- ANN-Posit uses from 48 to 50% less resources
- ANN-Posit consumes from 50 to 80% less power
- ANN-Posit is 13–74% faster.

Table 2 Comparison of ANN Implementations using Posit with IEEE 754

Author	Application	Accuracy obtained in Posit w.r.t. other representations	Resource utilisation in Posit w.r.t. other representations	Power consumption in Posit w.r.t. other representations	Time delay in Posit w.r.t. other representations
Nambi et al. [2]	Image	Comparable accuracy as obtained by IEEE 754-based software implementation	48% lesser LUTs utilised on FPGA	80% improvement over IEEE 754	Power-delay-product reduced by 74%
Cococcioni et al. [3]	Image	Comparable accuracy to IEEE 754	50% bit width reduction	Not given in paper	Not given in paper
Mishra et al. [4]	–	Posit offers 1.5 × more accuracy for the same bit width in comparison to bfloat16	Posit is 1.5 × less resource required	Posit—22.132 uW bfloat16—19.230 uW	Posit—126.998 ns; bfloat16—126.110 ns
Carmichael [5]	Image	Wisconsin breast cancer model—85.89% (Posit) versus 77.4% (IEEE 754); Iris model—98% (Posit) versus 96% (IEEE 754); Mushroom model—96.4% (Posit) versus 96.4% (IEEE 754)	For numerical precision n = 8—IEEE 754 = 1200 LUTs (approx.). Posit = 400 LUTs (approx.)	Energy delay product for 16-bit IEEE 754 = 2.0 (approx.); Energy delay product for 8-bit Posit = 1.9 (approx.)	Not given in paper
Lu et al. [6]	Image	Posit has an accuracy of 98.90% on MNIST Image Database	68% reduction in Silicon area; 75% reduction in memory capacity	51.37% reduction in power	Not given in paper

(continued)

Table 2 (continued)

Author	Application	Accuracy obtained in Posit w.r.t. other representations	Resource utilisation in Posit w.r.t. other representations	Power consumption in Posit w.r.t. other representations	Time delay in Posit w.r.t. other representations
Proposed ASIC Implementation	Water quality classification	99.8% accuracy	50% lesser resource utilisation	50% reduction in power	13% reduction in critical path delay
Proposed FPGA Impl	Water quality classification	99.5% accuracy	18.45% reduction in number of LUTs; 78.6% reduction in number of MUXs; 77.78% reduction in number of DSP blocks	57.47% reduction in power	Not available

7 Conclusions

ANN with Posit representation, of floating-point numbers, is a power, area, and speed-efficient architecture for the ASIC platform. The ANN-Posit design achieves 50% lesser resource utilisation, 50% lesser power consumption, and 13% lesser critical path delays. It achieves similar accuracy in comparison to ANN implemented with IEEE 754. Also, the design is more efficient than published ANN architectures.

References

1. Gustafon JL, Yonemoto IT (2017) Beating floating point at its own game: posit arithmetic. Supercomput Front Innov 4(2):71–86
2. Nambi S, Ullah S, Sahoo SS, Lohana A, Merchant F, Kumar A (2021) ExPAN(N)D: exploring posits for efficient artificial neural network design in FPGA-based systems. IEEE Access 9:103691–103708
3. Cococcioni M, Rossi F, Ruffaldi E, Saponara S (2019) Novel arithmetics to accelerate machine learning classifiers in autonomous driving applications. In: 2019 26th IEEE International conference on electronics, circuits and systems (ICECS), Genoa, Italy
4. Mishra SM, Tiwari A, Shekhawat HS, Guha P, Trivedi G, Jan P, Nemec Z (2022) Comparison of floating-point representations for the efficient implementation of machine learning algorithms. In: 32nd International conference radioelectronika (RADIOELECTRONIKA), Kosice, Slovakia
5. Han J, Orshansky M (2013) Approximate computing: an emerging paradigm for energy-efficient design. In: 2013 18th IEEE European test symposium (ETS), Avignon, France
6. Xu Q, Mytkowicz T, Kim NS (2016) Approximate computing: a survey. IEEE Des Test 33(1):8–22

Machine Learning-Based Stock Market Prediction

Risham Kumar Pansari, Akhtar Rasool, Rajesh Wadhvani, and Aditya Dubey

Abstract In the era of big data, various methods were utilized for forecasting the stock market established on standard time series data like real numbers, fuzzy time series data, and fuzzy sets design. Controlling semantic value data so that accurate values prediction can be generated is done by implementing fuzzy time series data in the stock market forecasting. The models are largely implemented for predicting the dynamic and non-linear datasets in various fields like the stock market and cryptocurrency market. Different modern and smart techniques are being used nowadays in forecasting stock-related data, such as soft computing techniques, back-propagation techniques, Neural Network (NN) techniques, and various other techniques. Artificial Intelligence (AI) methods like NN were developed to forecast a stock's price. Various models were developed which used the feed-forward (FF) NN technique for forecasting the stock market trend and checking different non-parametric and parametric models for predicting the stock market gains. Various soft computing techniques were used for proposing models that dispense with AI for presenting the judgement with the help of loss and profit criteria. The main purpose of the paper is to present a comprehensive review of several prediction techniques utilized for stock market price and return forecasting. This survey contains different forecasting techniques used in stock market forecasting. Also, the paper shows the comparisons between different models.

Keywords Neural network · ANFIS · ARIMA · Modified firefly algorithm

R. K. Pansari (✉) · A. Rasool · R. Wadhvani
Maulana Azad National Institute of Technology, Bhopal 462003, India
e-mail: pansari.risham96@gmail.com

A. Dubey
Madhav Institute of Technology & Science, Gwalior 474001, India

H. O. Bansal et al. (eds.), *Next Generation Systems and Networks*, Lecture Notes in Networks and Systems 641, https://doi.org/10.1007/978-981-99-0483-9_6

1 Introduction

There has been a brisk growth in stock market investors in last few years and predicting the trend of the stock market is now being considered a major task and is of great importance as an accurate prediction of these trends may mine great profits for investors [1, 2]. Stock market forecasting is a tough task because of the non-stationary, thunder-like, and disorganized data [3, 4]. Thus, forecasting the data related to the stock market becomes more challenging for the investors of the stock market to invest money and create profits [5]. There has been great research going on in the field of forecasting stock prices. The complications of the prices of stocks are due to market news, monthly or quarterly, or yearly earnings reports, political decisions, unforeseen conditions, and many more. The investors are dependent on these indicators which affect the prices of the stock and so, these are collected daily. Even though these features affect the price of the stock to greater extend. But, forecasting stock related data is a challenging task because of various other factors which influences the same.

Traditional methods present that the earnings from the stock market can be forecasted from historical stock gains or losses and various other financial features and macroeconomics [15]. The forecasting of returns from the stock market motivated the traders in the direction of exploring the reasons for prediction ability. Stock trend prediction is a cumbersome process as the stock prices are affected by several factors which include investor's expectations, circumstances involving financial conditions and news, administrative events, and various other situations related to market rage [16]. Furthermore, the prices of stock are normally effectual, complex, noisy, and non-linear by characteristics. So, the forecasting process for stock prices using stock TS data becomes a complicated task due to various complicated features, like fickleness, asymmetries, and changing market rages. There were various models registered for forecasting prices of the stock which are governed by models of time series that include the Generalized Auto Regressive Moving Average model, Auto Regressive Moving Average, and Auto Regressive Conditional Heteroscedastic model [17]. Additionally, these models require previous data and normal hypotheses for forecasting.

Various methods were utilized for forecasting the stock market based on standard time series data like real numbers, fuzzy time series data, and fuzzy sets design [18]. For controlling semantic value data so that accurate prediction can be generated is done by implementing fuzzy time series data in the stock market forecasting. The models are largely implemented for predicting the dynamic and non-linear datasets in a variety of fields such as the stock market and cryptocurrency market. Various modern and smart techniques are being used nowadays in forecasting stock-related data such as soft computing techniques, backpropagation (BP) techniques, NN techniques, and various other techniques for forecasting stock-related data. In [4], NN and time series model was integrated for proposing a system that can forecast the difference in stock price. AI methods like NN were developed for forecasting the prices of the stock. Various models were developed which used the technique of FF

NN for forecasting the trend of the stock market and checks various non-parametric and parametric methods to forecast the returns of the stock market. Numerous soft computing techniques are used for proposing models that dispense with AI for presenting the decisions with the help of loss and profit criteria. These models were mostly based on fuzzy-based models (FBM) [17], hybrid, and more.

The primary purpose of the work is to provide a comprehensive review of the variety of prediction techniques utilized for stock market price forecasting and return forecasting. The survey contains various forecasting techniques used in stock market forecasting. Further, the paper contains the discussion on the problem in Sect. 2. In Sect. 3, an analysis of numerous models based on prediction techniques was done. In Sect. 4, an analysis of numerous models based on classification techniques was done. In the last section, the conclusion of the paper and future work in the field is discussed.

2 Problem Discussion

All the stock market is attracting more and more traders day by day. So, traders need to have a good forecasting model that can provide them with some leads that could fetch more profits. But, the forecasting of trends in the stock market is affected directly by trades and investments done in stocks. There have been various studies that have developed tools that could be used for stock market forecasting. These tools can help traders in making better decisions. The market deals with different industrial stock data, which include the complete data related to the market. The traders or investors act according to the market trends. The traders make their buy or sell judgments based on the market status. Various factors influence the market status, such as the company's future income prediction, an announcement of dividends, quarterly, monthly, or yearly income reports, changes in management, political decisions, and more.

3 Analysis Based on Prediction Technique

Various techniques proposed under prediction techniques can be used for forecasting data related to the stock market. The following classification has subclasses of the models: NN, SVR, LSTM, and Hybrid. In each of these subclasses lies several methods under NN; a review of 6 different techniques was done. Similarly, three models were done for the SVR review: the LSTM review for two models and the Hybrid review for three. Under NN, there are models like Wolf optimization-Elman NN (GWO-ENN), Radial Basis Function NN (RBFNN), and Evolving Partially Connected NN (EPCNN). Similarly, SVR-related models were Support vector regression with chaos-based firefly algorithm (SVR-CBF), Support Vector Regression with

Non-Linear Independent component Analysis (NLICA-SVR), and Support Vector Regression with Modified Firefly Algorithm (MFA-SVR), and so on.

3.1 Neural Network

Just like human brains have neurons connected to each other via layers of the network. These neurons are known as nodes. NN can be thought of as a technique of data processing that maps information served as input to a stream of output after processing. It contains one input layer and one output layer, and there is no restriction on number of hidden layers. Different NNs may be utilized for various purposes like classification, pattern recognition, and predictive tasks. The NN is a non-parametric method; learning the market variables' pattern can predict the future price without strict assumptions. It records the correlations between a stock's price and performance and precisely determines features. Numerous sets of input parameters are used for stock prediction.

Grey Wolf optimization-Elman neural network ENN is a category of recurrent neural network (RNN) [3]. The output of the hidden layer in ENN is feedback to themselves through a context layer or recurrent layer. GWO is a type of Swarm Intelligence algorithm that copies the hunting behavior and hierarchy of the grey wolves. The model was tested with daily stock data of 8 companies such as Apple, Bank of America Corporation, Cognizant Technology Solutions Corporation, Goldman Sachs Group, Halliburton Company Common Stock, Microsoft Corporation, IOS, and Oracle Corporation common stock taken from the National Association of Securities Dealers Automated Quotations (NASDAQ) and New York Stock Exchange (NYSE) for analysis. Ten features were computed from the stock dataset, and these are normalized in the range from [0, 1]. These features are normalized using the min-max method. These scaled data were served as input to GWO-ENN, where training was done via ENN and optimization is done via GWO. Every search agent represents the initial solution. During training, the position of every search agent is manipulated by minimizing the MSE, which is the objective function of GWO. The GWO attempts to look for the best value of ENN based on the objective function and initial parameters value.

Radial Basis Function Neural Network (RBFNN) RBFNN was a 3-layered FF network that contained one input layer, one hidden layer, and one output layer [4]. The input layer was served with signal units from the source, and the hidden layer was the second layer. The number of elements in the hidden layer was taken as per requirement. The output layer, the third layer, responds to input data. There is a non-linear motion from the input layer toward the hidden layer and linear motion from the hidden layer toward the output layer. RB activation function was used inside the hidden layer. They applied the RBFNN procedure optimized via the Artificial Fish Swarm Algorithm (AFSA) to predict the trend of Shanghai Composite Indices.

The Algorithm comprises two parts. Classification of every input sample data and activation function center of units in the hidden layer and the width was the first step. The next part was training the connecting weights from the hidden layer toward the output layer after determining the center.

Evolving Partially Connected Neural Network (EPCNN) EPCNN is a combination of global search (GS) algorithms and NN that are partially connected and can be used to forecast the move of the stock price index [5]. The ability to express is more powerful than EPCNN than BP NN because the interconnection existing between neurons is probabilistic. EPCNN comprises six steps which start with identifying the number of layers and neurons. Then scale input variables, initialize weights, and confirm that the neurons remain interconnected. Then, the output value is calculated by serving input features to the network. Then search for optimal or near-optimal connections for the weights by Genetic Algorithms (GA). Then storing the weights that are optimized is done. The output and input of NN were the stock price index and technical indexes. For the experiment, Citigroup and Motors Liquidation Company stocks were selected. The stocks were selected based on earnings per share, monthly sales, daily transaction volume, capital size, and marginal capital cost. The dataset size for training and testing for Citigroup stock was 254 and 124. Similarly, for Motors Liquidation Company stock, the dataset size for training and testing was 253 and 124.

4 Support Vector Regression (SVR)

ANN has a problem of getting stuck into a local minimum, and determining the learning rate and number of hidden layers is difficult. Whereas SVR possesses a global optimum, and also prediction accuracy is high for SVR because it implements the principle of structural potential risk minimization, which considers the regression model capacity and training error.SVR is a type of supervised learning algorithm which is used for predicting discrete values. SVR and SVM use the same principle. SVM is a classification-based method that maps Independent variables of K samples into multidimensional space. It is used to classify data points by separating them via the most optimal hyper-plane. SVR is a kernel-based non-linear regression method that attempts to find a hyper-regression plane with minimal danger in a higher dimensional feature space. It owns better function generalization and approximation abilities. SVR is different from other regression methods as it not only tries to lower the error between the forecasted value and real value, but it also attempts to fit the best fit line within the boundary, which is the distance between the boundary-line and hyper-plane. Using the above capabilities, SVR predicts the future target values of the stock.

4.1 *Support Vector Regression with Chaos-Based Firefly Algorithm (SVR-CBF)*

The primary problem with SVR is determining its hyper-parameters which require prior experience [7]. SVR-CBF model consists of 3 stages. In the Preliminary stage, a method of embedding delay coordinate embedding is considered for reconstructing unknown phase space dynamics. In this stage, the first calculation for the mutual information for the stock time series dataset was done. Then considering the first optimal time delay, the minimum MI function value is obtained. Then the False Nearest Neighbor method was used to explore the least sufficient embedding dimension. Then, the time series phase space is constructed again to explore its unseen dynamics using the collected embedding dimension and optimum time delay. In the next stage, the chaotic firefly algorithm is engaged to optimize SVR hyper-parameters. In this stage, they start by generating the initial firefly position by the use of a chaotic mapping operator was done. Then evaluation for the forecasting error of each firefly (data point) is done. Then there is a chaotic movement of fireflies in which the data points which are less fit move toward the fittest data point, and their positions are updated. Then there is a chaotic movement of the fittest data-point in the solution space. Lastly, in the next stage, the optimized SVR can be utilized for forecasting the stock market price. The dataset used was daily stock market closing prices for Microsoft, Intel, and National Bank shares which were extracted from NASDAQ. Eighty percent of the dataset was utilized for training, and the rest was utilized for testing.

Support Vector Regression with Non-linear Independent component Analysis (NLICA-SVR) In practical life, the stock price is a mixture of various source signals (features), which might be non-linear [8]. So while modeling, these non-linear properties should be considered. So the proposed model integrated NLICA with the SVR technique to improve stock price forecasting. The model consists of 2 stages. The first stage is Applying NLICA to convert original variables into independent and separate components, i.e., smoothing the variables which will be used for forecasting. The next stage is integrating these independent components into the SVR model to develop the model for the stock price forecasting model. Radial Basis function kernel was used as it has a good performance in most of the previous forecasting models, and also it is the widely used activation function. The dataset collected for the experiment was daily stock closing prices for NIKKEI255 and SSEC. The technique of moving window (rolling window) is used for selecting training and testing data.

Support Vector Regression with Modified Firefly Algorithm (MFA-SVR) In the Initial Firefly algorithm, the movement of fireflies is established on a specified step size, which is ill-suited and might disturb the balance between local search and GS [9]. Therefore to overcome this issue, there is a need for a dynamically adjusting step value at different generations. So, the Opposition-based chaotic firefly algorithm

(OCFA) was used, which was developed by getting inspiration from the opposition-based learning approach. The first step is initializing all the fireflies, then fireflies (features) need to be sorted from largest (strongest) to smallest (weakest) as per the light intensities of each firefly. Now there is a need to determine poor fireflies with weak light intensity. Then for each iteration, a fixed number of fireflies is selected, their positions are updated, and their updated light intensities are evaluated. Lastly, the optimized SVR can be utilized for predicting stock market prices. The dataset selected was from Shanghai Stock Exchange. Six securities were selected for the training and testing purpose. The stock price was predicted as an output of the model.

4.2 Long Short-Term Memory (LSTM)

An LSTM is a variation of RNN that could handle long-term dependencies. LSTM model is designed for handling time series data. RNN establishes a directional loop, where a neuron's output can be applied as it is again to self at the later time stamp. This loop gives RNN an advantage in dealing with the problems before and after the input. To resolve the dependence of long term in NN and outburst and the disappearance of RNN model gradients, LSTM was introduced to better store and access information and serve as the basic method for other models. LSTM has connections that are feedback connections. An LSTM unit involves a cell and 3 gates: input, output, and forget. The gates are responsible for regulating the flow of facts in and out of the cell, and the cell is accountable for storing the values over time.

Sentiment and Empirical Mode Decomposition Based LSTM with Attention Mechanism (S-EMDAM-LSTM) In the proposed model, initially, investors' sentiment for the stock forecasting was included, which effectively can increase the model's forecasting accuracy [10]. The sentiment index was calculated via sentiment analysis on various comments on the stock market, which were categorized as either bearish or bullish. Based on this sentiment index, which reflects the shareholders' behavior directly, along with input as the past stock price to the prediction model, was adopted. Then modification to the above model was done by adopting Empirical Modal Decomposition to dig the trend terms of the sequence of stock price. After that LSTM was applied to forecast the closing price of the stock which might be affected due to various factors and highly presents non-linear properties and uncertainty. At last, an improvement to the LSTM was done using an attention mechanism. This final model was utilized for forecasting the stock market. The proposed model focuses on more important information. The experiment was done on an Apple stock dataset with five features: opening prices, highest price, lowest price, closing price, and traded quantity.

Multimodal Event-Driven LSTM Model (ME-LSTM) The problem with the LSTM model is that it might fail to remember the information absorbed from the previous news article before working on the second news event [11]. To resolve the mentioned issue, and Event-driven LSTM model was proposed, that extended an LSTM model to involve an event-driven storage mechanism in which an activating plan of action was imposed to fortify the event-based information found from the earlier stages. It further includes a tensor representation to record communications between different data modes. This was done by applying convolution operations so tensors could be processed using the Conv-LSTM model, which offers a special feature of temporally propagation interconnections via each Conv-LSTM state. Due to this, it is possible to process time series dataset representation as tensors. For the experiment, an extension to the CSI-100 stock dataset was used with additional news articles related to financial events extracted by the topic crawler.

4.3 Hybrid Model

Forecasting stock price tries to find out a company's stock value movement. The popularity of the stock market, that is, the number of investors in the stock market, has grown tremendously, and predicting accurate price movement would help investors to get more benefits from the investment. However, the toughest task is predicting stock movement because it includes factors that directly or indirectly impacts a stock's price like politics, economic growth, price of the currency in the global market, season, politics, etc. So to build a better prediction model for stock returns instead of using a single classifier with its disadvantages, combining various classifiers can be done so that the advantages of each of the used machine learning models can be used in a single hybrid model for prediction. In recent pieces of literature, the results of the different hybrid models were found to be promising. These hybrid models can also be used for forecasting price fluctuation over minutes and days.

Hybrid Red Deer-Grey Algorithm (HRDG) The proposed model is a mixture of concepts of two optimization models concepts: GWO and RDA [12]. The developed model consists of three major phases, which are feature extraction, optimal feature selection, and prediction. Firstly, statistical attributes like Variance, Standard deviation, mean, kurtosis, and skewness were mined from the gathered stock market data. Then the data gathered were used to calculate standard indicators like Exponential Moving Average, Average True Range, Rate of Change, and Relative Strength Index. Then from the selected features, the most optimal features were chosen using Red Deer Adopted Wolf Algorithm. Then the chosen data were utilized for forecasting the stock market via the ensemble learning technique. The model uses SVM, RF1, RF2, and an optimized NN classifier in the ensemble technique. The output was collected from the optimized NN. The input data to the model is stock inputs from the ALIMA, DAR ALARKAN, and SABIC with varying slot sizes from 7, 15, and 20, respectively. The output was the closing price values of the Saudi Stock on the 8th, 16th, and 21st day, respectively.

PSO-based integrated functional link net and interval type-2 fuzzy logic system (FLIT2FLS) The hybrid model utilizes Takagi-Sugano-Kang type fuzzy rules (FR), which contains a type-2 fuzzy set in the predecessor section and is generated from FLANN in the latter part [13]. Along with this, PSO and BP algorithms were independently utilized to optimize the model's parameters. FLIT2FNS has five layers. In the 1st layer, 7 points from the dataset and three features are selected as input. No calculations were required in the layer; it only transfers the data to a further layer. In the 2nd layer, every fuzzy set is expressed by a function of the Gaussian member. Each member trades in with two member functions, and outputs are computed. In the 3rd layer, nodes receive the degree of membership of the rules associated with the 2nd layer. In the 4th layer, nodes are called consequent nodes. The output produced by the 3rd layer and two local outputs generated from FLANN was used as input in the 4th layer, and two outputs were calculated. In the 5th layer, defuzzification is done, which is called the defuzzification layer. Finally, an output value was generated from the FLIT2FNS model after the data was passed from the above layers. For the experiment, the dataset used was of S&P500, BSE, and DJIA stock indices.

Integration of Genetic Fuzzy Systems and Artificial Neural Networks (GFS + ANN) The developed model integrates GA, and ANN approaches used to build an expert system of stock price forecasting [14]. The idea behind this approach is to attain the best prediction accuracy with the use of the minimum number of input data and an easy stock market system. Firstly, Stepwise Regression Analysis was utilized to figure out the features that have a high influence on the price of the stock and eliminate the factors which have a low influence on the stock price. Adding and removing factors is resolved by the statistical value of the F-test and by decreasing the squared error sum. Then the data is divided into n groups using Self-Organizing Map NN. This step reduces the complexity of the dataset into something of higher homogeny. Lastly, all the groups were served into independent GFS models, which can tune the dataset and extraction based on rules for predicting stock price. For the experiment, the dataset used was from 2 sectors: IT and Airlines. In the IT sector, IBM and Dell corporations are used, and British Airlines and Ryan Airlines are used in the airline sector.

5 Conclusion

This paper describes some methods or models for forecasting stock market data. The methods were clustered based on prediction techniques and classification techniques. The methods used for the forecasting included NN, SVR, LSTM, Fuzzy based methods, Hybrid methods, and optimization or filtering-based methods. The paper presents the issues and advantages of some currently existing models and presents a scope of future research into the area of stock price prediction. The method which was mostly utilized for forecasting the stock market is the fuzzy-based method and ANN. These methods could be effectively utilized for keeping track of the stock market. The primary issue of the forecasting stock price models notices is the use of historical data. This is because there are various historical data that affect the price of a stock. The other issue with some models is the computational cost for forecasting. This is an important aspect to be considered because the models should be less costly in terms of computation so that the model can be used by people who are planning for investment and are also investing and can get a better return. In most of the models, the attributes collected were not sufficient. The model should be made by considering more features like news, human sentiments, political decisions, decisions by higher management on specific issues or operations, and many more, as these also affect the stock price. So, further research can be done to generate a better forecasting model for the stock market-related data prediction.

References

1. Ticknor JL (2013) Energy A Bayesian regularized artificial neural network for stock market forecasting. Expert Syst Appl 40:5501–06
2. Rout AK, Biswal B, Dash P (2014) A hybrid FLANN and adaptive differential evolution model for forecasting of stock market indices. Int J Knowl-Based Intell Eng Syst 18:23–41
3. Chandar SK (2021) Grey Wolf optimization-Elman neural network model for stock price prediction. Soft Comput. 25:649–58
4. Shen W, Guo X, Wu C, Wu D (2011) Forecasting stock indices using radial basis function neural networks optimized by artificial fish swarm algorithm. Knowl-Based Syst 24:378–85
5. Chang PC, Wang D, Zhou C (2012) A novel model by evolving partially connected neural network for stock price trend forecasting. Expert Syst Appl 39:611–20
6. Asadi S, Hadavandi E, Mehmanpazir F, Nakhostin M (2012) Hybridization of evolutionary Levenberg-Marquardt neural networks and data pre-processing for stock market prediction. Knowl-Based Syst 35:245–58
7. Kazem A, Sharifi E, Hussain FK, Saberi M, Hussain OK (2013) Support vector regression with chaos-based firefly algorithm for stock market price forecasting. Appl Soft Comput 13:947–58
8. Kao LJ, Chiu CC, Lu CJ, Yang JL (2013) Integration of nonlinear independent component analysis and support vector regression for stock price forecasting. Neurocomputing 99:534–42
9. Zhang J, Teng YF, Chen W (2019) Support vector regression with modified firefly algorithm for stock price forecasting. Appl Intell 49:1658–74
10. Jin Z, Yang Y, Liu Y (2020) Stock closing price prediction based on sentiment analysis and LSTM. Neural Comput Appl 32:9713–29
11. Li Q, Tan J, Wang J, Chen H (2021) A Multimodal Event-Driven LSTM Model for Stock Prediction Using Online News. IEEE Trans Knowl Data Eng 33:3323–37
12. Alotaibi SS (2021) Ensemble technique with optimal feature selection for saudi stock market prediction: a novel hybrid red deer-grey algorithm. IEEE Access 9:64929–44
13. Chakravarty S, Dash PK (2012) A PSO based integrated functional link net and interval type-2 fuzzy logic system for predicting stock market indices. Appl Soft Comput 12:931–41
14. Hadavandi E, Shavandi H, Ghanbari A (2010) Integration of genetic fuzzy systems and artificial neural networks for stock price forecasting. Knowl-Based Syst 23:800–08
15. Boyacioglu MA, Avci D (2010) An adaptive network-based fuzzy inference system (ANFIS) for the prediction of stock market return: the case of the Istanbul stock exchange 37:7908–7912
16. Chen MY, Chen BT (2015) A hybrid fuzzy time series model based on granular computing for stock price forecasting. Inf Sci 294:227–41

17. Hassan MR (2009) A combination of hidden Markov model and fuzzy model for stock market forecasting 72(16):3439–46
18. Cheng CH, Chen TL, Wei LY (2010) A hybrid model based on rough sets theory and genetic algorithms for stock price forecasting. Inf Sci 180(9):1610–29

Nature-Inspired Optimization Algorithm-Assisted Optimal Sensor Node Positioning for Precision Agriculture Applications in Asymmetric Fields

Puneet Mishra and H. D. Mathur

Abstract Precision agriculture has recently become the need of the hour due to an unprecedented rate of world population increment. Through fine-tuned crop activities, the crop yield can be significantly increased for a given crop season from the same field. However, these fine-tuned crop activities majorly rely on the information received from the field through various sensing nodes. With proper integration and placement of sensing nodes in crop fields, a large proportion of pre-harvest activities can be dealt in a better manner. As the area of agricultural fields can be quite large, thereby identification of the optimal positioning of sensing nodes can become a cumbersome task. Improper placement of sensing nodes may result in suboptimal usage of available resources. To overcome this issue, the current work investigates the use of metaheuristic optimization algorithms to optimally place the sensing nodes on the field. This is done by minimizing an innovatively designed cost function which is an indication of the un-scrutinized field area and overlapping coverage area of the sensor nodes. Five different nature-inspired optimization algorithms are applied in this paper for the said task, and it was found that Particle Swarm Optimization provided the most optimal sensing node placement in comparison to the Cuckoo search algorithm, Salp swarm algorithm, Water cycle algorithm, and Life choice-based optimization algorithm.

Keywords Precision agriculture · Nature-inspired optimization · Sensor nodes

1 Introduction

In recent years, research toward technological development for sensor-enabled precision agriculture has become quite popular. With the advent of different sensors viz.

P. Mishra (✉) · H. D. Mathur
Department of Electrical and Electronics Engineering, Birla Institute of Technology and Science, Pilani Campus, Pilani, Rajasthan 333031, India
e-mail: puneet.mishra@pilani.bits-pilani.ac.in

H. D. Mathur
e-mail: mathurhd@pilani.bits-pilani.ac.in

H. O. Bansal et al. (eds.), *Next Generation Systems and Networks*, Lecture Notes in Networks and Systems 641, https://doi.org/10.1007/978-981-99-0483-9_7

soil moisture sensor, soil nutrient sensor, vision-based technologies for detection of crop diseases, etc., the cop productivity has been increased with tailor-made interventions using precision agriculture. It may be noted that such interventions including irrigation, delivery of fertilizers, and specific micro-nutrients are largely dependent on information received from various measurement means or sensor nodes placed in the field. Such sensor nodes are typically employed for cases where in-field measurement of crop health or soil health is desirable. Such in-field continuous soil health monitoring seems to be quite a promising solution; however, a typical problem is associated with such a scheme which is linked to the optimal positioning of the sensor nodes in the field. If the sensor nodes are not placed in the field optimally then it can lead to various disadvantages viz. suboptimal resource utilization, wrong inputs for precision agriculture actions, or even reduced crop yield, etc. In general, for practical use cases, placement of the sensor nodes at arbitrary locations can cause coverage problems in a network [1]. Hence, it is essential to optimize the position of sensor nodes to enhance the coverage rate of sensor nodes in networks [2–4]. Such networks formed by sensor nodes are often termed as wireless sensor networks (WSN), and many researchers have focused on the optimal positioning of sensor nodes for minimizing cost [5], maximizing lifetime [6], etc. A particle swarm optimization (PSO)-based node deployment is proposed in [7] which aims at energy consumption reduction and improving the coverage of nodes, as well. A recently proposed optimization algorithm, i.e., Glowworm swarm optimization was used in [8] for sensor node deployment which also helped in reducing energy requirements from the deployed sensor nodes.

A binary quantum-inspired gravitational search algorithm was proposed in [9], which targeted at the optimization of sensor nodes specifically for precision agriculture applications. This works aimed at finding out the operational mode of each sensor node for increased longevity of the WSN while considering communication-related limitations. It was claimed that the proposed algorithm could reduce energy consumption and hence was able to improve the life span of the WSN. Further, Kulkarni and Venayagamoorthy presented a critical review of the application of PSO in WSNs in [10]. This work not only reviews the applications of PSO on the optimal deployment of sensor nodes but also details their localization, clustering, and data aggregation. In [6], the optimal location of the sink node was found for a large area WSN. The performance of the proposed approach was compared against many other optimization algorithms, and it was found that Harris hawk optimization-based positioning of ink node could increase the lifetime of WSN by reducing energy consumption and localization error.

Based on the literature survey, it can be easily observed that a lot of work has been done in the area of optimal sensor node placement in WSN, however, this work has not been extended or tuned toward the area of precision agriculture. Since the application of WSN in precision agriculture poses other significant challenges such as arbitrary and asymmetric field geometry, it inhibits the use of existing methodologies to provide the optimal position of sensor nodes along with minimum overlap of sensing area in the region of interest (RoI). In view of this, the current work aims at proposing a novel scheme to place the sensor nodes, aiming at uniform and maximum coverage of asymmetric RoI, while minimizing the overlap to reduce the overall installation and operating cost of the WSN for precision agriculture applications.

Further, this paper is divided into 5 sections. Following a brief introduction in the current section, the details about the problem and the mathematical modeling of the proposed scheme for optimal sensor nodes in asymmetric fields are presented in Sect. 2. Section 3 presents a brief introduction to the considered optimization algorithms. Simulation results and the detailed comparative study on the obtained results are presented in Sect. 4. Finally, Sect. 5 concludes the paper.

2 Problem Formulation

It can be easily inferred from the literature survey that the majority of works in optimal sensor node placement usually deals with symmetric field structures, often having rectangular dimensions. However, this kind of uniform structure of the field is seldom possible in practical circumstances. Placement of sensor nodes in an asymmetrical field structure poses significant difficulties as the locations of sensor nodes also must be adjusted according to the field geometry, and placing nodes in symmetrical positions may not yield optimal coverage of the field under investigation. Further, the chances of overlapping the area coverage is also a vital issue. In this view, this paper presents an innovative way to optimally place the nodes while minimizing the overlapping coverage of the sensor nodes and maximizing the coverage of the RoI.

Assume $\mathcal{N}$ is the total number of nodes in the field and each node has a circular sensing area Φ with a radius of λ. Say, $\int = (\int_x, \int_y) : \int_x \in \mathbb{R}^{\mathrm{N}}$ and $\int_y \in \mathbb{R}^{\mathrm{N}}$ are the coordinates of the sensor nodes in the field. Assume the field coordinates are defined by $(\mathcal{X}, \mathcal{Y})$: $\mathcal{X} \in \mathbb{R}^{\Lambda}$ and $\mathcal{Y} \in \mathbb{R}^{\Lambda}$, where $\Lambda \in \mathrm{N}^{+}$, equal to the number of sides defining the arbitrary asymmetric field. The optimal locations of the sensor nodes are defined by $\int_x{}^*, \int_y{}^* \in \mathrm{R}^{\mathrm{N}}$ which will minimize the bare area of the field '$\psi(\int_x, \int_y)$' with minimum overlapping coverage area, say '$\mathcal{O}(\int_x, \int_y)$'.

Say the objective function to be minimized is

$$\mathcal{J}(\int_x, \int_y) := (\mathcal{O}(\int_x, \int_y) + \psi(\int_x, \int_y)) \tag{1}$$

then the optimization problem can be written as

$$(\int_x{}^*, \int_y{}^*) := \underset{\int_x, \int_y}{\operatorname{argmin}} \mathcal{J}(\int_x, \int_y) := \{\int_x, \int_y \in \mathbb{R} : \mathcal{J}(\int_x{}^*, \int_y{}^*) \leq \mathcal{J}(\int_x, \int_y) \forall \int_x, \int_y\} \tag{2}$$

subjected to the constraint that sensor node locations $(\int_x, \int_y)$ strictly lie inside or on the boundary of the polygon formed by the field coordinates, $(\mathcal{X}, \mathcal{Y})$.

The above constraint makes the solution to the optimization problem even more complex to be solved, as now the search space for the optimal node location will be of irregular shape and the optimization algorithm must satisfy the boundary constraint problem alongside the minimization problem. It is worth mentioning here that the current work uses five different nature-inspired optimization algorithms for the task of the minimization of the objective function $\mathcal{J}(\int_x, \int_y)$. Since the considered nature-inspired optimization algorithm is claimed to be free from the gradient-based approach to obtain the optimal solution, these are expected to provide the optimal solution within the boundary condition of the solution space with the most optimal cost function value.

It is worth mentioning here that in this work, the calculation of objective function $\mathcal{J}(\int_x, \int_y)$ is done in an innovative manner so that the solution of the developed optimization problem could be obtained with the help of any unconstrained optimization algorithm as well. Here, an unconstrained optimization algorithm means that the only constraint which is handled by the optimization algorithm is the generation of a solution within a specific lower and upper bound for the specified dimension of the optimization problem. The following pseudo-code is a generic way of obtaining the objective function value $\mathcal{J}(\int_x, \int_y)$ given the values of field coordinates $(\mathcal{X}, \mathcal{Y})$.

Indicative Pseudo-code for objective function calculation

Indicative Pseudo-code for objective function calculation

Get node locations $(\mathcal{s}_x, \mathcal{s}_y)$ *and field coordinates* $(\mathcal{X}, \mathcal{Y})$.
Check if all the node locations $(\mathcal{s}_x, \mathcal{s}_y)$ *are inside the polygon formed by field coordinates* $(\mathcal{X}, \mathcal{Y})$
  YES: Define boundary limits of field coordinates,
    $x_{min} := \min(\mathcal{X}), x_{max} := \max(\mathcal{X}), y_{min} := \min(\mathcal{Y}), y_{max} := \max(\mathcal{Y})$
    Divide the area into pixels (say 'p') as per the resolution (r) and initiate an array ϕ
    for all pixels, i = 1:p
      If pixel(i) is inside the field polygon
        *Calculate the Euclidean distance '**d**' of pixel(i) with sensor node locations* $(\mathcal{s}_x, \mathcal{s}_y)$
        *If **d** is lesser than node sensing radius,* λ
          Increment $\phi(i)$ *by* 1
        else
          Keep $\phi(i)$ *as it is*
      else
        set $\phi(i) = -1$
    Use $\phi(i)$ *to calculate* $\psi(\mathcal{s}_x, \mathcal{s}_y), \mathcal{O}(\mathcal{s}_x, \mathcal{s}_y)$, *and* $\mathcal{J}(\mathcal{s}_x, \mathcal{s}_y)$
  NO: Set $\mathcal{J}(\mathcal{s}_x, \mathcal{s}_y) = \gamma$
return $\mathcal{J}(\mathcal{s}_x, \mathcal{s}_y)$

The above pseudo-code is a mere indicative representation of the calculation of the objective function. Detailed calculations on $\boldsymbol{\phi}$ will be required to effectively calculate the value of $\mathcal{J}(\int_x, \int_y)$. However, the solution of using an unconstrained optimization algorithm for this constrained optimization (minimization) problem is straightforward. One may note that if the solution node positions are not lying in the intended field location, the cost function value, i.e., $\mathcal{J}(\int_x, \int_y)$ will assume a value of γ, a very high positive real number suitable for this problem. In this work, the value of γ is considered as 10^6, which is an appropriate value for this optimization problem.

As may be noted from the structure of the pseudo-code and nature of the optimization problem, this problem falls under the multi-modal category of problem with multiple solutions and plenty of maxima and minima values. Also, the discussed problem will be a high-dimensional problem as the dimension of the problem will be $2\mathcal{N}$. For the current work, the value of $\mathcal{N}$ is taken as 15, so the problem dimension will be 30, depicting the x- and y-coordinates of fifteen sensor nodes in the field. The formulated objective function may be difficult to be solved with the help of a conventional optimization algorithm, and hence, due to the above cited reasons, five different nature-inspired optimization algorithms are considered in this work to obtain the optimal node locations. The details of these algorithms are briefly discussed in the subsequent section.

3 Brief Discussion on Considered Optimization Algorithms

The current work uses five different recently proposed and well-established optimization algorithms viz. Water Cycle Algorithm (WCA) [11], Life Choice-Based Optimization (LCO) algorithm [12], Salp Swarm Algorithm (SSA) [13], Cuckoo Search Algorithm (CSA) [14], and Particle Swarm Optimization (PSO) [15]. All the mentioned algorithms are proven in solving numerical optimization problems and fall under the category of nature-inspired optimization algorithms. Out of all these algorithms, PSO is the oldest, and its first original version was proposed in the year 1995 by Kennedy and Eberhart. This version of PSO explained the use of PSO for the optimization of nonlinear functions using swarm behavior found in many living species. This work also explained about the relationship of PSO with artificial life and neural networks. Since then, the base paper [15] has been cited more than 73,000 times in the literature indicating its vast acceptance in the field of optimization. WCA is an iterative population-based nature-inspired algorithm which mimics the behavior of the water cycle process and models the movements of flowing water bodies, say rivers and their streams toward the oceans [11]. More details about WCA can be found in [11] with its code implementation in [16].

The cuckoo search algorithm is a bio-inspired optimization algorithm proposed by Gandomi et al. [14]. This algorithm is based on the popular behavior of cuckoos laying their eggs in other birds' nests. The stochastic nature of finding the egg by the other bird and then destroying it is typically modeled in the CSA. The concept of levy flight is also introduced in the implementation of the CSA for effective implementation exploration and exploitation in the optimization process [14]. Further, the salp swarm optimization algorithm is based on the swarm behavior of a sea creature, i.e., *salps* and was proposed by Mirjalili et al. [13]. The basic version of SSA tries to mathematically model the navigation and foraging behavior of the alp swarms in the sea. The salp swarms are often found in the form of chain and are believed to provide this creature better locomotion through coordinated changes in their positions. Various modifications and use cases of SSA have since been proposed proving its worth and efficacy in solving real-world optimization problems [17–19].

A recently proposed optimization algorithm, i.e., LCO [12], is also considered in this paper which falls under the population-based algorithm category. It must be mentioned here that it has been innovatively crafted by observing the critical decision-making aspects of humans and then optimizing the choices to achieve the best from life in terms of reaching one's goals. The algorithm was proposed by Khatri et al. in the year 2020. The performance of the LCO was tested for many CEC benchmark test functions, and it was found that it could outperform or perform equally well from its other counterparts. The current work is merely using the above-mentioned optimization algorithms for solving the problem of optimal sensor node positioning in an asymmetric field, and no attempt has been made to modify the structure or basic functionality of the considered optimization algorithms. A detailed explanation of these algorithms is omitted from this paper due to the sake of brevity, however, the same can be found from [11–15], for WCA, LCO, SSA, CSA, and PSO, respectively. Further, the next section discusses the simulation results and extensive comparative studies on the performance of the considered optimization algorithms.

4 Simulation Results and Discussion

In this work, the sensor node positioning optimization of 15 different nodes is carried out using the above-mentioned optimization algorithms. To test and maintain the integrity of the proposed approach, a highly asymmetric field has been considered with field coordinates as $(\mathcal{X}, \mathcal{Y}) = ((00), (110), (812), (148), (103), (3, 1))$, in the clockwise direction, where the first value in each pair indicates the x-coordinate and the second value indicates the y-coordinate of the different field coordinates. A pictorial representation of the field is shown in Fig. 1, from where it may be easily inferred that the considered field is quite asymmetric in nature.

The positioning of sensor nodes in the field is done with the help of the considered nature-inspired optimization algorithms. For the sake of uniformity and the same level of reference for the comparative analysis of the considered algorithms' performance, the number of solution candidates in each iteration and the maximum number of iterations are kept uniform with values of 20 and 200, respectively. Further, for the sake of balancing out the stochastic nature of each algorithm, the optimization process was run independently 50 times and the best solution was stored. Figure 2 shows the convergence curves of the objective function $\mathcal{J}(\int_x, \int_y)$ with respect to iterations for all the considered algorithms. It is worth mentioning here that only the best convergence curves are shown in Fig. 2, which are chosen from the 50 individual independent runs as mentioned above. It may be noted that the cost function value $\mathcal{J}(\int_x, \int_y)$, achieved by CSA, is worst, standing at a value of 1931 and the value obtained by PSO is the least, at a value of 569 only. Please note that the cost function values indicated here are for pixels as shown in the pseudo-code. For a better

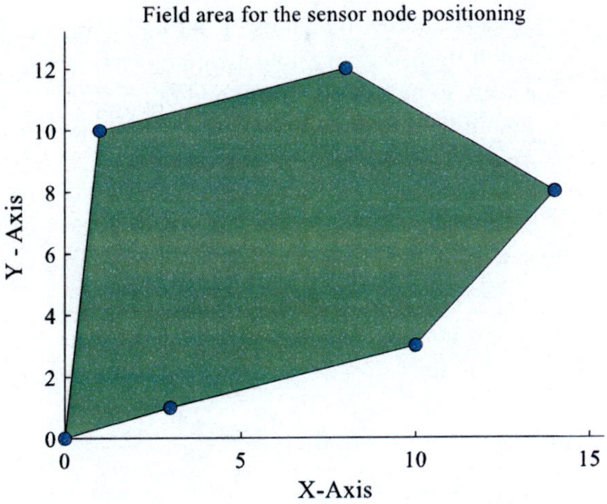

Fig. 1 Pictorial representation of the field under consideration

understanding of the performance of the optimization algorithms in the placement of sensor nodes in the field, more realistic performance indices are calculated, which include the percentage area of the field covered by the sensor nodes, percentage overlapping area of the sensor node coverage of the field, and the percentage uncovered area of the field. It needs a mention here that the covered area is indicating the area which is sensed by a single sensor node. The overlapping area is not included in the covered area performance index to essentially showcase the effectiveness of the proposed approach and optimization algorithm. One can easily observe that the lesser the overlapping area, the better will be optimal placement of the sensor nodes in the field.

Table 1 shows the obtained optimal values of the sensor node positions in the considered asymmetric field from their best runs of all the optimization algorithms.

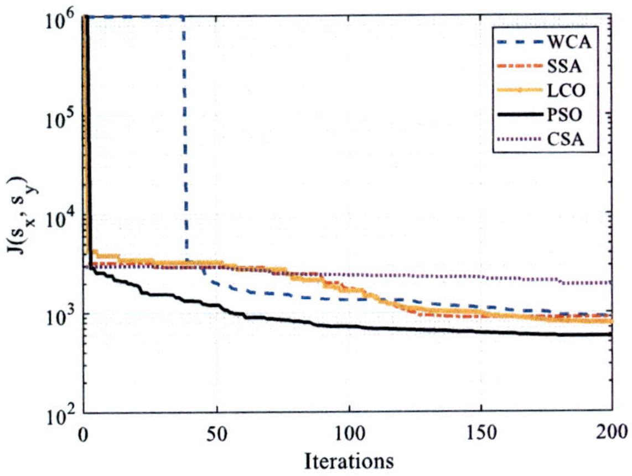

Fig. 2 Best convergence curves for all the considered optimization algorithms

Figure 3 shows the locations of the sensor nodes ($\mathbf{N_P}$) in the field for better visualization. It may be easily noted from this figure that the node positions as obtained by PSO seem to be more uniform and optimally sparse in the entire field, covering the maximum field with the minimum overlapping area.

Table 1 Optimal sensor node positions obtained by considered optimization algorithms

Node #	CSA		SSA		WCA		LCO		PSO	
	f_x	f_y	f_x	f_y	f_x	f_y	f_x	f_y	f_x	f_y
1	8.62	8.35	4.75	6.56	7.90	7.71	1.92	6.90	12.13	7.79
2	5.00	4.56	4.68	9.49	8.80	10.28	9.68	5.72	6.68	8.10
3	11.07	8.31	6.73	5.30	11.63	7.73	11.69	7.88	1.68	8.34
4	7.04	2.18	3.08	3.65	5.20	8.04	8.59	2.97	6.18	2.55
5	6.56	5.99	6.92	10.64	3.88	10.60	4.85	8.17	9.25	6.58
6	10.70	4.87	10.46	5.98	6.43	10.62	4.17	5.14	3.65	3.62
7	6.08	8.08	1.84	6.25	2.45	1.33	7.87	8.05	8.87	3.68
8	2.23	9.45	1.97	9.00	6.71	5.24	0.42	1.34	7.00	10.81
9	8.72	5.22	11.40	8.77	2.17	8.95	2.31	9.12	9.37	9.46
10	10.28	6.62	9.50	10.27	9.73	5.33	1.27	4.30	1.41	1.70
11	1.43	3.50	5.90	2.77	1.74	3.76	5.65	10.81	11.45	4.86
12	6.45	10.15	8.62	3.84	8.11	2.78	8.48	10.65	3.98	6.66
13	2.46	6.92	13.18	7.03	4.94	2.91	2.96	1.99	4.27	9.83
14	8.76	3.74	0.86	1.44	1.08	6.33	5.73	2.60	1.39	5.23
15	2.63	1.72	7.91	7.86	3.86	5.85	6.81	5.35	6.48	5.38

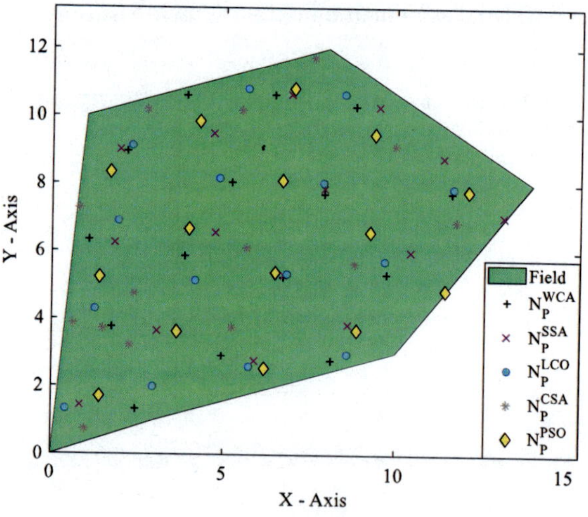

Fig. 3 Optimal sensor node position in the asymmetric field obtained after optimization

Further, from the bar graph shown in Fig. 4, it can be observed that CSA-optimized node positions are far from optimal positioning and only 58.49% area is under coverage by a single sensor node. The overlapping area is approximately 16.44%, indicating the coverage of the field by more than one sensor node, which hints at larger wastage of the available resources in the form of sensor nodes. On the other hand, remaining optimization algorithms viz. WCA, SSA, LCO, and PSO could reach near the optimal node positioning in their best runs, out of which PSO stood out best covering 86.77% of the field area by single nodes and the overlap area was decreased to a level of only 1.39% of the total field. Further, the uncovered area was also decreased to the minimum of all the considered algorithms at the level of 10.83% only. A doughnut representation for all the considered optimization algorithms is also shown in Fig. 5, which aids in a quick overview of the optimizing performance of the algorithms.

Further, it is necessary to visualize the sensor node coverage placed at optimal locations as obtained by all the considered algorithms. It may be noted from Fig. 6e that the PSO-optimized sensor nodes have the least overlapping, indicated using yellow color pixels in the field. Further, green-colored pixels cover almost all of the field. However, optimal node locations obtained by CSA are gathering toward the right side of the field and form regions having overlap from 3 sensor nodes, as shown by a dark red color. For the other algorithms (shown in Fig. 6b–e), there are no

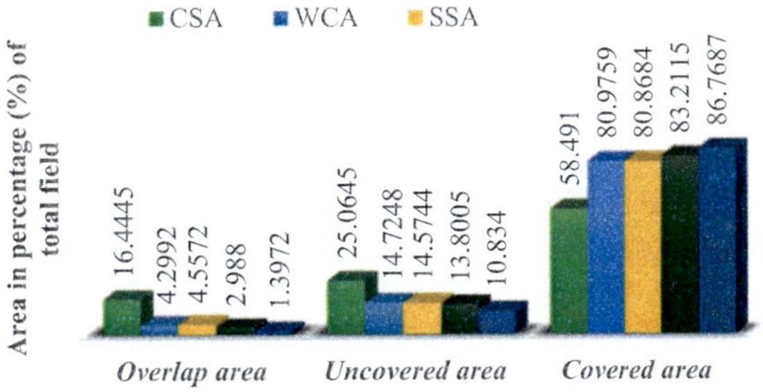

Fig. 4 Bar chart comparison of the performance of considered optimization algorithm

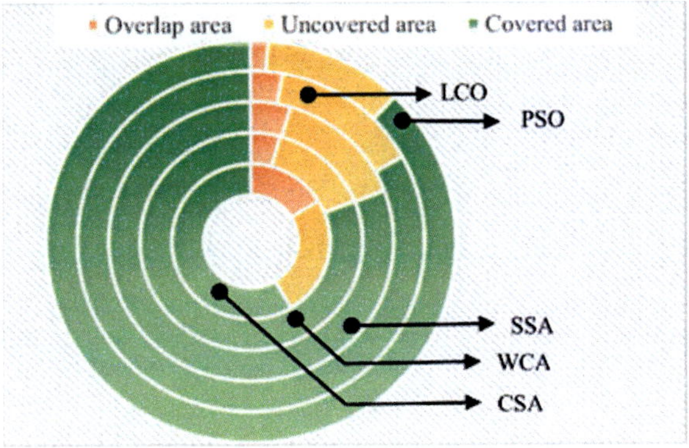

Fig. 5 Donut chart comparison of the performance of considered optimization algorithm

regions on the field having overlap from more than two sensor nodes, however, the overlapped region is higher in comparison to optimal sensor node locations obtained by PSO.

Using the statistical analysis, it could be easily inferred that PSO again emerged as the winner with a minimum mean value of the optimal cost function value and minimum standard deviation among all the runs. Further, for the algorithms SSA, LCO, and PSO, the success rate was 100%, i.e., the optimization could deploy all the sensor nodes optimally inside the field boundaries, or in other words the constrained optimization could be solved efficiently. However, for WCA and CSA, the success rate was 36% and 14%, respectively, i.e., the optimization was converged 18 and 7 times only out of 50 independent runs. Table 2 depicts various indices related to the statistical analysis of the considered optimization algorithms. The high values of standard deviation and mean value for WCA and CSA indicate that for the majority of the runs the optimization could not be converged, and the function value remained at 1E6 at the last of non-converged iterations.

5 Conclusion

In this work, a novel scheme for sensor node placement in the asymmetric field for precision agriculture applications is presented. The proposed scheme encompasses an innovatively designed objective function which aims at optimum resource utilization by maximizing the sensor node coverage in the field and minimizing the overlapped monitoring area of sensor nodes. Five different optimization algorithms, viz. CSA, WCA, SSA, LCO, and PSO were used for solving the proposed optimization problem, and the results were obtained for a polygon-shaped field defined by straight lines connecting six field coordinates. Detailed comparative studies were drawn after extensive simulation studies which included 50 independent runs of each algorithm. It was observed that PSO outperformed all the other considered optimization algorithms in terms of maximum sensor node coverage area, minimum overlapping area, minimum standard deviation, and minimum mean value with a 100% success rate.

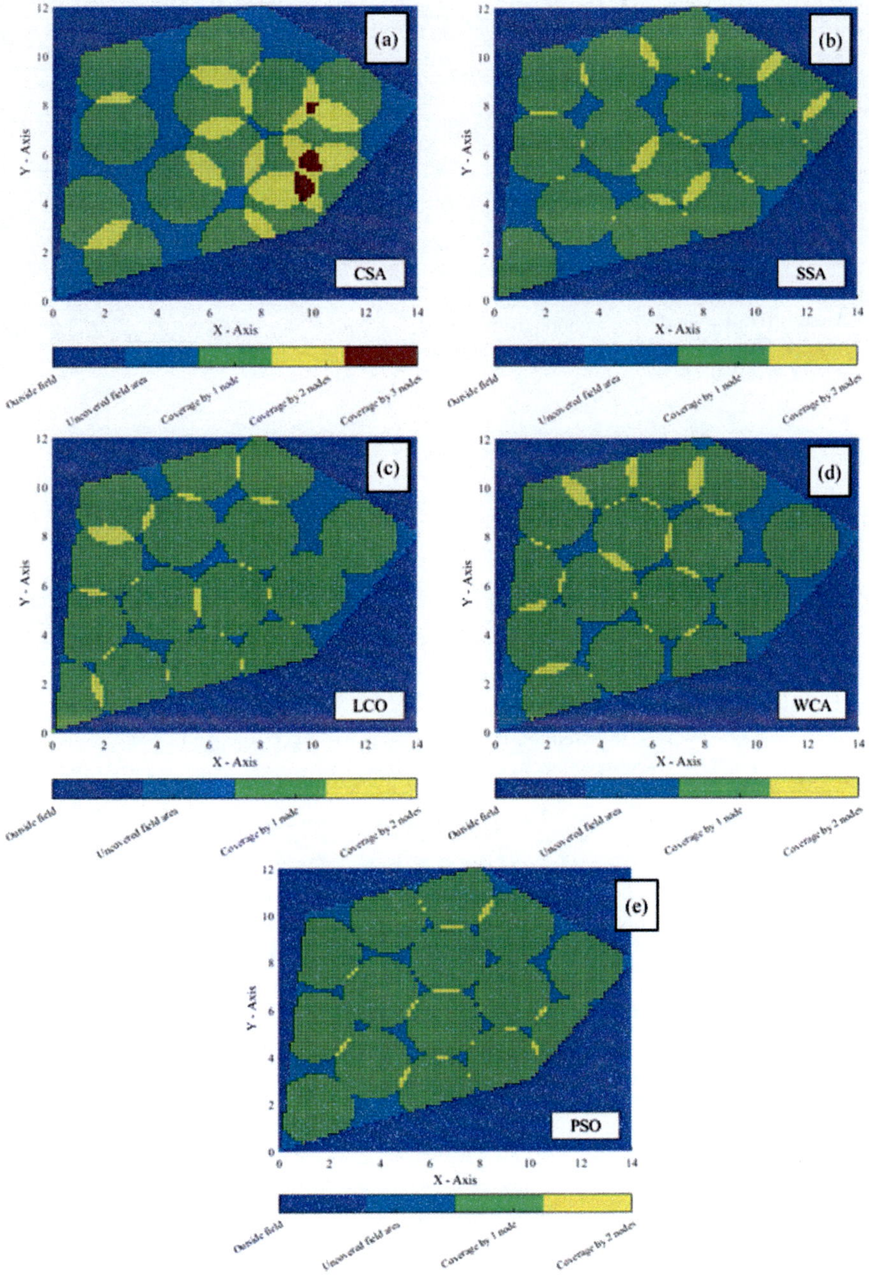

Fig. 6 Sensing node coverage area in field **a** CSA, **b** SSA, **c** LCO, **d** WCA, and **e** PSO

Table 2 Statistical analysis of the considered optimization algorithms

	Standard deviation	Best	Mean	Success rate (%)
CSA	349,679	1931	860,331	14
WCA	484,273	885	640,445	36
SSA	265.22	890	1291.92	100
LCO	301.36	781	1265.72	100
PSO	143.31	569	787.24	100

Acknowledgements The authors thankfully acknowledge the support received from Department of Biotechnology, Ministry of Science and Technology, Govt. of India, through grant BT/PR36167/NNT/28/2019.

References

1. Zhang H, Liu C (2012) A review on node deployment of wireless sensor network. Int J Comput Sci Issues 9(6):378–383
2. Wang G, Cao G, Porta TFL (2006) Movement-assisted sensor deployment. IEEE Trans Mob Comput 5(6):640–652
3. Zou Y, Chakrabarty K (2004) Uncertainty-aware and coverage oriented deployment for sensor networks. J Parallel Distr Comput 64(7):788–798
4. Aitsaadi N, Achir N, Boussetta K, Pujolle G (2011) Artificial potential Feld approach in WSN deployment: cost, QoM, connectivity, and lifetime constraints. Comput Netw 55(1):84–105
5. Mohar SS, Goyal S, Kaur R (2021) Optimized sensor nodes deployment in wireless sensor network using bat algorithm. Wirel Pers Commun 116(4):2835–2853
6. Houssein EH, Saad MR, Hussain K, Zhu W, Shaban H, Hassaballah M (2020) Optimal sink node placement in large scale wireless sensor networks based on Harris' hawk optimization algorithm. IEEE Access 8:19381–19397
7. Li Z, Lei L (2009) Sensor node deployment in wireless sensor networks based on improved particle swarm optimization. In: Proceeding of IEEE international conference on applied superconductivity and electromagnetic devices, Chengdu, China
8. Liao WH, Kao Y, Li YS (2011) A sensor deployment approach using glowworm swarm optimization algorithm in wireless sensor networks. Expert Syst Appl 38:12180–12188
9. Mirhosseini M, Barani F, Nezamabadi-pour H (2017) Design optimization of wireless sensor networks in precision agriculture using improved BQIGSA. Sustain Comput: Inform Syst 16:38–47
10. Kulkarni RV, Venayagamoorthy GK (2010) Particle swarm optimization in wireless-sensor networks: a brief survey. IEEE Trans Syst Man Cybernet Part C (Appl Rev) 41(2):262–267
11. Eskandar H, Sadollah A, Bahreininejad A, Hamdi M (2012) Water cycle algorithm—a novel metaheuristic optimization method for solving constrained engineering optimization problems. Comput Struct 110:151–166
12. Khatri A, Gaba A, Rana KPS, Kumar V (2020) A novel life choice-based optimizer. Soft Comput 24(12):9121–9141
13. Mirjalili S, Gandomi AH, Mirjalili SZ, Saremi S, Faris H, Mirjalili SM (2017) Salp swarm algorithm: a bio-inspired optimizer for engineering design problems. Adv Eng Softw 114:163–191
14. Gandomi AH, Yang XS, Alavi AH (2013) Cuckoo search algorithm: a metaheuristic approach to solve structural optimization problems. Eng Comput 29(1):17–35

15. Eberhart R, Kennedy J (1995) Particle swarm optimization. In: Proceedings of the IEEE international conference on neural networks, vol 4, pp 1942–1948. IEEE
16. Sadollah A, Eskandar H, Lee HM, Kim JH (2016) Water cycle algorithm: a detailed standard code. SoftwareX 5:37–43
17. Tubishat M et al (2021) Dynamic salp swarm algorithm for feature selection. Expert Syst Appl 164:113873
18. Mishra AK, Mishra P, Mathur HD (2022) Enhancing the performance of a deregulated nonlinear integrated power system utilizing a redox flow battery with a self-tuning fractional-order fuzzy controller. ISA Trans 121:284–305
19. Mishra AK, Mishra P, Mathur HD (2021) Design of a dual-layered tilt fuzzy control structure for interconnected power system integrated with DFIG. Int Trans Electr Energy Syst 31(9):e13015

Performance Analysis of All-Optical Relaying for UWOC Over EGG Oceanic Turbulence

Ziyaur Rahman, Rohan Aynyas, and Syed Mohammad Zafaruddin

Abstract In this paper, we consider an amplify-and-forward (AF) optical relaying to enhance the performance of the underwater wireless optical communication (UWOC) system with a negligible line-of-sight (LOS) link over the oceanic turbulence with pointing errors. We analyze the relay-assisted system's end-to-end performance, which consists of complicated probability distribution functions. We derive analytical expressions of the outage probability, average signal-to-noise ratio (SNR), and ergodic rate in terms of UWOC system parameters. We also develop an exact integral-form expression of these performance metrics using the half-harmonic mean of individual SNRs to validate the tightness of the derived analytical expressions. The numerical and simulation analysis shows that the proposed dual-hop relaying has significant performance improvement comparing the direct transmission over considered channel impairments. The considered system also provides a significant gain in the average SNR and ergodic rate for practical scenarios of UWOC deployment.

Keywords Exotic channels · Oceanic turbulence · Underwater optical wireless communication · Performance analysis · Outage probability · Pointing errors · Relaying · SNR

1 Introduction

The underwater wireless optical communication (UWOC) system transmits data through an unguided underwater wireless channel in the unlicensed optical spectrum [1, 2]. However, signal transmission at small wavelengths encounters different channel impairments, such as oceanic turbulence (due to the scintillation effect of light

Z. Rahman (✉) · R. Aynyas · S. Mohammad Zafaruddin
Birla Institute of Technology and Science, Pilani Campus, Pilani, India
e-mail: p20170416@pilani.bits-pilani.ac.in

R. Aynyas
e-mail: h20210086@pilani.bits-pilani.ac.in

S. Mohammad Zafaruddin
e-mail: syed.zafaruddin@pilani.bits-pilani.ac.in

H. O. Bansal et al. (eds.), *Next Generation Systems and Networks*, Lecture Notes in Networks and Systems 641, https://doi.org/10.1007/978-981-99-0483-9_8

propagation) and pointing errors (caused by the misalignment between the transmitter and receiver). The pointing errors caused by the dynamic wind loads, weak earthquakes, and thermal expansion have a statistical effect on the signal quality and present a major challenge in the UWOC deployment [3–6].

The use of relaying to improve the performance of UWOC systems has been extensively studied under the effect of oceanic turbulence and/or pointing errors [7–13]. Hybrid RF/UWOC systems, where relays act as an interface between RF and optical links, have been studied in [7–10]. Considering a dual-hop transmission with a single relay and ignoring the direct transmission, the authors in [7–9] have analyzed the UWOC performance under turbulence and pointing errors. The use of multiple relays in parallel with active relay selection is shown to improve the reliability of a UWOC system in [10].

Relaying for all-optical UWOC systems has recently gained considerable research interest [11–15]. In the seminal work [11], multi-hop and cooperative relaying coupled with amplify-and-forward and decode-and-forward modes have been considered only for turbulence channels. The authors in [13] provided asymptotic expressions of the outage probability and average bit-error-rate (BER) under oceanic turbulence and pointing errors by considering a single amplify-and-forward (AF) relay with fixed and variable gain. Multi-hop relaying with AF and decode-and-forward (DF) protocol under the Gamma–Gamma turbulence and pointing errors have been considered in [12].

In the aforementioned and related research, the statistical effect of exponentiated generalized Gamma (EGG)-distributed oceanic channels with pointing errors has not been considered. Following this direction, the authors in [16] have analyzed the impact of oceanic turbulence without pointing errors on UWOC using various performance measures such as average signal-to-noise ratio (SNR), ergodic rate, outage probability, and BER. However, combining the effect of pointing errors with oceanic turbulence shows high degradation in performance [17]. The authors in [17] have considered three techniques to mitigate this effect, those being multi-hop relay systems using DF, active laser selection, and parallel RF/UWOC links. Multi-hop relaying, on the other hand, requires channel state information (CSI) at each relay to decode the signal, which can be challenging in practice.

It is noted that there are no analyses of average SNR, ergodic rate, and diversity order for the performance improvement of a relay-assisted UWOC system over the oceanic turbulence with pointing errors. Our main contributions to this paper are as follows:

- We analyze the performance of a relay-assisted UWOC system over the oceanic turbulence with pointing errors. We consider a single optical relay with AF relaying, assuming no direct transmission to the destination.
- Using an upper bound approximation of the end-to-end SNR, we analyze the closed-form expressions for average SNR, ergodic rate, and outage probability.
- The analysis shows that the proposed dual-hop relaying significantly improves the performance of the UWOC system.

- We also provided an integral-form expression of the probability density function (PDF) for the end-to-end SNR to validate the tightness of the derived analytical expressions.
- Numerical evaluation is presented to validate the results of the analytical formulae, demonstrating the improvement in performance compared to the direct transmission.

2 System Model

We consider a UWOC system using intensity modulation/direct detection (IM/DD). It consists of a single-aperture transceiver system with a negligible line-of-sight (LOS) link, under oceanic turbulence and pointing errors. The signal received at the receiver aperture, y, is given as

$$y = h_l h_t h_p x + w, \tag{1}$$

where x is the transmitted signal, w is the additive white Gaussian noise (AWGN) with variance σ_w^2, and $h_l = e^{-\alpha l}$ **is the oceanic path loss coefficient (with link distance** l **(in m) and extinction attenuation coefficient** α) [17]. The terms h_t and h_p are the random states of the oceanic turbulence channel and pointing errors, respectively.

We define $\gamma = \gamma_0 |h|^2$ as the SNR, where $h = h_p h_t$, $\gamma_0 = 2P_t^2 R^2/\sigma_w^2$, and P_t is the average optical transmitted power. The PDF of SNR for the UWOC system can be easily derived using the distribution of $|h|^2$ [18]:

$$f(\gamma) = \frac{\omega\rho^2}{2\gamma} G_{1,2}^{2,0}\left(\begin{matrix} \rho^2+1 \\ 1, \rho^2 \end{matrix} \middle| \frac{1}{\lambda A_0} \left(\sqrt{\frac{\gamma}{\gamma_0}}\right) \right)$$
$$+ \frac{(1-\omega)\rho^2}{2\Gamma(a)\gamma} G_{1,2}^{2,0}\left(\begin{matrix} \frac{\rho^2}{c}+1 \\ a, \frac{\rho^2}{c} \end{matrix} \middle| \frac{1}{b^c A_0^c} \left(\sqrt{\frac{\gamma}{\gamma_0}}\right)^c \right), \tag{2}$$

where ω **represents the mixture coefficient of the distributions** (i.e., $0 < \omega < 1$), λ, a, b, and c are the exponential distribution parameters [19]. The parameter $\rho = \omega_{z_{eq}}/2\sigma_s$ is the ratio of the equivalent beam radius and the standard deviation of the pointing errors (due to misalignment between the transmitter and receiver) displacement at the receiver [17]. The parameter A_0 is the fraction of collected power given as $A_0 = (\text{erf}(\upsilon))^2$ where $\upsilon = \sqrt{\pi/2}\, a/\omega_z$, and a is the receiver aperture radius.

The cumulative distribution function (CDF) of EGG-distributed oceanic turbulence channel with pointing errors [18]:

$$F(\gamma) = \omega\rho^2 G_{2,3}^{2,1}\left(\begin{matrix}1, \rho^2+1\\ 1, \rho^2, 0\end{matrix}\,\middle|\,\frac{1}{\lambda A_0}\left(\sqrt{\frac{\gamma}{\gamma_0}}\right)\right)$$
$$+\frac{(1-\omega)\rho^2}{c\Gamma(a)} G_{2,3}^{2,1}\left(\begin{matrix}1, \frac{\rho^2}{c}+1\\ a, \frac{\rho^2}{c}, 0\end{matrix}\,\middle|\,\frac{1}{b^c A_0^c}\left(\sqrt{\frac{\gamma}{\gamma_0}}\right)^c\right). \tag{3}$$

For the case of relayed transmission, the expressions for signals received at the relay and destination are

$$y_r = h_{l1}h_{t1}h_{p1}Rx + w_1 \tag{4}$$

$$y_d = h_{l2}h_{12}h_{p2}Gy_r + w_2, \tag{5}$$

where h_{l1}, h_{t1}, h_{p1} and h_{l2}, h_{t2}, h_{p2} are oceanic path loss, oceanic turbulence, and pointing error states between source–relay and relay–destination, respectively, each having w_1 and w_2 as AWGNs. $h_1 = h_{l1}h_{t1}h_{p1}$ is the combined channel between source and relay, and $G = \sqrt{\frac{P_t}{h_1^2 P_t^2 R^2 + \sigma_w^2}}$ is the antenna gain. Here, x and y_r are transmitted from the source and relay, respectively.

Taking $P_t = P_r$, we get the instantaneous SNRs of signals received at the relay and receiver as $\gamma_1 = \gamma_0 h_1^2$ and $\gamma_2 = \gamma_0 h_2^2$, respectively. Using these, the expression for the end-to-end SNR is given as [20]

$$\gamma_{e2e} = \frac{\gamma_1\gamma_2}{\gamma_1 + \gamma_2 + 1}. \tag{6}$$

To derive tractable analytical expressions using the well-known half-harmonic mean expression, we ignore the 1 (since the PDF value is small for $\gamma < 1$) in the numerator to get

$$\gamma_{e2e} = \frac{\gamma_1\gamma_2}{\gamma_1 + \gamma_2} \leq \min(\gamma_1, \gamma_2). \tag{7}$$

We assume similar oceanic turbulence conditions (i.e., same ω, λ, a, b, and c) for both links. This assumption is practical since oceanic turbulence does not change over distances of a few meters, over which a majority of UWOC systems work. However, we get different values for ρ and A_0 depending on the individual link lengths. We also consider that the relay is situated the mid-way between the source and destination since this results in optimum performance as demonstrated in the numerical section. This causes the parameters A_0, w_z, ρ, ω, λ, a, b, and c are the same for both links (i.e., a symmetric condition), facilitating insightful analytical expressions.

3 Performance Analysis

In this section, we present an integral-form expression to determine the exact outage probability, average SNR, and ergodic rate. Considering the intractability of the integration, we use the upper bound approximation of the harmonic mean and derive closed-form analytical bounds of these parameters. The derived expressions show the system behavior in a relay-assisted environment under the combined effect of pointing errors and oceanic turbulence.

3.1 Distribution Function of End-to-End SNR

We first present an exact expression (with the exception of ignoring 1 in the denominator of (6)) of the PDF of end-to-end SNR and then represent a tractable PDF for the performance analysis. Denoting the PDF of γ_1 and γ_2 as $f_1(\gamma)$ and $f_2(\gamma)$, respectively, we evaluate the PDFs of $1/\gamma_1$ and $1/\gamma_2$ using the inverse distribution and employ the PDF of their sum [21, Eq. 5.55] to get resulting PDF of γ_{e2e} as

$$f(\gamma) = \gamma \int_{t_{min}}^{t_{\max}} f_1\left(\frac{\gamma}{t}\right) f_2\left(\frac{\gamma}{1-t}\right) \frac{dt}{t^2(1-t)^2}. \tag{8}$$

This representation is used to evaluate the UWOC performance numerically as a benchmark for our derived analytical expressions. Since the integral-form expression in (8) is intractable for the PDF in (8), we use an upper bound approximation $\gamma_{e2e} = \min(\gamma_1, \gamma_2)$, to get the CDF and PDF [20]:

$$F(\gamma) = F_1(\gamma) + F_2(\gamma) - F_1(\gamma)F_2(\gamma) \tag{9}$$

$$f(\gamma) = f_1(\gamma) + f_2(\gamma) - f_1(\gamma)F_2(\gamma) - f_2(\gamma)F_1(\gamma), \tag{10}$$

where $f_1(\gamma)$, $f_2(\gamma)$, $F_1(\gamma)$, and $F_2(\gamma)$ can be obtained from (2) and (3) using $l_1 = l_2 = l/2 = l_r$, where l_1 is the distance between source to relay and l_2 is the distance from the relay to the destination.

3.2 Outage Probability

Outage probability is a performance metric to demonstrate the effect of fading channels. It is defined as the probability of failing to reach a specified quality of service (QoS), for example, an SNR threshold value γ_{th}. Using the PDF in (8), an exact integral representation of the outage probability is specified as

$$P_{out} = P(\gamma < \gamma_{th}) = \int_{0}^{\gamma_{\text{th}}} f(\gamma) d\gamma.$$

Further, we can use the CDF in (9) with $\gamma = \gamma_{\text{th}}$ to get an upper bound. In what follows, we develop an asymptotic expression of the outage probability.

Proposition 1 *If $\{\omega, \lambda, a, b, c\}$ are the parameters of oceanic turbulence, A_0 and ρ are the parameters of pointing errors with $l = 2l_r$ as the transmission link length for a relay-assisted UWOC system, then an asymptotic expression for the outage probability at high SNR $\gamma_0 \to \infty$ is given as*

$$P_{out}^{\infty} = 2P'_{out} - (P'_{out})^2, \tag{11}$$

where

$$P'_{out} = \omega\rho^2 \sum_{k=1}^{2} \frac{\prod_{j=1, j\neq k}^{2} \Gamma(\mathcal{B}_j - \mathcal{B}_k)\Gamma(1-\mathcal{A}_j+\mathcal{B}_k)}{\Gamma(\mathcal{A}_j-\mathcal{B}_k)\Gamma(1-\mathcal{B}_j+\mathcal{B}_k)} \left(\frac{\sqrt{\gamma_{th}}}{\lambda A_0}\right)^{\mathcal{B}_k} \gamma_0^{-\frac{\mathcal{B}_k}{2}} +$$
$$\frac{(1-\omega)\rho^2}{c\Gamma(a)} \sum_{k=1}^{2} \frac{\prod_{j=1, j\neq k}^{2} \Gamma(\mathcal{D}_j - \mathcal{D}_k)\Gamma(1-\mathcal{C}_j+\mathcal{D}_k)}{\Gamma(\mathcal{C}_j-\mathcal{D}_k)\Gamma(1-\mathcal{D}_j+\mathcal{D}_k)} \left(\frac{\sqrt{\gamma_{th}}}{bA_0}\right)^{\frac{\mathcal{D}_k}{c}} \gamma_0^{-\frac{\mathcal{D}_k}{2c}}, \tag{12}$$

where $\mathcal{A}_j = \mathcal{A}_k = \{1, \rho^2 + 1\}$, $\mathcal{B}_j = \mathcal{B}_k = \{1, \rho^2, 0\}$, $\mathcal{C}_j = \mathcal{C}_k = \{1, \frac{\rho^2}{c} + 1\}$, and $\mathcal{D}_j = \mathcal{D}_k = \{a, \frac{\rho^2}{c}, 0\}$.

Proof sing $\gamma = \gamma_{th}$ in (3) and applying the identity [22, Eq. 07.34.06.0006.01], we get the term in (12), and thus (11).

It can be easily seen from the outage probability expression that the exponent of the SNR γ_0 is $\{\frac{1}{2}, \frac{a}{2}, \frac{\rho^2}{2}\}$. Thus, the diversity order, $M = \min\{\frac{1}{2}, \frac{a}{2}, \frac{\rho^2}{2}\}$.

3.3 *Average SNR and Ergodic Rate*

Exact expressions of the average SNR $\bar{\gamma}^{\text{exact}}$ and ergodic rate $\bar{\eta}^{\text{exact}}$ are

$$\bar{\gamma}^{\text{exact}} = \int_{0}^{\infty} \gamma f(\gamma) d\gamma, \quad \bar{\eta}^{\text{exact}} = \int_{0}^{\infty} \log_2(1+\gamma) f_\gamma(\gamma) d\gamma, \tag{13}$$

where $f(\gamma)$ is given in (8). Note that we have assumed that relay requires negligible time to relay the data while computing the ergodic rate. Considering the intractability of the above integral, we use the PDF in (10) to derive closed-form approximations. Since the oceanic turbulence and pointing error parameters are the same for the source to relay and relay to the destination, expressions for the average SNR and ergodic rate are

$$\bar{\gamma} = 2\int_{0}^{\infty} \gamma\,[f(\gamma) - f(\gamma)F(\gamma)]\,d\gamma \tag{14}$$

$$\bar{\eta} = 2\int_{0}^{\infty} \log_2(1+\gamma)\,[f(\gamma) - f(\gamma)F(\gamma)]\,d\gamma. \tag{15}$$

Theorem 1 *If* $\{\omega, \lambda, a, b, c\}$ *are the parameters of the oceanic turbulence channel,* A_0 *and* ρ *are the parameters of pointing errors with* $l = 2l_r$ *as the transmission link length for a relay-assisted UWOC system, then an approximation of average SNR and ergodic rate is given in (16) and (17).*

$$\begin{aligned}
\bar{\gamma} = 2\rho^2\lambda^2A_0^2\gamma_0 \Bigg(& 2\omega + \tfrac{2b^2(1-\omega)\Gamma(\frac{2}{c}+a)}{\lambda^2\Gamma(a)(2+\rho^2)} - \omega^2\rho^2 G_{4,4}^{2,3}\left(\begin{matrix} 1, -2, -\rho^2-1, \rho^2+1 \\ 1, \rho^2, -\rho^2-2, 0 \end{matrix}\,\middle|\, 1\right) \\
& - \tfrac{\omega(\omega-1)\rho^2}{c\Gamma(a)} H_{4,4}^{2,3}\left[\begin{matrix} (1,1), (-2,c), (-\rho^2-1,c), (\frac{\rho^2}{c}+1,1) \\ (a,1), (\frac{\rho^2}{c},1), (-\rho^2-2,c), (0,1) \end{matrix}\,\middle|\, \tfrac{\lambda^c}{b^c}\right] - \tfrac{\omega(\omega-1)\rho^2}{\Gamma(a)} \\
& H_{4,4}^{3,2}\left[\begin{matrix} (-2,c), (-\rho^2-1,c), (-1,c), (\frac{\rho^2}{c}+1,1) \\ (a,1), (\frac{\rho^2}{c},1), (-2,c), (-\rho^2-2,c) \end{matrix}\,\middle|\, \tfrac{\lambda^c}{b^c}\right] - \tfrac{(\omega-1)^2\rho^2b^2}{\lambda^2(\Gamma(a))^2} \\
& G_{4,4}^{2,3}\left(\begin{matrix} 1, 1-\frac{2}{c}-a, 1-\frac{2-\rho^2}{c}, \frac{\rho^2}{c}+1 \\ a, \frac{\rho^2}{c}, \frac{-\rho^2-2}{c}, 0 \end{matrix}\,\middle|\, 1\right)\Bigg)
\end{aligned} \tag{16}$$

$$\begin{aligned}
\bar{\eta} = {} & 1.44\omega\rho^2 H_{3,4}^{4,1}\left[\begin{matrix} (0,\frac{1}{2}), (1,\frac{1}{2}), (\rho^2+1,1) \\ (1,1), (\rho^2,1), (0,\frac{1}{2}), (0,\frac{1}{2}) \end{matrix}\,\middle|\, \tfrac{1}{\lambda A_0\sqrt{\gamma_0}}\right] + \tfrac{1.44(1-\omega)\rho^2}{\Gamma(a)} \\
& H_{3,4}^{4,1}\left[\begin{matrix} (0,\frac{c}{2}), (1,\frac{c}{2}), (\frac{\rho^2}{c}+1,1) \\ (a,1), (\frac{\rho^2}{c},1), (0,\frac{c}{2}), (0,\frac{c}{2}) \end{matrix}\,\middle|\, \tfrac{1}{\lambda^c A_0^c(\sqrt{\gamma_0})^c}\right] - 2.88\omega^2\rho^4 \\
& H_{2,1:2,2;2,3}^{0,2:1,2;2,1}\left[\begin{matrix} (0;2,1), (1-\rho^2;2,1) : (1,1), (1,1); (1,1), (\rho^2+1,1) \\ (-\rho^2;2,1) : (1,1), (0,1); (1,1), (\rho^2,1), (0,1) \end{matrix}\,\middle|\, \begin{matrix} \lambda^2A_0^2\gamma_0 \\ 1 \end{matrix}\right] \\
& -2.88\tfrac{\omega(1-\omega)\rho^4}{c\Gamma(a)} H_{2,1:2,2;2,3}^{0,2:1,2;2,1} \\
& \left[\begin{matrix} (0;2,1), (1-\rho^2;2,1) : (1,1), (1,1); (1,1), (\frac{\rho^2}{c}+1,1) \\ (-\rho^2;2,1) : (1,1), (0,1); (1,1), (\frac{\rho^2}{c},1), (0,1) \end{matrix}\,\middle|\, \begin{matrix} \lambda^{2c}A_0^{2c}\gamma_0^c \\ 1 \end{matrix}\right] - \\
& 2.88\tfrac{\omega(1-\omega)\rho^4}{c\Gamma(a)} H_{2,1:2,2;2,3}^{0,2:1,2;2,1} \\
& \left[\begin{matrix} (1-a;\frac{2}{c},\frac{1}{c}), (1-\frac{\rho^2}{c};\frac{2}{c},\frac{1}{c}) : (1,1), (1,1); (1,1), (\rho^2+1,1) \\ (-\frac{\rho^2}{c};\frac{2}{c},\frac{1}{c}) : (1,1), (0,1); (1,1), (\rho^2,1), (0,1) \end{matrix}\,\middle|\, \begin{matrix} b^2A_0^2\gamma_0 \\ \frac{b}{\lambda} \end{matrix}\right] - \\
& 2.88\tfrac{(1-\omega)^2\rho^4}{c(\Gamma(a))^2} H_{2,1:2,2;2,3}^{0,2:1,2;2,1}
\end{aligned}$$

$$\left[\begin{array}{c} (1-a; \frac{2}{c}, \frac{1}{c}), (1-\frac{\rho^2}{c}; \frac{2}{c}, \frac{1}{c}) : (1,1), (1,1); (1,1), (\frac{\rho^2}{c}+1, 1) \\ (-\frac{\rho^2}{c}; \frac{2}{c}, \frac{1}{c}) : (1,1), (0,1); (1,1), (\frac{\rho^2}{c}, 1), (0,1) \end{array} \middle| \begin{array}{c} b^2 A_0^2 \gamma_0 \\ (b A_0 \sqrt{\gamma_0})^{1-c} \end{array} \right]. \tag{17}$$

Proof To find the approximate expression for the average SNR, we substitute (2) and (3) in (14) with $\sqrt{\gamma} = t$, $(\sqrt{\gamma})^c = u$ and apply the identities [22, Eqs. 07.34.21.0009.01, 07.34.21.0011.01, 07.34.21.0012.01]. Similarly, for ergodic rate we substitute (2) and (3) in (15) with $\sqrt{\gamma} = t$, $(\sqrt{\gamma})^c = u$ and apply the identities [22, Eqs. 07.34.21.0012.01, 07.34.21.0009.01] with the definition of Meijer G-function [23, Eq. 2.9.1] and multivariate Fox's H-function [24, Appendix: A1].

4 Numerical Analysis

In this section, we use numerical analysis and Monte Carlo simulation (averaged over 10^6 channel realizations) to demonstrate the average SNR, ergodic rate, and outage probability performance of the relay-assisted UWOC system under the combined effect of oceanic turbulence and pointing errors. We provide a comparison between the performance of direct and relay-assisted transmissions at various link distances. We use the different oceanic turbulence conditions and consider the experimental values of oceanic turbulence channel parameters $\{\omega, \lambda, a, b, c\}$ [19]. We consider the range for transmitted power as $P_t = 0$ dBm to 40 dBm, receiver with AWGN power $\sigma_w^2 = 10^{-14}$. Since the experimental values of pointing error parameters ρ and A_0 are available in [25], we use Gaussian beam optics to determine these parameters for other link distances for a better estimate of performance.

First, we analyze the optimal location of the relay by changing the source–relay distance from 20 to 80 m for a UWOC system with the source-to-destination distance of 100 m. It can be seen from Fig. 1a that the minimum outage performance occurs at $d_r = 50$ m. Thus, we see that the symmetric analysis considered in the paper is justified giving a near-optimal performance for the UWOC system. Note that a similar conclusion holds when we analyze the optimal relay location using average SNR and ergodic rate.

In Fig. 1b, we demonstrate the diversity order of the system by analyzing the outage probability at two transmission link lengths ($d = 60$ m and $d = 120$ m) by considering the threshold SNR $\gamma_{th} = 6$ dB. The figures show that the relay-assisted channel provides a significant gain in the outage performance comparing the single-link channel. It can be seen that direct transmission requires almost 100 times of transmit power to achieve the same outage probability (i.e., $P_{\text{out}} = 0.170$ at 30 dBm) with relaying at a transmit power of 10 dBm. It can also be seen that the derived approximation excellently matches the simulation results.

In Fig. 2a, we plot the average SNR at different transmitted powers for three transmission link lengths ($d = 80$ m, $d = 120$ m, and $d = 200$ m). As expected, the performance degradation with link length is evident here, being ≈ 19 dB from

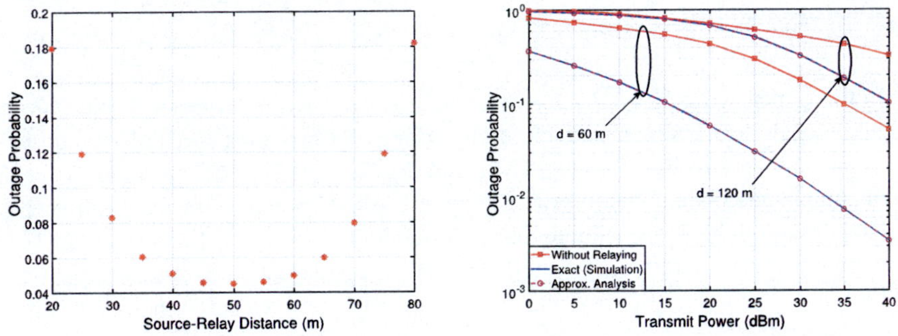

(a) P_{out} versus d_r at $P_t = 15$ dBm and $d = 100$ m.

(b) Outage probability.

Fig. 1 Outage performance of relay-assisted UWOC system under oceanic turbulence channel with pointing errors

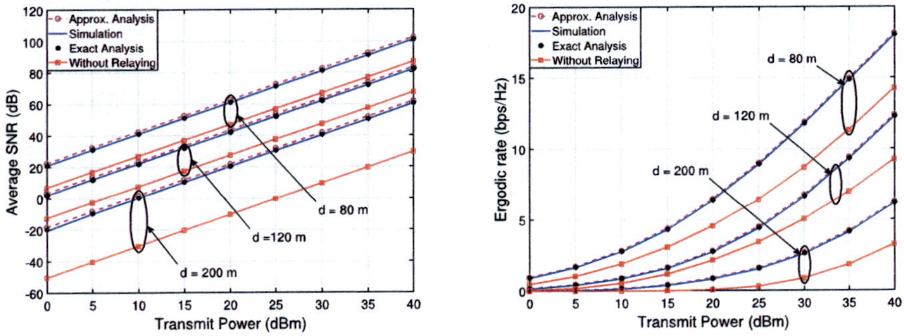

(a) Average SNR.

(b) Ergodic rate.

Fig. 2 Average SNR and ergodic rate performance of relay-assisted UWOC system under oceanic turbulence channel with pointing errors

80 to 120 m and ≈ 17 dB from $d = 120$ to $d = 200$ m. It can be seen that relay and single-link SNRs differ by 14 dB (approximately 25 times in linear scale) at $d = 80$ m, which is a significant performance improvement in the SNR. However, compared to the average SNR, an enhanced performance improvement for ergodic rate with relaying can be observed in Fig. 2b. We also notice the distance-dependent benefit in ergodic rate performance for the relay-assisted channel compared to direct transmission. It can be seen that gain with relaying is larger for $d = 80$ m than for $d = 120$ m and $d = 200$ m. It is noted that we have assumed that relay consumes negligible time to relay the data. It can also be seen that the derived approximation for average SNR and ergodic rate is very close, comparing the exact numerical analysis and Monte Carlo simulations.

5 Conclusions

In this paper, we presented the performance of an AF relay-assisted UWOC system over oceanic turbulence with pointing errors. We provided a detailed analysis of the outage probability, average SNR, and ergodic rate in order to show the benefit of the proposed dual-hop relaying for the UWOC system with weak direct links. Numerical analysis and Monte Carlo simulations show that the derived analytical bounds are close to the complicated integral expressions and thus can be implemented for real-time tuning of the system parameters for optimized performance. It was also demonstrated that the relay-assisted system shows better performance than the single-link transmissions. The extended version of the work will involve a consideration of asymmetric situations with different channel statistics at the relay and destination.

References

1. Zeng Z, Fu S, Zhang H, Dong Y, Cheng J (2017) A survey of underwater optical wireless communications. IEEE Commun Surv Tutor 19(1):204–238
2. Gussen C, Diniz P, Campos M, Martins W, Costa F, Gois J (2016) A survey of underwater wireless communication technologies. J Commun Inf Syst 31(1)
3. Elamassie M, Miramirkhani F, Uysal M (2019) Performance characterization of underwater visible light communication. IEEE Trans Commun 67(1):543–552
4. Jamali MV, Mirani A, Parsay A, Abolhassani B, Nabavi P, Chizari A, Khorramshahi P, Abdollahramezani S, Salehi JA (2018) Statistical studies of fading in underwater wireless optical channels in the presence of air bubble, temperature, and salinity random variations. IEEE Trans Commun 66(10):4706–4723
5. Rahman Z, Wariya S, Zafaruddin SM (2021) Performance analysis of mixed underwater and terrestrial owc systems over foggy weather. In: 2021 3rd IEEE Middle East and North Africa communications conference (MENACOMM), pp 152–157
6. Rahman Z, Tailor NV, Zafaruddin SM, Chaubey VK (2022) Unified performance assessment of optical wireless communication over multi-layer underwater channels. IEEE Photonics J 14(5):1–14
7. Jamali MV, Akhoundi F, Salehi JA (2016) Performance characterization of relay-assisted wireless optical CDMA networks in turbulent underwater channel. IEEE Trans Wirel Commun 15(6):4104–4116
8. Wang C, Yu H-Y, Zhu Y-J (2016) A long distance underwater visible light communication system with single photon avalanche diode. IEEE Photonics J 8(5):1–11
9. Sun X, Kang CH, Kong M, Alkhazragi O, Guo Y, Ouhssain M, Weng Y, Jones BH, Ng TK, Ooi BS (2020) A review on practical considerations and solutions in underwater wireless optical communication. J Light Technol 38(2):421–431
10. Akhoundi F, Jamali MV, Hassan NB, Beyranvand H, Minoofar A, Salehi JA (2016) Cellular underwater wireless optical CDMA network: potentials and challenges. IEEE Access 4:4254–4268
11. Zhang H, Dong Y (2016) Impulse response modeling for general underwater wireless optical MIMO links. IEEE Commun Mag 54(2):56–61
12. Gabriel C, Khalighi M-A, Bourennane S, Leon P, Rigaud V (2013) Monte-carlo-based channel characterization for underwater optical communication systems. IEEE/OSA J Opt Commun Netw 5(1):1–12
13. Jamali MV, Chizari A, Salehi JA (2017) Performance analysis of multi-hop underwater wireless optical communication systems. IEEE Photonics Technol Lett 29(5):462–465

14. Rahman Z, Zafaruddin SM, Chaubey VK (2022) Direct air-to-underwater optical wireless communication: Statistical characterization and outage performance. IEEE Trans Veh Technol 1–6
15. Das S, Rahman Z, Zafaruddin SM (2022) Optical wireless transmissions over multi-layer underwater channels with generalized gamma fading. In: 2022 IEEE 95th vehicular technology conference: (VTC2022-Spring), pp 1–6
16. Jamali MV, Salehi JA, Akhoundi F (2017) Performance studies of underwater wireless optical communication systems with spatial diversity: MIMO scheme. IEEE Trans Commun 65(3):1176–1192
17. Agheli P, Beyranvand H, Emadi MJ (2021) UAV-assisted underwater sensor networks using RF and optical wireless links. J Light Technol 1
18. Yang L, Zhu Q, Li S, Ansari IS, Yu S (2021) On the performance of mixed FSO-UWOC dual-hop transmission systems. IEEE Wirel Commun Lett 10(9):2041–2045
19. **Zedini E, Oubei HM, Kammoun A, Hamdi M, Ooi BS, Alouini M-S (2019) Unified statistical** channel model for turbulence-induced fading in underwater wireless optical communication systems. IEEE Trans Commun 67(4):2893–2907
20. Yang L, Gao X, Alouini M-S (2014) Performance analysis of relay-assisted all-optical FSO networks over strong atmospheric turbulence channels with pointing errors. J Light Technol 32(23):4613–4620
21. Soong TT (2004) Fundamentals of probability and statistics for engineers. Wiley, New York
22. Wolfram function. https://functions.wolfram.com/hypergeometric-functions/meijerg. Accessed 10 Feb 2022
23. Kilbas A (2004) Analytical methods and special functions, H-transforms: theory and applications. Taylor and Francis, New York, NY, USA
24. Mathai A, Saxena R, Haubold H (2009) The H-function: theory and applications. Springer, New York
25. Esmail MA, Fathallah H, Alouini M-S (2016) Outdoor FSO communications under fog: attenuation modeling and performance evaluation. IEEE Photonics J 8(4):1–22

Stock Price Prediction Using GRU and BiLSTM Models

Yash Thorat, Arya Talathi, Kirti Wanjale, and Abhijit Chitre

Abstract The stock market has become a very important part of the world's economy, and the accurate prediction of stock prices has gained significant attention over the years. Predicting accurate stock prices can help an individual make sound financial investment decisions and reduce the risk involved in investing in stock markets. This paper proposes a Recurrent Neural Network based stock price prediction framework using two very popular deep learning models, namely the Gated Recurrent Unit (GRU) and the Bidirectional Long Short-Term Memory (BiLSTM) models, and compares the performances of both of these models in predicting future stock price [1]. Experimental results conducted by us show that both the GRU and BiLSTM models can be used to accurately predict future stock prices. The RMSE and MAE evaluation metrics for both the GRU and BiLSTM models were calculated by tuning the hyperparameters and comparing the performances of both models to predict future stock prices. The experiments were carried out using a freely available dataset of stock market, which contained the stock prices of various companies. It was observed that the BiLSTM model performed better than the GRU model. However, GRU was almost twice as fast as the BiLSTM model in predicting stock prices due to its simpler architecture. Both models provide accurate results and can be very effective in predicting future stock market prices.

Keywords Stock market · Stock price prediction · Gated recurrent unit · Bidirectional LSTM

Y. Thorat · A. Talathi · K. Wanjale (✉) · A. Chitre
Vishwakarma Institute of Information Technology, Pune, Maharashtra, India
e-mail: kirti.wanjale@viit.ac.in

Y. Thorat
e-mail: yash.21920004@viit.ac.in

A. Talathi
e-mail: arya.21810346@viit.ac.in

A. Chitre
e-mail: abhijit.chitre@viit.ac.in

H. O. Bansal et al. (eds.), *Next Generation Systems and Networks*, Lecture Notes in Networks and Systems 641, https://doi.org/10.1007/978-981-99-0483-9_9

1 Introduction

Stock price prediction is a difficult and complex task. Several factors affect the value of a stock, and to accurately predict the price of a stock, many alternative approaches can be taken, such as performing a fundamental or technical analysis of a company's stock or using various machine learning and deep learning approaches. Various methods have been used to perform stock predictions, which have provided good results, such as the Autoregressive Integrated Moving Average (ARIMA) model, Artificial Neural Networks (ANNs), and Convolutional Neural Networks (CNNs). However, the deep learning-based models outperform the traditional ARIMA-based models and are able to predict stock prices more accurately. Numerous models based on Recurrent Neural Networks exist. RNNs have a tendency to encounter vanishing and exploding gradient problems. These problems resulted in poor accuracy, and therefore, the LSTM and GRU models were designed to overcome these problems faced by a traditional RNN. These two models were capable of learning from past data and were able to model the relationships between the input and output data more effectively.

As these newer deep learning-based models performed better than the ANN, CNN, and ARIMA-based models, we propose the use of the GRU and the extension of LSTM called Bidirectional LSTM models to predict future stock prices and perform a comparative study of both of these models to find which of these models performs better when forecasting a company's stock prices by fine-tuning the various hyperparameters of the models to produce the best results.

2 Related Works

A lot of work involving time series data has been published by researchers using GRU and BiLSTM models. The methods and approaches used in other papers, such as the use of ANN and CNN models to predict stock prices, are decent and can be used to predict stock prices. However, deep learning-based models are considered to be generally better than these models and are used more often. Chen et al. [1] proposed a method based on Convolutional Neural Network to predict the stock prices of the Chinese stock market and found the performance of the model to be decent. Devadoss et al. [2] proposed an Artificial Neural Network (ANN) model to predict the stock prices of the Bombay Stock Exchange. Verma et al. [3] proposed the application of multiple ANN models for financial time series prediction. Di Persio et al. [4] compared different artificial neural network methods like MLP, CNN, and RNN, to predict stock market prices. Khashei et al. [5] proposed a combination of both artificial neural networks and ARIMA model in order to overcome the limitations of ANNs and provide a more accurate forecasting model.

Some papers have presented a comparison of GRU, LSTM, and BiLSTM models and found the performance of these models to be almost similar, with BiLSTM having better accuracy than LSTM and LSTM having slightly better accuracy than GRU. Research study by Gao et al. [6] showed a comparative study of both LSTM and GRU for stock market prediction. Their experimental results show that both LSTM and GRU models provide similar results and that one model is not more effective than the other. Siami-Namini et al. [7] compared the LSTM and BiLSTM models to forecast stock market prices. Their results showed that BiLSTM models performed better than the regular LSTMs when it came to the accuracy of the prediction. Research study by Sunny et al. [8] compared both the LSTM and BiLSTM models and showed that both models can be used for stock price prediction with proper adjustment of different hyperparameters of the models. Chen et al. [9] presented a case study of Chinese stock market in 2015 and predicted their stock returns using the LSTM model. Althelaya et al. [10] evaluated and compared different LSTM architectures for stock market prediction.

A comparative study of GRU with the newer BiLSTM model in predicting stock prices could help determine if the newer BiLSTM model performs only slightly better than the traditional GRU and LSTM or provides a significant difference in terms of the accuracy of the predicted values. As per our knowledge, no comparisons have been made between the GRU and BiLSTM models, and therefore, we propose the use of GRU and BiLSTM models to predict stock prices and compare their performance in forecasting stock prices.

3 Methodology

3.1 Gated Recurrent Unit (GRU)

Gated Recurrent Unit, or GRU, is a kind of recurrent neural network which was designed to solve the limitations of a basic Recurrent Neural Network by Cho et al. [11] in 2014. GRU incorporates gating mechanisms and there are two gates in a GRU, update gate and reset gate.

1. Update gate: The update gate is used to conclude what amount of previous data is to be passed to the model. The equation for the update gate is shown in Eq. (1) as follows:

$$z_t = \sigma(W_z * [h_{t-1}, x_t]) \tag{1}$$

2. Reset gate: The reset gate is used to determine what past information should be removed. The equation for the reset gate is shown in Eq. (2) as follows:

$$r_t = \sigma(W_r * [h_{t-1}, x_t]) \tag{2}$$

The candidate status value is updated using Eq. (3) and the final output is represented by Eq. (4).

$$\tilde{h}_t = \tanh(W_{\tilde{h}} * [r_t * h_{t-1}, x_t]) \tag{3}$$

$$h_t = (1 - z_t) * h_{t-1} + z_t * \tilde{h}_t \tag{4}$$

3.2 Bidirectional Long Short-Term Memory (BiLSTM)

Bidirectional LSTM, or BiLSTM, proposed by Hochreiter and Schmidhuber [12] in 1997, is a RNN-based model in which two LSTMs are applied to the input data. One LSTM takes the input fed into the model in the forward direction, and the other LSTM takes the input in the backward direction. This arrangement of two LSTMs improves the model's learning ability to understand complex problems and understand the underlying context much better. The following gates are used in a BiLSTM:

1. Forget gate: The forget gate is used to conclude what data needs to be removed from the LSTM. The equation for forget gate is shown in Eq. (5) as follows:

$$f_t = \sigma(W_f * [h_{t-1}, x_t] + b_f) \tag{5}$$

2. Input gate: The input gate is used to decide if the new information should be added to the network or not. The equation for the input gate is shown in Eqs. (6)–(8). Equation (6) decides which values need to be updated in the network. Equation (7) adds a vector of new candidate values into the LSTM network and Eq. (8) is used for updating the current value in the network.

$$i_t = \sigma(W_i * [h_{t-1}, x_t] + b_i) \tag{6}$$

$$\tilde{c}_t = \tanh(W_c * [h_{t-1}, x_t] + b_c) \tag{7}$$

$$c_t = f_t * c_{t-1} + i_t * \tilde{c}_t \tag{8}$$

3. Output gate: The output gate is used to determine the subsequent hidden state value and carries information on previous data inputs to the network. The equation for the output gate is shown in Eqs. (9) and (10) as follows:

$$o_t = \sigma(W_o * [h_{t-1}, x_t] + b_o) \tag{9}$$

$$h_t = o_t * \tanh(c_t) \tag{10}$$

4 Research Method

To compare and evaluate the performances of both the GRU and BiLSTM models, we performed the following tasks.

4.1 Dataset

The dataset used for this study was gathered from Kaggle, which contained stock market data for all stocks in the NIFTY 50 index from the National Stock Exchange, India. All datasets have data for every single day from the past 20 years, with pricing and trading values stored in multiple files for each company's stock. The data consists of attributes such as Date, Open, High, Low, Close, etc.

4.2 Training and Testing Dataset

The closing price of a stock was selected to be fed into both models. The datasets were split into two sets, a training set with 75% of the stock data, and 25% of the stock data for testing the predictions of the models.

4.3 Evaluation Metrics

The evaluation metrics used to evaluate the performance of the model were the Root Mean Squared Error (RMSE) and Mean Absolute Error (MAE). These metrics are widely used for evaluating the accuracy of a model.

4.4 Hyperparameter Tuning

Performing the proper hyperparameter tuning for both models was essential to compare the performances of the models. The number of epochs, batch size, hidden layers, dense layers, etc., affect the performance of the model and hence need to be properly tuned in order to achieve the best results.

Varying the number of epochs for both models gave the following values for the evaluation metrics as shown in Tables 1 and 2. After analyzing the results, we can see that the model with 80 epochs gave much better values (lower is better) compared to other model configurations in both GRU and BiLSTM. The graphs shown in Figs. 1

and 2 illustrate the predicted stock prices by both models with respect to the original stock prices.

We compared the batch sizes for both models. Table 3 shows the values for the evaluation metrics for GRU, and Table 4 shows the values for BiLSTM. We found that a batch size of 64 gave better results for GRU and a batch size of 32 gave better results for the BiLSTM model.

By comparing the number of hidden layers in both models, we found that increasing the number of hidden layers increased the complexity of the model and the time it took the model to train. However, increasing the number of hidden layers helped the model understand the context better and increased the accuracy of the models. According to the data in Tables 5 and 6, we concluded that the three hidden layer GRU model performed slightly better than the two and four hidden layer

Table 1 Comparison of the number of epochs (GRU)

Epochs (GRU)	Training RMSE	Testing RMSE	Training MAE	Testing MAE
10	20.4936	37.3998	15.7943	28.3055
50	16.8381	32.3155	13.4823	22.6770
80	12.0560	21.2440	10.4083	15.7192

Table 2 Comparison of the number of epochs (BiLSTM)

Epochs (BiLSTM)	Training RMSE	Testing RMSE	Training MAE	Testing MAE
10	16.2211	26.7665	12.0496	19.5524
50	10.8623	17.2313	7.9946	12.3111
80	10.4912	15.8088	7.9097	11.6974

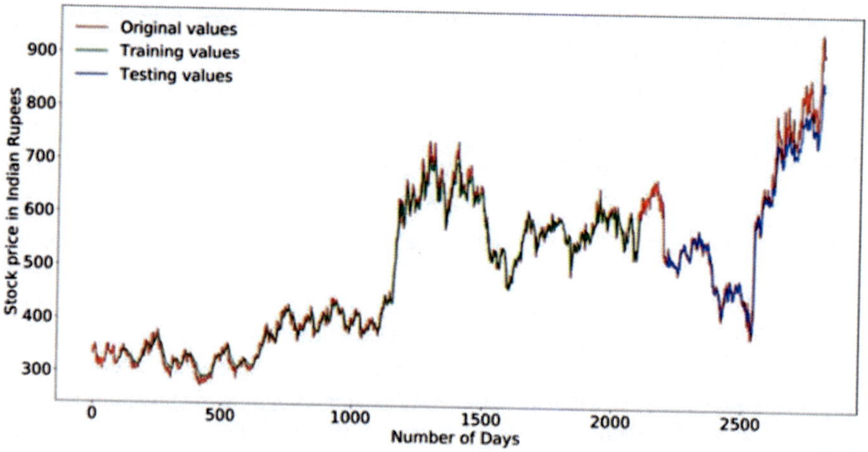

Fig. 1 Prediction for 80 epochs by GRU model

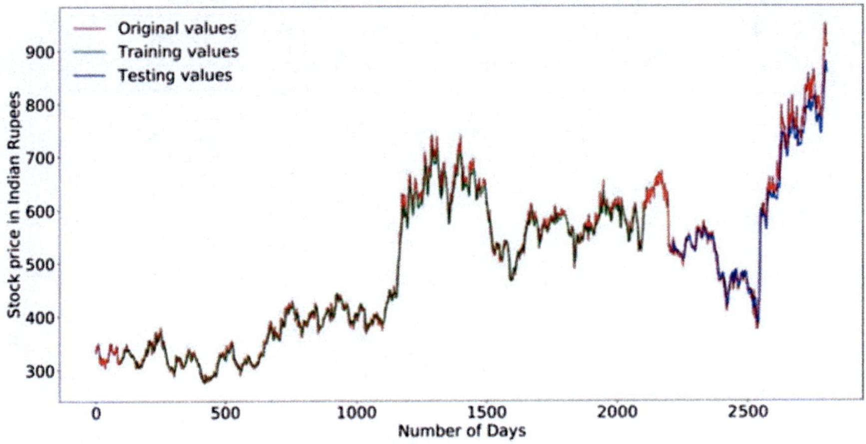

Fig. 2 Prediction for 80 epochs by BiLSTM model

Table 3 Comparison of batch sizes (GRU)

Batch size (GRU)	Training RMSE	Testing RMSE	Training MAE	Testing MAE
32	15.5039	29.6641	11.8367	22.5880
64	13.3890	25.7286	10.3920	17.9577
128	14.8747	26.6752	10.3409	20.3611

Table 4 Comparison of batch sizes (BiLSTM)

Batch size (BiLSTM)	Training RMSE	Testing RMSE	Training MAE	Testing MAE
32	10.3561	13.8308	8.1096	10.0895
64	10.6672	15.2392	8.0589	11.2867
128	12.6355	18.7229	9.6056	14.6265

GRU models. Also, the BiLSTM model with two hidden layers performed the best. Figures 3 and 4 show the prediction of the stock prices of GRU and BiLSTM with respect to the original values.

Table 5 Comparison of the number of hidden layers (GRU)

No. of hidden layers (GRU)	Training RMSE	Testing RMSE	Training MAE	Testing MAE	Time (min)
2 hidden layers	16.2688	31.4856	12.3768	23.4606	5
3 hidden layers	14.9456	27.4537	10.9567	19.7525	6
4 hidden layers	14.8730	26.4054	11.8479	18.3868	9

Table 6 Comparison of the number of hidden layers (BiLSTM)

No. of hidden layers (BiLSTM)	Training RMSE	Testing RMSE	Training MAE	Testing MAE	Time (min)
2 hidden layers	10.7144	15.3641	8.3729	11.2902	10
3 hidden layers	10.5925	19.1465	7.8588	14.1821	17
4 hidden layers	11.0786	34.1767	8.0937	25.5415	21

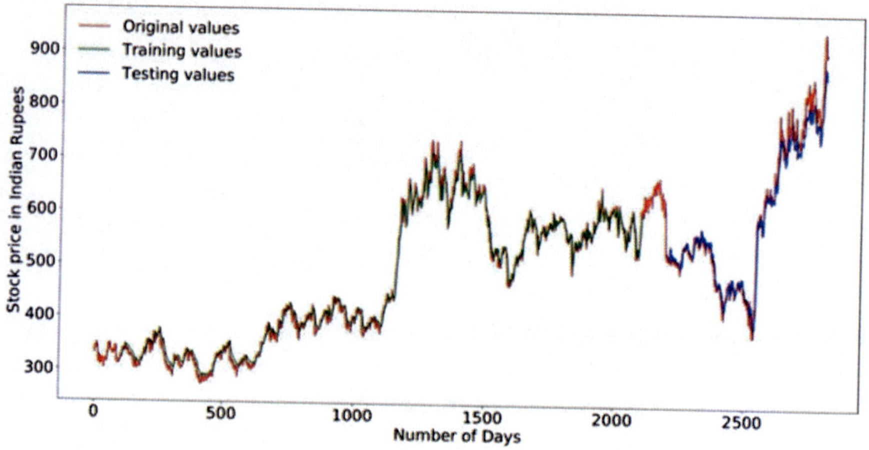

Fig. 3 Prediction for 80 epochs and three hidden layers by GRU model

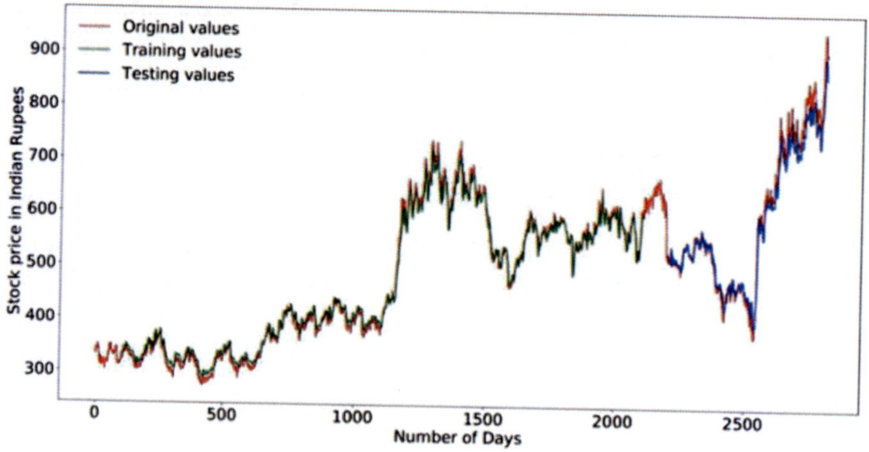

Fig. 4 Prediction for 80 epochs and two hidden layers by BiLSTM model

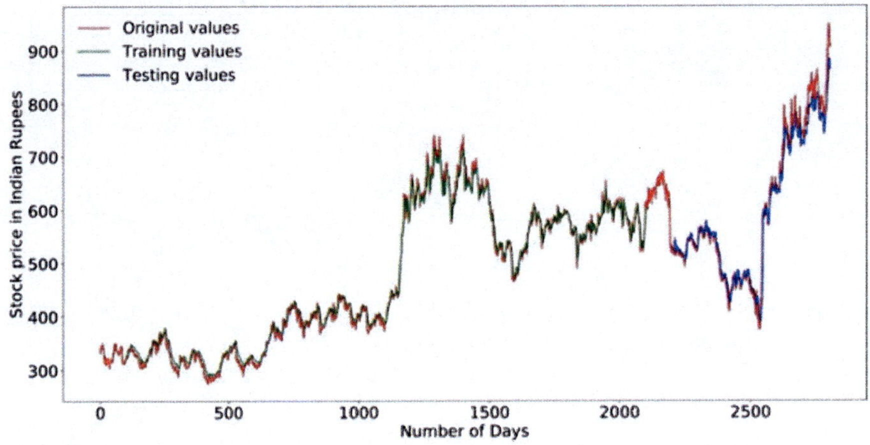

Fig. 5 GRU model with lowest RMSE and MAE values

5 Result Analysis

By modifying and analyzing the different configurations for both models, we configured both models with hyperparameters that resulted in the best values for the RMSE and MAE evaluation metrics.

We observed that the following configurations gave the best results for respective models:

1. GRU model with 80 epochs, 64 batch size, 3 hidden layers, 1 dense unit, dropout of 0.4, and ReLU activation function.
2. BiLSTM model with 80 epochs, 32 batch size, 2 hidden layers, 1 dense unit, dropout of 0.2, and ReLU activation function.

Analyzing Figs. 5 and 6 and the performance metrics of both models in Table 7, we can observe that BiLSTM performed much better on test data compared to the GRU model and was able to understand the underlying context much better. However, GRU performed well enough and was able to train almost twice as fast as the BiLSTM model. Both models gave excellent results and were almost evenly matched in terms of accuracy.

6 Conclusion

Through this research study, we performed a comparative analysis of both the GRU and BiLSTM models and evaluated the performance of each of these models. We tuned the hyperparameters of both models properly, and through our experiments,

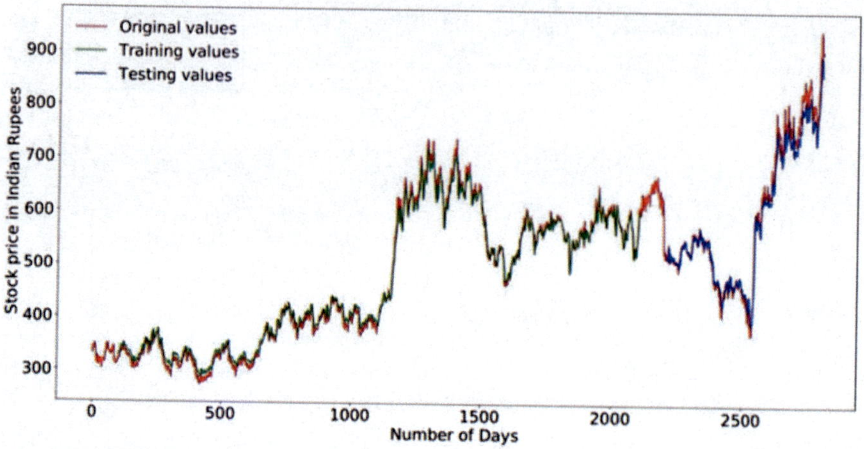

Fig. 6 BiLSTM model with lowest RMSE and MAE values

Table 7 Comparison of the configuration of both GRU and BiLSTM models

Model	Training RMSE	Testing RMSE	Training MAE	Testing MAE	Time (min)
GRU	14.1482	28.1178	10.9747	19.8770	6
BiLSTM	10.6100	16.5757	7.6995	11.8718	10

we concluded that the GRU model is slightly less accurate than BiLSTM. The difference is very minimal when both models are trained on smaller datasets, but is more noticeable when trained on larger datasets. Although BiLSTM was more accurate, the time taken by each model to train on a particular dataset should also be taken into consideration. Our experimental analysis shows that GRU was able to train on datasets much faster than BiLSTM because of its much less complex architecture, and thus was able to produce results almost twice as fast as the BiLSTM model. Thus, we conclude that both the GRU and BiLSTM models are capable of predicting stock prices accurately.

References

1. Chen S, He H (2018) Stock prediction using convolutional neural network. IOP Conf Ser Mater Sci Eng 435(1):012026. IOP Publishing
2. Devadoss AV, Ligori TAA (2013) Stock prediction using artificial neural networks. Int J Data Min Tech Appl 2(1):283–291
3. Verma SA, Thampi G, Rao M (2017) Inter-comparison of artificial neural network algorithms for time series forecasting: predicting Indian financial markets. Int J Comput Appl 162(2):1–13
4. Di Persio L, Honchar O (2016) Artificial neural networks architectures for stock price prediction: Comparisons and applications. Int J Circ Syst Signal Process 10(2016):403–413

5. Khashei M, Bijari M (2011) A novel hybridization of artificial neural networks and ARIMA models for time series forecasting. Appl Soft Comput 11(2):2664–2675
6. Gao Y, Wang R, Zhou E (2021) Stock prediction based on optimized LSTM and GRU models. Scientific programming, vol 2021, Article ID 4055281, 8 p. https://doi.org/10.1155/2021/4055281
7. Siami-Namini S, Tavakoli N, Namin AS (2019) The performance of LSTM and BiLSTM in forecasting time series. In: 2019 IEEE international conference on big data (Big Data), pp 3285–3292. IEEE
8. Sunny MAI, Maswood MMS, Alharbi AG (2020) Deep learning-based stock price prediction using LSTM and bidirectional LSTM model. In: 2020 2nd novel intelligent and leading emerging sciences conference (NILES), pp 87–92. IEEE
9. Chen K, Zhou Y, Dai F (2015) A LSTM-based method for stock returns prediction: A case study of China stock market. In: 2015 IEEE international conference on big data (big data), pp 2823–2824. IEEE
10. Althelaya KA, El-Alfy ESM, Mohammed S (2018) Evaluation of bidirectional LSTM for short- and long-term stock market prediction. In: 2018 9th international conference on information and communication systems (ICICS), pp 151–156. IEEE
11. Cho K, Van Merriënboer B, Bahdanau D, Bengio Y (2014) On the properties of neural machine translation: encoder-decoder approaches. arXiv:1409.1259
12. Hochreiter S, Schmidhuber J (1997) Long short-term memory. Neural Comput 9(8):1735–1780. https://doi.org/10.1162/neco.1997.9.8.1735

Evaluation of BERT Model for Aspect-Based Sentiment Analysis

Jaspreet Singh, Deepinder Kaur, and Parminder Kaur

Abstract Thanks to digital evaluations and ratings on E-commerce sites, people can now purchase items more effortlessly. Purchasers may share their stories on e-commerce giants like Amazon, Flipkart, and others, giving prospective buyers concrete proof of the product's effects. One of the most difficult challenges in Sentiment Analysis (SA) is determining a customer's fine-grained opinion on a single component of a product. This research offers a machine learning model based on Bidirectional Encoder Representation Transformer (BERT) for performing Aspect-Based Sentiment Analysis (ABSA) of user evaluations. This paper proposes a BERT-based model for SA of three different datasets of user reviews obtained from Amazon, IMDB, and Tripadvisor, respectively. The proposed BERT model outperformed other classifiers when compared by varying learning rates, dropouts, and sentence lengths of user texts. The proposed model beats three baseline models in terms of accuracy of the classification and achieves state-of-the-art results.

Keywords Sentiment analysis · Aspect based · Product reviews · Sentiment classification · BERT

1 Introduction

The success of any product is directly proportional to its consumers; if a product wins the heart of a customer then it defines the success of the product, otherwise,

J. Singh (✉) · D. Kaur
Department of Computer Science and Engineering, Chandigarh University, Chandigarh, India
e-mail: jaspreet.e10279@cumail.in

D. Kaur
e-mail: deepinder.e9237@cumail.in

P. Kaur
Department of Computer Science, Guru Nanak Dev University, Amritsar, India
e-mail: parminder.dcse@gndu.ac.in

H. O. Bansal et al. (eds.), *Next Generation Systems and Networks*, Lecture Notes in Networks and Systems 641, https://doi.org/10.1007/978-981-99-0483-9_10

the product loses its repute. These days, to identify the reputation of a product, e-commerce companies evaluate its user reviews and apply the processes in research and development of that product to win more buyers in the online market [1]. The real emotions of a customer who buys a certain product lie in the reviews and ratings of a product. Therefore, evaluation of sentiments has been performed on reviews and ratings of products in order to perform sentiment analysis of customers which indirectly leads to the optimization of a product [2].

Customers may offer feedback on things or services on most online platforms, including Amazon, by giving both subjective and objective feedback choices [3]. Subjective feedback comprises dialogues, drop-downs, and text spaces where customers may freely submit a review, whereas objective feedback includes a star rating, numerical rating, and happiness index, among other things. Sentiment analysis will be used for customer reviews and ratings in order to extract key data such as the customer's happiness with things, a description of negative remarks, and so on [4]. After utilizing a product for a certain amount of time, customers can contribute rating points for the things or services they purchased. For example, Amazon.com uses a numerical rating scale ranging from 1 to 5, with 1 being the worst and 5 being the best. Because each individual rating is required to calculate the average ratings, reviews with a range of rating scores are invited to participate [5, 6].

The fundamental purpose of SA is to figure out if a customer loves or hates something based on their review data. Although SA is a text-centered analysis, there are real challenges in detecting a statement's exact mood. SA is divided into three levels: (i) aspect level, (ii) phrase level, and (iii) document level [7].

Despite the fact that the SA aids in the process of recognizing the emotion hidden in the text evaluations, a Computer program considers identifying the sentiment to be a challenging task since anticipating the sentiments involves multiple components [8, 9]. It is easier to extract good feelings from a text than it is to extract negative sentiments. The reason for this is that optimism is communicated more overtly in the review language than criticism. Understanding affirmative (positive) sentiment in review language, for example, is much simpler than interpreting negative sentiment. Controlling the 'component' of the statement that has been labeled as an unpleasant sensation is also crucial [10].

It's conceivable that a customer writes a negative review due to a delivery delay, even if he loves the goods otherwise. In this case, the negative review triumphs over the positive evaluation. As a result, the ABSA enters the picture. The ABSA is a strategy that examines the language linked with the aspects and associates a customer's sentiments with each. It is also true that components of the text, such as irony and indirect references, are occasionally obscured. As a result, determining the aspect of a sentence in a review language is a difficult process [11].

A language model of Natural Language Processing (NLP) is used to locate aspects of a phrase, particularly a pre-trained language model for providing a context for sentences. It's difficult to extract context at the sentence and paragraph levels from an unannotated review text. In the year 2019, Google researchers produced a language model called BERT for ABSA. In a sentence, this model goes to both sides of a contextual word [1, 11]. It gathers word connections from both the left and right sides of a contextual word and aids in determining the precise feeling of a phrase or paragraph. BERT can also do tasks such as automatic question answering, complicated word identification, SA, named entity recognition, and word sense disambiguation, to mention a few.

Embedding from a Language Model (ELM) is a pre-trained language model for extracting contextualized information from a text. In bi-directional language models, this model aids in the extraction of contextual terms from a text (BERT) [12].

Despite the fact that SA and ABSA are both text-level and sentence-level tasks, one phrase or contextual word might relate to many objects, and sentence-level tasks cannot address multiple targets. As a result, the targeted ABSA (TABSA) task was developed, which attempts to estimate fine-grained opinion polarity toward a particular attribute associated with a specific target. This assignment is broken down into two parts: (1) Determine the polarity of aspects associated with a certain target; (2) Determine the many aspects of a single target word [13].

2 Literature Review

Recently, many Machine Learning algorithms are gaining popularity for performing aspect-based sentiment analysis. The most famous are SVM, NB, DT, CNN, and LSTM. However, the BERT-based model that has been introduced by researchers at Google lately in 2019 has proved state-of-the-art performance for SA and its different allied tasks. The reason for the fame of BERT lies in the fact that it provides the pre-trained model with automated transformers. However, the BERT-based ABSA has gained more attention due to its accuracy of sentiment classification especially, extracting aspects from contextual words. The following are some latest publications where BERT is used for determining aspects from text followed by sentiment analysis (Table 1).

3 Statistics of Datasets

Numerous kinds of user reviews are available over the most renowned online platforms like Amazon, IMDB, and Tripadvisor. The reviews available on these platforms are free from any prejudiced context. Hence, the chances of spam reduce to almost negligible [6]. Consequently, sentiment analysis from user reviews obtained from these platforms has been considered for our proposed model and other baseline classifiers. Around ten thousand paragraphs have been collected in three text files for the respective three platforms, whereas the metadata from these platforms have been collected in JSON format [14]. The CSV files are built with the help of Python's Natural Language Toolkit; these CSV files contain segregated sentences per their aspects and sentiments through pre-processing techniques. Table 2 represents the statistics of the three datasets at both the sentence as well as document level.

Table 1 Summary of recent literature

Author(s)	Year	Methods	Dataset	Results
Mickel et al.	2019	BERT	Restaurant, Hotel, and Laptop reviews	89.5%
Zhengxuan Wu and Desmond C. Ong	2021	Two variants of BERT (CG-BERT), Quasi attention (QACG-BERT)	Sentihood, SemEval 2016, and SemEval-2014	93.1% for CG-BERT and 93.7% for QACG-BERT
Guangyao Pang et al.	2021	Feature location-based model	Laptop, Restaurant, and Twitter reviews	3.1% accuracy hike and F1 by 6.56%
Geetha and Karthika Renuka	2021	Fine-tuned BERT	Amazon product reviews	88.4%
Zhiguo Wang et al.	2018	BERT for automatic question answering	1. OpenSQuAD 2. TriviaQA 3. Quasar T 4. Search QA	4.7% of hike in accuracy
Lysimachos Maltoudoglou et al.	2020	BERT for sentiment analysis	IMDB movie reviews	92.28%
Fredrico Dias Souza et al.	2022	Pre-tuned and fine-tuned BERT	UTLC movie reviews	98.1% for pre-tuned and 98.4% for fine-tuned
Chi Sun et al.	2019	Constructed auxiliary sentences and utilized BERT for SA	SentiHood dataset	79.8%
Methew E Peters et al.	2018	Word embeddings	ELM (Embeddings for language models)	4.7% of hike
Chao Chen et al.	2020	BERT for fine grained SA	Chinese film reviews	94.57% precision
Hongchan Li et al.	2021	Amalgamated BERT and Deep learning for SA	Weibo text for Covid-19	91.2%

4 BERT Model

The proposed model deploys two variants of pre-trained BERT models viz., BERT-base uncased and BERT-base cased. Each model contains 12 layers, 768 hidden neurons, 12 heads, and 110 mega parameters. The BERT models are downloaded from Google APIs (googleapis.com) as zipped files using the WEGET function followed by extraction of the zipped file into Python's variable [15]. Both BERT-cased and uncased models use word embeddings of each sentence to predict the exact sentiment of the text. Pandas and NumPy libraries are used to showcase the results of prediction.

Table 2 Statistics of three different datasets

Dataset		Paragraphs	Target aspects	Targets with tags	Targets without tags
Sentence level	Amazon product reviews	2071	15	1465	408
	Hotel reviews	2186	79	1853	199
	IMDB movie reviews	2914	82	1752	280
Document level	Amazon product reviews	245	15	985	57
	Hotel reviews	291	79	567	91
	IMDB movie reviews	347	82	358	42

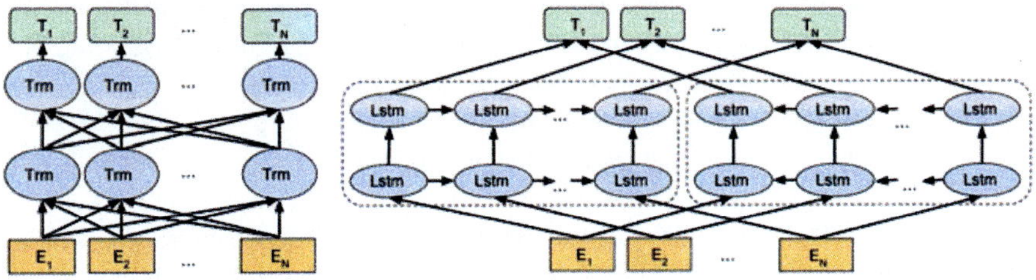

Fig. 1 BERT architecture used for sentiment prediction

The above Fig. 1 represents the architecture of the BERT model, where the top layer holds the contexts of words. The middle layer contains transformers which are heavily connected to collect the influence of left and right contexts of words in a sentence. The lower layer takes word embeddings from processed texts obtained after pre-processing user reviews.

5 Results and Discussions

Table 3 shows the performance evaluation of the BERT model for different epochs. It is observed that the accuracy of classification stabilizes at higher epochs ranging from 30 to 50. Almost similar trends have been observed in the case of precision, recall, and F1-score.

Table 3 Iterative experiment by increasing number of epochs

Epoch	Accuracy (%age)	Precision (%age)	Recall (%age)	F1-score (%age)
1	86.2	89.1	82.5	85.8
5	88.3	89.5	81.5	85.2
10	87.5	89.5	81.5	85.2
15	90.1	89.6	81.6	85.6
20	90.4	90.6	86.5	85.4
25	90.5	91.6	86.4	87.1
30	90.4	88.4	86.9	88.6
35	90.3	88.5	90.1	86.5
40	90.4	89.2	86.4	85.4
45	90.4	89.2	86.4	85.4
50	90.4	89.2	86.3	85.4

Below is Table 4 where the performance of both the BERT variants have been taken by varying sentence lengths of texts. A counter variable holds the count of words; four Python lists have been populated for different counter values ranging from 8, 9, 10 to 12. The classifiers are trained for all the sentence lengths. It is observed that the maximum accuracy of both the BERT variants has been found for higher sentence lengths (SL = 12 here).

The high accuracy for higher sentence length is due to the fact that the middle layer in the BERT model is heavily connected to 768 transformers which take care of both sides of the contextual words. Both the left and right contexts of a target word are considered by transformers. In case of higher sentence lengths, a sufficient number of words on both sides of the contextual word are taken for defining the exact sentiment of a sentence.

Table 5 represents the comparison of performance parameters by varying values of dropouts. Dropouts are varied to overcome the overfitting of the BERT models. This is because our datasets are limited to only a few hundreds of sentences and a few thousands of contextual words. The best results are obtained at 0.4 of a dropout value which reflects good accuracy and F1-score.

The best and stable values of the parameters of the BERT model are taken by successfully running 50 epochs of the experiment. Table 6 shows eight variables of the BERT model for which the suitable values for the three datasets are collected.

Table 4 Performance of proposed BERT classifier at different sentence lengths

Sentence length	Accuracy (%age)	Precision (%age)	Recall (%age)	F1-score (%age)
8	86.4	85.4	83.4	84.5
9	86.3	88.6	82.3	83.3
10	89.5	87.4	80.5	85.5
12	92.4	90.6	81.3	86.8

Table 5 Performance of BERT model by varying dropouts

Dropouts	Accuracy (%age)	Precision (%age)	Recall (%age)	F1-score (%age)
0.1	80.3	90.1	83.4	84.7
0.2	84.4	89.2	83.5	88.3
0.3	82.5	91.5	83.6	87.3
0.4	90.2	88.6	83.7	86.4
0.5	91.3	91.4	87.6	86.5
0.6	92.4	90.4	87.5	86.4
0.7	89.4	89.5	85.9	85.5

Table 6 Description of model parameters

Parameter	Value
Number of epochs	20
Dropouts	0.4
Average sentence length	9
Dimensions of word vector	512
# of neurons in hidden LSTM layer	64
# of neurons in hidden CNN layer	64
Convolutional kernel	3 × 3
Learning rate	0.0004

Table 7 shows the performance parameters of the experiment conducted with two **BERT methods for different neural network classifiers. The experimental results of** BERT and other neural network models are compared in Table 7 below. The BERT method outperforms the deep learning models (CNN, LSTM, and biLSTM).

The following Fig. 2 shows the performance comparison of traditional deep learning models with the BERT-based model. The green colored bar represents the performance parameters of the BERT method. Since the BERT method is a pre-trained bidirectional encoder representation for transformers, therefore the performance parameters have shown an extra hike over the performance of existing neural network models.

Table 7 Performance comparison of different neural network classifiers with BERT

Classifier	Accuracy (%age)	Precision (%age)	Recall (%age)	F1-score (%age)
CNN	78.3	82.1	82.4	85.4
LSTM	80.4	84.2	84.5	84.3
biLSTM	81.5	85.5	83.6	85.3
CNN + LSTM	89.3	84.6	82.7	85.4
CNN + biLSTM	89.8	89.4	84.5	86.2
BERT	94.3	91.4	89.6	88.5

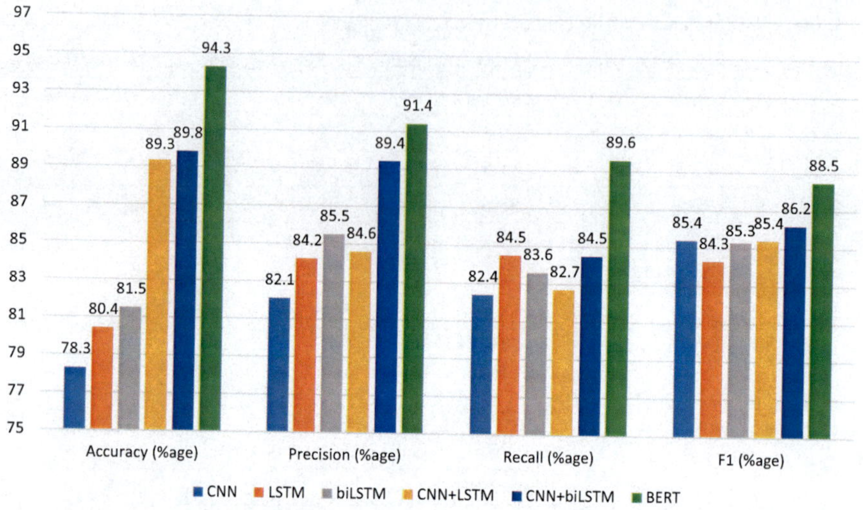

Fig. 2 Performance graph of different Neural Network classifiers with BERT

We vary the learning rate of the classifier from 0.00001 to 0.00008; it has been observed that the BERT model was most accurate at the learning rate of 0.00004. The maximum recall of the model has been observed at the learning rate of 0.00003 as shown in Table 8.

We have conducted the experiment with other baseline classifiers obtained from SVM and NB (supervised). The precision and recall of these baselines are not that much good as that of CNN and BERT models. The performance of model also depends upon the epochs; for the pre-trained BERT and CNN, almost 50 epochs are performed due to which the models stabilized quite quickly as compared to other supervised counterparts (Table 9).

Figure 3 contains the performance graph for comparing SVM and NB with CNN, LSTM, and BERT. The four different performance parameters reveal that the topmost performer is BERT, while CNN takes second followed by LSTM in third place among

Table 8 Performance of proposed model by varying learning rates

Model	Learning rate	Precision (%age)	Recall (%age)	F1-measure (%age)	Accuracy (%age)
BERT	1.00E-05	84.25	87.16	88.65	91.74
	3.00E-05	83.75	89.53	87.42	90.64
	4.00E-05	84.12	87.52	88.14	**95.25**
	7.00E-05	84.09	87.34	88.34	91.35
	1.00E-06	84.56	87.12	88.67	91.24
	8.00E-06	84.82	87.68	88.89	91.78

Table 9 Performance comparison of different classifiers with BERT

Model	Precision (%age)	Recall (%age)	F1-measure (%age)	Accuracy (%age)
SVM	68.32	63.46	69.45	81.43
NB	52.84	50.35	55.28	69.45
CNN	80.41	75.24	65.12	68.21
LSTM	72.46	71.26	77.54	82.65
BERT	84.25	87.16	88.65	91.74

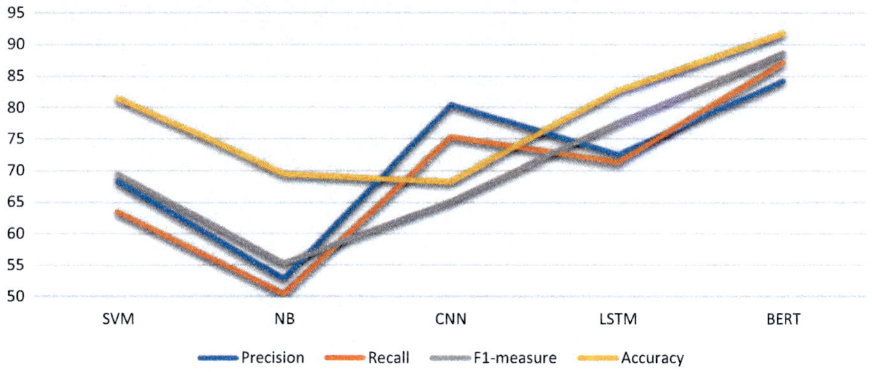

Fig. 3 Performance comparison of SVM, NB, CNN, LSTM, and BERT

the five classifiers. The performance of SVM and NB was below the baselines for all three datasets.

6 Conclusion

This study presents a BERT-based model for sentiment analysis of Amazon product reviews, Tripadvisor hotel reviews, and IMDB movie reviews. The BERT model is fine-tuned for two transformer block and output layer combinations. From the first baseline for the NB classifier, the fine-tuned and pre-trained BERT model deployed on pre-processed text exhibits a 32% increase in sentiment classification accuracy. The other two baselines, on the other hand, show improvements of roughly 13 and 11%, respectively. The pre-processing of datasets, fine-tuning of the model for phrase pairs, and many runs for refinement of the proposed model are the reasons for the BERT model's high performance. The suggested model might be used to help automated Chatbots with the sentiment analysis job of Question Answering.

References

1. Rietzler A, Stabinger S, Opitz P, Engl S (2020) Adapt or get left behind: Domain adaptation through BERT language model finetuning for aspect-target sentiment classification. In: LREC 2020—12th International conference on language resources and evaluation, conference proceedings, 2020. pp 4933–4941
2. Sun C, Huang L, Qiu X (2019) Utilizing BERT for aspect-based sentiment analysis via constructing auxiliary sentence. In: NAACL HLT 2019 - 2019 Conf North Am Chapter Assoc Comput Linguist Hum Lang Technol—Proc Conf, vol. 1. pp 380–385
3. Chakriswaran P, Vincent DR, Srinivasan K, Sharma V, Chang CY, Reina DG (2019) Emotion AI-driven sentiment analysis: A survey, future research directions, and open issues. Appl. Sci. (Switzerland) 9(24). MDPI AG. https://doi.org/10.3390/app9245462.
4. Shaikh MAM, Prendinger H, Ishizuka M (2008) Sentiment assessment of text by analyzing linguistic features and contextual valence assignment. Appl Artif Intell 22(6):558–601https://doi.org/10.1080/08839510802226801
5. Singh J, Singh G, Singh R (2017) Optimization of sentiment analysis using machine learning classifiers. Human-centric Comput Inf Sci. 7(1):32. https://doi.org/10.1186/s13673-017-0116-3
6. Glorot X, Bordes A, Bengio Y (2011) Domain adaptation for large-scale sentiment classification: a deep learning approach. In: Proceedings of the 28th international conference on machine learning no. 1. pp 513–520. Accessed 19 Sep 2018. [Online]. Available http://www.cs.jhu.edu/.
7. Pang G et al. (2021) Aspect-level sentiment analysis approach via Bert and aspect feature location model. Wirel Commun Mob Comput 2021. https://doi.org/10.1155/2021/5534615
8. Minsky ML (2006) The emotion machine: Commensense thinking, artificial intelligence, and the future of the human mind. Simon & Schuster
9. Maltoudoglou L et al. (2020) BERT-based conformal predictor for sentiment analysis. Proc Mach Learn Res 128:1–16
10. Xu H, Liu B, Shu L, Yu PS BERT Post-training for review reading comprehension and aspect-based sentiment analysis. https://doi.org/10.48550/arXiv.1904.02232
11. Jin Z, Lai X, Cao J (2020) Multi-label sentiment analysis base on BERT with modified TF-IDF. https://doi.org/10.1109/ISPCE-CN51288.2020.9321861
12. Rehman AU, Malik AK, Raza B, Ali W (2019) A hybrid CNN-LSTM model for improving accuracy of movie reviews sentiment analysis. Multimed Tools Appl 78(18):26597–26613 https://doi.org/10.1007/s11042-019-07788-7
13. Tao J, Fang X (2020) Toward multi-label sentiment analysis: a transfer learning based approach. J Big Data 7(1):1–26 https://doi.org/10.1186/s40537-019-0278-0
14. Wang Z, Ng P, Ma X, Nallapati R, Xiang B (2020) Multi-passage BERT: A globally normalized BERT model for open-domain question answering. In: EMNLP-IJCNLP 2019—2019 Conf Empir Methods Nat Lang Process 9th Int Jt Conf Nat Lang Process Proc Conf, no 2. pp 5878–5882. https://doi.org/10.18653/v1/d19-1599
15. Li H, Ma Y, Ma Z, Zhu H (2021) Weibo text sentiment analysis based on bert and deep learning. Appl Sci 11(22). https://doi.org/10.3390/app112210774

Solidly Mounted BAW Resonator for 5G Application on AlN Thin Films

Poorvi K. Joshi and Meghana A. Hasamnis

Abstract Development in 5G technology and other wireless standard poses new challenges in RF filtering than the traditional acoustic filters. Evolution in Bulk Acoustic Wave (BAW) filters can address new challenges in RF filtering. BAW filters are efficient at higher frequencies. In this paper, a solidly mounted resonator of piezoelectric aluminium nitride is designed for 5G Application. Finite element simulations in COMSOL multiphysics software are used to evaluate the properties of the designed SMR (Solidly Mounted Resonator). Simulation results show that the designed device resonates at 5.34 GHz with good return loss and Q-factor of 3326. FBAR (Thin Film Bulk acoustic wave Resonator) devices with high resonance frequency are beneficial for 5G mobile technology.

Keywords MEMS · Piezoelectric · SMR · BAW · 5G

1 Introduction

Significant new RF filtering issues are brought on with an increase in the growth of the Internet of Things (IoT) and the global rollout of 5G. To address these issues, there is a significant advancement in bulk acoustic wave (BAW) filters. BAW filters are pivotal by enabling higher frequencies and enhanced bandwidth considering the 5G and Wi-Fi band extension, and as a result, play essential roles in wireless devices and infrastructure. They focus on providing isolation between an increasing number of frequency bands while also providing low insertion loss to support device performance requirements. However, the expansion of other wireless technologies, such as Wi-Fi, and the global rollout of 5G present new filtering problems that go beyond the scope of conventional acoustic filters. To address these new challenges, major advancements in BAW technology were required. A BAW resonator is an electromechanical device that uses an electrical signal to generate a standing acoustic wave

P. K. Joshi (✉) · M. A. Hasamnis
Department of Electronics Engineering, Shri Ramdeobaba College of Engineering and Management, Nagpur, India
e-mail: joshipk@rknec.edu

H. O. Bansal et al. (eds.), *Next Generation Systems and Networks*, Lecture Notes in Networks and Systems 641, https://doi.org/10.1007/978-981-99-0483-9_11

in the piezoelectric material. A piezoelectric material—typically quartz, aluminium nitride, or zinc oxide—is placed between two metallic electrodes to form a device. In other words, the structure of the resonator is made up of two thin-film metal electrodes and a piezoelectric thin-film layer. To achieve the appropriate operating frequency, design characteristics such as material thickness and natural frequency are used. An acoustic wave is excited by the electrical field or voltage present between the two electrodes. Inverse relationships exist between the resonant frequency and the material characteristics and film thickness.

We need to explore beyond conventional RLC circuits in order to create a high Q narrow band-pass filter that incorporates all of the desired features of electronic equipment. BAW filters work efficiently at higher frequencies in the range from 2 to 16 GHz [1]. The market for low frequency (below 1.5 GHz) applications has been dominated by SAW filters, despite the fact that the production of RF filters and duplexers for wireless electrical devices is where BAW resonators are mainly used. Research on SMR so far is discussed below.

Chengzhang Han et al., solidly mounted resonator (SMR) fabrication which can also function as a biological molecule sensor [2]. Mohamed Mabrouk et al., for ladder BAW, a new customised interface tool for identifying parameters of Modified Butterworth-Van Dyke models [3]. R. Aigner et al., a summary of current BAW trends for high frequency bands, broad bandwidth filters, miniaturisation and thermal management. The development of complicated RF modules and their challenges are covered, as well as the evolution of RF content [4]. Villa-López et al. emphasises the development of Solidly Mounted Resonator (SMR) devices for the detection of particulate matter in order to produce an intelligent, reasonably priced particle sensor for air quality that can work at either 870 MHz or 1.5 GHz resonant frequency. [5]. Wei-Che Shih et al. determined that the corresponding circuit alteration based on the Mason model is how the actual device measurements are achieved. On resonator performance, the effects of various film conditions, material properties, and geometrical variations are investigated [6]. D. Ekeom, B. Dubus, and A. Volatier, for two-dimensional SMR simulation, presented the finite element–boundary element (FEMBEM) technique. The ATILA finite element programme is constructed with a discretised elasto-dynamic half-space Green function to characterise the substrate [7]. J. Olivares et al. explains how to create low acoustic impedance SiO2 films that can be used as a layer in Bragg mirrors for BAW resonators. Infrared absorption spectroscopy is used to evaluate the material's composition and structure [8]. According to K. M. Lakin et al., Material interfaces are required for acoustic resonators, and these interfaces are commonly created at the electrodes using an air or vacuum interface. Another approach is to attach the resonator to a series of quarter-wavelength-thick layers to a substrate to create a solid for making low insertion loss GPS filters (SMR) [9]. M. A. Duboisa et al., a solidly mounted resonator with a 10-layer acoustic reflector was created. They feature a strong resonance peak with a centre frequency of 2 GHz, and they have a high Q factor with a k2 coupling coefficient of 1% [10]. K. M. Lakin et al. describes how to use high mechanical impedance electrodes to boost the effective coupling coefficient of bulk acoustic wave (BAW) resonators in SMR or FBAR designs. A better electrode arrangement that raises

Ke^2 (Effective coupling coefficient) for resonators has been developed, enabling the construction of broader bandwidth filters than would have been possible with electrodes made of solely aluminium [11]. Roman Lanz et al. created bulk acoustic wave (BAW) resonators and filters with frequency ranges of 7–8 GHz, quality factors of 400–600, and coupling factors of 4–5.5%. [12]. S. Marksteiner et al., the acoustic mirror layer stack optimization using a quantitative technique was developed. Using this methodology, the effect of the mirror structure on the trapping of both longitudinal and shear wave energy was investigated [13]. Gayathri Pillai et al. suggested a composite thin-film bulk acoustic wave resonator (c-FBAR) based on the apodization approach to permit the containment of displacement and strain energy at the resonator's centre section [14]. An in-depth analysis of resonators based on MEMS (Micro Electro Mechanical Systems) technology is discussed [15–17]. Cassella et al. demonstrated a brand-new class of monolithic integrated RF passive parts based on recently created MEMS cross-sectional Lamé-mode resonator (CLMR) technology made of aluminium nitride (AlN) [18] The study of Ruby et al. discusses FBARs' most recent modification, the temperature-compensated zero-drift resonator (ZDR) [19].

2 BAW Resonators

There are two typical arrangements for BAW resonators, namely Thin Film Bulk Acoustic Wave Resonator (TFBAR or FBAR) and Solidly Mounted Resonator (SMR).

FBARs are often created in one of two ways: composite or edge-supported. FBAR is shown in Fig. 1. above. An edge-supported structure's back is totally etched away from the substrate, whereas another dielectric material is positioned beneath the primary structure in a composite design to serve as a supporting layer. In an SMR Fig. 2, additional layers are positioned beneath the primary structure to serve as Bragg reflectors and as structural support. Acoustic wave leakage from the bulk structure is reduced using Bragg reflectors.

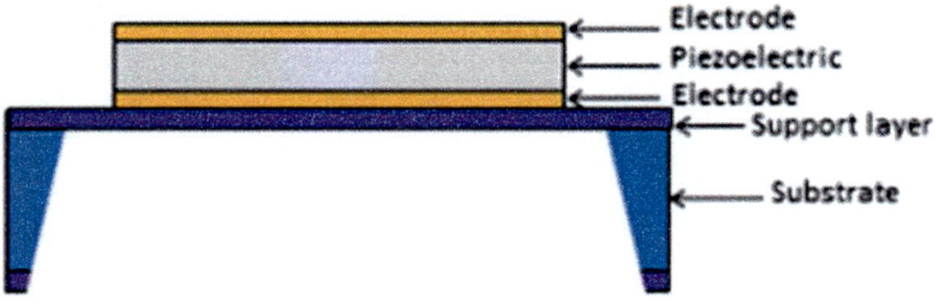

Fig. 1 Thin Film Bulk Acoustic Wave Resonator (FBAR)

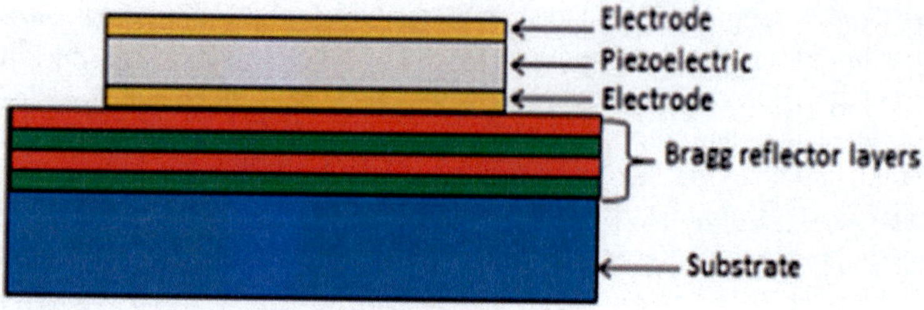

Fig. 2 Solidly Mounted Resonator (SMR)

Table 1 SMR design dimensions

Parameter	Value
Top/Bottom electrode dimension	
Length	900 nm
Height of top electrode Height of bottom electrode	100 nm 200 nm
Width	350 nm
Piezoelectric material dmension	
Length	900 nm
Height	200 nm
Width	350 nm
Bragg reflector dimension(2 stage)	
SiO2	100 nm
Tungsten(W)	100 nm

3 Proposed SMR Resonator Design

SMR is designed using aluminium nitride (AlN) as piezoelectric material. The SMR is designed which is as shown in Fig. 2. Aluminium top electrode with a thickness of 100 nm and a bottom electrode with a thickness of 200 nm were used. The design makes use of a piezoelectric layer with a thickness of 0.2 μm and an active area of $0.9 \times 0.35\ \mu m^2$. For the Bragg reflector's high-low acoustic impedance layer, a 100 nm thick (Tungsten) W and SiO2 (Silicon Dioxide) layer is created.

The SMR design with specifications is given in Table 1.

4 FEM Simulation Result

The simulations are performed in COMSOL Multiphysics Tool. First, 2D work planes with predetermined dimensions are used to design the resonator. The 3D model of the device, i.e. SMR is designed. All of the geometry's domains are given solid mechanics

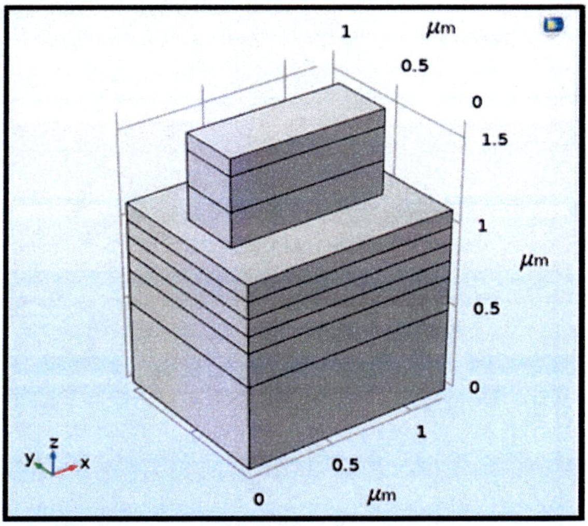

Fig. 3 3D model of SMR in COMSOL

physics assignments, while piezoelectric materials are given electrostatic physics assignments. The bottom and top Al electrodes function as the resonator's grounded and potential terminals, as shown in Fig. 2. The geometry starts to resonate at a predefined frequency as soon as 1 mV of electric potential is applied to the top electrode to activate the resonator. This frequency was calculated using eigenfrequency analysis. Figure 3 shows a 3D Model of SMR in COMSOL.

Figure 4 depicts the mode shape and induced stress of the first eigenmode for AlN material. The Quality factor and Admittance graph versus frequency are shown in Fig. 5a and b.

Calculating the Quality factor and admittance is necessary for examining the resonator's performance. The frequency domain analysis in simulation is performed with the range of frequency 5–6 GHz. Table 2 shows the Comparison of the quality Factor of Recent SMR with our proposed SMR through simulator validation.

An additional connector is attached to the top electrode to give input power for estimating the S11 and S22 parameters. For 1mW input power supply to the resonator, the S11 and S22 are obtained for both piezoelectric material as shown in Fig. 6 a and b.

5 SMR Micro-Fabrication Process Steps

Three steps make up the SMR fabrication process: Deposition of Bragg reflectors, fabrication of AlN thin films, and top Al electrode deposition. The Bragg reflector's minimum surface roughness is one of the most crucial elements in minimising wave energy loss into the substrate. In the investigation, the Si substrate is coated with W and SiO2 thin films using a Bragg reflector. The bottom electrodes on the Bragg

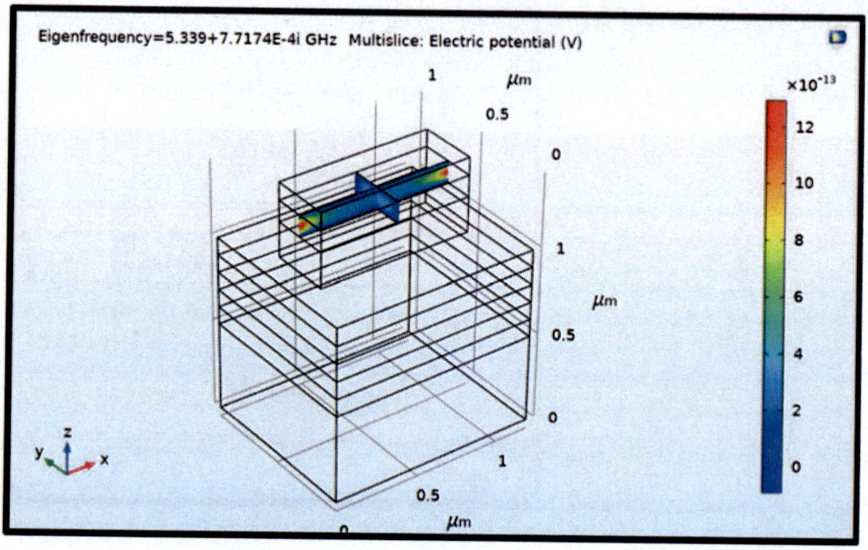

Fig. 4 Mode shape at eigenfrequency

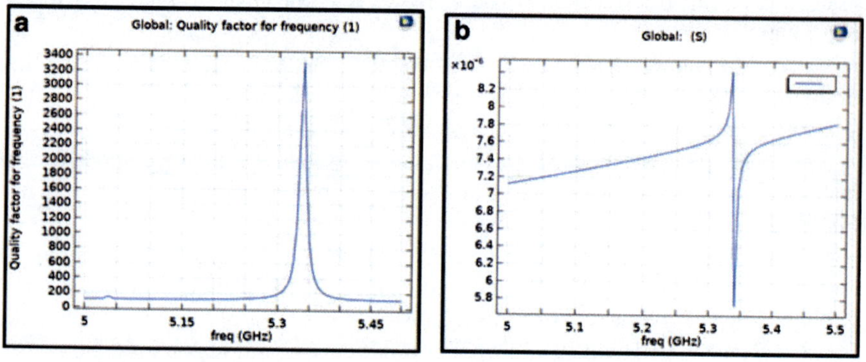

Fig. 5 **a.** Frequency versus quality factor **b.** Admittance versus Frequency

Table 2 Comparison of quality Factor of Recent SMR with our proposed SMR through simulator validation [13, 14, 18]

	Frequency	Quality factor
Proposed Work	5.34 GHz	3326
Gayathri Pillai et al. [14]	3.264 GHz	2507
Cristian Cassella et al. [18]	920-MHz	1750
Marksteiner et al. [13]	1.9 GHz	200

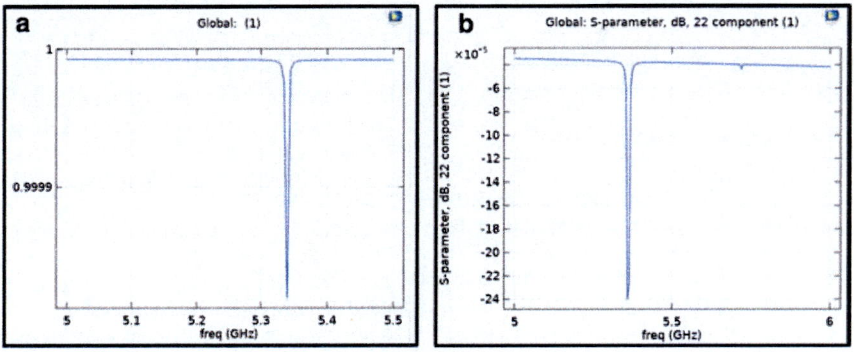

Fig. 6 **a.** S parameter **b.** S_{22} parameter

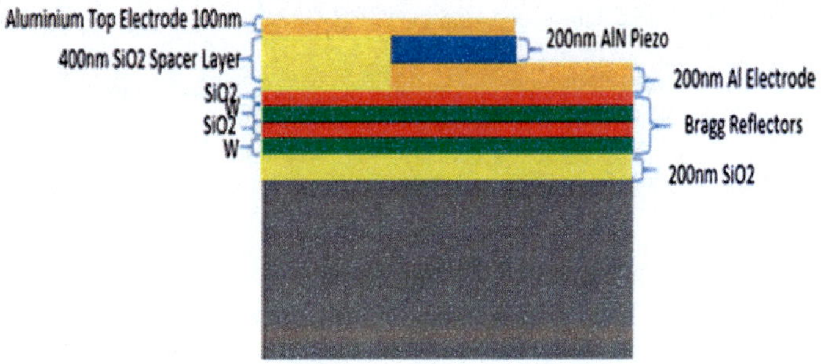

Fig. 7 Fabrication of the proposed SMR

reflector are implemented after it had been finished. The bottom electrode is then covered in the AlN piezoelectric thin films. Finally, the AlN piezoelectric thin films were built up with a top electrode. The top and bottom electrodes need a spacer layer to be separated, and this completes the SMR device. The SMR fabricated device is shown in Fig. 7.

6 Conclusion

This paper proposes an optimised form of SMR device which is highly feasible for mobile applications. The suggested SMR device employs a SiO_2 and W film-based Bragg reflector. The SMR device is found to resonate at 5.34 GHz frequency with good return loss and a high Q factor around 3326. Due to their compact size, lightweight, high resonance frequency, and superior resonance characteristics, SMR devices are advantageous for use in mobile applications.

References

1. Joshi PK, Latwe KR, Hasamnis MA (2022) Analysis and enhancement of Q-factor in thin-film bulk acoustic wave resonator (FBAR). J Phys: Conf Ser. 2273 012010. https://doi.org/10.1088/1742-6596/2273/1/012010
2. Han C et al. (2019) Solidly mounted resonator sensor for biomolecule detections. RSC Adv 9(37):21323–21328. https://doi.org/10.1039/C9RA01695C
3. Mabrouk M et.al. (2016) Flexible engineering tool for radiofrequency parameter identification of RF-MEMS BAW filters. ETRI J 38(5):988–995. https://doi.org/10.4218/etrij.16.0115.1075
4. Aigner R et.al. (2018) BAW filters for 5G bands. In: IEEE International Electron Devices Meeting (IEDM). pp 14.5.1–14.5.4. https://doi.org/10.1109/IEDM.2018.8614564
5. Villa Lopez FH et.al. (2015) Design and modelling of solidly-mounted resonators for low-cost particle sensing. Meas Sci Technol 27(2). https://doi.org/10.1088/0957-0233/27/2/025101
6. Shih W-C. et.al. (2017) Simulation of solidly mounted resonator using mason model and its implementation. Sens Mater 29(4):405–410. https://doi.org/10.18494/SAM.2017.1522
7. Volatier A, Dubus B, Ekeom D (2006) P1J-7 solidly mounted resonator (SMR) FEM-BEM simulation. IEEE Ultrason Symp 2006:1474–1477. https://doi.org/10.1109/ULTSYM.2006.371
8. Olivares J et al. (2009) Sputtered SiO2 as low acoustic impedance material for Bragg mirror fabrication in BAW resonators. In: 2009 IEEE International frequency control symposium joint with the 22nd European frequency and time forum. pp 316–321. https://doi.org/10.1109/FREQ.2009.5168193
9. Lakin KM, McCarron KT, Rose RE (1995) Solidly mounted resonators and filters. In: 1995 IEEE Ultrasonics Symposium. Proceedings. An international symposium, vol 2. pp. 905–908. https://doi.org/10.1109/ULTSYM.1995.495711.
10. Dubois M, Muralt P, Matsumoto H, Plessky V (1998) Solidly mounted resonator based on aluminum nitride thin film. In: 1998 IEEE Ultrasonics symposium. Proceedings (Cat. No. 98CH36102), vol 1. pp. 909–912. https://doi.org/10.1109/ULTSYM.1998.762291
11. Lakin KM, Belsick J, McDonald JF, McCarron KT (2001) Improved bulk wave resonator coupling coefficient for wide bandwidth filters. In: 2001 IEEE Ultrasonics symposium. Proceedings. An international symposium (Cat. No.01CH37263), vol.1. pp. 827–831. https://doi.org/10.1109/ULTSYM.2001.991848
12. Lanz R, Dubois M, Muralt P (2001) Solidly mounted BAW filters for the 6 to 8 GHz range based on AlN thin films. In: 2001 IEEE Ultrasonics symposium. Proceedings. An international symposium (Cat. No.01CH37263), vol.1. pp. 843–846. https://doi.org/10.1109/ULTSYM.2001.991851
13. Marksteiner S, Kaitila J, Fattinger GG, Aigner R (2005) Optimization of acoustic mirrors for solidly mounted BAW resonators. IEEE Ultrason Symp 2005:329–332. https://doi.org/10.1109/ULTSYM.2005.1602861
14. Pillai G, Zope AA, Tsai JM-L, Li S-S (2017) Design and optimization of SHF composite FBAR resonators. IEEE Trans Ultrason Ferroelectr Freq Control 64(12):1864–1873. https://doi.org/10.1109/TUFFC.2017.2759811
15. Abdolvand R, Bahreyni B, Lee JEY, Nabki F (2016) Micromachined resonators: A Review. Micromachines, 7(9):160(2016). https://doi.org/10.3390/mi7090160
16. Wang L, Wang C, Wang Y, Quan A, Keshavarz M, Madeira BP, Zhang H, Wang C, Kraft M (2022) A review on coupled bulk acoustic wave MEMS resonators. Sensors (Basel) 22(10):3857. https://doi.org/10.3390/s22103857
17. Pillai G, Li S.S (2021) Piezoelectric MEMS resonators: A review. IEEE Sens J 21(11):12589–12605. https://doi.org/10.1109/JSEN.2020.3039052
18. Cassella C, Chen G, Qian Z, Hummel G, Rinaldi M (2017) RF Passive components based on aluminum nitride cross-sectional lamé-mode MEMS resonators. IEEE Trans Electron Devices 64(1):237–243. https://doi.org/10.1109/TED.2016.2621660

19. Ruby R et al (2012) Positioning FBAR technology in the frequency and timing domain. IEEE Trans Ultrason Ferroelectr Freq Control 59(3):334–345. https://doi.org/10.1109/TUFFC.2012.2202

Design and Investigation of PMUT Sensor for Medical Imaging Applications

Sujay Krishnan Subramanian and Soumendu Sinha

Abstract Piezoelectric Micromachined Ultrasonic Transducers (PMUTs) are adaptable ultrasonic sensors and transducers utilised in a number of applications. In this work, we have designed PMUT cells made of PZT piezoelectric material and simulated its performance using COMSOL™. We compared the obtained results with the analytical model and found them to be closely matching. The designed cell uses a circular unimorph diaphragm with a high piezoelectric coefficient and a low dielectric constant. We also performed parametric studies for membrane dimensions and thickness of piezoelectric layer and it was found that the resonant frequency is inversely related to membrane diameter, whereas the resonant frequency increases with dielectric thickness.

Keywords Ultrasound · Piezoelectric thin films · PMUT · COMSOL · Sensors

1 Introduction

Ultrasound is a mechanical vibration of a frequency range greater than 20 kHz [1]. It is capable of traversing any physical medium. A device called an ultrasonic transducer is used to transform another kind of energy into an ultrasonic vibration. Contrary to traditional piezocomposite ultrasonic transducers, piezoelectric micromachined ultrasonic transducers (PMUTs) are miniaturised ultrasonic transducers that are manufactured utilising MEMS process technology [2]. Piezoelectric materials convert the changes in pressure, acceleration, strain, force or temperature into an electrical signal in the form of resonance frequency shift, charge density or voltage [3]. Miniaturised piezoelectric devices have become an essential element for integrated systems, as

S. K. Subramanian (✉)
Birla Institute of Technology and Science (BITS), Hyderabad 500078, Telangana, India
e-mail: f20200472@hyderabad.bits-pilani.ac.in

S. Sinha
CSIR - Central Electronics Engineering Research Institute (CEERI), Pilani 333031, Rajasthan, India
e-mail: soumendu@ceeri.res.in

H. O. Bansal et al. (eds.), *Next Generation Systems and Networks*, Lecture Notes in Networks and Systems 641, https://doi.org/10.1007/978-981-99-0483-9_12

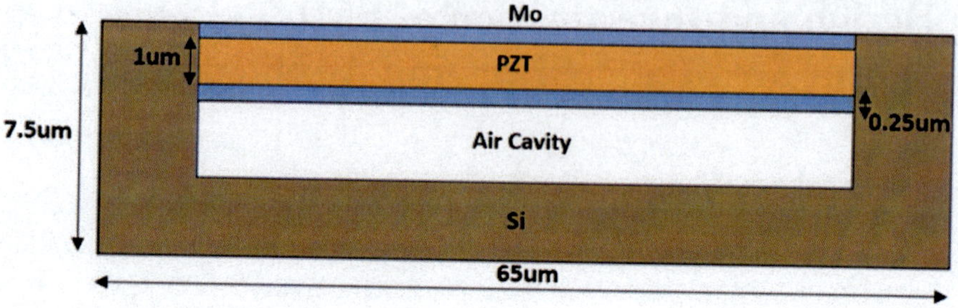

Fig. 1 Cross sectional view of PMUT device

smaller device dimensions provide advantages such as reduced power consumption, lower noise levels, multi-functionality and batch production. These characteristics are the primary drivers behind the semiconductor device industry's adoption of MEMS technology for the creation of advanced sensors and actuators [4]. MEMS is a process technology used to fabricate miniature integrated mechanical and electrical devices and systems [5, 6]. The size of these devices can vary from a few micrometres to a few millimetres and they are fabricated utilising integrated circuit (IC) batch manufacturing techniques. These devices (or systems) are capable of micro-scale sensing, control and actuation as well as generate effects on the macro-scale.

PMUT device consists of a piezoelectric material sandwiched between two electrodes. Figure 1 depicts the cross-sectional schematic of a PMUT device. The basic working principle is that the piezoelectric material vibrates and produces ultrasonic waves when an AC signal is delivered to the two electrodes. Acoustic sound waves are created when the strain caused by AC signals at a specific frequency causes vibrations in the piezoelectric material layer. Ultrasound is used in a number of ways, including for sensing, actuation and imaging, all of which call for small and low power transducers, which can be made with the use of a PMUT. A circular unimorph diaphragm with a high piezoelectric coefficient and a low dielectric constant is used in this work [7].

The main goal of a PMUT design is to change the device parameters such as electrode size, material composition and layer thickness in order to increase the coupling coefficient. This is because a PMUT with a high coupling coefficient can generate wide mechanical and electrical bandwidths, leading to short pulse length.

Rangefinders, intravascular ultrasound (IVUS) and fingerprint sensors are just a few applications of a PMUT sensor. IVUS is a diagnostic procedure that uses ultrasound waves to view inside blood arteries. This device has an ultrasonic transducer which produces ultrasonic waves and scans the inner wall of blood vessels [2, 8]. Rangefinders are utilised as object detectors, non-destructive inspection tools and detection sensors in automobiles. It often uses an AlN-based pMUT since fabrication is less complicated and there is no need for bias voltage [9, 10]. A typical fingerprint scanner is a capacitive fingerprint scanner that can recognise finger ridges and valleys. An ultrasound based fingerprint sensor is preferred over a standard capacitive type sensor as it can detect fingerprints even on oily or wet surfaces [2, 11].

The paper is organised as follows: Sect. 1 introduces the PMUT device and its applications. Section 2 presents the design and simulation of the device in COMSOL. Results and discussion have been presented in the Sect. 3. Finally, Sect. 4 provides the conclusion of the work.

2 Design and Simulation of PMUT

The design and simulation of PMUT has been carried out using COMSOL Multiphysics tool. For the design, a 2-D axis symmetric model has been used. This ensures that the 2-D model would revolve around the Y-axis and create the desired 3-D structure. For the PMUT design, a circular diaphragm was used. The top membrane should ideally have a thin supporting layer due to its high degree of flexibility. However, practically a layer that is too thin, i.e. a thickness less than 0.1 µm, may break during the operation. A 7.5 µm standard silicon supporting layer is used. The geometry of designed PMUT is summarised in Table 1.

The piezoelectric layer vibrates on the application of a potential difference, acting as a driver. In this study, a 1 µm PZT was used due to its fabrication convenience. Voltage is applied to the top electrode while the bottom electrode is grounded.

We utilise Solid Mechanics and Electrostatics modules in our simulations. In the Solid Mechanics module, the bottom surface is kept fixed, and the rest is left free for deformation. The left side is kept axis symmetric. In the Electrostatics module, a floating potential is applied to the bottom side of the top electrode and the top side of the bottom electrode is grounded. Additionally, the piezoelectric layer is subjected to charge conservation. The Multiphysics involved is the piezoelectric effect. A finer tetrahedral mesh is applied.

Table 1 Details of the geometry of designed PMUT

Layer details	Radius (in µm)	Thickness (in µm)	Material
Substrate	32.5	7.5	Silicon
Cavity	22.5	3	Air
Piezoelectric layer	22.5	1	PZT
Electrodes	22.5	0.25	Molybdenum
Surrounding medium	32.5	12.5	Water

3 Results and Discussion

We have performed three studies for the PMUT device:

1. Stationary studies
2. Eigenfrequency studies
3. Eigenfrequency with parametric sweep

3.1 *Stationary Studies*

There are two parts in this study. In the first part, a 1 V potential is applied across the electrodes, and we study the generated stress. In the second study, we apply an external force to the piezoelectric layer to emulate an impinging ultrasound on the PMUT diaphragm and study the generated electric potential. This demonstrates the capability of the PMUT device to be used as a sensor as well as a transducer. During the transmission operation, an AC voltage is supplied to the top electrode to generate ultrasound signals, which are useful for medical device applications. The reflected signal impinges upon the receiving PMUT device, which converts the external stress into a measurable electrical potential.

1. For the first case, the obtained stress is shown in Fig. 2. A surface stress of 15 MN/m^2 is obtained when an electric potential of 1 V was applied.
2. For the second case, a total external force of 1 mN is applied in the downward direction to emulate an impinging ultrasound on the PMUT diaphragm and the obtained electric potential is shown in Fig. 3. An electric potential of 0.15 V is generated when a total force of 1 mN is applied in the downward direction.

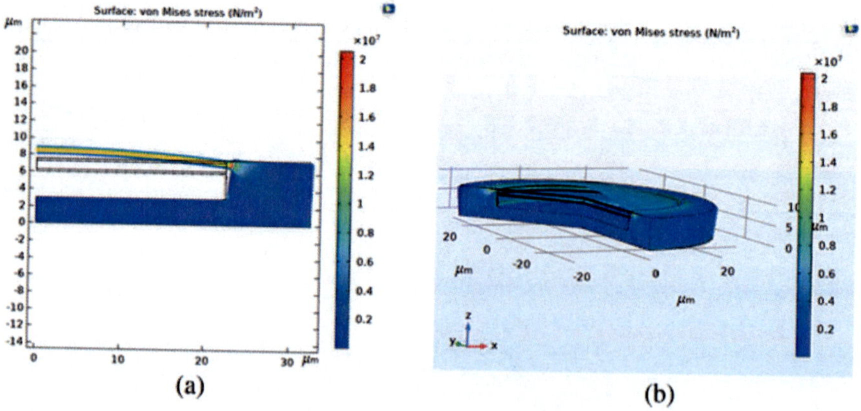

Fig. 2 **a** A surface stress of 15 MN/m^2 is obtained when an electric potential of 1 V was applied; and **b** the surface stress obtained in 3D

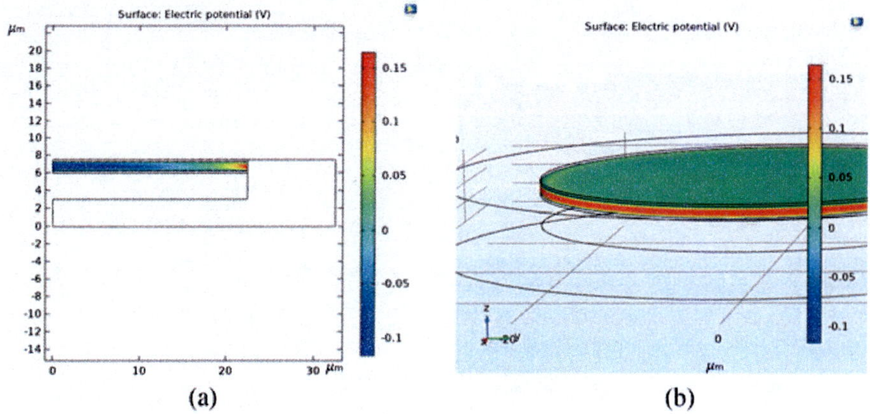

Fig. 3 **a** An electric potential of 0.15 V is generated when a total force of 1 mN is applied in the downward direction, and **b** the electric potential obtained in 3D

3.2 Eigenfrequency Studies

Eigenfrequency or resonance frequency is the natural frequency of a device, where the membrane vibrates at the highest amplitude. Eigenfrequency depends on various factors, as discussed in Eq. 1. The analytical model to obtain the eigenfrequency is as follows [10]:

$$f_o = \frac{0.47t\sqrt{E'/\rho_m}}{r^2} \tag{1}$$

where the layer's total thickness is t, the average plate modulus is $E = (E_p t_p + E_s t_s)/(t_p + t_s)$, $E'_{(p,s)} = E_{(p,s)}/(1 - \nu^2)$, Young's modulus of the PZT and Si layers are E_p (57~80 GPa) and E_s (168 GPa), respectively. ν is the Poisson's ratio. Poisson's ratio of PZT and Si are 0.32 and 0.064, respectively. The average density and radius of the PMUT diaphragm is given by ρ_m and r, respectively. Table 2 lists the values of different parameters.

The analytical calculations are done using MATLAB^{TM}, where we obtained the eigenfrequency as 6.747 MHz for the designed PMUT using PZT as the piezoelectric layer. After performing eigenfrequency study, the simulated eigenfrequency obtained in COMSOL was 6.781 MHz, as shown in Fig. 4. We realise that the error between the analytical and simulated eigenfrequency is ~0.034 MHz and a deviation of 0.5% from the desired frequency.

We performed detailed modal studies for the PMUT device. We have considered six modal eigenfrequencies and obtained the PMUT oscillation modes when excited with an AC signal. The harmonic oscillations at different eigenfrequencies observed in this study are listed in Table 3. Figures 5, 6 and 7 depict a few of the higher eigenfrequency vibrational modes for the PMUT device.

Table 2 List of parameters used for analytical calculations

Variables	Values
E_p	6.00e10 N/m^2
E_s	1.68e11 N/m^2
ν_p	3.20e–01
ν_s	6.40e–02
E'_p	6.68e10 N/m^2
E'_s	1.69e11 N/m^2
t_p	1.00e–06 m
t_s	7.50e–06 m
r	2.25e–05 m
E	1.57e11 N/m^2
ρ_p	7.75e03 kg/m^3
ρ_s	2.33e03 kg/m^3
ρ_m	2.97e03 kg/m^3
f_o	6.747 MHz

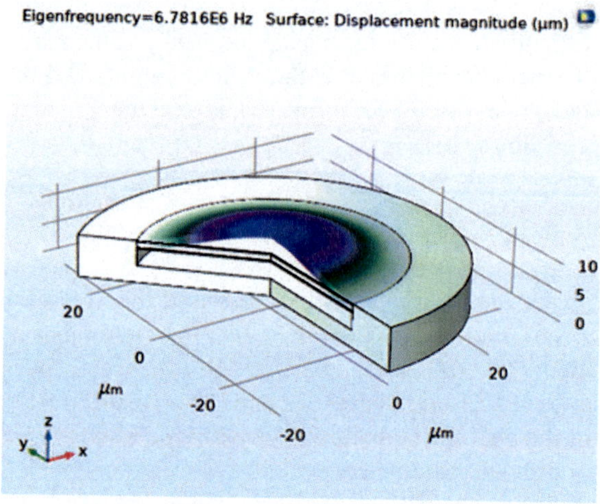

Fig. 4 Results of eigenfrequency study: fundamental mode

Table 3 Modal frequencies for the designed PMUT

Mode	Frequency (MHz)
Mode 1	6.781
Mode 2	24.832
Mode 3	51.28
Mode 4	82.522

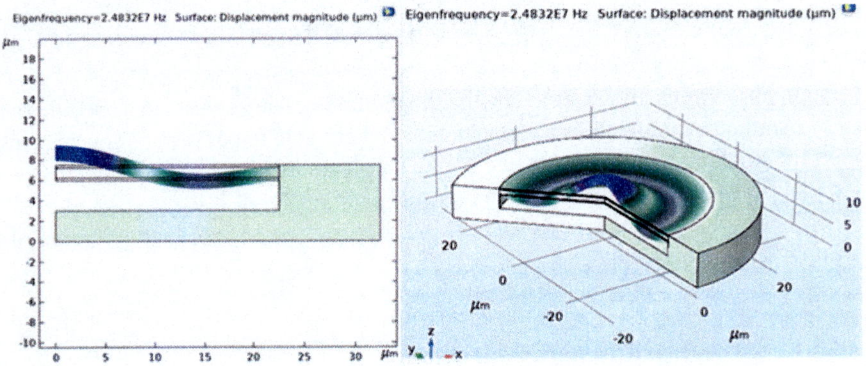

Fig. 5 Harmonic oscillation at an eigenfrequency of 24.832 MHz (Mode 2)

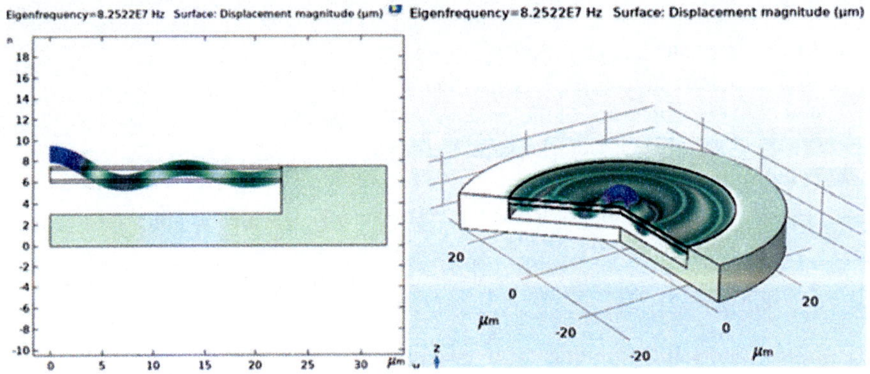

Fig. 6 Harmonic oscillation at an eigenfrequency of 51.28 MHz (Mode 3)

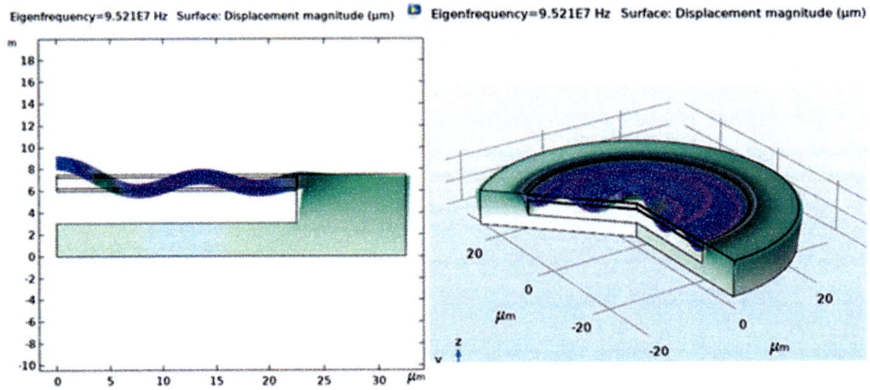

Fig. 7 Harmonic oscillation at an eigenfrequency of 82.522 MHz (Mode 4)

3.3 *Eigenfrequency with Parametric Sweep*

Parametric sweep allows us to change the parameter values through a specific range. By parametrizing the global parameters, we were able to perform parametric sweep on the thickness and radius of the piezoelectric and device layers. Table 4 lists the parametric study variables used in this study.

The resonance frequency of the PMUT depends on several factors. Some of the important factors include

3.3.1 Piezoelectric Material Variation

We have studied two different piezoelectric materials popularly used in PMUT devices.

a. Lead Zirconate Titanate (PZT)
   PZT provides high PMUT transceiver efficiency due to low dielectric constant and high piezoelectric coefficient ($e_{31,f} \approx 16$–$24\ \mathrm{C/m^2}$). The resonance frequency obtained using PZT material is 6.7816 MHz.
b. Aluminium Nitride (AlN)
   AlN is a CMOS compatible piezoelectric material, which enables cost-effective device fabrication. The resonance frequency obtained using AlN material is 9.6393 MHz. However, PZT is a preferred material due to superior piezoelectric constant [2].

We can observe that AlN has a higher eigenfrequency in comparison to PZT due to much higher Young's modulus (273–346 GPa) compared to PZT (57–80 GPa). As we can see from Eq. 1, as Young's modulus increases, the resonant frequency increases as well.

Table 4 Parameters used for performing the parametric sweep

Parameter	Expression	Value (m)
thk_mo	0.25 μm	2.5e–7
thk_pzt	1.00 μm	1e–6
thk_air	thk_si-(2*thk_mo)-thk_pzt- 3.00 μm	3e–6
thk_water	12.5 μm	1.25e–5
thk_si	7.5 μm	7.5e–6
rad_air	22.5 μm	2.25e–5
rad_si	rad_air + 10 μm	3.25e–5

3.3.2 Radius of Piezoelectric Material

Further, a parametric sweep was performed to obtain the resonance frequency at different diaphragm radii. The obtained results are shown in Fig. 8a. As observed from the graph, as the radius of the piezoelectric layer increases, the resonance frequency decreases. Thus, we obtain an inverse relation between radius and the resonant frequency. We can also correlate this result from the analytical model (Eq. 1).

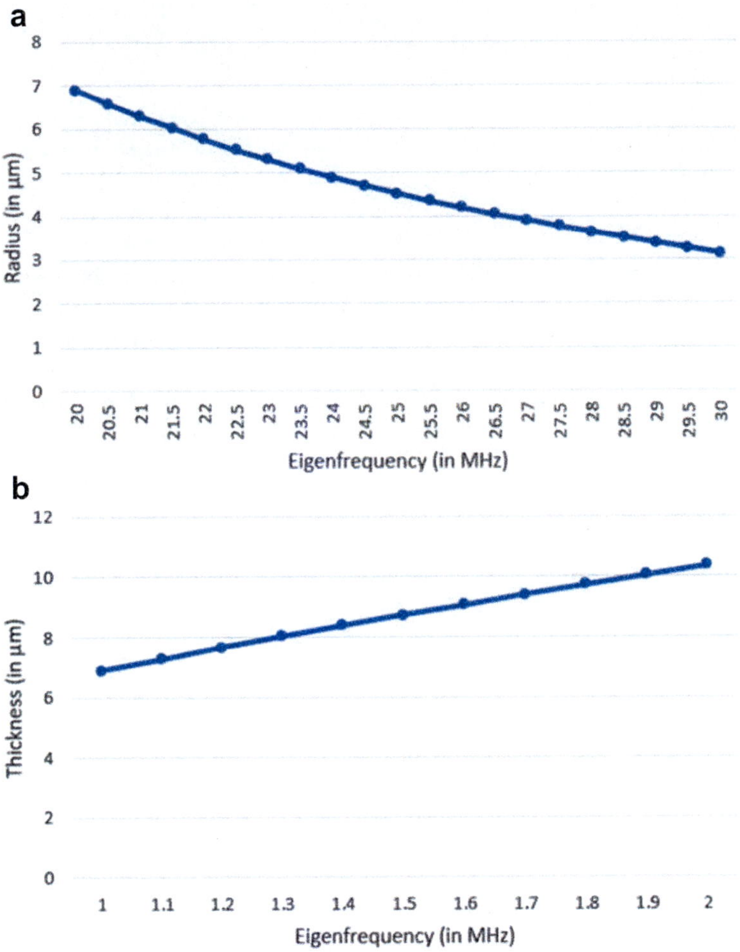

Fig. 8 **a** Parametric study of the eigenfrequency (in MHz) versus radius (in µm), **b** Parametric study of the eigenfrequency (in MHz) versus thickness (in μm).

3.3.3 Thickness of Piezoelectric Layer

Another factor on which the resonance frequency depends on is the thickness of the piezoelectric layer. We have again used parametric sweep to find the resonant frequency at different thickness. We then plotted the results as shown in Fig. 8b. As observed from the figure, as the thickness of the piezoelectric layer increases, the resonance frequency also increases (almost linearly). Again, from Eq. 1, we can confirm that the frequency and thickness have a direct relationship.

4 Conclusion

In this work, PZT-based PMUT was designed, simulated and analysed for ultrasonic sensing applications. The displacement of the PZT layer was studied using COMSOL, when a voltage was applied across the electrodes of the PMUT. A stress of 15 MN/m^2 was generated when a voltage of 1 V was applied. Moreover, an electric potential of 0.15 V was developed when an external total force of 1 mN was applied to the PMUT device. A resonance frequency of 6.6781 MHz was obtained after performing eigenfrequency simulations. Analytical calculations were also performed using MATLAB, and the results obtained closely match the simulation results. The PMUT resonance frequency was studied by varying several device parameters, such as material, thickness and radius of the piezoelectric layer. The obtained results show that the diaphragm radius had an inverse relation with the resonance frequency, while the thickness had a direct relation.

Acknowledgment The authors are thankful to the Director, CSIR-CEERI, Pilani for his generous support for this work. They also express their gratitude to the scientific/technical staff members of Semiconductor Devices Area for their kind support.

References

1. Sadeghpour S, Zilonova E, D'Hooge J, Kraft M (2021) A novel 6 MHz phased array piezoelectric micromachined ultrasound transducer (pMUT) with 128 Elements for medical imaging. In: 2021 IEEE International Ultrasonics Symposium (IUS). IEEE, pp 1–4. https://doi.org/10.1109/IUS52206.2021.9593633
2. Joontaek J, Wonjun L, Woojin K, EunJung S, Jungho R, Hongsoo C (2017) Review of piezoelectric micromachined ultrasonic transducers and their applications. J Micromechanics Microengineering 27. https://doi.org/10.1088/1361-6439/aa851b
3. Qiu Y, Gigliotti J, Wallace M, Griggio F, Demore C, Cochran S, Trolier-McKinstry S (2015) Piezoelectric Micromachined Ultrasound Transducer (PMUT) arrays for integrated sensing, actuation and imaging. Sensors. 15:8020–8041. https://doi.org/10.3390/s150408020
4. Roy K, Kalyan K, Ashok A, Shastri V, Jeyaseelan A, Mandal A, Pratap R (2021) A PMUT integrated microfluidic system for fluid density sensing. J Microelectromechanical Syst 1–8. https://doi.org/10.1109/JMEMS.2021.3091651
5. An Introduction to MEMS (Micro-electromechanical Systems). Lboro.ac.uk. (2002)

6. Mukhiya R, Garg M, Gaikwad P, Sinha S, Singh AK, Gopal R (2020) Electrical equivalent modeling of MEMS differential capacitive accelerometer. Microelectron J 99:104770. https://doi.org/10.1016/j.mejo.2020.104770
7. Roy K, Gupta H, Shastri V, Dangi A, Pratap R (2018) Fluid density sensing using PMUTs, pp 1–4. https://doi.org/10.1109/ICSENS.2018.8589662
8. Ma T, Yu M, Chen Z, Fei C, Shung KK, Zhou Q (2015) Multi-frequency intravascular ultrasound (IVUS) imaging. IEEE Trans Ultrason, Ferroelectr, Freq Control 62:97–107. https://doi.org/10.1109/TUFFC.2014.006679
9. Richard P, Igor I, Mitchell K, Bernhard B, Stefon S (2010) An ultrasonic rangefinder based on an ALN piezoelectric micromachined ultrasound transducer. Proc IEEE Sens 2417–2421. https://doi.org/10.1109/ICSENS.2010.5690777
10. Luo G-L, Kusano Y, Horsley D (2020) Airborne piezoelectric micromachined ultrasonic transducers for long-range detection. J Microelectromechanical Syst 1–9. https://doi.org/10.1109/JMEMS.2020.3037298
11. Tang H-Y, Lu Y, Fung S, Tsai J, Daneman M, Horsley D, Boser B (2015) Pulse-echo ultrasonic fingerprint sensor on a chip. https://doi.org/10.1109/TRANSDUCERS.2015.7181013
12. Akhbari S et al (2016) Equivalent circuit models for large arrays of curved and flat piezoelectric micromachined ultrasonic transducers. IEEE Trans Ultrason, Ferroelectr, Freq Control 63:432–447. https://doi.org/10.1109/TUFFC.2016.2525802

Intelligent Cost Estimation Engine for EV Charging Stations: A Deep Learning-Based Approach

Antarip Giri, Akash Dixit, Sidheshwar Harkal, Uday Satya Kiran Gubbala, and Niranjan Mehendale

Abstract The potential of automation possibility and enhanced user experience for electric vehicle stakeholders is discussed in this paper. Electric vehicles (EV) have acquired development chances in recent years because of the pressures of the energy crisis and environmental pollution and awareness within users. With the large-scale construction of charging stations, exploded EV buying rate, it results in increased power demand on local grids. Further, various direct, indirect, static, and dynamic factors involved in the usage and total charging cost estimation make it difficult for charging station owners, utility, and overall tedious user experience. In the future, EVSE load forecast and associated cost estimation, particularly a short-term forecast, will play a critical role in grid load planning, and the safe functioning of the electric power system while enhancing end user experience. AI-ML-driven and deep learning-based intelligent cost estimation engine proposed here will leverage the IOT connectivity for EV to forecast the load demand at charging station and efficient price estimation for EV charging. EV will provide details like battery state parameters and user historical data, etc. whereas Electric Vehicle Supply Equipment (EVSE) details, base and peak tariff details will be captured from charging station through connected platform. This engine is to provide dynamic and optimized tariff forecast up to next 5 h based on the historical usage of the specific user, past charging station usage, etc. Intelligent price forecast engine can efficiently estimate more accurate total charging cost based on the customer demand. User will have flexibility to choose the best tariff based on the different time slots and this engine will provide the hassle-free estimation.

Keywords Intelligent cost estimate · Time-series forecast engine · IoT-Connected vehicle platform · Electric Vehicles (EV) · EVSE · AI-ML · Deep learning · State parameters

---

A. Giri (✉) · A. Dixit · S. Harkal · U. S. K. Gubbala · N. Mehendale
Cognizant Technology Solutions, Chennai, India
e-mail: Antarip.Giri@cognizant.com

N. Mehendale
e-mail: Niranjan.Mehendale@cognizant.com

H. O. Bansal et al. (eds.), *Next Generation Systems and Networks*, Lecture Notes in Networks and Systems 641, https://doi.org/10.1007/978-981-99-0483-9_13

1 Introduction

The foundation of EV development is infrastructure, such as charging stations, which is one of the most significant components in EV market exploitation. With the large-scale building of charging stations, the widespread usage of electric vehicles will result in a quick increase in the power load, resulting in peak electric power in local areas and periods of time when everyone is charging. In a study conducted in [1] in Beijing, it was found that the demand for EV charging stations would account for 3% of its energy demand of the city if all their buses were converted to EVs.

Currently, a large majority of EV owners do not have access to private charging stations, indicating the critical need for public charging stations as the EV market increases. Moreover, directives like AFI (Alternative Fuels Infrastructure) by the European Union (EU) recommend one public access charger for 10 EVs [2]. This creates a major issue of grid integration of the charging stations. Methods like smart charging [3] is considered an important step towards achieving the successful grid integration all the while creating a profitable business for the stakeholders. Apart from smart charging, dynamic pricing methods are also being adopted by stakeholder as an attempt to create a profitable business [4]. In countries like the USA, the increase in EV has laid to various pricing model including time-based along with idle charges, kWh-based fees, and session-based fees on top of the usage-based fees, monthly/annual subscription-based models [5, 6]. In an analysis done in [7], it is found that most charging stations in the US are based on traditional time-of-use (TOU) structure, which is found to incur higher charging costs in certain scenarios such as fast charging stations with lower utilization like in remote places.

The EV charging station load is more intricate and variable than the standard electricity load because of the flexible user behavior. In contrast to traditional electricity load, the current EV charging station load curve is more strongly tied to human behavior than to weather. In addition, models should consider the required flexibility for individual user/vehicle requirements.

This paper proposes a real-time solution in the form of a smart price estimation engine, based on the user profile using the deep learning techniques. This solution is developed using the open-source tools like Python, deep learning and HTML. This solution handle dynamic price fluctuation due to the dynamic change in charging load demand. This paper discussed about the architecture design of the proposed solution and its working operation.

The main contributions of this paper can be summarized as follows:

- The proposed price estimation engine, based on deep learning approaches, helps in estimating the price for charging the EV vehicles. This study diversifies the forecasting models, such as LSTM.
- A SMART estimate Engine handles the personalized profile of each individual user.
- This SMART estimate engine will also have the user historical data about the charging summary and the travel miles.

The remainder of this document is structured as follows: Sect. 2 describes the main structure followed for pipeline creation with main blocks involved, and describes the IoT layer, and the ML/DL model layer in detail. Data understanding, feature engineering and model development stages are discussed in detail. It further presents and analyzes the findings of LSTM, GRU and RNN. Section 3 describes SMART Cost Forecast Engine, which utilizes complex rules developed to consider dynamic nature of the tariff, and user data. SMART web-app development and operational step are explained Sect. 4. Finally, results are discussed in Sect. 5, and Sect. 6 shows final conclusions on the entire work. In addition, future scope is added in last Sect. 7 for quick reference.

2 Main Structure and Model Pipeline Development

The complete architecture and data pipeline proposed in this paper are shown in Fig. 1. The architecture contains four main parts: the IoT layer, the model development layer, the smart cost forecasting engine (SCFE) layer and the front-end layer. The IoT layer will connect the two main components of the pipeline, i.e., the vehicle and the charging station (EVSE) through wired and/or wireless connections. Further, this layer will gather relevant data from both through cloud/edge services and feed the information to the appropriate databases. Next, the model development layer, which gets data from historians for EVSE, is fed first through the data preparation block, which consists of all the necessary blocks for making the dataset ready for ML/DL models. The middle core layer named smart cost forecasting engine (SCFE) layer, which connects to all other 3 layers, gets relevant information from IoT layer, gets input from the deployed ML model and the instantaneous user inputs from the front-end layer (web-app). The front-end layer shows the various static data parameters of EV user and the EVSE along with forecasted prices. The sub-sections below describe the IoT layer, and the ML/DL model layer in detail while SCFE and front-end layer web application are discussed in separate respective sections.

The current scope of paper does not consider any details about IoT layer and assumes it will be suitably established at ground based on the specific requirements and feasibility.

2.1 ML Model Development

The second component of the model pipeline is the model development layer. It consists of three sub-parts, namely, the data layer, the AI/ML engine layer and the deployment layer. Figure 2 represents a drill down of the model development layer with its various components. The data layer consists of the historian data of the charging station logs. For this paper, the dataset used is considered in this layer. The AI/ML engine again is broken down into two sub-parts: the data preparation layer

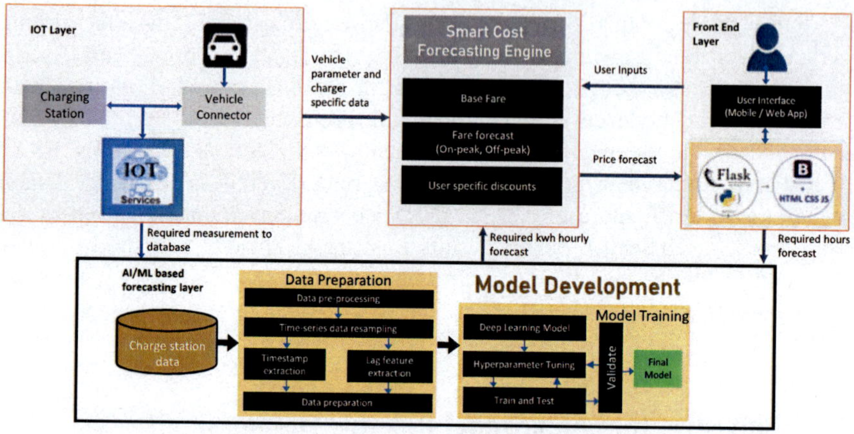

Fig. 1 Architecture/design of the proposed solution

and the model training layer. It is to be noted that this is the case for deep learning-based models, which is considered for the final forecasting pipeline. The final layer is the deployment layer, this takes the trained final DL model and creates an API (flask API in this case) which is called from the web-app. A detailed discussion of all the sub-part of the model development layer is discussed in this section.

Dataset Details/Description

The dataset used in this paper is openly available at https://ev.caltech.edu/dataset. The supporting paper for this dataset [8] talks about the various aspects of the charging station set-up, data acquisition methods, and data analysis methods such as statistical methods, use predictive methods, adaptive charging and Duck Curve smoothing. The

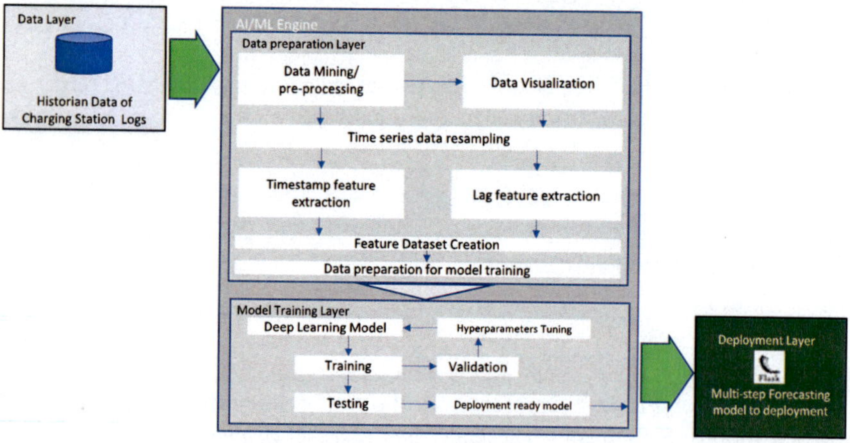

Fig. 2 AI/ML engine architecture of the proposed solution

data, we are using, are from commercial workplace charging data. The complete ACN (Adaptive Charging Network) data contain data from two charging sites, Caltech, and JPL. In this paper, we have focused on JPL data as it has no fast DC charger and contains 52 EVSEs unlike the other location. We have taken 2-year data starting from January 1, 00:00 am 2019 to January 01, 2021, 00:00 am. This is one of the many openly available EV charging station datasets [9, 10], which are provided for related works like building indicators through statistical modeling for planning and evaluating purposes [11, 11]. There has been other EV charging data collected from residential complex's EV chargers to specifically study the impact of them on the distribution grid [12].

Feature Engineering

The dataset we are using in this paper is described in the section above. The dataset is given in a nested json format as shown in Fig. 3.

This nested json is extracted in the form of a dataframe. Further processing steps are followed to create a final dataframe. This dataframe is further processed to create a dataset with appropriate typesetting for each column. The dataset consists of minute data which is resampled to hourly data to create a dataset for our requirement. Fig. 3 shows that there are many parameters that are getting captured and stored for the charging station. Although, these parameters are important for a smooth operation of the EVCS, keeping in mind our problem statement, we extract the univariate data of the kWh delivered from the dataframe of all the parameters and create a dataframe with only the timestamps and the corresponding kWh delivered. An hourly distribution of the dataset is shown in Fig. 4a. It compares the hourly kWh delivered for the each "Thursday" of the month of January 2019. Further, Fig. 4b. compares the

Fig. 3 json format used for data extraction


```
{
  "_id": "5c412c1df9af8b12cb56c27d",
  "clusterID": "0039",
  "connectionTime": "Tue, 01 Jan 2019 18:09:17 GMT",
  "disconnectTime": "Wed, 02 Jan 2019 02:39:32 GMT",
  "doneChargingTime": "Tue, 01 Jan 2019 20:16:10 GMT",
  "kWhDelivered": 12.534,
  "sessionID": "2_39_79_379_2019-01-01 18:09:16.991864",
  "siteID": "0002",
  "spaceID": "CA-327",
  "stationID": "2-39-79-379",
  "timezone": "America/Los_Angeles",
  "userID": "000000558",
  "userInputs": [
    {
      "WhPerMile": 222,
      "kWhRequested": 17.76,
      "milesRequested": 80,
      "minutesAvailable": 514,
      "modifiedAt": "Tue, 01 Jan 2019 18:09:19 GMT",
      "paymentRequired": true,
      "requestedDeparture": "Wed, 02 Jan 2019 02:43:17 GMT",
      "userID": 558
    }
  ]
},
```

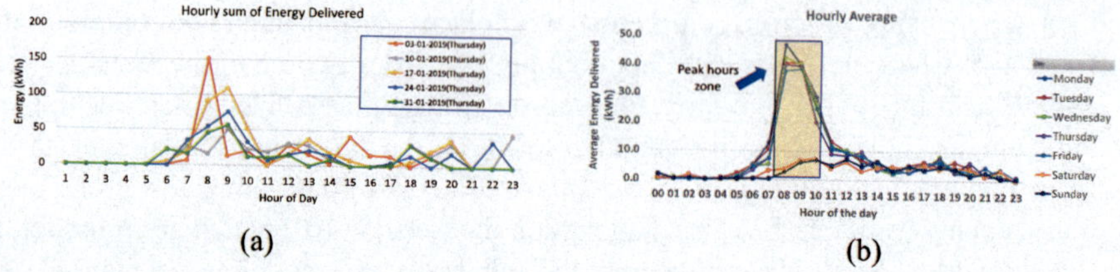

Fig. 4 **a**. Hourly total kWh delivered for a particular day (e.g., Thursday in the month of Jan 2019), **b** Average kWh delivered for hour of the day

average kWh delivered for various days of the week for a particular month (March 2019). It is evident that there is no specific identifiable pattern emerging for the various days of the week. This further infers that there are other factors that can impact the EVCS kWh delivered.

Figure 5a and b shows a relation between the average energy delivered for the months with the average connection time for an EV in those months. We can see that there is a significant correlation between them. Further, we observe in Fig. 5b that the cumulative kWh delivered over the months for years 2019, 2020 shows a significant dip due to the COVID-19 pandemic. Thus, we can see that there are many factors which can be socio-economic and can alter the usage pattern of the specific EVCS.

Statistical and ML-based forecast models

Before creating a dataset with extracted features, several statistical time-series forecasting techniques are performed on the univariate data itself. These techniques are used as base models to ensure that the deep learning model we create outperforms them in forecasting the kWh delivered. For our paper, we explore techniques like, moving average with a window size of 6, exponential smoothing with various alpha values, double exponential smoothing with various combination of alpha and beta values, triple exponential smoothing with different combination of alpha, beta, and gamma values (Holt-Winters). Figure 6 shows a moving average filter (window =

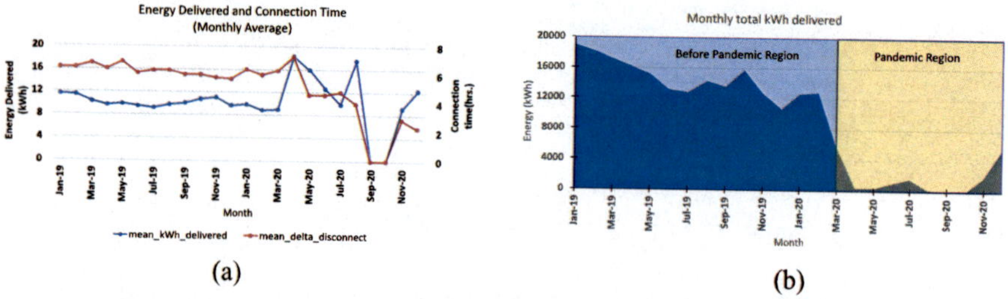

Fig. 5 **a** Relation between monthly average kWh delivered and average charging time for a specific charging station; **b** Monthly cumulative kWh delivered for specific charging station—showing steep change before and during pandemic time frame

6) applied on the kWh delivered, it can be observed that the moving average doesn't track the sudden peaks in the delivered energy and smoothens out the peaks significantly. Figure 7b, c shows exponential, double exponential smoothing applied on the original data, we can observe although exponential smoothing with alpha = 0.7 tracks the kWh delivered well but fails to provide accurate tracking. Whereas the double exponential smoothing shows negative values, which is not practically true. These inferences depict that these statistical methods are not much suitable in forecasting kWh delivered for this dataset.

Further, linear regression, ridge regression and lasso regression model are applied after creating feature dataset with lagged value and date-time features. Lastly, XGBoost regressor is applied on the feature dataset created from the univariate data. The ML algorithms are compared based on their mean absolute error (MAE) metrics and the XGBoost is taken as the base model with MAE of 8.34%.

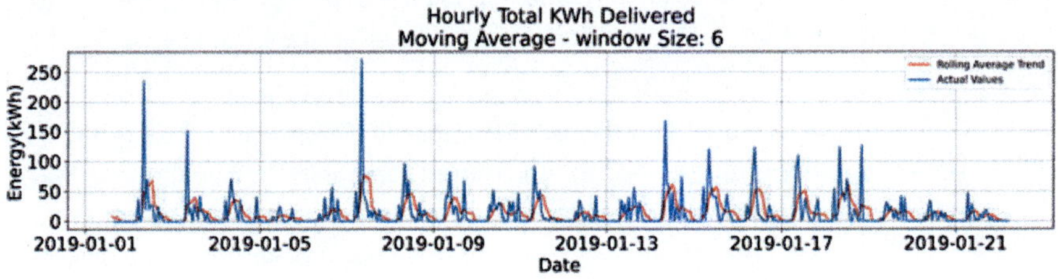

Fig. 6 Moving average filter (window = 6) applied on hourly total kWh delivered

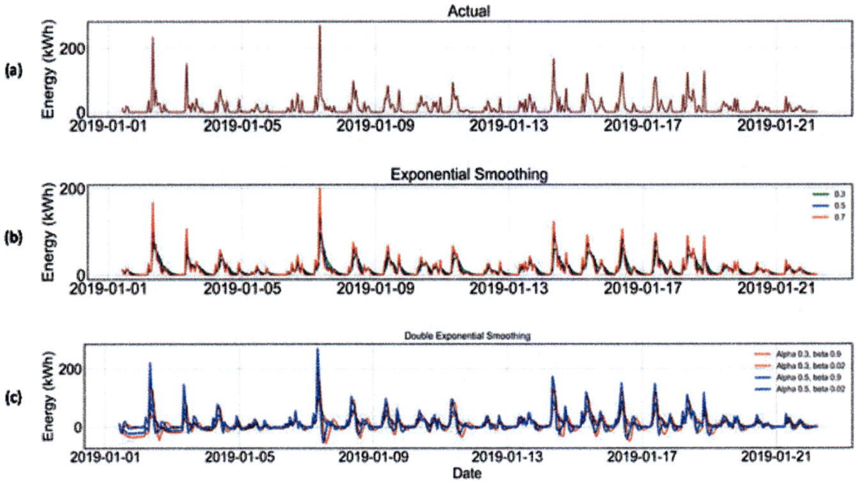

Fig. 7 **a** Original signal; **b** Exponential smoothing applied on hourly kWh delivered data (various alpha values), **c** Double exponential smoothing applied on hourly kWh delivered data (various combinations of alpha and beta values)

Table 1 Hyperparameters used for DL models

Hyperparameters	Values
Input dimension	462
Output dimension	6
Layer dimension	3
Batch size	64
Dropout	0.2
Learning rate	1e-6
Weight decay	1e-4

Deep learning-based forecast models

In this paper, we have explored three such deep learning models, namely, gated recurrent unit (GRU), long short-term memory (LSTM) and recurrent neural networks (RNN). A detailed understanding for each model can be found in [13].

The hyperparameters used in all the three models are given in Table 1.

The input dimension used is same as the number of training data rows used. As we are performing multi-step forecasting, the output dimension is equal to the number of steps that needs to be forecast, i.e., 6 for this paper. The hidden layer dimension is taken as 3, which has been found to work well from our trial-and-error experiments.

The deep learning models are tasked to forecast n (where maximum n is 6) steps of kWh delivered by the charging station. To create a training and test dataset for the model to be trained on, the univariate time-series data is processed with various feature extraction methods applicable for time-series data. The feature dataset was created by taking lag features of 24*7 which was found by trial-and-error method to be the most suitable for our forecast models. Further, hour, day, month, day of the week, week of the year were the features that were extracted from the time stamp of the univariate data. The holiday feature was added to the dataset and was found to be useful in improving the performance of the model. Finally, all the categorical features were further expanded to form one-hot encoded features for the whole feature dataset. All this feature extraction created a dataset with 295 features.

For time-series data, a cut-off point-based train and test split was performed to create train, validation and test dataset from the complete feature set. Once the dataset was split, normalization technique of MinMax scaling was used to create the dataset ready for the deep learning model.

The RNN model was found to be performing best for our problem statement. The loss functions used were mean squared error (MSE), smooth L1 loss and Huber loss. Out of the loss function used smooth L1 loss was found to be suitable for our model's performance. Three metrics were calculated for all the deep learning models, namely, mean absolute error (MAE), root mean square error (RMSE), R-squared error (R^2). A comparison of the three metrics for ML models and the combination of DL model with three loss functions are shown in Table 2, which shows a significant and drastic drop in metric values from ML to DL models.

Table 2 Performance evaluation for various ML/DL models

Sl. No	ML/ DL model names	MAE	RMSE
1	Ridge regression w CV	24.393	32.523
2	Lasso w CV	23.137	31.329
3	Linear regression (w encoded features)	13.769	21.093
4	Linear regression	9.465	16.449
5	XGBoost Regressor	9.412	22.587
6	LSTM w Huber loss	0.070	0.087
7	LSTM w MSE loss	0.069	0.086
8	LSTM w smooth L1 loss	0.067	0.085
9	GRU w Huber loss	0.067	0.080
10	GRU w smooth L1 loss	0.063	0.077
11	GRU w MSE loss	0.058	0.072
12	RNN w Huber loss	0.048	0.063
13	RNN w MSE loss	0.047	0.062
14	RNN w smooth L1 loss	0.056	0.061

It was observed that, for some of the models, R^2 score is negative, which is due to lack of data samples as compared to the features extracted leading to overfitting or due to the distribution of test data is completely different from the training data distribution. However, for error metrics (MAE, RMSE), we can see that the DL models are performing better than the ML models and hence we go for the best-performing DL model.

3 Smart Cost Forecasting Engine

The deep learning model for forecasting kWh delivered by the charging station is one of the building blocks of our smart forecasting engine. The SCFE block diagram shown in Fig. 8 depicts the various building blocks required. The deep learning model requires charging historical data of the charging station. This is used to extract features and forecast the kWh delivered. The number of forecasts required is calculated through a calculation engine, which is specific for each vehicle that comes in and connects. If the vehicle needs to be charged for 'n: number of hours', it needs few details from the vehicle user such as battery and BMS details, the vehicle range details and required miles. Once 'n' is calculated, it is fed to the ML/DL-based forecast block, which forecasts the kWh delivered by the charging station for next 'n' hours. These values are fed to the cost forecasting engine, which considers further user's profile data and charging station data to set various dynamic fares such as the base fare, peak tariff charges and the discount rate depending upon the user. These inputs are fed to the data to estimate the cost for that user for that requirement.

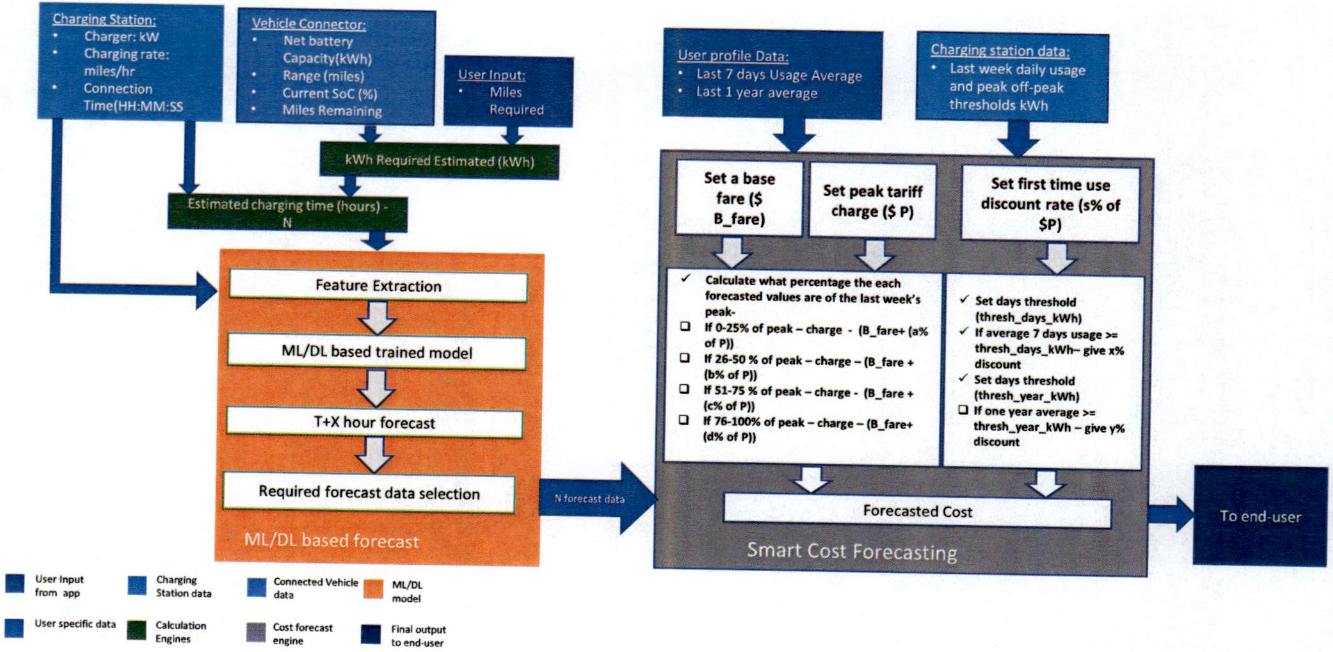

Fig. 8 Various internal steps and inputs to create smart cost forecasting engine

The SCFE is a smart rule-based engine used for calculating the current and future tariff to the end-user. This is carried out using two steps of rule-based calculation. The charging station authority needs to provide the SCFE three values: a base tariff (in \$B), a peak charge tariff (in \$P) and a first-time user discount (% of peak tariff). Further, the engine considers the peak charge for the last 7 days, which is used to check with the current demand in kWh. Depending upon whether the demand is x% of the peak demand, the initial tariff is calculated based on the formula (a% of \$P + \$B), where 'a' can be values that can be set by the charging station owner. In the next step of the rule engine, the calculated tariff is taken and checked whether the user is a first time, or a regular user. At the end of this step, the tariff, or the cost to be charged to the user is output and this is fed to the web-app, which is discussed later in this paper. The complete SCFE is wrapped with a web-app, which considers different user and charging station-centric scenarios. The SMART web app is explained in detail below.

4 Smart Web-App

A web-app is designed to demo and create use cases for the SCFE described above. It also shows information of the past charging station usage and forecast for the next 4 h.

Fig. 9 shows the flowchart for the user journey, which depicts the various steps the user will follow arriving at the charging station.

Once the user enters the charging station, the battery charge state and other necessary parameters are captured, and the user is prompted to connect the vehicle to a charging station.

In the SMART App, the first page is used to login, where a user can login by entering its credentials (username and password). The various key pages available are shown in Fig. 10 and briefly explained next.

4.1 Working Principle

Assuming user profile is already created, once user opens the web-app, it will prompt for login credentials on main page and after successful authentication user will be allowed to use the SMART app.

After login user is redirected to first page (refer Fig. 10b) where specific details corresponding to vehicle and EVSE are shown on left side pane. Vehicle details and EVSE details are fetched from connected platform and displayed for quick reference. Usage history, average miles requested, average charging cost paid and average charging time utilized by user in last 7 charges are shown in bottom pane. The tool automatically calculates and prompts user with the required miles for vehicle

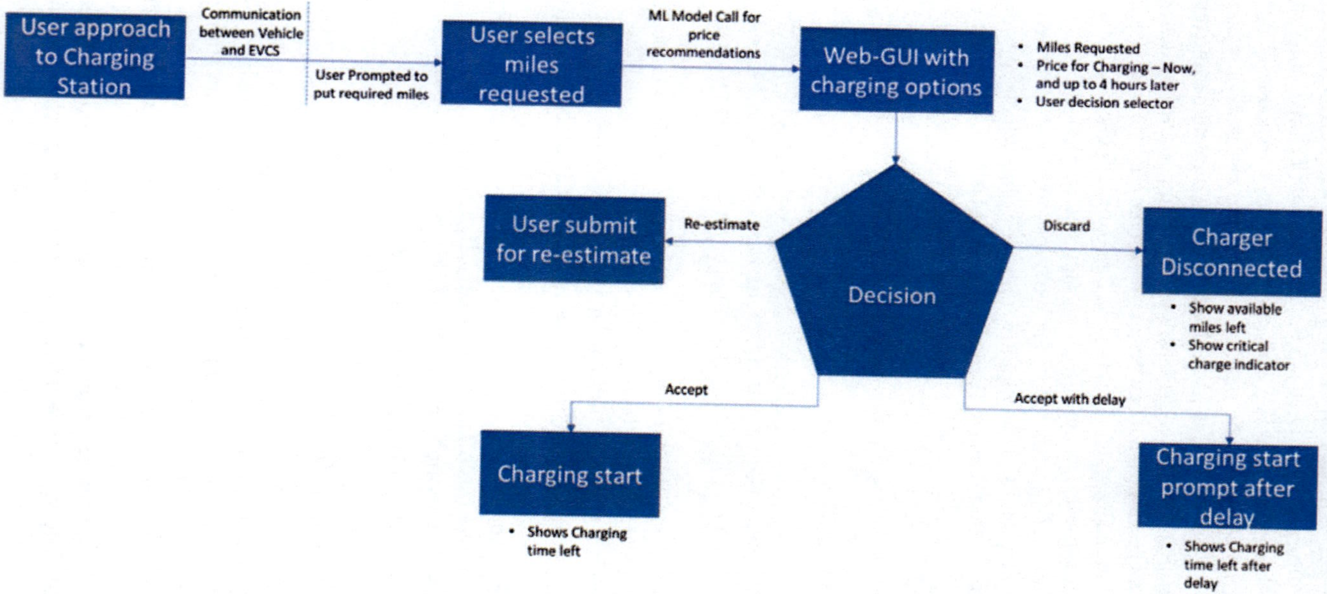

Fig. 9 Flowchart of the interaction between the user and the charging station web-app

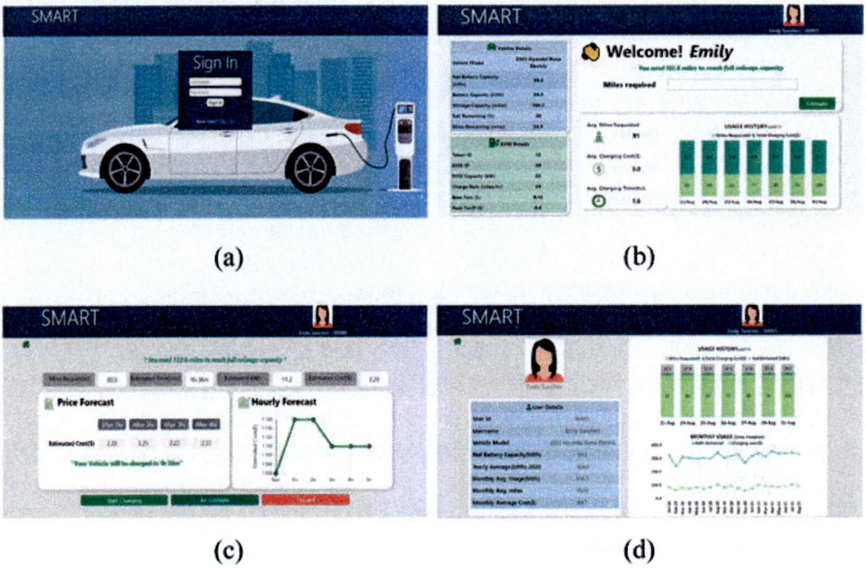

Fig. 10 Screenshots of SMART web-app: **a** Home page with log in, **b** User front page, **c** Cost forecast page, **d** User profile page

to attain full charge in top middle section. User can input required miles for battery and submit for total charging cost forecast.

Once the user clicks on estimate button, then total charging cost, time required based on the charging station c-rate capacity, estimated kWh and few hourly forecasts for total cost are computed, and user is redirected to a page where all these details are displayed in neat manner (refer Fig. 10c). User can see the hourly forecast for the next 4 h and the respective total estimated price considering charging time.

User can accept the total charging cost and start charging. In that case, user will be thanked with message showing charging status and approximate charging time. In case user wants to start charging later which may have lower charging cost it can be selected directly on screen and then user will be intimated accordingly that the vehicle is on hold for respective hours and when it will complete the required charging for requested miles. In another scenario, user may completely discard the charging and message will be displayed with left miles capacity of vehicle.

Further, a detail user profile page (refer Fig. 10d) is created where more details such as User Id, Username along with usage pattern are displayed for better understanding. User can see yearly average in kwh, monthly average usage, monthly average miles, and monthly average cost. Usage history for last week and monthly usage since inception could be seen in the form of charts and graphs.

5 Result and Discussion

The EV charging station data are highly randomized in terms of monthly, weekly, hourly and daily usage. This makes it a great challenge to successfully predict single step let alone multi-step predictions. It becomes very difficult for conventional statistical time-series models to learn important dynamic features from the data. When the statistical, ML and DL models are compared on common metrics such as MAE, RMSE and R^2, it can be observed that the ML models although perform better than statistical models, they are still significantly underperforming as compared to DL models with appropriate loss functions. A graphical comparison shown in Fig. 11 indicates significant drop in error metrics from ML to DL models.

Appropriate DL model (RNN with smoothL1 loss) is selected and saved for online inference to be used via web-app created for enhanced user experience. It can be observed that a dynamic pricing is estimated depending on multiple factors: one being the predicted kWh from the DL models, the other being multiple inputs from the charging station authority and the various BMS and user-specific data which are gathered from the vehicle data.

This solution can forecast the price of the charging of EV vehicles for up to next 4–6 h. Fig. 12 shows forecasts (6 h) for various test data fed through the final selected model and entire pipeline. The inference pipeline takes batches of data forecasts for each row of that batch and forecasts 6 h of energy delivered.

It can be observed that the predicted forecasts can capture the overall trend of the actual kWh. This can be beneficial in creating an appropriate price forecasting engine using DL-based models. However, for sudden drastic fall or rise in the actual kWh demand, the DL model fails to predict and can be further explored to capture these sudden changes kWh demands. The cost forecast for one of the test cases, or response to real-time model API call when fed into the SCFE, produces a price forecast.

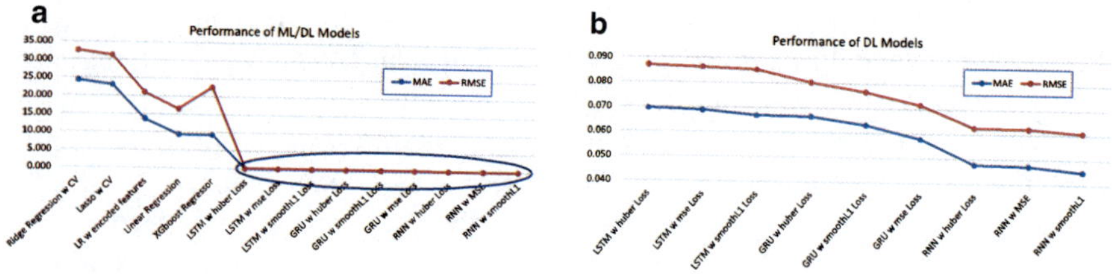

Fig. 11 Performance of ML/DL models: **a** All models considered in this work, **b** zoom-in view for only DL models

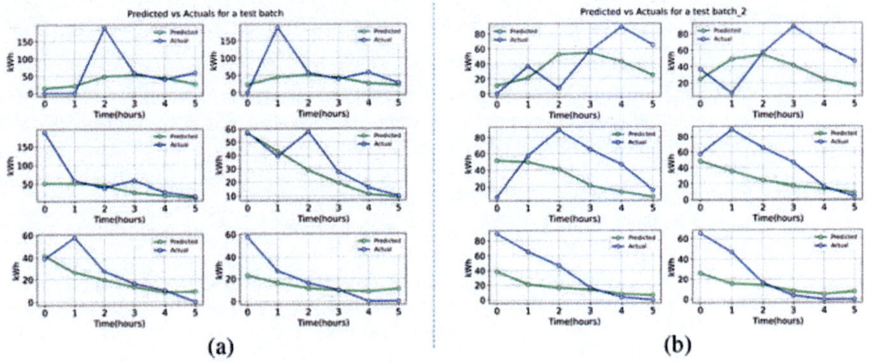

Fig. 12 Forecast of kWh for test batches using DL model: **a** Batch-1, **b** Batch-2

6 Conclusion

With the purpose of research on EV charging station load forecasting, this paper provides a solution that can forecast the load of the charging station and the estimated price for charging the EV vehicle. With the large-scale installation of charging stations, research on EV charging station load forecasting is the next trend, and it will play a key role in local electric grid dispatching automation.

This paper describes an effective ETL method used to extract relevant information on the EV charging data. Various parameters provided in the dataset are explored and understood. Once, the univariate time-series data of the kWh delivered are segregated, various statistical models are applied and model performance on the data is observed. The data being highly random, non-generalized, full of uncertainty due to seasonality, trends, and other unpredictable elements, it is indicated that the normal statistical models perform poorly. Next, several ML models trained and tested on the data observed that these perform better than the statistical model but still poor in forecasting. Lastly, DL models are trained and tested with extra features extracted from the data show improved performance. Next, the SCFE is devised for creating a dynamic cost forecasting engine. The rules built in are considering multiple aspects and dynamics of data, tariff, and user profile to give most accurate forecast and total charging cost. This complete pipeline is wrapped with a web-app with elegant page designs, which is the front-end to this solution and gives enhanced user experience. Maximum possible details are considered during web app development, and appropriate emphasis is given for improved user experience. The SMART web-app provides an easy and interactive way for the user to forecast engine to choose the charging cost and charging method for requested miles according to their convenience.

Acknowledgements Authors wish to thank their organization (Cognizant Technology Solutions, India), management, teammates and all individuals for their support in this technical work and publication.

References

1. Xie F, Huang M, Zhang W, Li J (2011) Research on electric vehicle charging station load forecasting. International Conference on Advanced Power System Automation and Protection 2011:2055–2060. https://doi.org/10.1109/APAP.2011.6180772
2. International Energy Agency (2018) Global EV Outlook 2018; International energy agency: Paris. France. https://doi.org/10.1787/9789264302365-en
3. Wang Q, Liu X, Du J, Kong F (2016) Smart charging for electric vehicles: a survey from the algorithmic perspective. In: IEEE Communications Surveys & Tutorials, vol. 18, no. 2. pp 1500–1517, Second quarter 2016. https://doi.org/10.1109/COMST.2016.2518628
4. Limmer S (2019) Dynamic Pricing for Electric Vehicle Charging—A Literature Review. Energies 12:3574. https://doi.org/10.3390/en12183574
5. Berman B (2018) The ultimate guide to electric car charging networks. Available online: https://www.plugincars.com/ultimate-guide-electric-car-charging-networks-126530.html Accessed on 27 July 2022
6. Archive V (2022) Who pays for electric car charging stations? 2019. Available online https://evarchive.com/who-pays-for-electric-car-charging-stations/. Accessed on 27 July 2022
7. Muratori M, Kontou E, Eichman J (2019) Electricity rates for electric vehicle direct current fast charging in the United States. Renew Sustain Energy Rev 113: 109235. ISSN 1364-0321. https://doi.org/10.1016/j.rser.2019.06.042.
8. Lee ZJ, Johansson D, Low SH (2019) ACN-Sim: An open-source simulator for data-driven electric vehicle charging research. In: 2019 IEEE International conference on communications, control, and computing technologies for smart grids (SmartGridComm) pp 1–6. https://doi.org/10.1109/SmartGridComm.2019.8909765
9. Develder C, Sadeghianpourhamami N, Strobbe M, Refa N (2016) Quantifying flexibility in EV charging as DR potential: Analysis of two real-world data sets. IEEE International Conference on Smart Grid Communications (SmartGridComm) 2016:600–605. https://doi.org/10.1109/SmartGridComm.2016.7778827
10. Flammini MG, Prettico G, Julea A, Fulli G, Mazza A, Chicco G (2019) Statistical characterisation of the real transaction data gathered from electric vehicle charging stations. Electr Power Syst Res 166:136–150. ISSN 0378-7796. https://doi.org/10.1016/j.epsr.2018.09.022
11. Lucas A, Prettico G, Flammini M, Kotsakis E, Fulli G, Masera M (2018) Indicator-based methodology for assessing EV charging infrastructure using exploratory data analysis. Energies 11(7):1869. https://doi.org/10.3390/en11071869
12. Chung Y, Khaki B, Chu C, Gadh R (2018) Electric vehicle user behavior prediction using hybrid kernel density estimator. IEEE International Conference on Probabilistic Methods Applied to Power Systems (PMAPS) 2018:1–6. https://doi.org/10.1109/PMAPS.2018.8440360
13. Zargar s (2021) Introduction to sequence learning models: RNN, LSTM, GRU. https://doi.org/10.13140/RG.2.2.36370.99522

DNA Sequencing: The Future Perspective

Kshatrapal Singh, Manoj Kumar Gupta, and Ashish Kumar

Abstract Surprises are unavoidable. Indeed, it's feasible that more of the world's data (currently saved on hard disks as well as in the clouds) will be saved in DNA decades over the next, but also that the primary factor of DNA sequencing will be our insatiable hunger for data storage rather than our effort to combat disease. We have presented our best estimations for the future of DNA sequencing in this publication. Furthermore, public-health experts are debating how to analyze the DNA of all bacteria found in waste-water discharges across urban centers in order to detect disease epidemics faster. Marine biologists are also looking at using methodical meta-genomic investigations to monitor the health of the seas. With smoke detectors and thermostats, DNA-sequencing devices could be the next 'smart' or 'integrated' products in the house. One commenter even suggested using the toilets to check basic health using real-time DNA sequencing.

Keywords DNA analyses · Meta-genome · Gene-expression · Crime lab · Data storage

1 Introduction

Two studies published 40 years ago presented the very first workable algorithms for detecting the sequence of chemical bases in DNA segments. Molecular methods could only sequence fragments before these 1977 findings. DNA sequencing has progressed at a breakneck speed from such early procedures to today's advanced technologies [1, 2]. In the last decade, nearly 30 years of exponentially growing data

K. Singh (✉)
Krishna Engineering College, Ghaziabad 201007, India
e-mail: mekpsingh1@gmail.com

M. K. Gupta
Shri Mata Vaishno Devi University, Katra J&K, 182320, India

A. Kumar
I.T.S Engineering College, Greater Noida 201308, India

H. O. Bansal et al. (eds.), *Next Generation Systems and Networks*, Lecture Notes in Networks and Systems 641, https://doi.org/10.1007/978-981-99-0483-9_14

production have giving rise to extremely growth. The resulting data has produced revolutionary uses in basic biology and beyond, including archaeology, criminal probe, and pregnancy diagnoses.

Optimists are frequently incorrect in their predictions of which technology—or, more significantly, what applications—seems to be the most revolutionary. Few imagined that e-mail would become so widespread in the era of the internet. Likewise, Wall Street traders and Silicon Valley capitalists missed the fact that games, net streaming video, as well as social networking would come to dominating the utilization of today's computing capacity as well as available bandwidth.

We would certainly do just as well projecting the future of sequencing technology. However, we provide a foundation for understanding it [3]. Our main point would be that killer applications, not killer technologies, would accelerate DNA sequencing advances.

Technology advancements could either raise or decrease costs. The latter was widely noted by Microsoft co-founder Bill Gates: since radial tyres were much more reliable than previous designs, the demand for tyres decreased, as well as the tyre industry shrunk. DNA analysis, unlike computation and imaging, we believe, will follow the trend of tyres. Implementations will increase as they get faster and more affordable, and consumption will rise [4, 5]. As DNA sequencing moves beyond laboratory and into therapeutic, commercial, and other areas, the adage of "more supply equals larger market" will become increasingly true.

The demand for Analysis of DNA is overwhelming among scientists. The prospect of sequencing a human genome looked intimidating in the 1990s. Evolutionary biologists now want Gene sequences from each cell in tissue at every developmental phase (with epigenetic alterations), in healthy or sickness, for everyone on the planet. They'd also like to sequence the equivalent Copies of DNA of messenger RNA molecules to obtain full gene-expression profiles. Likewise, researchers are starting to piece together the passage of DNA across ancestral communities, much like they did with linguistic, cultural traditions, and material things. Biologists, environmentalists, biochemists, as well as evolutionary biologists are all attempting to decipher the nucleotides among all alive or dead species, as well as entire ecosystems. Figure 1 depicts the past, present, and future of genetics and genomics technology for hereditary anaemia diagnosis and study [6–8]. The graphic represents the current DNA sequencing analytical techniques, as well as the fundamental distinctions between them. WGS stands for whole-genome analysis; SNV is for single nucleotide variant; CNV stands for copy number variation.

Furthermore, if there is a constant demand for content, the massive cataloguing activities must provide actual comprehension. The analysis and interpretation of all DNA-sequence data is currently the bottleneck. However, we expect that huge DNA-seq data sets combined along phenotypic details will empower scientists to conclude the biological behaviours encoded within nucleotide sequences, just like new informatics methodologies as well as large data sets have enhanced substantially language support / image classification.

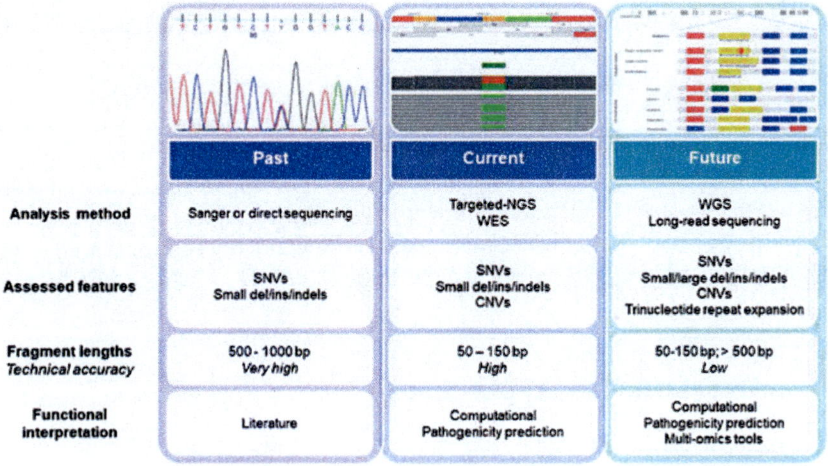

Fig. 1 Past, present, and future of genetics

Moreover, most of the underlying science required to analysis the information has already been established for a rising number of practical uses (like as good-quality reference sequences of bacterial genes, rule by which particular gene networks work in well persons). They include everything from finding microbial Nucleotide sequence in impartial surveys of environment or available scientific to discovering genomic changes linked to specific physiological outcomes.

2 DNA Sequencing Approaches

The technologies for Genetic analysis have been updated numerous times during the last 40 years. Just about all the DNA sequence was done with the Sanger or dideoxy chain-termination approach by 1985; reaction items typically tagged using radiom-nucleotides, segregated on acrylamide slab gels, as well as identified along autoradiography (the application of X-ray or photo film to identify radioactively tagged sample). Till year 2000, the 4-color fluorescent approach had taken over; reaction items were tagged with chain-terminating nucleotides equivalents, isolated electrophonetically in capillaries loaded with a jelly-like medium, as well as monitored along energy-transfer fluorescent dye. By 2010, the methods become more diverse [9, 10]. Parallel processing analysis of DNA 'polonies' as well as sequencing-by-synthesis chemicals were the leading tools (these depend on reversible chain-terminator).

The criteria for every DNA-sequencing system will indeed be determined by the application. In oncology as well as clinical genetics, the objective is frequently to properly identify each base and specify every variation of many versions of genomic regions. Whenever a yes or no 'decesion' is needed, such as in recognition of species, the power to execute testing rapidly as well as effortlessly in the fields may be highly critical than precision.

A further element that is likely to shift is the comparative importance of centralized vs decentralized Sequencing technology. In order to identify what virus has afflicted a specific hamlet in actual environments, an epidemiologist may require inexpensive, transportable instruments. However, for those producing large data sets, it would be much highly power full as well as cheaper to transfer samples to centralized business enterprises, particularly when labs must meet strict quality control as well as sample monitoring criteria, as in medical application.

3 Transformative Applications

The technologies for DNA sequencing have evolved substantially over the years. However, the trajectories of previous methods with apparently insatiable greed—cell phones, the Internet, electronic imaging—propose that now the true disruptors will be the services that come from the technological advances, not the new technologies themselves.

Healthcare is one area wherein we believe DNA sequencing will be genuinely transformational. Prenatal test for the existence of aberrant set of chromosomes, like as trisomy, that produces Down syndrome, is today's modern 'breakout' process of effective of DNA seq in conditions of the absolute volume of test performed. This testing now focuses on detecting a trace amount of cell-free foetal DNA in the mother's blood [11]. This has been hailed like "the quickest increasing genetic test in healthcare records," something that was not even envisioned at the conclusion of the Genomics. The future applications of genomics are shown in Fig. 2.

In reality, experts predict around 4.5 million to 6.5 million pregnant ladies globally receive the testing every year, with the count expected to rise to 15 million inside a decade. Several future uses of DNA sequencing in general practice are likely to have some of the test's characteristics: it really is non-invasive, simple to execute, as well as has minimal nucleotide-level precision requirements.

DNA sequencing has already been used regularly in high-income nations to screen kids with ill-defined congenital disorders. In roughly 30% of situations, analysis of the resultant sequences can show the disease-causing variants, a percentage that will only climb as the capacity to understand the data improves [12, 13]. The ensuing diagnosis have led to significant advances in clinical management in several cases. They usually help both households and doctors by bringing an end to a diagnostic saga and offering clinical clarification.

Oncology is investing much in the creation of fluid biopsies. It's possible to visualize a sequence-centric cancer diagnostic, similar to Pap smears and colonoscopies, becoming a standard tool for screening. Fluid biopsies could eventually lead treatment methods even when tumours are proven to occur only from DNA-sequence signature contained in sample taken, thanks to the emergence of cancer medicines which targets individual mutations instead of tumour categories.

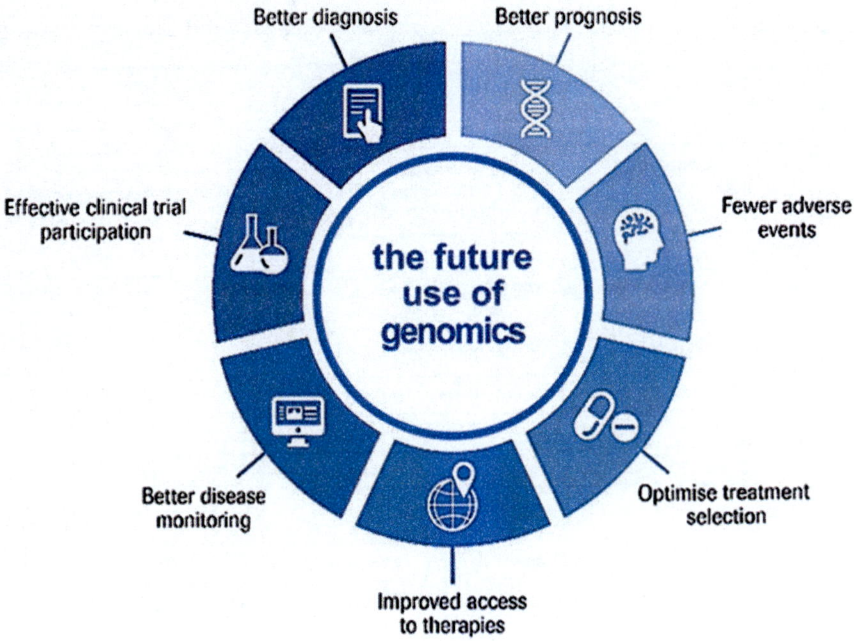

Fig. 2 The future use of genomics

Beyond the health centre, several uses, especially for hand-held DNA sequencers, might be imagined. Such devices might be used by epidemiologists or even caretakers in rural areas to analyze air, groundwater, food, animal as well as insect vectors, as well as person throat swabs and liquids. Actually, in low- and middle-income nations, cheap availability to DNA-sequencing tools really is allowing programmes like the Global Virome Project [14–16]. This project seeks to sequence a large sample size of wildlife DNA in order to discover a major percentage of the viruses that can infect people and cause sickness.

Similarly, public-health experts are debating how to analyze the DNA over all bacteria found in wasted-water outflows across all towns in order to detect disease epidemics faster. Marine biologists are also looking at using comprehensive meta-genomic examinations to assess the health of the seas. Figure 3 describes the effect of genomics over the future of medicines and healthcare.

Handheld equipment might take DNA analysis out from the forensics lab and put it in the hands of police officers on the street. Authorities may be capable to 'read' people's DNA in the same way they examine automobile license plates and identification cards now. Indeed, human-rights organizations have recently expressed alarm about the extent to which economical as well as accessible DNA seq sets up opportunities for bulk observation.

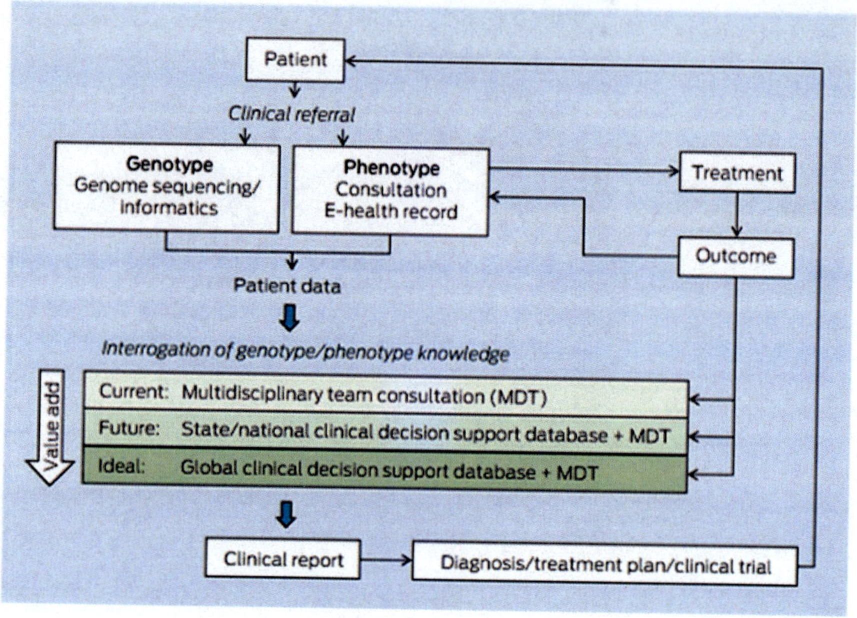

Fig. 3 The impact of genomics on the future of medicine and health

4 Beyond the Ordinary

The basic goal of bringing biological information regarding cells to meaningful application has shifted from an informative to a meta-informational in just forty years.

Consider the medical potential of genome-sequence information. DNA sequencing may eventually be regularly used to analyze body fluids acquired for any diagnostic purposes. However, only a massive amount of well-organized data about millions of people's multi-year medical histories will give the meta-information required to determine when to disregard such information and when to respond on it. Figure 4 explains where genomics is going next in detailed way.

In terms of medical, we agree with advisory committees like the Precision Medicine Committee of the US National Research Council about the need to develop a huge "knowledge commons." The microbe DNA sequences of millions of persons would be overspread with clinical as well as molecular information. The UK Bio-bank repository and the US All of Us Research Group are two of this kind population-scaled endeavors now ongoing. Surprises are unavoidable [17, 18]. Indeed, it's feasible that much of the world's information will be saved in DNA decades from now and that the key factor of DNA seq will be our voracious hunger for information storage rather than our effort to combat illnesses.

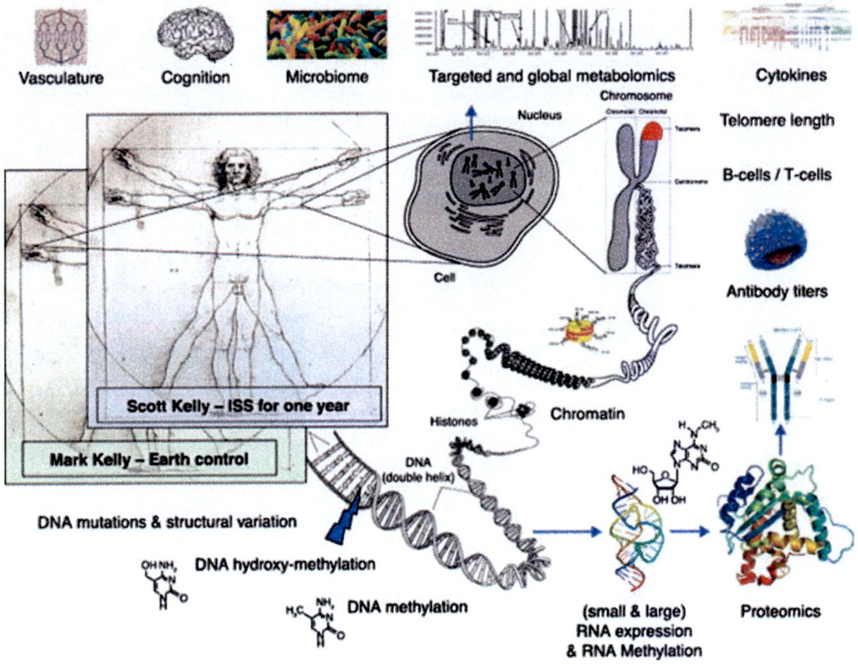

Fig. 4 Where genomics is going next

5 Conclusion

DNA sequencing is now having the largest effect in classifying cancer, describing inborn illness, and anticipating a patient's possibly treatment outcomes. In the long term, genomics will be applied to diagnose and predict particular disease. Liquid biopsy for earlier disease diagnosis will be a tremendous step forward in Medicare, and I predict it will be possible in the coming time. Many genomic instruments were created for basic research. DNA microarray technique, commonly familiar as the DNA chip, is the newly advancement in nanotechnology, allowing scientists to study the DNA Sequencing at a maximum efficiency. It could be applied for gene expression profiling, which permits researchers to analyze which genes are up or down-regulated. In healthcare, DNA sequencing has been utilized for disease diagnosis and treatment, as well as epidemiology investigations.

References

1. Findlay GM, Daza RM, Martin B, Zhang MD, Leith AP, Gasperini M et al (2018) Accurate classification of BRCA1 variants with saturation genome editing. Nature 562:217–222
2. Martincorena I, Fowler JC, Wabik A, Lawson ARJ, Abascal F, Hall MWJ et al (2018) Somatic mutant clones colonize the human esophagus with age. Science 362:911–917
3. Xing D et al (2021) Accurate SNV detection in single cells by transposon-based whole genome amplification of complementary strands. Proc Natl Acad Sci USA 118:e2013106118
4. McIntyre ABR, Rizzardi L, Yu AM, Alexander N, Rosen GL, Botkin DJ et al (2016) Nanopore sequencing in microgravity. NPJ Microgravity. 2:16035
5. Schneider VA, Graves-Lindsay T, Howe K, Bouk N, Chen HC, Kitts PA et al (2017) Evaluation of GRCh38 and de novo haploid genome assemblies demonstrates the enduring quality of the reference assembly. Genome Res 27:849–864
6. Lander ES, Linton LM, Birren B, Nusbaum C, Zody MC, Baldwin J et al (2021) Initial sequencing and analysis of the human genome. Nature 409:860–921
7. Sedlazeck FJ, Lee H, Darby CA, Schatz MC (2018) Piercing the dark matter: bioinformatics of long-range sequencing and mapping. Nat Rev Genet 19:329–346
8. Nattestad M, Goodwin S, Ng K, Baslan T, Sedlazeck FJ, Rescheneder P et al (2018) Complex rearrangements and oncogene amplifications revealed by long-read DNA and RNA sequencing of a breast cancer cell line. Genome Res 28:1126–1135
9. Computational Pan-Genomics Consortium (2018) Computational pan-genomics: status, promises and challenges. Brief Bioinform 19:118–135
10. Church DM, Schneider VA, Steinberg KM, Schatz MC, Quinlan AR, Chin CS et al (2021) extending reference assembly models. Genome Biol 16:13
11. Shendure J et al. (2017) Nature. http://dx.doi.org/10.1038/nature24286
12. Paxton A (2017) CAP today; available at go.nature.com/2hoipsp
13. Wang P, Xiao X, Chou K-C (2019) NR-2L: A two-level predictor for identifying nuclear receptor subfamilies based on sequence-derived features. PLoS ONE 6(8):e23505
14. Eldomery MK et al (2017) Genome Med. 9:26
15. Worthey EA et al (2019) Genet Med 13:255–262
16. Bainbridge MN et al. (2020) Sci Transl Med 3, 87re3 (2020)
17. Alix-Panabières C, Pantel K (2019) Cancer Discov 6:479–491
18. National Research Council. Toward Precision Medicine: Building a Knowledge Network for Biomedical Research and a New Taxonomy of Disease; available at go.nature.com/2fmz99 (National Academies Press, 2021).

DAGE: A Deviation Assessment-Based Grey-Hole Detection Method for Ad Hoc Wireless Networks

Nenavath Srinivas Naik, Athota Kavitha, L. Nirmala Devi, and B. Vijender Reddy

Abstract Ad hoc wireless networks need cooperation between nodes, like multi-hop forwarding of neighbouring nodes, for the proper function of the network. Some nodes intentionally drop the packets in the real environment to disrupt the network's performance. Grey-hole nodes drop packets with some probability. Since it drops the packet selectively, it is challenging to identify such types of nodes. We propose a grey-hole detection mechanism DAGE, using the node's behaviour deviations to identify the grey-hole node. We integrate it with ad hoc on-demand distance vector and show that DAGE successfully identifies the grey-hole nodes in the network and establishes the grey-hole-free route from source to destination.

Keywords Grey hole · Wireless networks · MANETs · AODV

1 Introduction

Mobile ad hoc networks are connected by wireless links to form a network. A node may join and leave at any time from the network. Because of the openness, MANETs can be attacked by internal nodes. A packet drop attack is an attack where the malicious node drops the packets with some probability. If the probability is 1, the malicious node drops all the packets passed through which is called a black-hole attack. A grey-hole attack is a special type of black-hole attack where the malicious node

N. Srinivas Naik (✉)
International Institute of Information Technology (IIIT),Naya Raipur, India
e-mail: srinu@iiitnr.edu.in

A. Kavitha
JNTUH College of Engineering, JNTUH University, Hyderabad, Telangana, India

L. N. Devi
University College of Engineering, Osmania University, Hyderabad, India

B. V. Reddy
Advanced Data Processing Research Institute, Secunderabad, Telangana, India
e-mail: vijender@adrin.res.in

H. O. Bansal et al. (eds.), *Next Generation Systems and Networks*, Lecture Notes in Networks and Systems 641, https://doi.org/10.1007/978-981-99-0483-9_15

drop packets selectively disrupt the communication. Identifying a grey-hole attack is challenging because the node shows capricious behaviour and drops packets selectively.

Apart from the malicious packet drop, a node may drop packet [9] due to MAC buffer overflow, IEEE 802.11 protocol drop because of RTS frame reaching the maximum allowed number and packet corrupted due to wireless channel. But our work is focussed only on identifying the grey-hole attack nodes.

Route establishment is an important task in MANETs to forward the data from source to destination. Routing protocols in MANETs are categorized as reactive and proactive based on the route establishment phase. Ad hoc On-Demand Distance Vector (AODV) [6] and Dynamic Source Routing (DSR) [6] are most popular reactive routing protocols and Optimized Link State Routing (OLSR) [6] is the popular proactive routing protocol. Since routing is an important service, most attacks on routing protocols DAGE can be embedded into any routing protocol.

The network's performance improves when the routing protocol selects a good path. The routing protocol should differentiate malicious nodes from legitimate ones to identify a good path. We integrate our model into AODV to evaluate the performance of the proposed model. DAGE can be integrated into any routing protocol.

The main contributions of this paper are

- We propose a model to identify the deviation of node behaviour by considering the frequency of deviation occurrences.
- We integrate the proposed model to the AODV routing protocol in Mobile Ad hoc Networks (MANETs).
- We do analysis and simulations in the presence of grey-hole nodes in the network. The model successfully identifies the grey-hole attack nodes.

The rest of the paper is organized as follows: Sect. 2 briefly explains previous research on packet drop attacks. Section 3 shows the proposed grey-hole detection mechanism, DAGE. Section 4 shows the performance evaluation of DAGE. Finally, we conclude the paper in Sect. 5.

2 Related Work

This section briefly describes some of the packet drop attacks' detection mechanisms. Watchdog and pathrater [7] use forwarding ratio to detect the malicious node. The watchdog mechanism observes the neighbour to calculate the behaviour. The calculation is based on packet forwarding. This type of solution identifies a malicious node if it drops more than 50% of packets.

Deng et al. [3] proposed a method to detect black-hole attacks by verifying the following hop route to forward RREP. The verification is done by using an alternate route because routing overhead is high.

Yu and Xiao [1] proposed a malicious node detection using a multi-hop acknowledgement scheme. The intermediate node uses the downstream and upstream of the flow to identify malicious nodes. Upstream node uses ACKs, which travels multi-hop to detect malicious and downstream node uses Packet IDs. If ACKs are less than downward packets, it sends an alarm message, and the node is declared a malicious node. It is difficult to identify malicious nodes if a malicious node can modify the packet IDs or send ACKs without forwarding them.

Mohammad et al. [5] propose that the node waits for more than one Route Reply, then it uses two paths to send with different id and seq. no if two paths contain one common node. This solution gives extra processing of nodes and doubles the traffic.

Shila and Anjali [2] proposed a solution to identify selective forwarding attacks in WMNs. It uses a packet counter and threshold to discover the attack. Control ACK packet is used to detect the attacker. These extra messages give an additional burden.

Thaier et al. [10] use many RTS, CTS and Signal-to-Noise Ratio (SNR) to identify grey-hole nodes. This method estimates the channel errors with statistics, and when a loss occurs, it calculates the RTS and CTS and based on SNR, it identifies the grey-hole node.

Djhal et al. [9] survey on packet dropping attacks. The proposals include cryptographic approaches, Feedback methods using ACKs and observing the neighbours using promiscuous mode. Cryptographic procedures consume a lot of battery power.

Tripo [4] uses micropayments to stimulate the rational packet droppers. Tripo identifies malicious nodes by measuring the frequency of dropping packets based on payment receipts instead of overhearing. A third-party server is required to generate payment receipts, which is impossible for distributed kinds of ad hoc wireless networks.

Pal et al. [8] propose a mechanism to identify grey-hole attacks on synchrophasor data using inter-delay between the packets. The grey-hole node is detected using the delay slope, calculated before and after the packet loss. This is proposed on wired networks, but delays may not always be uniform, even though the network's grey holes are not present.

Reddy et al. [11] proposed a novel similarity model that is used to compute recommendation credibility instead of using direct trust as the weight to compute indirect trust. Their proposed model mitigates packet dropping, collusion and badmouthing attack in IoT.

In [12], the authors, to calculate the indirect trust values, used weighted Dempster–Shafer (D-S) theory for Wireless Sensor Networks (WSNs). The forwarding ratio with the cos(x) function is used to calculate direct trust as it reduces more for greater values than lower values, which mitigates on–off attacks. It used the weighted D-S theory to aggregate the recommendations. Their proposed mechanism is shown to mitigate collusion and badmouthing attacks and performs better than other methods like weighted average.

Ghasempouri et al. [13] proposed a new extensive framework for specifying various robust properties conceived in Trust and Reputation Systems (TRS) and verifying them. This is required as, even though there are many schemes or plans for verification of TRSs against attacks, none are extensive enough. This proposed framework is thorough because it provides a reason for the robustness of property specification and algorithms for quantitative and probabilistic model checking. The proposed framework experiments on real-world case studies.

3 System Model and Assumptions

3.1 Deviation Assessment-Based Grey-Hole Detection (DAGE)

In mobile ad hoc networks, nodes must be cooperative and forward other nodes' packets like routers. If a node is not forwarding the packets, it deviates from routine behaviour. The proposed DAGE mechanism uses deviation of a node to identify the grey-hole attack. It observes the neighbour's activity to calculate the deviation. This mechanism runs on each node independently and can be embedded into any routing protocol like AODV, DSR and OLSR.

A grey-hole attack is a selective forwarding type of Denial-of-Service (DoS) attack. The attacker drops the packets with probability.

Nodes should collaborate to establish a network. A legitimate node behaviour is always consistent, but grey-hole node behaviour changes, like selectively dropping the packet to degrade the network's performance or a particular node. DAGE identifies the node's behaviour change by verifying the past behaviour with the present one. Deviating the node's behaviour is one of the ways to identify malicious activity. We assume that all nodes are homogeneous and packet forwarding is the contribution of a node to the network.

Each node observes the neighbour node in the promiscuous mode, whether the neighbour is forwarding the packets or not. The packet forward update function is defined as

$$PF_n = k \times PF_{cur} + (1 - k) \times PF_{n-1} \tag{1}$$

where $PF_0 = 0$. PF_{cur} is the present packet forwarding value which is about the most recent packet forward and is defined as

$$PF_{cur} = \begin{cases} 1 & \text{If forwarded successfully} \\ 0 & \text{Otherwise} \end{cases} \tag{2}$$

The weight k is defined as follows:

$$k = 0.25 + p \times \frac{D_n}{1 + \lambda_n} \tag{3}$$

We add 0.25 to prevent k from saturating to a fixed value. The initial value of k is 1. The deviation of the node's behaviour is identified from the difference between the

present packet forwarding and the past packet forwarding.
Deviation D_n is

$$D_n = |PF_{cur} - PF_{n-1}| \tag{4}$$

and accumulated deviation λ_n

$$\lambda_n = p \times D_n + (1 - p) \times \lambda_{n-1} \tag{5}$$

p is the constant, which gives weight to the recent error. If the p-value is more, then the recent deviation is more important than the accumulated deviation and vice versa. Initial value of p is 0.9.

The accumulated deviation is used to identify the grey-hole node in the network. δt is the observed time of the node's behaviour. If the node's behaviour deviates η times in δt, then the node is marked as a grey-hole node. δt and η values are decided based on the network's application and environment.

3.2 DAGE Integration with AODV

To explain the working of DAGE, we integrate it with the Ad hoc On-demand Distance Vector (AODV) routing protocol as a case study. AODV is a reactive routing protocol for MANETs. A brief description of this protocol is given below.

The route establishment takes only a node (i.e. source) that wants to send data to another node (i.e. destination). Source broadcasts Route Request (RREQ) message. Once the neighbour node receives RREQ, if it is not the destination and does not have a route, it rebroadcasts the RREQ message and stores the path to the source in its routing table. Once the destination receives the RREQ. It unicasts the Route Reply (RREP) message to the source. After the source receives RREP, it forwards the data to the destination. If any intermediate node has a route to the destination, then the intermediate node also unicasts RREP to the source.

The communication will be disrupted if any malicious node is present in the network. We assume that each node overhears the neighbour node's activity by using promiscuous mode. According to the neighbour's activity, PF_{cur} value is set and calculated accumulated deviation of the neighbour node. Suppose the variations are more than η within δt time, then the node-set neighbour node as grey-hole node. The node broadcasts the grey-hole node message (GN). Once the source receives the GN message, it initiates the route establishment phase. If any RREP or RREQ receive from the grey-hole node, then the legitimate node drops that message. So the grey-hole node is avoided from the route to the destination.

A MANET example is shown in Fig. 1. Source S wants to send data to destination D. Node S broadcasts the RREQ packet. Nodes A, E, G receive RREQ packets and rebroadcast them if these do not have route information about node D. Similarly, nodes C, $BandF$ also do the same. Once node D receives RREQ, it unicasts RREP

Fig. 1 MANET example

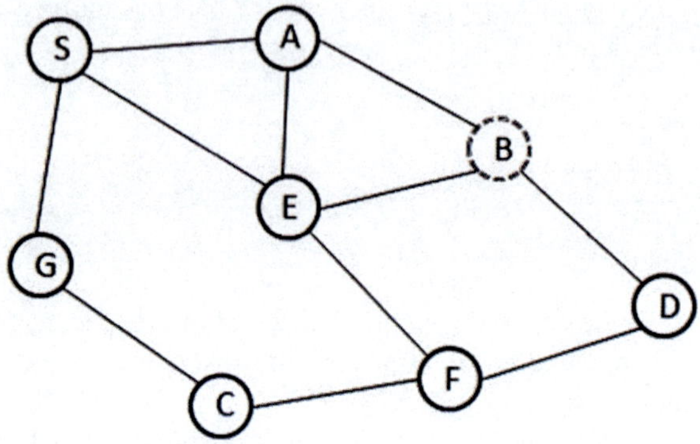

to source S. RREP follows B, A nodes to reach source S. After RREP reaches source S, it forwards the data to destination D

Suppose B is a malicious node and drops the packets like a grey-hole node. Node A overhears node B's communication in promiscuous mode. Node A identifies node B as malicious using the λ value. Node B broadcasts grey-hole node message (GN). When source S receives the GN message, it initiates the route discovery process. No legitimate node forward RREQ or RREP message receive from the malicious node. So the malicious node avoids the path to the destination. From Fig. 1, the new path from S to D is S, E, F, D.

4 Performance Evaluation

We have implemented the DAGE to evaluate the λ value. The input to the DAGE is given as a bitmap, showing packet drop and forward of neighbour nodes. Bit map contains 0 and 1's which shows drop and forward, respectively. The λ value for two cases is given in Figs. 2 and 3.

1 The grey-hole node drops the packets with probability(Pr) 0.5. Figure 2 shows the λ versus time. Because of high occurrences of deviation between 200 and 400, the λ value is high. At 200, the λ value oscillates very high and finally converges to one value.
2 The grey-hole node drops the packets with probability(Pr) 0.2. The λ is shown in Fig. 3. The deviations occur very frequently; hence λ value fluctuates.

We integrate the DAGE into the AODV routing protocol in the ns-2 simulator. Simulation run with 50 nodes spread randomly over 670×670 m^2 area. Constant Bit Rate (CBR) traffic packets flow between source and destination. Simulation run about 600 s. The λ value in Fig. 5 shows the packet droppings of the neighbour node. DAGE successfully identifies the grey-hole node with δt is 3 s. Figure 4 shows the λ value without grey-hole nodes. Since there is no packet drop, the λ value is 0. Figures

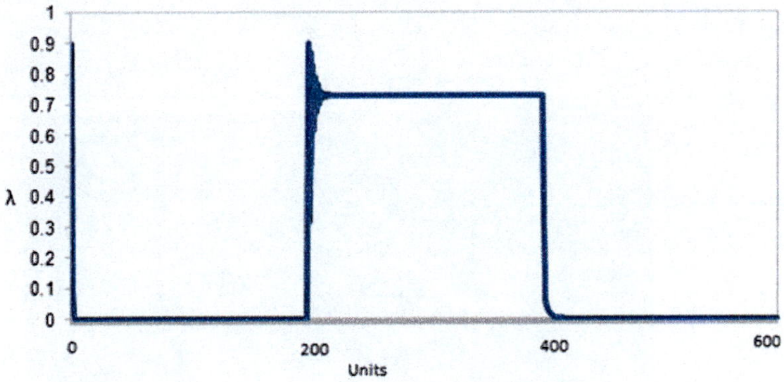

Fig. 2 λ value with Pr = 0.5

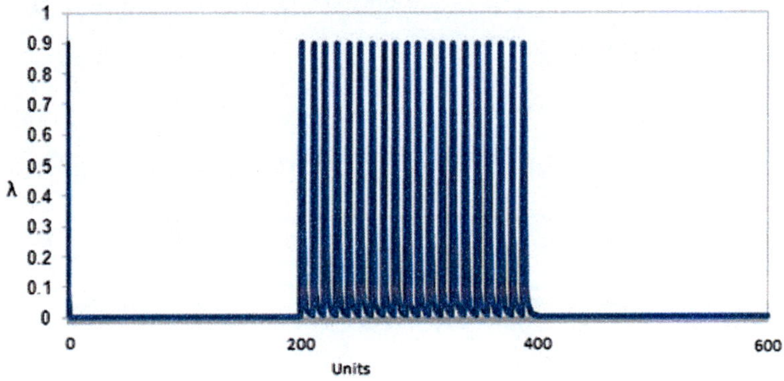

Fig. 3 λ value with Pr = 0.2

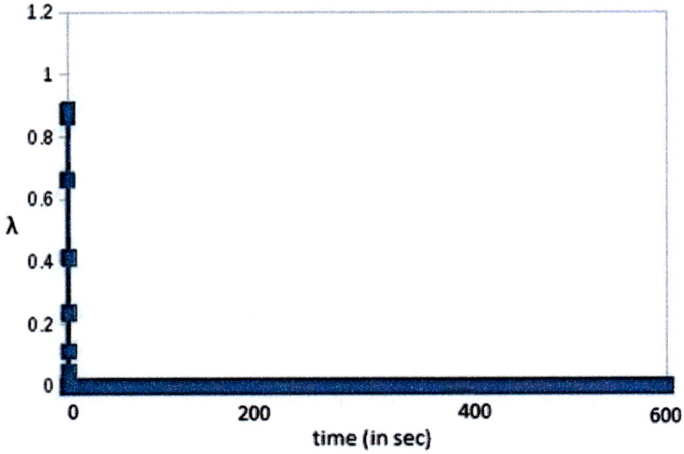

Fig. 4 λ value without grey-hole node

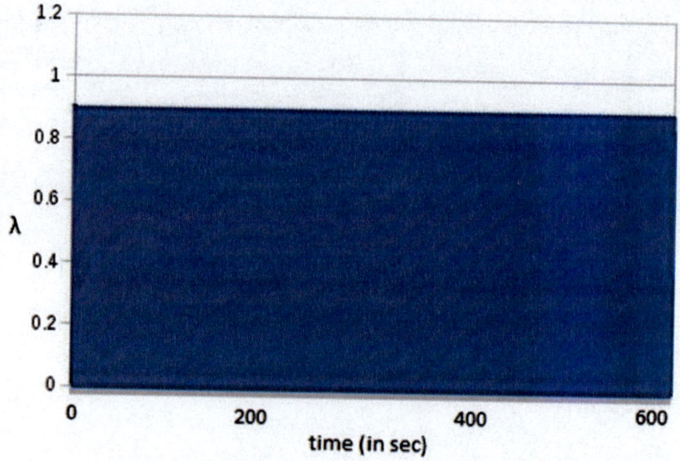

Fig. 5 λ value with grey hole and Pr = 0.1

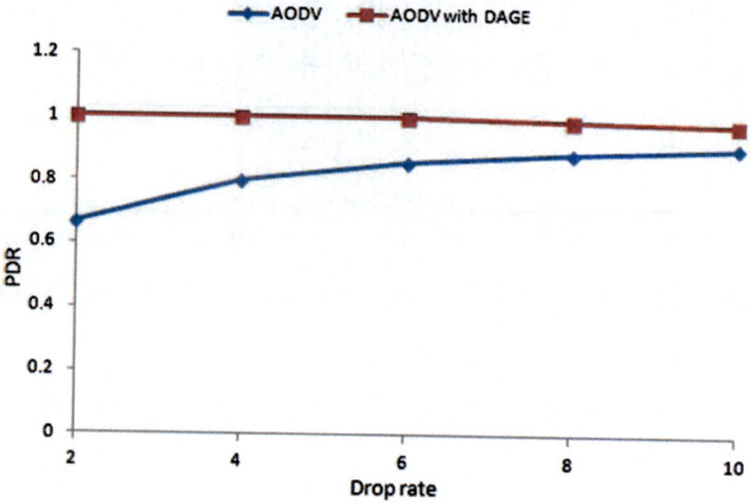

Fig. 6 PDR versus grey-hole node drop rate

4 and 5 clearly show that the packet losses are due to grey-hole nodes. In Fig. 5, the λ value fluctuates when a malicious node drop packets. So, the λ value significantly identifies the grey-hole attacks.

Figure 6 shows the grey-hole drop rate versus Packet Delivery Ratio (PDR). The drop rate is the rate at which one packet is dropped. A lower drop rate reduces PDR because of frequent packet loss, whereas DAGE successfully identifies a grey-hole node, and due to the GN message, the node establishes a grey-hole-free route to the destination. As the drop rate increase, the packet losses will decrease, so the PDR increases. Drop rate 2, 4, 6, 8 and 10 means 50, 25, 16, 12 and 10% of packets loss, respectively.

5 Conclusion and Future Work

This paper presents a DAGE model to identify the grey-hole node. It uses the node's deviations to identify the node's malicious activity. We integrate the proposed algorithm into AODV to evaluate the performance and show that DAGE successfully identifies the grey-hole node.

Apart from the malicious packet drop, the node may drop the packet because of buffer flow and radio errors. As part of future work, we will find a mechanism to differentiate malicious packet drops from other packet losses.

References

1. Yu B, Xiao B (2006) Detecting Selective forwarding attacks in wireless sensor networks. In: 20th IEEE international conference on parallel and distributed processing symposium (IPDPS)
2. Shila DM, Anjali T (2008) Defending selective forwarding attacks in WMNs. In: IEEE International conference on Electro/Information technology (EIT). Ames, IA
3. Deng H, Li H, Agrawal DP (2002) Routing security in wireless ad hoc networks. IEEE Commun Mag 40(10):70–75
4. Mohamed Elsalih Mahmoud, X (2011) An integrated stimulation and punishment mechanism for thwarting packet dropping attack in multihop wireless networks. IEEE Trans Veh Technol 60(8):3947–3962
5. Mohammad Al-Shurman, S-M, Park S (2008) Blackhole attack in mobile ad hoc network. In: ACM-SE 42nd annual southeast regional conference. NY, USA
6. Siva Ram Murthy C, Manoj BS (2004) Ad hoc wireless networks: architectures and protocols. Prentice Hall
7. Marti S, Guili T, Lai K, Baker M (200) Mitigating routing misbehaviour in mobile ad hoc networks. In: 6th annual international conference on Mobile computing and networking (MOBICOM), pp 255–265
8. Pal S, Li H, Sikdar B, Chow J (2014) A mechanism for detecting gray hole attacks on synchrophasor data. In: IEEE international conference on communications (ICC), pp 4131–4136
9. Djahel S, Nait-Abdesselam F, Zhang Z (2011) Mitigating packet dropping problem in mobile ad hoc networks: proposals and challenges. IEEE Commun Surv Tutor 13(4):658–672
10. Hayajneh T, Krishnamurthy P, Tipper D, Kim T (2009) Detecting malicious packet dropping in the presence of collisions and channel errors in wireless ad hoc networks. In: IEEE international conference on communications (ICC), pp 1–6
11. Reddy VB, Negi A, Venkataraman S, Venkataraman VR (2019) A similarity based trust model to mitigate badmouthing attacks in internet of things (IoT). In: 2019 IEEE 5th world forum on internet of things (WF-IoT), pp 278–282
12. Reddy VB, Venkataraman S, Negi A (2017) Communication and data trust for wireless sensor networks using DS theory. IEEE Sens J 17(12):3921–3929
13. Ghasempouri SA, Ladani BT (2020) Model checking of robust- ness properties in trust and reputation systems. Futur Gener Comput Syst 108:302–319

Microexpression Analysis Using a Model Based on CNN and Facial Expression Recognition

B. H. Pansambal and A. B. Nandgaokar

Abstract Emotions don't have correct definition and are abstract in nature. Though emotions are easily understandable, it is hard to process and know the expression within fraction of second. Therefore, machine learning algorithms and methods help recognizing the correct expressions especially the ***microexpressions***. These facial expression recognition models can be categorized under the *supervised learning* where a model is already trained and remaining data (images/videos) can be processed to get the expressions. We have used CNN and FER (facial expression recognition) to identify the expressions, microexpressions and their correlation in the given images or video. It helps in developing a model that accurately predicts the expressions on the face. Basic emotions like anger, sadness, happiness, neutral, surprise and disgust are considered to categorize the images based on the scores obtained by executing our proposed model. There are few layers in our proposed model like CNN layer, dropout layer, etc. We strongly believe with the help of this analytic approach one can easily identify the emotions/microexpression either using a web camera or processing a video file. In the field of robotics application, computer vision these machine learning-based models will bring the revolution. Multi-cascade CNN and HAAR cascade classifier helps to processing the streaming data and identifying the expression and their relations for the images.

Keywords CNN · FER (facial expression recognition) · Machine learning · Emotions · HAAR classifier

1 Introduction

AI, ML are achieving new milestones and new applications are being developed using these technologies. Computer vision, robotics, cognitive science, etc. are the few research areas where face detection, emotion analysis plays an important role. It became possible to predict the emotions by processing images, videos, live streaming

B. H. Pansambal (✉) · A. B. Nandgaokar
Department of Electronics and Telecommunications Engineering, Dr. Babasaheb Ambedkar Technological Univesity, Lonere, India
e-mail: bpansambal@gmail.com; abnandgaonkar@dbatu.ac.in

H. O. Bansal et al. (eds.), *Next Generation Systems and Networks*, Lecture Notes in Networks and Systems 641, https://doi.org/10.1007/978-981-99-0483-9_16

and scientists are working on finding the 'microexpressions' and stimuli behind the same. May facial expression recognition (FER) systems are developed to predict the emotions/sentiments by processing the speech (audio), text and gestures i.e. multimodalities. This approach falls under the supervised learning category, and learnable model can be developed using application of techniques such as CNN and classification. Model can be defined for the entire data as follows

$$Model_m = f(x) \tag{1}$$

2 Review of Literature

A very popular non-verbal way of communication by expressing emotions like happiness, anger, neutral reaction, etc. It's an emerging research area and one can use recent technologies, methods from CNN, deep learning etc. Authors [1] have developed a model which extracts features from given images and trained the model to process the upcoming features automatically. Authors have focused on important aspect of the emotion recognition, i.e. working on static images and dynamic images, i.e. real-time images or images appearing from the video. Model processes the video as sequences of different frames and each frame consists of images. Model processes the features extracted from given images automatically by introducing three steps as follows–

1. *Image acquisition*
2. *Feature extraction*
3. *Expression recognition.*

Model developed by the authors [1] mainly deals with the few emotional states such as *happy, surprise, angry, disgust, fear, sad* and *neutral*. Conversation among human plays a very important role in daily activities. Authors [2, 12] have focused on these conversations of different people. Authors analyzed multi-round conversation of people to extract emotions and named it as 'sequence mining' for emotion recognition. It is obvious that one has to use natural language processing (NLP) to process these conversations. Authors have used LSTM algorithm to find different emotion patterns and changes occurring in forthcoming patterns. LSTM helps in acquiring different patterns from the sequences and gives good results in recognizing the emotions, validating the correctness. Authors [2] also suggested one can use ***SVM, Naïve Bayes, lexicon methods*** to recognize emotions from different conversations. Hence, model generates different vectors according to the different conversations. Model helps in identifying sentiments of people and knowing the depth of conversation. Model could also help in analyzing different styles of conversations, time required for certain conversation, etc. Authors [3] have developed a PEIS model based on positive emotion incentive system. This model achieved 79% accuracy in recognizing emotions. Authors have introduced robots and embedded this PEIS model to process the real-life images. In this model, robot extracts images from camera, video, etc. and analyzes different frames. FERW model—facial expression recognition in the wild processes the features and gives corresponding feedback.

Authors [4, 19] have used CNN and auto-encoders to extract different features from images for classification purpose. Authors have proposed six architectures out of which two are trained on Japanese female facial expressions and four with Berlin database. Two architectures dealt with image-based emotions and four with speech recognition. CNN architecture consists of three convolution layers, three *pooling layers*, three *dense layers with output layer*. These convolution and pooling layers help in extracting different features from emotions. Auto-encoder helps in reducing the dimensions. Authors have concluded one can introduce two or more feature extractors to improve the classification accuracy. Emotions play an important role in decisions and recognizing different emotions either from images or speech became very necessary. Therefore, emotion recognition is very popular research area among the researchers. Authors [5, 6] suggested that ML machine learning and methods especially from the deep learning techniques can handle complex feature extraction from images and also helps in classifying them accordingly. 'Gabor Filter' is applied to analyze the textures, edge detection, and extracting different features from the images. Authors pointed good response in edge detection can be obtained after application of the Gabor filter. Hence, the proposed algorithm consists of two Gabor filters. Once the Gabor filter processes the images and features, output of this phase is given as input to the CNN. Authors [7, 13] have focused on six elementary emotions like happy, sad, fear, anger, surprise and disgust. Using MATLAB, deep learning methods are implemented to extract and process the features from images. Model processes the JAFEE dataset, which consists of six elementary emotions. It consists of 213 images and image format is '.tiff'. Architecture of the model consists of different layers such as–

1. *Input layer*
2. *CNN layer*
3. *Activation*
4. *Pooling layer*
5. *Fully connected layer*

Pooling layer processes the image and divides it into sub-regions. According to this division, there could be max-pooling and mean-pooling. System processes large volume of data. However, data could be labeled or unlabeled. Authors [8, 11, 18] have proposed a model to process unlabeled data. A hybrid label-less learning is proposed to automatically process the unlabeled data. Main advantage of this approach is human intervention and is almost reduced and model automatically labels the data. Authors have proposed 'LLEC' label less learning for emotion cognition. Authors suggested when labeling the unlabeled data prediction probability plays an important role. Entropy is used as a measure to assess the prediction uncertainty. IoT systems are becoming integral part of human lives. These devices can track humans and mainly used in healthcare applications. To realize the emotions of the human one can use IoT devices. Devices can help in detecting the feelings of the persons. Microcontrollers, motion sensors and camera are the basic things required to implement this system.

Authors [9] have proposed a model using IoT devices which captures the images of the human and passes to the system. Model uses motion sensor to detect the motion of human. Camera attached with the module gets activated once the motion sensor detects the motion of human and captures the image using ***HAAR—feature-based cascade classifier***. Next phase is cropping—system crops the image and selects the face of human. Deep learning methods are applied on these cropped face images to get different emotions. Once the system generates classified results, same will be stored in emotion database. Authors [10, 11, 13] have proposed a comparative approach for different deep learning architectures available in 'Keras' for emotion recognition. Pre-trained models like VGC-16, ResNet152V2, Inception V3 and Xception, etc. Authors have used two datasets Cohn-Kanade dataset and JAFFE Japanese female facial emotion. All the pre-trained models are evaluated on these datasets and their accuracies are measured. First layer of the proposed architecture consists of few steps as follows–

1. *Converting the image in gray scale image*
2. *Detecting the face using HAAR cascade classifier*
3. *Cropping*
4. *Resizing*

Second layer consists of application of these pre-trained models, extracting features and classifying these features. Output layer shows different probabilities for the emotions like angry, fear, disgust, neutral, happy, sad and surprise. A three-class annotation approach [12] is proposed to get the stimuli behind emotions. Authors have developed emotion classification framework using LSTM. Framework also consists of thresholding scheme and categorizes emotions in different segments like anxiety, anger, disgust, sadness, joy, etc. and ratings are given in the range from 1 to 9. Authors have also used EEG signals and facial videos to extract the features from images. Authors also concluded that EEG feature-based classifier can give better accuracy when used with thresholding scheme. Authors [14, 19] have considered five emotions and developed a model based on deep convolutional neural network. Model is based on few layers like convolutional layer, dropout, fully connected layer and features are extracted from images. Using convolution filter helps is representing the features of interests. Back-propagation algorithm can be embedded in the framework to reduce the errors. Activation function is introduced—**ReLu (rectified linear unit)** to make negative values zero. If any overfitting of the data is observed then dropout layer can be used to reduce the same. An interesting topic 'microexpressions' in the emotion recognition is presented by the authors [15] using deep learning architecture. Authors have used FERC-2013 database to extract the different features to identify the microexpressions. Model is based on cross-entropy loss function and uses Adam optimizer. This optimizer helps in training and providing best results and also reduces the losses. To handle the overfitting of data dropout layer can be introduced which reduces the overfitting of the model. Using confusion matrix different emotions and their scores are represented. Authors [16] have proposed a LDL label distribution learning model with conditional probability function to reduce the entropy. A framework EDL-LBCNN is implemented using CNN on the s-JAFFE dataset. Basic emotions like happiness, sadness, fear, anger, surprise and disgust are accurately evaluated. Framework consists of four convolutional layers and single LBC (local binary convolutional layer. Initially, framework extracts the features from grayscale images. Once the features are extracted, next phase is to concatenate them and pass to the next phase. LBC (local binary convolutional layer) consists of few trainable parameters and fixed size of filters. Output of this layer is generation of feature map and in the next layer, i.e. FC layer (fully connected layer) concatenates the feature maps generated by CNN and LBC layer. Authors [17, 18] have used LSTM and CNN to learn speech emotion features. Interesting approach is proposed by the authors by combining facial and speech expression. Advantage of this fusion approach is it gives information of audiovisual features. High level of features is extracted using deep neural networks. Audio features are extracted using pyAudioAnalysis an open-source library. Local features of data and global training features can be extracted using CNN—convolutional neural network. A multi-modal fusion is carried out on different features, also known as early fusion or feature level fusion. Model also consists of decision level fusion known as late-fusion. There is score level fusion implemented in the model to calculate the classification scores. Authors have combined happy and excite category into excite and works on remaining basic emotion like excite, sad and neutral. Authors [20] have implemented a pose-guide estimation model where features are extracted using few methods like pyramid histogram orientation gradient method (PHOG), edge histogram descriptor (EHD) and local binary pattern (LBP). Canny detector is introduced to extract the edge of the image. LBP helps in converting image into integer array. Authors have evaluated their proposed model on the four datasets—CK+, JAFFE, CASIA and AR dataset. Model works on three steps—target pose estimation, template generation and target matched.

3 Proposed System

As shown in Fig. 1, images/videos can be given as an input to the system. If the video is input then model splits the same into number of frames. Model also analyzes each frame and gets emotions for the same. Pandas data frame analyzes the values and plots the dataframe. Main objective of research is to get the analysis of different microexpressions. To do the same model must capture the expressions within microseconds. Once the model captures the microexpressions, it can be processed to get the score. *Output of this stage is cumulative score of the images which can be used to correlate the future scope of this research work, i.e. will be used for identification of stimuli behind appropriate expression.* Cumulative score helps in data analysis as shown in Tables 2 and 3 and Sect. 4.3 with the help of scatter plots. Analysis of different microexpressions has compared in Sect. 4.3 with the help of scatter plot and Pearson's heatmap. Increase in iteration will also increase the running time and have an impact on good precision and recall. Section 4 of this research work elaborates the processing of video file. Model processed the video file in terms of images and calculates the individual scores.

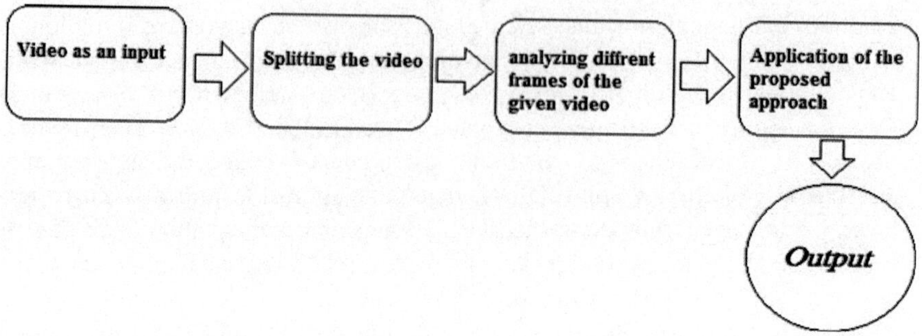

Fig. 1 Architecture of the proposed system

3.1 Algorithmic Details

(a) Input image/s or video
(b) Model processes the same
(c) If the input is—video
   a. Video (stream) will be divided into frames
   b. Each frame will be processed to identify the emotions
   c. Emotions are categorized into following types
      i. *Sadness*
      ii. *Happy*
      iii. *Surprise*
      iv. *Neutral*
      v. *Angry*
      vi. *Fear*
      vii. *Disgust*
(d) Score comparisons

3.2 MTCNN

It is good to execute the MTCNN (Multi-task CNN) using GPU (Graphics Processing Unit) to boost the speed of processing no. of frames /second. It is not suggested to execute the file (video) with the CPU because it might take no. of hours to process the 'n' frames. The video file used consists of 1123 frames and it takes ~1.4 s to process one frame. Hence with the normal CPU (Central Processing Unit) it takes minutes to process the video file.

No. of frames * time '*t*' required to process one frame = no. of seconds and one can convert it into the minutes.

Output of this stage is one can calculate the time required to process the video and images as shown in Eq. (2). Different points on the image (face), i.e. nose, eye, mouth can be detected using MTCNN method (Fig. 2).

Fig. 2 Processing of image using MTCNN

$$\boldsymbol{Total\ time\ to\ process\ the\ video_v = Frames_f * time_t\ for\ one\ frame} \quad (2)$$

Shape of image/array: (225, 225, 3)

There are three stages of the MTCNN method and are as follows

1. *Classifying the face*
2. *Bounding box regression*
3. *Landmark localization (eyes, nose and mouth)*

3.3 HAAR Cascade Classifier

As an integral part of the openCV library, HAAR cascade classifier detects the objects from given image(s) or video. It collects the features from given image by processing it. HAAR feature collector traverses through the given image. The algorithm uses edge or line detection features proposed by Viola and Jones in their research paper 'Rapid Object Detection using a Boosted Cascade of Simple Features' published in 2001. Working of HAAR feature mapping is quite similar to convolutional neural networks. It puts rectangle once the eyes, nose ears, etc. are detected and represents it in the following format—(x, y, width 'w', height 'h') and draws a rectangle around the same.

Output of this classifier helps 'mapping' of different objects from 'n' images. Local and global features will be extracted, categorized and a feature vectored map can be formed using this HAAR cascade classifier (Fig. 3).

Fig. 3 Detecting objects in the image

3.4 Getting Emotions from the Images/Video

Few images are given as an input and their individual scores (analysis) can be calculated, e.g. sadness score, happiness score, etc.

As shown in Table 1, basic emotions are listed like anger, sad, happy, etc. and once the model processes the images/videos scores will be calculated. Model calculates the score and applies the scale of 0 to 1, i.e. 1 means 100%. Few example images are processed as shown in Table 1 and their appropriate score is listed. E.g. happy 0.75 (75%), neutral 0.53 (53%), happy 1.0 (100%), surprise 0.44 (44%). We can calculate the average score for individual emotions as shown in Fig. 4.

$$\boldsymbol{Meanscore} = (\boldsymbol{sum\ of\ total\ scores\ obtained}) / \boldsymbol{total\ scores} \tag{3}$$

Output of this stage is calculating the 'Meanscore' for different images with different emotions. Once the model successfully processes the given image/video a data file (.csv) can be generated with the obtained score.

4 Results and Discussions

Model processes the given images/video(s) and extracts the features. Model analyzes all the frames and stores the data in separate file, which can be used for future analysis.

Table 1 Image analysis and scores obtained

	*[{'box': [81, 46, 82, 113], 'emotions': {'angry': 0.06, 'disgust': 0.0, 'fear': 0.02, 'happy': 0.75, 'sad': 0.01, 'surprise': 0.11, 'neutral': 0.04}}]* **happy 0.75**
	*[{'box': [101, 24, 86, 109], 'emotions': {'angry': 0.2, 'disgust': 0.0, 'fear': 0.0, 'happy': 0.06, 'sad': 0.22, 'surprise': 0.0, 'neutral': 0.53}}]* **neutral 0.53**
	*[{'box': [15, 38, 115, 155], 'emotions': {'angry': 0.0, 'disgust': 0.0, 'fear': 0.0, 'happy': 1.0, 'sad': 0.0, 'surprise': 0.0, 'neutral': 0.0}}]* **happy 1.0**
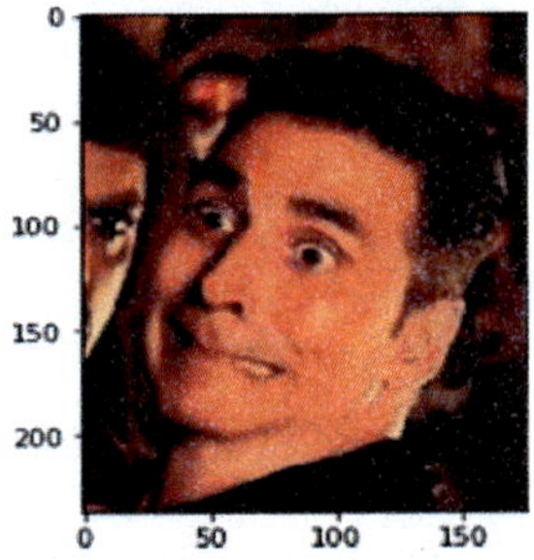	*[{'box': [29, 46, 107, 147], 'emotions': {'angry': 0.02, 'disgust': 0.0, 'fear': 0.1, 'happy': 0.18, 'sad': 0.03, 'surprise': 0.44, 'neutral': 0.22}}]* **surprise 0.44**

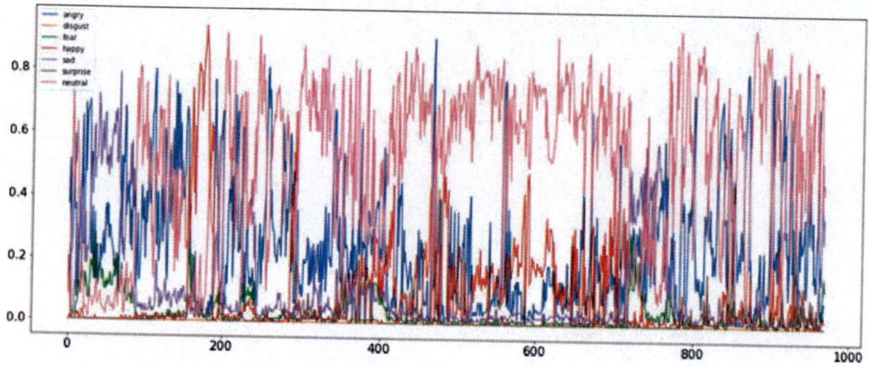

Fig. 4 Processing of the frames from given video

A video is given as an input to the propsoed model which divides it into 'n' frames and analyzes them independently.

For each frame, model calculates the score in terms of sad, anger, happy, etc. Cumulative emotional value from the given video will be shown as mentioned in Table 2. Processing of the images from the video file is shown in Fig. 4.

With the help of Fig. 5 and Table 2, one can conclude that, in a given video file, there is a maximum score for the neutral images. Next to the neutral image, score of the angry images is 234.45. Hence, one can process any streaming file, video and calculates the scores of different emotions. Figure 6 elaborates the average values of the different expressions.

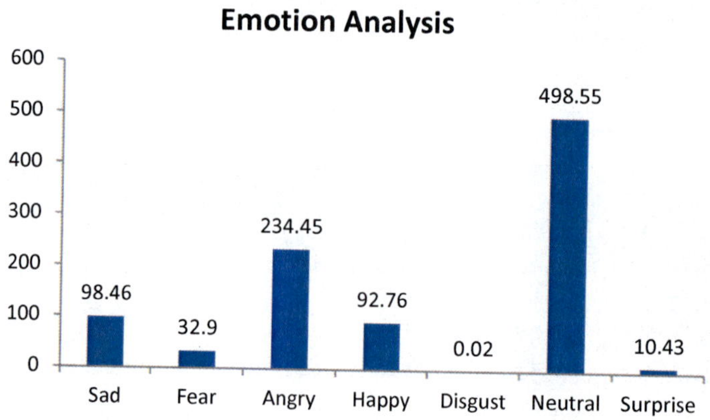

Fig. 5 Scores obtained for the 'n' frames

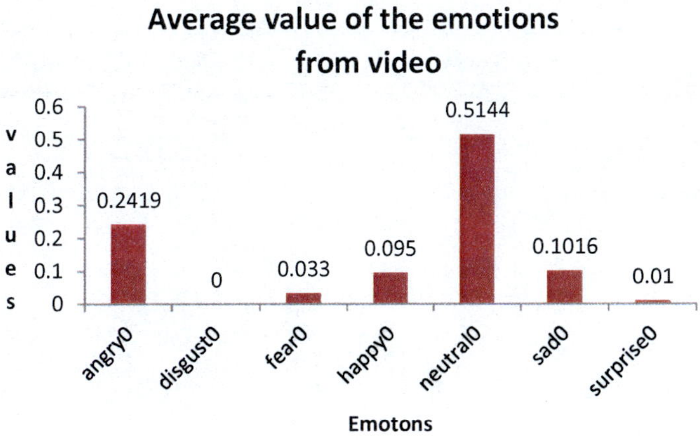

Fig. 6 Average value of the emotions from video

Table 2 Emotional values from the video

Count	Emotions	Emotion value from video
0	Sad	98.46000000000056
1	Fear	32.90000000000039
2	Angry	234.45000000000036
3	Happy	92.7600000000005
4	Disgust	0.02
5	Neutral	498.54999999999944
6	Surprise	10.429999999999957

4.1 Correlational Analysis (Angry, Disgust, Fear)

This section focuses on data analysis, i.e. once the model captures the microexpressions and calculates the individual and average scores one can go for getting the correlational analysis between them. Pearson's correlational value lies between -1 and 1. '-1' denotes the negative correlation and '+1' denotes the positive correlation.

4.2 Scatter Plots

As discussed in previous section, Pearson's correlation can be calculated for different variables. Different variables will have different correlational value. This correlation between variables can be shown using scatter plot like Figs. 7 and 8.

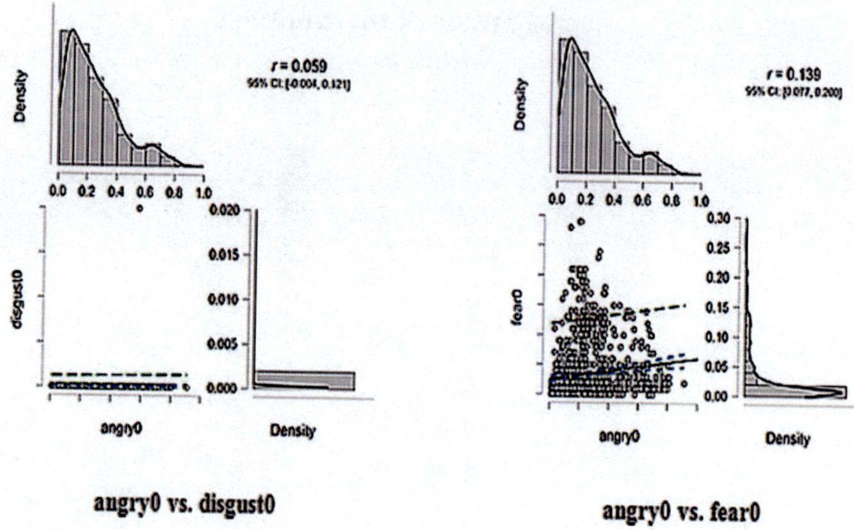

Fig. 7 Scatter plot

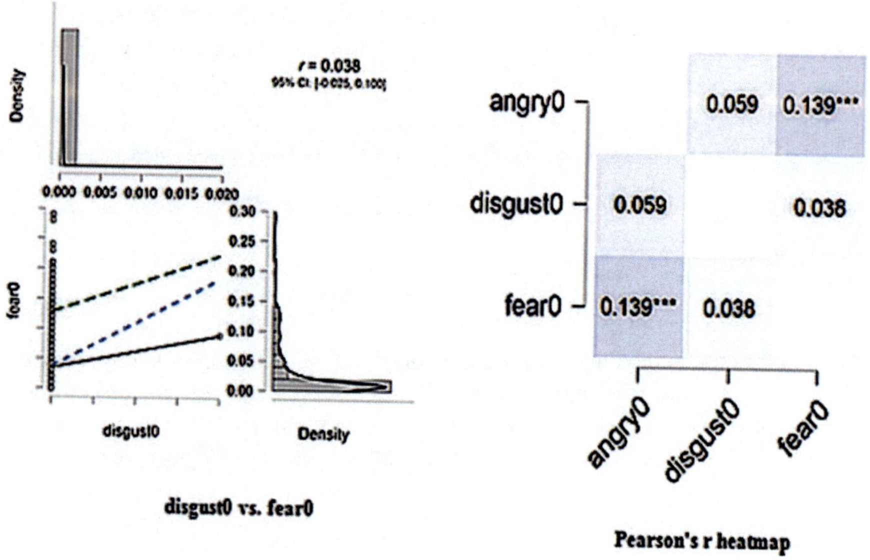

Fig. 8 Scatter plot and Pearson's heatmap

As shown in previous section, three variables are considered '**angry**', '**disgust**' and '**fear**'. Their correlational analysis is shown in Table 3 and Figs. 7 and 8. There is positive correlation between the pairs—'angry-fear' and 'disgust-angry'.

Table 3 Pearson's correlation

		n	Pearson's r		p	Lower 95% CI	Upper 95% CI	VS-MPR†
Angry0	Disgust0	969	0.059		0.067	−0.004	0.121	2.039
Angry0	Fear0	969	0.139	***	< .001	0.077	0.200	2336.880
Disgust0	Fear0	969	0.038		0.242	−0.025	0.100	1.071

5 Conclusion

These systems will be very useful in many applications such as detecting the objects, monitoring the traffic, robotics, computer vision, etc. It helps in detecting the emotions, microexpressions and in future work stimuli behind the same. Model detects the emotions and saves the scores obtained accordingly. Later on, one can process this data for analysis purpose as shown in Figs. 5 and 6 where scatter plots can be drawn. One can also correlate the scores and find out the association among different emotions. In this way, one can get the rid of specific emotion and try to find out the stimuli behind the same. Such systems will be very useful in predictive analytics.

Acknowledgements This work is supervised by Dr. A. B. Nandgaokar, Department of Electronics & Telecommunications, Dr. Babasaheb Ambedkar University, Lonere. Author thanks him for continuous help & guidance provided to complete this research work. Author also thanks Director of the Institution, Head of Department and other technical staff of the University for help and providing necessary infrastructure.

Conflict of Interest The authors declared that they have no conflicts of interest to report regarding the present study.

References

1. Mohammadpour M, Khaliliardali H, Hashemi SMR, AlyanNezhadi MM Facial emotion recognition using deep convolutional networks. In: IEEE 4th international conference on knowledge-based engineering and innovation (KBEI), pp 0017–0021. IEEE, Tehran, Iran
2. Wang L, Song Y, Chen J, Sun G, Li H (2018) Emotion recognition using sequence mining. In: IEEE International Conference on Progress in Informatics and Computing (PIC), pp 129–133. IEEE, Suzhou, China
3. Chen H, Gu Y, Wang F, Sheng W (2018) Facial expression recognition and positive emotion incentive system for human-robot interaction. In: 13th World Congress on Intelligent Control and Automation (WCICA), pp 407–412. IEEE, Changsha, China
4. Kattubadi IB, Garimella RM (2019) Emotion classification: novel deep learning architectures. In: 5th International Conference on Advanced Computing & Communication Systems (ICACCS), pp 285–290. IEEE, Coimbatore, India
5. Zadeh MMT, Imani M, Majidi B (2019) Fast facial emotion recognition using convolutional neural networks and Gabor filters. In: 5th Conference on Knowledge Based Engineering and Innovation (KBEI), pp 577–581. IEEE, Tehran Iran

6. Hua W, Dai F, Huang L, Xiong J, Gui G (2019) HERO: Human emotions recognition for realizing intelligent Internet of Things. IEEE Access 7: 24321–24332
7. Harshitha S, Sangeetha N, Shirly AP, Abraham CD (2019) Human facial expression recognition using deep learning technique. In: 2nd International Conference on Signal Processing and Communication (ICSPC), pp 339–342. IEEE, Coimbatore, India
8. Chen M, Hao Y (2019) Label-less learning for emotion cognition. IEEE Trans Neural Netw Learn Syst 31.7:2430–2440
9. Yokoo K, Atsumi M, Tanaka K, Haoqing W, Lin M (2020) Deep learning based emotion recognition iot system. In: International Conference on Advanced Mechatronic Systems (ICAMechS), pp 203–207. IEEE, Hanoi, Vietnam
10. Kondaveeti HK, Goud MV (2020) Emotion detection using deep facial features. In: IEEE international conference on advent trends in multidisciplinary research and innovation (ICATMRI), pp 1–8. IEEE, Buldhana, India
11. Begaj S, Topal AO, Ali M (2020) Emotion recognition based on facial expressions using convolutional neural network (CNN). In: International Conference on Computing, Networking, Telecommunications & Engineering Sciences Applications (CoNTESA), pp 58–63. IEEE, Tirana, Albania
12. Zhang S, Guan C (2020) Emotion recognition with refined labels for deep learning. In: 42nd Annual international conference of the ieee engineering in medicine & biology society (EMBC), pp 108–111. IEEE, Montrreal, QC, Canada
13. Jaiswal A, Raju AK, Deb S (2020) Facial emotion detection using deep learning. In: International Conference for Emerging Technology (INCET), pp 1–5. IEEE, Belgaum, India
14. Pranav E., Suraj, Kamal S, Chandran C, Supriya MH (2020) Facial emotion recognition using deep convolutional neural network. In: 6th International conference on advanced computing and communication Systems (ICACCS), pp 317–320. IEEE, Coimbatore, India
15. Yadahalli SS, Rege S, Kulkarni S (2020) Facial micro expression detection using deep learning architecture. In: International Conference on Smart Electronics and Communication (ICOSEC), pp.167–171. IEEE, Trichy, India
16. Almowallad A, Sanchez V (2020) Human emotion distribution learning from face images using CNN and LBC features. In: 8th International Workshop on Biometrics and Forensics (IWBF), pp.1–6. IEEE, Porto, Portugal
17. Cai L, Dong J, Wei M (2020) Multi-modal emotion recognition from speech and facial expression based on deep learning. In: Chinese Automation Congress (CAC), pp.5726–5729. IEEE, Shanghai, China (2020).
18. Choi DY, Song BC (2020) Semi-supervised learning for continuous emotion recognition based on metric learning. IEEE Access 8:113443–113455
19. Subramanian RR, Niharika CS, Rani DU, Pavani P, Syamala KPL (2021) Design and evaluation of a deep learning algorithm for emotion recognition. In: 5th International conference on intelligent computing and control systems (ICICCS), pp. 984–988. IEEE, Madurai, India
20. Liu J, Feng Y, Wang H (2021) Facial expression recognition using pose-guided face alignment and discriminative features based on deep learning. IEEE Access 9: 69267–69277

Performance Analysis of Outdoor THz Wireless Transmission Over Mixed Gaussian Fading with Pointing Errors

Pranay Bhardwaj and S. M. Zafaruddin

Abstract This paper presents the performance analysis for THz transmissions by considering a novel channel fading model for outdoor THz transmission with the combined effect of zero-boresight pointing error. We consider the recently proposed Gaussian mixture fading for outdoor THz transmissions to model the small-scale fading. We derive the probability density function of the channel fading with zero-boresight pointing errors. Using the derived PDF, we derived the outage probability and the average bit error rate (BER) of the considered system. We validate the derived analytical expressions with Monte Carlo simulations and demonstrate the effect of various system and channel parameters on the performance for single THz wireless communications.

Keywords Outage probability · Performance analysis · Pointing error · THz transmissions

1 Introduction

Terahertz (THz) wireless systems are believed to be a pivotal enabling technology for the next generation of wireless communications as it can provide enormous amount of unlicensed bandwidth for high bandwidth applications such as virtual reality, autonomous driving, and holographic imaging [1–3]. The THz band can provide ubiquitous connectivity as it can be utilized as wireless backhaul/fronthaul, mobile ad hoc backhaul in difficult-to-access terrains, replacing the need for wireline backhaul/fronthaul. THz communication is line-of-sight communication that requires high gain antennas to mitigate the relatively higher path loss which is caused by molecular absorption [4, 5]. Apart from the short-term fading and higher path loss,

P. Bhardwaj (✉) · S. M. Zafaruddin
Birla Institute of Technology and Science, Pilani Campus, Pilani, Rajasthan, India
e-mail: p20200026@pilani.bits-pilani.ac.in

S. M. Zafaruddin
e-mail: syed.zafaruddin@pilani.bits-pilani.ac.in

H. O. Bansal et al. (eds.), *Next Generation Systems and Networks*, Lecture Notes in Networks and Systems 641, https://doi.org/10.1007/978-981-99-0483-9_17

THz transmissions are also susceptible to random antenna misalignment errors if there is a misalignment between the transmitting and receiving antennas.

Recently, a significant amount of research has been done on the dual-hop and multi-hop THz relaying systems [6–19]. More specifically, the authors in [6] proposed an optimal relaying distance to maximize the network throughput for THz-band communication. In [7], a relay selection approach was suggested to mitigate the impact of antenna misalignment and shadowing due to the human blockage in a multi-relay setup. The authors in [8] considered a reconfigurable intelligent surface (RIS)-assisted multi-hop THz system over Rician fading to mitigate the signal attenuation using deep reinforcement learning (DRL)-based beam-forming technique. The authors in [10, 12] employed the decode-and-forward (DF) protocol to integrate THz and radio frequency (RF) technologies by considering the generalized α-μ fading combined with stochastic pointing errors. The DF relaying was studied for a dual-hop THz-THz link in [9] using multi-antenna transceivers. The authors in [19] presented the performance analysis for multi-hop THz transmission by considering generalized α-μ for the THz backhaul and also considered the effect of shadowing by considering generalized-k shadowed fading [20] in the access link.

Lately, the researchers have shown keen interest to model the fading characteristics of THz communications [21–25]. The authors in [22] proposed a Gamma mixture channel model for short ($<$ 1m) link THz transmissions. The authors in [23] show that the α-μ fading model is more apt for indoor THz transmission with link distances up to 50 m at 152 GHz. Recently, the authors in [25] have presented a channel fading model for outdoor THz transmissions consisting of Gaussian Mixture (GM) distribution. The authors have shown that the analytical distributions such as Rayleigh, Rice, Nakagami-M, Weibull, and α-μ distribution effectively analyze single peak fading. Whereas the GM distribution can accommodate multiple peaks fading characterization of the channel by including more Gaussian mixture components.

In this paper, we analyze the performance of a single-link THz transmissions by considering a novel channel fading model for outdoor THz transmission with the combined effect of zero-boresight pointing error. We consider the recently proposed Gaussian mixture fading for outdoor THz transmissions to model the small-scale fading. We derive the combined probability density function (PDF) of the channel fading with zero-boresight pointing errors. Using the derived PDF, we present the analysis for outage probability and average bit error rate (BER) of the considered system. Finally, We validate the derived analytical expressions with Monte Carlo simulations and demonstrate the effect of various system and channel parameters on the performance of single antenna THz wireless communications.

2 System Model

The system model consists of a single antenna point-to-point wireless transmission between source to destination in an outdoor environment. To model the short-term fading of the wireless channel, we consider the GM model, while zero-boresight pointing error is considered to model the antenna misalignment errors. The received signal y at the destination is given by

$$y = h_l h_{fp} s + w, \tag{1}$$

where h_l denotes the path gain; $h_{fp} = |h_f| h_p$ denotes the channel coefficient with combined effect of small-scale fading (h_f) and pointing errors (h_p), respectively; s denotes the transmitted signal with power P; and w_i is the additive white Gaussian noise with a variance σ_w^2. The deterministic path gain h_l is given by [26]

$$h_l = \frac{c\sqrt{G_t G_r}}{4\pi f d} \exp\left(-\frac{1}{2}k(f, T, \psi, p)d\right), \tag{2}$$

where G_t and G_r are the gains of the transmitting and receiving antenna, respectively. The terms f, c, and d depict the transmission frequency, the speed of light, and transmission distance. The molecular absorption coefficient is denoted by the term $k(f, T, \psi, p)$, which depends on the atmospheric pressure p, relative humidity ψ, and atmospheric temperature T given by

$$\begin{aligned} k(f, T, \psi, p) =& \frac{r_1 v(r_2 v + r_3)}{(r_4 v + r_5)^2 + (\frac{f}{100c} - r_1)^2} \\ &+ \frac{r_6 v(r_7 v + r_8)}{(r_9 v + r_{10})^2 + (\frac{f}{100c} - p_2)^2} + g_1 f^3 + g_2 f^2 + g_3 f + g_4, \end{aligned} \tag{3}$$

where $v = \frac{\psi}{100} \frac{p_w(T,p)}{p}$ with $p_w(T, p)$ represents the saturated water vapor partial pressure at temperature T, which can be calculated utilizing Buck's equation. The other parameters are provided in Table II of [27]. We use the zero-boresight model for pointing errors h_p [28]:

$$f_{h_{pi}}(h_p) = \frac{\rho^2}{S_0^{\rho^2}} h_p^{\rho^2 - 1}, 0 \leq h_p \leq S_0, \tag{4}$$

where $S_0 = \text{erf}(\upsilon)^2$ with $\upsilon = \sqrt{\frac{\pi}{2}} \left(\frac{r_1}{\omega_z}\right)$ and ω_z denotes the signal beam width, $\rho = \frac{\omega_{z\text{eq}}}{2\sigma_s}$ with $\omega_{z_{\text{eq}}}$ as the equivalent beam width at the receiving antenna, which is given as $\omega_{z_{\text{eq}}}^2 = \frac{\omega_z^2 \sqrt{\pi} \text{erf}(\upsilon)}{2\upsilon \exp(-\upsilon^2)}$, and σ_s^2 is the variance of pointing errors displacement characterized by the elevation and horizontal sway [28]. To model the short-term fading ($|h_f|$), we use the GM distributed channel with PDF [25]:

$$f_{|h_f|}(x) = \sum_{i=1}^{K} w_i \frac{\exp\left(-\frac{(x-\mu_i)^2}{2\sigma_i^2}\right)}{\sqrt{2\pi}\sigma_i}. \tag{5}$$

The parameters μ_i is the mean and σ_i is the standard deviation of the i-th GM component. K denotes the number of Gaussian mixture (GM) components and w_i is the weight of the i-th mixture component with $w_i \in [0, 1]$ and $\sum_{i=1}^{K} w_i = 1$. We denote signal-to-noise ratio (SNR) as $\gamma = \gamma_0 |h_{fp}|^2$, where the term $\gamma_0 = \frac{P|h_l|^2}{\sigma_w^2}$ denotes the SNR without the channel fading.

3 Performance Analysis

In this section, we carry out the closed-form analysis for the PDF and CDF of the single THz link for the GM fading combined with zero-boresight pointing errors. We derive closed-form expressions of outage and average BER for the THz link in terms of standard Mathematical functions, Gamma functions, and bivariate Fox's H-functions.

Proposition 1 *If $|h_{fp}| = |h_f||h_p|$ is the combined effect of GM fading with pointing errors, then the PDF and CDF of the single-link SNR $\gamma_i = \gamma_0 |h_{fp}|^2$ are given by*

$$f_\gamma(\gamma) = \frac{\rho^2 \gamma^{-\frac{1}{2}}}{\sqrt{8\pi} S_0 \gamma_0^{\frac{1}{2}}} \sum_{i=1}^{K} \frac{w_i e^{\frac{-\mu_i^2}{2\sigma_i^2}}}{\sigma_i}$$

$$\times H^{0,1;0,1;0,1}_{1,1;1,0;1,0}\left[\begin{matrix}\{(2-\rho^2; 2, 1)\} : (0,1); (0,1) \\ \{1-\rho^2; 2, 1\} : -; -\end{matrix}\middle| \frac{2\sigma_i^2 S_0^{2\rho^2}\gamma_0}{\gamma}, \frac{\sigma_i^2 S_0^{\rho^2}\gamma_0^{\frac{1}{2}}}{-\mu\gamma^{\frac{1}{2}}}\right]. \quad (6)$$

$$F_\gamma(\gamma) = \frac{-\rho^2 \gamma^{\frac{1}{2}}}{\sqrt{8\pi} S_0 \gamma_0^{\frac{1}{2}}} \sum_{i=1}^{K} \frac{w_i e^{\frac{-\mu_i^2}{2\sigma_i^2}}}{\sigma_i}$$

$$\times H^{0,2;0,1;0,1}_{2,2;1,0;1,0}\left[\begin{matrix}\{(2-\rho^2; 2, 1), (\frac{3}{2}; 1, \frac{1}{2})\} : (0,1); (0,1) \\ \{(1-\rho^2; 2, 1), (\frac{1}{2}; 1, \frac{1}{2})\} : -; -\end{matrix}\middle| \frac{2\sigma_i^2 S_0^{2\rho^2}\gamma_0}{\gamma}, \frac{\sigma_i^2 S_0^{\rho^2}\gamma_0^{\frac{1}{2}}}{-\mu\gamma^{\frac{1}{2}}}\right]. \quad (7)$$

Proof ubstituting the limits of (4) and (5), the joint PDF of $|h_{fp}| = |h_f|h_p$ can be expressed as [29]

$$f_{h_i}(x) = \int_0^{S_o} \frac{1}{y} f_{h_f}\left(\frac{x}{y}\right) f_{h_p}(y) \mathrm{d}y. \quad (8)$$

Substituting the respective PDFs from (4) and (5) in (8) and simplifying, we get

$$f_{h_{fp}}(x) = \frac{\rho^2}{S_0^{\rho^2}\sqrt{2\pi}} \sum_{i=1}^{K} \frac{w_i e^{\frac{-\mu_i^2}{2\sigma_i^2}}}{\sigma_i} \int_0^{S_o} y^{\rho^2-2} e^{\left(\frac{-x^2}{y^2 2\sigma_i^2}\right)} e^{\left(\frac{\mu x}{y\sigma_i^2}\right)} \mathrm{d}y. \tag{9}$$

Writing the exponential terms into their Meijer's G equivalent and converting $S_1 = -S_1$ and $S_2 = -S_2$, where S_1 and S_2 are complex variables in contour $\mathcal{L}_1$ and $\mathcal{L}_2$, we can rewrite the equation as

$$f_{h_i}(x) = \frac{\rho^2}{S_0^{\rho^2}\sqrt{2\pi}} \sum_{i=1}^{K} \frac{w_i e^{\frac{-\mu_i^2}{2\sigma_i^2}}}{\sigma_i} \left(\frac{1}{2\pi j}\right)^2 \int_{\mathcal{L}_1} \Gamma(0+S_1)\left(\frac{2\sigma_i^2}{x^2}\right)^{S_1} dS_1$$
$$\int_{\mathcal{L}_2} \Gamma(0+S_2)\left(\frac{\sigma_i^2}{-\mu x}\right)^{S_2} dS_2 \times I_1 \tag{10}$$

The inner integral I_1 can be simplified to

$$\int_0^{S_o} y^{\rho^2-2+2S_1+S_2}\mathrm{d}y = \frac{S_0^{\rho^2+2S_1+S_2-1}}{\rho^2+2S_1+S_2-1} = \frac{S_0^{\rho^2+2S_1+S_2-1}\Gamma(\rho^2+2S_1+S_2-1)}{\Gamma(\rho^2+2S_1+S_2)}. \tag{11}$$

Substituting back the inner integral in (10) and utilizing the definition of bivariate Fox's H-function, we represent the PDF of the channel as

$$f_{h_i}(x) = \frac{\rho^2}{S_0\sqrt{2\pi}} \sum_{i=1}^{K} \frac{w_i e^{\frac{-\mu_i^2}{2\sigma_i^2}}}{\sigma_i}$$
$$\times H^{0,1;0,1;0,1}_{1,1;1,0;1,0}\left[\begin{matrix}\{(2-\rho^2;2,1)\}:\{(0,1)\};\{(0,1)\}\\ \{1-\rho^2;2,1\}:\{-\};\{-\}\end{matrix}\middle| \frac{2\sigma_i^2 S_0^{2\rho^2}}{x^2}, \frac{\sigma_i^2 S_0^{\rho^2}}{-\mu x}\right]. \tag{12}$$

Transforming the random variables the integral in (12) using $\frac{1}{2\sqrt{\gamma\gamma_0}} f_{|h_{fp}|^2}\left(\sqrt{\frac{\gamma}{\gamma_0}}\right)$, we derive the PDF of SNR in (6) of Proposition 1. To derive the CDF we integrate the PDF, i.e., $F_{\gamma_i}(\gamma) = \int_0^{\gamma} f_{\gamma_i}(\gamma)d\gamma$

$$F_{\gamma_i}(\gamma) = \frac{\rho^2}{\sqrt{8\pi}S_0\gamma_0^{\frac{1}{2}}} \sum_{i=1}^{K} \frac{w_i e^{\frac{-\mu_i^2}{2\sigma_i^2}}}{\sigma_i} \left(\frac{1}{2\pi j}\right)^2 \int_{\mathcal{L}_1} \Gamma(0+S_1)\left(2\sigma_i^2\gamma_0\right)^{S_1} dS_1$$
$$\times \int_{\mathcal{L}_2} \Gamma(0+S_2)\left(\frac{\sigma_i^2\gamma_0^{\frac{1}{2}}}{-\mu}\right)^{S_2} dS_2 \frac{S_0^{\rho^2+2S_1+S_2-1}\Gamma(\rho^2+2S_1+S_2-1)}{\Gamma(\rho^2+2S_1+S_2)} \times I_2 \tag{13}$$

The inner integral I_2 can be simplified as

$$\int_0^{\gamma} \gamma^{-S_1-\frac{S_2}{2}-\frac{1}{2}} d\gamma = -\frac{\gamma^{-S_1-\frac{S_2}{2}+\frac{1}{2}}\Gamma\left(S_1+\frac{S_2}{2}-\frac{1}{2}\right)}{\Gamma\left(S_1+\frac{S_2}{2}+\frac{1}{2}\right)}. \tag{14}$$

Substituting the inner integral back in (13) and using the definition of bivariate Fox's H-function, we get the CDF in (7) of Proposition 1 to conclude the proof.

3.1 Outage Probability

Outage probability is a measure of link reliability and is measured by the probability of instantaneous SNR to become less than a threshold value of the SNR and is given as $P_{\text{out}} = P(\gamma < \gamma_{\text{th}})$ where γ_{th} is the threshold SNR. Thus, the outage probability is expressed by $P_{\text{out}} = F_\gamma(\gamma_{\text{th}})$, where $F_\gamma(\gamma)$ is presented in (7).

3.2 Average BER

Average BER is a measure of occurrence of error in the transmitted bits and is an important performance parameter for a communication systems, defined as [30]

$$\bar{P}_e = \frac{q^p}{2\Gamma(p)} \int_0^{\infty} \gamma^{p-1} e^{-q\gamma} F_\gamma(\gamma) d\gamma, \tag{15}$$

where parameters p and q specify different modulation techniques and $F_\gamma(\gamma)$ is the CDF from (7).

$$\bar{P}_e = \frac{-\rho^2 q^p}{\sqrt{32\pi} S_0 \gamma_0^{\frac{1}{2}} \Gamma(p)} \sum_{i=1}^{K} \frac{w_i e^{\frac{-\mu_i^2}{2\sigma_i^2}}}{\sigma_i} \left(\frac{1}{2\pi j}\right)^2 \int_{\mathcal{L}_1} \Gamma(0+S_1)\left(2\sigma_i^2\gamma_0\right)^{S_1} dS_1$$

$$\int_{\mathcal{L}_2} \Gamma(0+S_2)\left(\frac{\sigma_i^2\gamma_0^{\frac{1}{2}}}{-\mu}\right)^{S_2} dS_2 \frac{S_0^{\rho^2+2S_1+S_2-1}\Gamma(\rho^2+2S_1+S_2-1)}{\Gamma(\rho^2+2S_1+S_2)}$$

$$\times \frac{-\Gamma\left(S_1+\frac{S_2}{2}-\frac{1}{2}\right)}{\Gamma\left(S_1+\frac{S_2}{2}+\frac{1}{2}\right)} \times I_3. \tag{16}$$

The inner integral I_3 can be simplified using the identity ([31], Eq. (3.381/4)) to get

$$\int_0^\infty \gamma^{p-S_1-\frac{S_2}{2}-\frac{1}{2}} e^{-q\gamma} d\gamma = q^{-p+S_1+\frac{S_2}{2}-\frac{1}{2}} \Gamma\left(p+S_1+\frac{S_2}{2}+\frac{1}{2}\right). \tag{17}$$

Substituting the inner integral in (16) and utilizing the definition of bivariate Fox's H-function, we express the average BER as

$$\bar{P}_e = \frac{-\rho^2 \gamma^{\frac{1}{2}}}{\sqrt{8\pi} S_0 \gamma_0^{\frac{1}{2}}} \sum_{i=1}^{K} \frac{w_i e^{\frac{-\mu_i^2}{2\sigma_i^2}}}{\sigma_i} H^{0,3;0,1;0,1}_{3,2;1,0;1,0}$$
$$\times \left[\begin{matrix} \{(2-\rho^2;2,1),(\frac{3}{2};1,\frac{1}{2}),(\frac{1}{2}-p;1,\frac{1}{2})\}:(0,1);(0,1) \\ \{(1-\rho^2;2,1),(\frac{1}{2};1,\frac{1}{2})\}:-;- \end{matrix}\middle| 2\sigma_i^2 S_0^{2\rho^2}\gamma_0, \frac{\sigma_i^2 S_0^{\rho^2}\gamma_0^{\frac{1}{2}}}{-\mu}\right]. \tag{18}$$

4 Simulation and Numerical Results

In this section, we illustrate the performance of the outdoor THz link and verify the derived analytical expressions with Monte Carlo simulations. We consider the distance of the THz link 30–80 m, the gain of transmitter and receiver antennas as 40 dBi, and carrier frequency 275 GHz. For the THz link, we calculate the path loss by considering the temperature, atmospheric pressure, and relative humidity as 296 K, 101325 Pa, and 50%, respectively. To compute the pointing error parameters ρ and S_0, We utilize [28] by changing the beam width and the jitter variance 5–20 cm. The AWGN noise power is considered as −94.2 dBm.

In Fig. 1, we depict the outage probability of the considered system for varying pointing errors and link distance. Figure 1a shows the effect of pointing errors parameter ρ on the outage performance of the considered THz system with $S_0 = 0.054$ and a SNR threshold (γ_{th}) of 4 dB where S_0 signifies the severity of the pointing error (S_0=1 denotes no pointing error). It is clearly visible from the plots that the out-

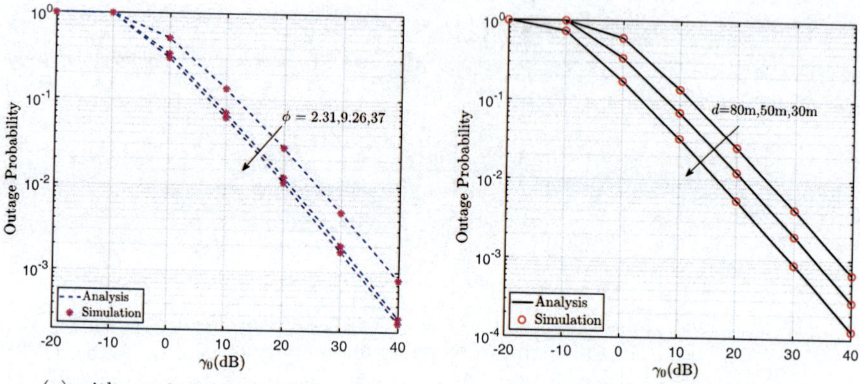

(a) with varying pointing errors (ϕ).

(b) for different link distances (d).

Fig. 1 Outage probability of THz wireless link

age probability of the THz link decreases with the increase in the value of ρ as an increase in the value of ρ results in reduced pointing errors. It can be observed that there is distinguishable difference, when we change the value of ρ from $\rho = 2.31$ and $\rho = 9.26$. Afterward for $\rho = 9.26$ and $\rho = 37$ the change is marginal. It can be inferred from the plots that at the average SNR of 30 dB, the outage performance improves by three times when we increase ρ from 2.31 to 9.26. Figure 1b depicts the effect of link distance d on the outage probability of the considered system. We have considered the link distance in the range of 30–80 m. The outage performance of the THz link degrades as we increase the link distance because the SNR at the receiver decreases.

In Fig. 2, we show the average bit error probability of the considered system with varying antenna gains. It can be observed from the plots that the bit error probability

Fig. 2 Average BER of the THz wireless link with varying antenna gains (G_t, G_r)

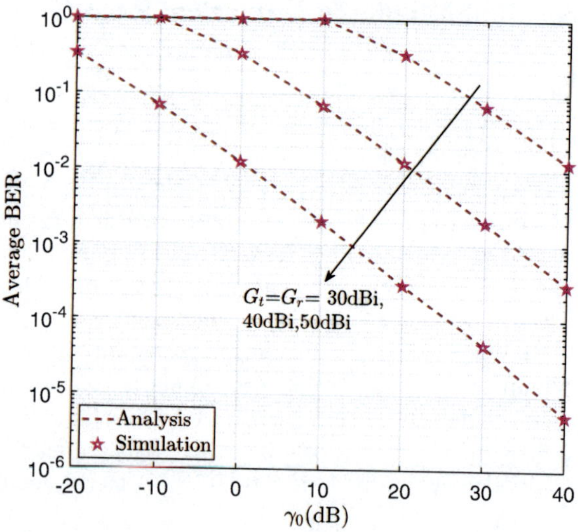

reduces when we increase the antenna gain. An improvement of almost 50 times is visible if we increase the antenna gain from 30 to 40 dBi for an average SNR of 30 dB.

5 Conclusions

In the above work, we carried out the performance analysis for an outdoor THz wireless link by considering the Gaussian mixture fading model, which is experimentally proven to be the best fit to model the short-term fading for outdoor THz wireless channels. First, we derived the PDF of combined channel fading considering the effect of statistical zero-boresight pointing errors. Using the derived PDF, we derived two primary performance metrics for the considered system, i.e., outage probability and average BER. Finally, using the Monte Carlo simulations, we validated our derived analytical expressions and analyzed the effect of various system parameters on the system's performance, which were demonstrated with the help of various plots.

References

1. Koenig S et al (2013) Wireless sub-THz communication system with high data rate. Nat Photon 7:977–981
2. Wang C et al (2014) 0.34-THz wireless link based on high-order modulation for future wireless local area network applications. IEEE Trans THz Sci Technol 4(1):75–85
3. Elayan H et al (2020) Terahertz band: The last piece of RF spectrum puzzle for communication systems. IEEE Open J Commun Soc 1:1–32
4. Kokkoniemi J et al (2018) Simplified molecular absorption loss model for 275–400 Gigahertz frequency band. In: 12th the European association on antennas and propagation (EuCAP), pp 1–5
5. Kokkoniemi J et al (2020) Impact of beam misalignment on THz wireless systems. Nano Commun Netw 24:100302
6. Xia Q, Jornet JM (2017) Cross-layer analysis of optimal relaying strategies for Terahertz-band communication networks. In: 2017 IEEE 13th international conference on wireless and mobile computing, networking and communications (WiMob), pp 1–8
7. Stratidakis G et al (2020) Relay-based blockage and antenna misalignment mitigation in THz wireless communications. In: 2020 2nd 6G wireless summit (6G SUMMIT), pp 1–4
8. Huang C et al (2021) Multi-hop RIS-empowered Terahertz communications: a DRL-based hybrid beamforming design. IEEE J Sel Areas Commun 39(6):1663–1677
9. Boulogeorgos A-AA, Alexiou A (2020) Outage probability analysis of THz relaying systems. In: 2020 IEEE annual international symposium on personal, indoor and mobile radio communications, pp 1–7
10. Boulogeorgos AA, Alexiou A (2020) Error analysis of mixed THz-RF wireless systems. IEEE Commun Lett 24(2):277–281
11. Bhardwaj P, Zafaruddin SM (2021) Performance of dual-hop relaying for THz-RF wireless link. In 2021 IEEE 93rd vehicular technology conference (VTC2021-Spring), pp 1–5
12. Bhardwaj P, Zafaruddin SM (2021) Performance analysis of dual-hop relaying for THz-RF wireless link over asymmetrical $\alpha - \mu$ fading. IEEE Trans Veh Technol 1–1

13. Bhardwaj P, Zafaruddin SM (2021) Fixed-gain AF relaying for RF-THz wireless system over α-κ-μ shadowed and α-μ channels. IEEE Commun Lett 1–1
14. Rong Z et al (2017) Relay-assisted nanoscale communication in the THz band. Micro Nano Lett 12(6):373–376
15. Bhardwaj P, Zafaruddin SM (2022) Performance analysis of cooperative relaying for multi-antenna rf transmissions over thz wireless link. In: 2022 IEEE 95th vehicular technology conference: (VTC2022-Spring), pp 1–5
16. Bhardwaj P, Zafaruddin SM (2022) Performance of hybrid THz and multi-antenna RF system with diversity combining. [Online]. Available: https://arxiv.org/abs/2202.08515
17. Abbasi QH et al (2017) Cooperative In-Vivo nano-network communication at Terahertz frequencies. IEEE Access 5:8642–8647
18. Mir T et al (2020) Hybrid precoding design for two-way relay-assisted Terahertz massive MIMO systems. IEEE Access 8:222 660–222 671
19. Bhardwaj P, Zafaruddin SM (2022) On the performance of multihop THz wireless system over mixed channel fading with shadowing and antenna misalignment. IEEE Tran Commun 1
20. Ye J et al (2021) Modeling co-channel interference in the THz band. IEEE Trans Veh Technol 70(7):6319–6334
21. Ekti AR et al (2017) Statistical modeling of propagation channels for terahertz band. In: IEEE conference standards for communication network (CSCN), pp 275–280
22. Kursat T et al (2021) Modeling and analysis of short distance sub-terahertz communication channel via mixture of gamma distribution. IEEE Trans Veh Technol 70(4):2945–2954
23. Papasotiriou EN et al (2021) A new look to THz wireless links: fading modeling and capacity assessment. In: 2021 IEEE 32nd annual international symposium on personal, indoor and mobile radio communications, pp 1–6
24. Du H, Zhang J, Guan K, Ai B, Kürner T (2020) Reconfigurable intelligent surface aided terahertz communications under misalignment and hardware impairments. arXiv:2012.00267
25. Papasotiriou EN et al (2022) Outdoor THz fading modeling by means of gaussian mixture distribution. Nat, Sci Rep. In Press
26. Boulogeorgos AA, Alexiou A (2019) Analytical performance assessment of THz wireless systems. IEEE Access 7:11 436–11 453
27. Boulogeorgos AA et al (2018) Performance evaluation of THz wireless systems operating in 275–400 GHz band. In: 2018 IEEE 87th vehicular technology conference (VTC Spring), pp 1–5
28. Farid AA, Hranilovic S (2007) Outage capacity optimization for free-space optical links with pointing errors. J Lightw Technol 25(7):1702–1710
29. Papoulis A, Pillai S (2002) Probability, random variables, and stochastic processes, 4th edn. McGraw Hill, Boston
30. Ansari IS et al (2011) A new formula for the BER of binary modulations with dual-branch selection over generalized-k composite fading channels. IEEE Trans Commun 59(10):2654–2658
31. Gradshteyn IS, Ryzhik IM (2000) Table of integrals, series, and products, 6th edn. Academic, San Diego, CA

Detecting and Analyzing Malware Using Machine Learning Classifiers

Vishal Dhingra, Jaspreet Singh, and Parminder Kaur

Abstract Nowadays, the security of systems is one of the prime focuses of any organization. Malware detection techniques can be in static form or dynamically identify malware present inside network communications that affect the performance of systems. The malware detection is the field of network security where different techniques for detecting malicious software are used to protect a network from any harmful consequences. This paper presents a method for detecting malware using machine learning classifiers. The proposed methodology deploys six machine learning classifiers over three publicly available authentic datasets. Around 1.2 million features of malware are extracted from these datasets using feature engineering. The static, dynamic, hybrid, and URL features are also considered in the proposed methodology. The performance of classifiers is compared in terms of detection rate and false positive rate. The SVM classifier with hyper-parameter 0.01 and threshold of 0.02 was found more accurate among the six classifiers. The proposed work provides insights into large networked systems by classifying software as malicious or benign.

Keywords Malware detection · Machine learning · Static analysis · Dynamic analysis · Behavior analysis

1 Introduction

In the era of Internet, there are increasingly many threats to the security and safety of Internet users connected with each other through various computing devices like desktop, laptops, mobiles phones, tablets, etc. Nowadays, Internet users can avail online services anywhere–anytime through Internet. With the growth of good use of

V. Dhingra · P. Kaur
Guru Nanak Dev University, Amritsar, Punjab 143005, India
e-mail: parminder.dcse@gndu.ac.in

J. Singh (✉)
Chandigarh University, 140413, Mohali, Punjab, India
e-mail: jaspreet.e10279@cumail.in

H. O. Bansal et al. (eds.), *Next Generation Systems and Networks*, Lecture Notes in Networks and Systems 641, https://doi.org/10.1007/978-981-99-0483-9_18

Internet, malicious activities and threats have also increased exponentially over past few years. These malicious activities include Malicious URL's attack, Intrusions in enterprise network system or home systems, Spread of malware files that are responsible for privacy leaks, denial of services, theft of valuable data, corruption of data, etc., Antivirus software (such as Norton, McAfee, Avast, Kaspersky, AVG, and Bitdefender) is a major line of defense against malware attacks. Traditionally anti-virus software used signature-based methods for malware detection. But these techniques cannot provide protection against zero-day attacks. Also with malware generation toolkits like Zeus, thousands of variants of the same malware could be generated using different obfuscation techniques.

Antivirus companies and researchers in this field are continuously looking for alternative techniques that are more effective in malware detection, intrusion detection, malicious URL's detection, etc. to prevent users from being exploited. In recent years, anti-malware companies and researchers have reported using machine learning and deep learning techniques for detecting maliciousness in computer systems. The Sect. 2.1 gives insights into static, dynamic, and hybrid analysis of malware detection, while Sect. 2.2 explains the three approaches based on machine learning viz., supervised, unsupervised and hybrid learning followed by literature review in the Sect. 3 of the manuscript. The subsequent Sects. 4 and 5 represents the proposed framework and pseudocode for malware detection, respectively. The paper ends with results and discussion and conclusion in the respective Sects. 6 and 7.

2 Malware Detection Methods

The existing literature over malware detection methods using machine learning techniques can be broadly classified in two ways.

2.1 Feature Detection-Based Method

For training a machine learning model, features from data need to be extracted. Features relating to maliciousness or being benign of software can be extracted using one of the following five methods:

Static Analysis

In static analysis, discriminatory attributes are collected without the execution of code. Various static features like op-codes, sequence of windows API calls, strings, size of program, header information, fingerprinting, packer detection, strings and import functions, etc. are recorded to create feature vector. Similarly in [3] binary files were disassembled to assembly code using Radre2 to generate n-grams features vector using TF-IDF metrics. In [4] Windows PE files were viewed using PEview tool and 78 features were extracted. Out of these 78 features, only 28 relevant features were selected for model training. In [2] n-gram op-codes were extracted from binary code. Out of these features, malware classifier model trained on snort-sub signature features.

Dynamic or Behavior-based Analysis

In dynamic analysis, binary programs are executed in a safe environment to generate behavior reports of malwares. Various dynamic features like API Function calls, their parameters, files created or deleted, website or ports accessed, etc. are recorded to create feature vector for model training. In [5] Windows PE files were executed in Cuckoo sandbox to generate malware behavior report in JSON File. Features representing dynamic behavior of malware were extracted for model training. In [6] binary programs were executed in sandbox for logging the behavior of program. These behavior reports were converted into 1-g feature vector. Sparse vector models were created for each XML report file and Attribute-Relation File Format (ARFF) files were created.

Network Analysis

Malware writers try to evade detection by obfuscation, encryption, packing techniques in static analysis. Similar techniques like code injections, API hook, dynamic link library hook, etc. are used to evade detection in dynamic analysis. The necessity of communication between infected hosts and attackers makes it difficult to evade network-behavior-based methods. In [9] various network-based features were extracted with dynamic analysis and given as input to RNN. The maliciousness of HTTP and HTTPS communications is determined using the malicious URL detection method which is discussed in [10].

Graphics Analysis

In addition to the above ML-based malware analysis methods, there is a completely different approach in which the binary code of executables is reshaped into the matrix and viewed as an image called a malware image. Then using deep learning approach malware files are identified using image recognition or related approaches. Deep neural networks are so trained to identify malware images even with some modifications in the malware image, thus detecting unknown malwares of similar behavior. In [11], author created Gray images [256 X 256] from binary files extracted from dataset and applied CNN to classify malwares. Similarly, in [12] training phase of t-GAN, malware images with slight modifications are generated by the generator module and discriminator to distinguish real data from generated data.

Hybrid Analysis

Both static analysis and dynamic analysis-based methods have their pros and cons. To better utilize these techniques, sometime mixtures of static and dynamic features are extracted for analysis and better accuracy. In [13] both static and dynamic features were extracted from android apk files for proposed DBN-GRU model training. In [10] both static and dynamic features were extracted for malicious URL detection.

2.2 *Machine Learning Methods for Malware Detection*

Detecting malware using machine learning approaches is categorized into the following three methods:

Supervised Learning

In supervised learning, machine learning model is trained on labeled dataset to detect malware. This labeled dataset could be downloaded from online repositories or we can download sample malware files and benign files to create our own labeled data mine based on various feature extractions techniques as discussed above. In [14] and [15] experimentation was done or pre-labeled datasets. However, in other research papers that we surveyed, data mines are created by using feature extraction techniques.

Unsupervized Learning

In this approach clusters of unlabeled malware are formed based on behavior similarities between them. In [16] author extracted malware features from computer registry data and applied K-mean clustering approach to detect malware.

Hybrid Learning

In hybrid learning approach, a mixture of supervised and unsupervised learning methods is used. In [17], author proposed an incremental framework based on clustering and classification techniques to detect malware.

3 Literature Review

This section considers recent submissions on malware detection using machine learning based on statistics features, dynamic features, network-based features, graphic or image-based features, and hybrid approaches for the malware detection (Table 1).

3.1 *Insights Obtained from Literature*

- Although the behavior analysis-based identification of malware is well established and yields good results [10, 19]. But the malicious activities in the network as well as URL-based attacks are left unearthed.
- Applying network security concerns through cloud or handling the attacks using signature-based methods also not fulfilling the quench of the problem. These techniques are time consuming

Table 1 Summary of recent literature

Paper	Description	Methodology	Results	Dataset
Behavior-based MD [8]	MD using five ML classifiers (KNN, NB, J48 DT, SVM, and MLP)	Anubus, XML processing, and ARFF files	J48 DT outperformed	220 malware and 250 benign files
Behavior analysis [17]	Incremental approach of clustering and classification	System calls, Sequential reports and embedded vectors generation	Improved space and time complexity	3131 instances of 'Malheur' dataset
Securing network systems [7]	Malware Analysis using ML Techniques	Feature Vectors, API calls, Address tables and model training	Proposed Method achieved higher accuracy	4500 instances of 'Virus dataset'
Signature generation [6]	Un-supervised approach to generate fixed sized malware signature	Behavior reports to Uni-gram vector, feature extraction and model training	Deep learning yields satisfactory performance	1800 malware of six different categories from C4 Security
Preventing Zero Days attack [18]	Malware detection using hybrid approach	Cuckoo Sandbox, JSON files, features extraction and model training	Random Forest outperformed	3130 windows PE binary files
Network behavior analysis [9]	Communication change and common latent function	Features extracted from communications and training of RNN	Proposed method covers 97.9% URL as compared to baseline	29,562 malware extracted from VirusTotal
Detecting Malware using CNN [11]	MD from malware images	Gray images [256 X 256] from binary files extracted then CNN	CNN outperformed	12,000 'Hauri' and 'Kaist' Cyber Security Research Center'
Analysis of Virtual Memory access pattern [19]	Hardware assisted MD	Virtual Memory accesses used to detect Malware	99% of detection rate achieved	Patterns of memory accesses
MD using t-GAN [12]	Automated classification t-GAN without feature extraction	Transfer Learning applied to GAN to overcome instability	t-GAN performed better than baselines	Kaggle Microsoft Malware

(continued)

Table 1 (continued)

Paper	Description	Methodology	Results	Dataset
MD using ML and DL [1]	Auto-encoders Vector Threshold technique	ASM file, Opcode frequency, dimension reduction, and model training	Random Forest outperformed	55 malware from *'Malicia Project'*
Accuracy comparison of MLT for MD [5]	Rule based, Tree based, and other learning techniques	Binaries, features extraction and model training	Instance based technique outperformed	13,286 malware and benign files from '*VirusTotal*' dataset
n-gram with Snort, feature selection [2]	ML and feature selection techniques	Binaries, n-grams, feature selection and model training	Proposed method shows high accuracy	3622 files 'VXHeavens'
Text analytics for MD [3]	MD using text processing in assembly program	Binaries, tfidf metrics, feature extraction and model training	Proposed model is 81% accurate	5000 files from 'Virus Total'
MD using MLT [4]	Deployed three ML Algorithms	Feature extraction then model training for DT, KNN and SVM	DT outperformed	541 files from '*Virus Share*', '*Virus Total*' and '*VX Heaven*'
Clustering for MD [16]	Behavior analysis and registry data	Binaries, registry features and model training	96.41 detection rate	Malware from www.malware.com
Attribute-based MD from Windows [14]	Malware analysis using AI	Feature selection and model training	Around 5.7% of improvement for Random Forest	57 files from Brazilian malware dataset
MD from Android platform using DL [13]	Deep Belief Network[DBN] and Gate recurrent unit [GRU]	Features extraction and model training	2.4% of improvement in accuracy	7000 files from 'Google Play' and 'APK'
Behavior and attributes based MD from URL's [10]	Malicious URL's detection	URL cleaning, feature extraction and model training	10.28% of FP rate and 92.37% of TP rate detected	4,70,000 URLs from kaggle.com and VirusTotal
Ensemble Learning [15]	MD using Ensemble Learning	Features selection and reduction and model training	Ensemble method outperformed over five baselines	19,611 Windows PE Files 'Kaggle' dataset

- In [1, 2] binary files were disassembled into assembly code (.asm) files using GNU Bin utility package and feature vector created using op-code count. Firstly 1600 unique op-codes were extracted. To overcome the problem of curse of dimensionality, techniques like Variance Threshold, Auto-Encoders (AE-1L, AE-3L) used over these static features.
- In [7] Windows Portable Executables were run and various API Calls were stored in data mine with the help of an import address table. For dynamic analysis in [8] Binary files were submitted to Anubis tool and reports were generated in XML Format.
- Various network-based features extracted include the number of IP addresses corresponding to the domain, countries corresponding to the resolved IP addresses, application protocols such as DNS or HTTP, ratio of the numerical characters to the domain length, status code, file type, depth of path, number of bytes of communication sent and received by malware, the maliciousness of communication as content-based features, etc.

4 Proposed Work

One of the biggest challenges of this work is to acquire relevant datasets for the training of any proposed model. This work devised a proposal where the framework would acquire the data from various sources and would be manually annotated (Fig. 1).

Authors of the malware detection try to evade detection by obfuscation, encryption, packing techniques in static analysis. Similar techniques like code injections, API hook, dynamic link library hook, etc. are used to evade detection in dynamic analysis. The proposed framework considers a diversified set of features that are based on static analysis, dynamic analysis, network-based analysis, graphic or image-based analysis, and URL-based analysis.

As the number of features increases, machine learning models suffer from the curse of dimensionality resulting increase in space and time requirements. The proposed framework adopts various feature selection techniques to keep the total number of features as low as possible without affecting the performance much. In addition to malware detection, proposed framework also incorporates a malicious URL detection module so as to avoid visiting or downloading any malicious file. To improve overall detection accuracy, an ensemble of various machine learning models is considered in the proposed framework.

5 Pseudocode for Proposed Model

The input data was obtained from three well-authentic datasets namely DrebinRed, VirusTotal, and Kaggle. The dataset contains entries for malicious files as well as

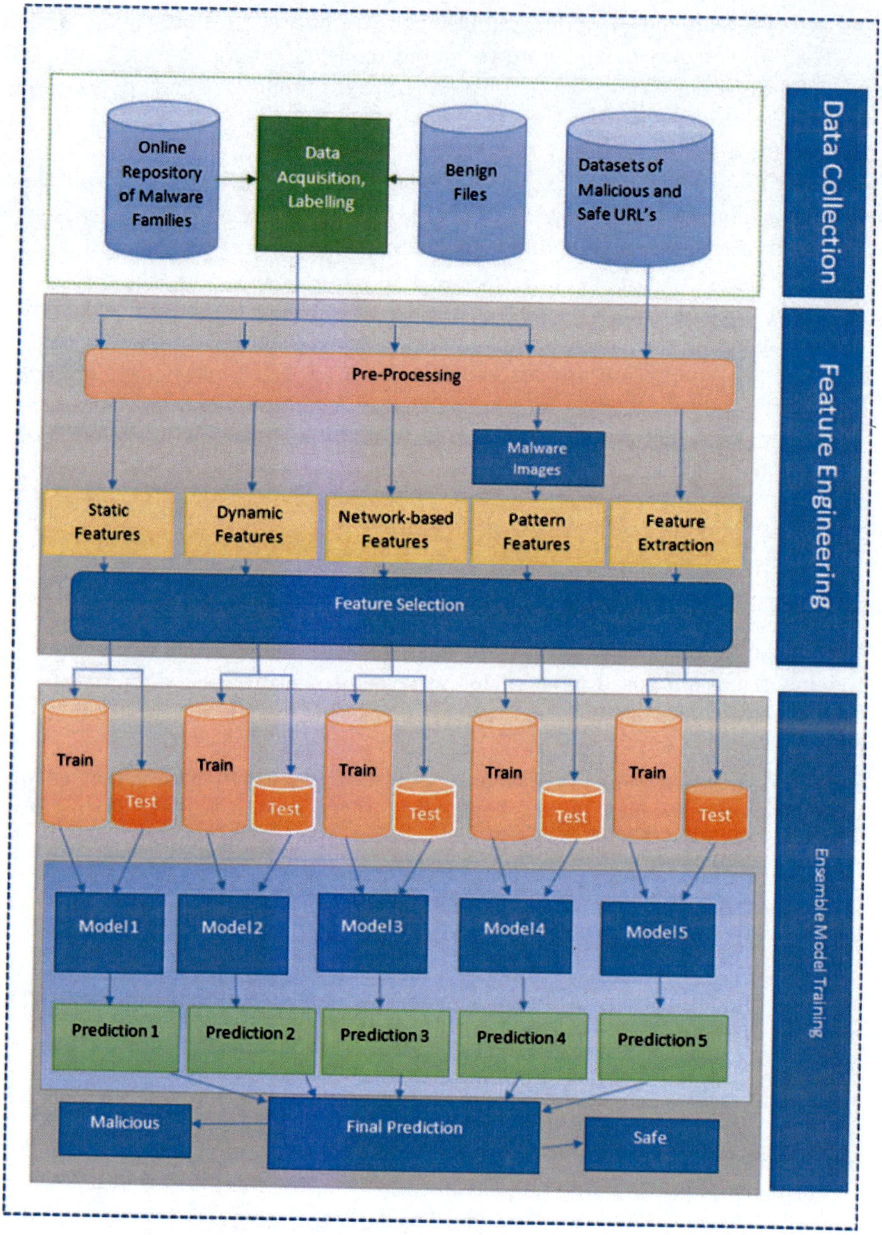

Fig. 1 Proposed framework for malware detection

for benign files. Around 12 thousand benign files and 550 malicious files have been extracted from the datasets. The feature engineering of SECML is applied on datasets to generate features, around 1.2 million features are extracted from the dataset.

Six machine learning classifiers are trained with collected datasets. The performance evaluation of classifiers is done based on detection rate, F1 score and accuracy of classification. The performance of classifiers is compared, where SVM is found most accurate among all the classifiers. Also the detection rates of Linear and RF classifiers are also good.

```
Step-1: Import SECML from gitlab, and RE
Step-2: Import File Management and File downloading utilities from
SECML
Step-3: Download DrebinRed, VirusTotal, and Kaggle datasets from
gitlab.com
Step-4: Obtain Malicious and Benign files using PICKLE utility
Step-5: Extract features from datasets
Step-6: Split the dataset into train and test sets in ratio 80:20
using CTrainTestSplit utility of SECML
Step-7: Set the hyper-parameter (C=0.1, 0.01, and 0.001) for three
runs of the model
Step-8: Train and Test SVM (CClassifierSVM, CClassifierLinear,
CClassifierDecisionTree,   CClassifierKNN,   CClassifierRandom-
Forest, CClassifierDNN of SECML)
Step-9: Performance Evaluation
Step-9.1: Obtain the Detection Rate at FPR= 2% (0.02 FPR threshold)
Step-9.2: Print F1 score
Step-10: GOTO Step-7 for next value of hyper-parameter
Step-11: Plot the graph of Malware Detection rate versus False Posi-
tive Rate
```

6 Results and Discussion

The reason for the good performance of SVM is due to the pre-processing of datasets. The feature extraction as well as the importance of features was done using the feature extraction functions of SECML. Another reason for good performance is due to the threshold and cross-validation of the model. The tenfold cross-validation adds to the model training. Also for two class problems, the performance of SVM is always above other models in each epoch. Table 2 shows the comparison of the performance of all six classifiers.

Figure 2 shows a comparison of malware detection rate versus false positive rate at the threshold of 2%. The false positive rate further rises when the detection rate rises. This shows that more number of samples processed, the more are the chances of getting false positives, which indirectly decreases the accuracy of the classifier. Figure 2 is the average case of proposed methodology when the threshold for the false positive rate is fixed as 0.02, the accuracy of classification is 99.07 and the malware detection rate is 95.04%.

Table 2 Performance evaluation of classifiers

Classifier	Detection rate	F1 Score	Accuracy
Linear	91.54	82.45	80.25
SVM	**95.04**	**89.09**	**99.07**
DT	85.16	80.05	88.25
KNN	78.34	79.15	82.41
RF	92.75	85.27	85.07
DNN	88.45	84.33	91.13

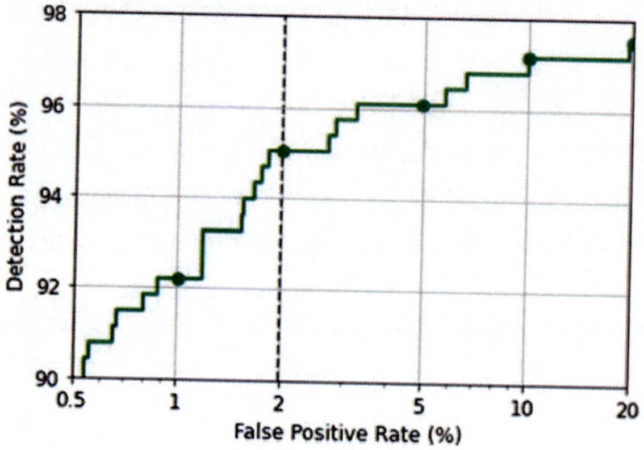

Fig. 2 Malware detection rate versus FPR for proposed methodology

7 Conclusion

The three runs of each of the six classifiers for the different values of hyper-parameter are averaged and the mean results are plotted for malware detection. The SVM classifiers were found most accurate among the six classifiers in terms of accuracy of classification at 99.07%. The trailing classifier was DNN, where the RF was second most effective in terms of detection rate when compared with SVM. The reason for the global maxima of SVM is that the training of the model is done after applying the state-of-the-art feature engineering of SECML library. The proposed method can be extended by considering IOT-based networks to assist smart homes, smart cities, and the smart healthcare industry by securing their systems from malware.

References

1. Rathore H, Agarwal S, Sahay SK, Sewak M (2018) Malware detection using machine learning and deep learning. Lecture notes in computer science (including Subseries. Lecture notes in artificial intelligence. Notes bioinformatics), vol 11297 LNCS, pp 402–411. https://doi.org/10.1007/978-3-030-04780-1_28

2. Khammas B (2018) Malware detection using sub-signatures and machine learning technique. J Inf Secur Res 9(3):96. https://doi.org/10.6025/jisr/2018/9/3/96-106
3. Cunningham E, Boydell O, Doherty C, Roques B, Le Q (2019) Using text classification methods to detect malware. CEUR Workshop Proc 2563:95–103
4. Selamat NS, Ali FHM (2019) Comparison of malware detection techniques using machine learning algorithm. Indones J Electr Eng Comput Sci 16(1):435–440. https://doi.org/10.11591/ijeecs.v16.i1.pp435-440
5. Zhao H, Li M, Wu T, Yang F (2018) Evaluation of supervised machine learning techniques for dynamic malware detection. Int J Comput Intell Syst 11(1):1153–1169. https://doi.org/10.2991/ijcis.11.1.87
6. David OE, Netanyahu NS (2015) DeepSign: deep learning for automatic malware signature generation and classification. In: Proceedings of the international joint conference on neural networks, 2015-Septe, 2015. https://doi.org/10.1109/IJCNN.2015.7280815
7. Singhal P (2012) Malware detection module using machine learning algorithms to assist in centralized security in enterprise networks. Int J Netw Secur Its Appl 4(1):61–67. https://doi.org/10.5121/ijnsa.2012.4106
8. Firdausi I, Lim C, Erwin A, Nugroho AS (2010) Analysis of machine learning techniques used in behavior-based malware detection. In: Proceedings of 2010 2nd international conference on advances in computing, control, and telecommunication technologies ACT 2010, pp 201–203. https://doi.org/10.1109/ACT.2010.33
9. Shibahara T, Yagi T, Akiyama M, Chiba D, Yada T (2016) Efficient dynamic malware analysis based on network behavior using deep learning. In: 2016 IEEE global communications conference GLOBECOM 2016 - Proceedings, 2016. https://doi.org/10.1109/GLOCOM.2016.7841778
10. Do Xuan C, Nguyen HD, Nikolaevich TV (2020) Malicious URL detection based on machine learning. Int J Adv Comput Sci Appl 11(1):148–153. https://doi.org/10.14569/ijacsa.2020.0110119
11. Choi S, Jang S, Kim Y, Kim J (2017) Malware detection using malware image and deep learning. In: 2017 international conference on information and communication technology convergence (ICTC), pp 1193–1195. https://doi.org/10.1109/ICTC.2017.8190895
12. Kim JY, Bu SJ, Cho SB (2017) Malware detection using deep transferred generative adversarial networks. In: Lecture notes in computer science (including Subseries. Lecture notes in artificial intelligence. Notes bioinformatics), vol 10634 LNCS, pp 556–564. https://doi.org/10.1007/978-3-319-70087-8_58
13. Lu T, Du Y, Ouyang L, Chen Q, Wang X (2020) Android malware detection based on a hybrid deep learning model. Secur Commun Netw ii. https://doi.org/10.1155/2020/8863617
14. Kumar A et al (2020) Malware detection using machine learning, pp 61–71
15. Azeez NA, Odufuwa OE, Misra S, Oluranti J, Damaševičius R (2021) Windows PE malware detection using ensemble learning. Informatics 8(1). https://doi.org/10.3390/informatics8010010
16. Rosli NA, Yassin W, Faizal MA, Selamat SR (2019) Clustering analysis for malware behavior detection using registry data. Int J Adv Comput Sci Appl 10(12):93–102. https://doi.org/10.14569/ijacsa.2019.0101213
17. Rieck K, Trinius P, Willems C, Holz T (2011) Automatic analysis of malware behavior using machine learning. J Comput Secur 19(4):639–668. https://doi.org/10.3233/JCS-2010-0410
18. Gandotra E, Bansal D, Sofat S (2016) Zero-day malware detection. In: 2016 sixth international symposium on embedded computing and system design (ISED), pp 171–175. https://doi.org/10.1109/ISED.2016.7977076
19. Xu Z, Ray S, Subramanyan P, Malik S (2017) Malware detection using machine learning based analysis of virtual memory access patterns. In: Proceedings of 2017 design, automation, and test in Europe DATE 2017, pp 169–174. https://doi.org/10.23919/DATE.2017.7926977

Sonar Signal Prediction Using Explainable AI for IoT Environment

Tanisshk Yadav, Parikshit Mahalle, Saurabh Sathe, and Prashant Anerao

Abstract In recent years, we have witnessed rapid advancements in IoT and IIoT industries giving a huge momentum to computational technologies like data analytics, artificial intelligence, visual computing, etc. Data science had its influence in such a powerful way that its application became omnipresent in every category of applied science. With the help of predictions and interpretations made by AI and ML algorithms, there urged a need to verify and explain these predictions. Hence, explainable AI was then introduced in computing which fulfilled this need and gave extra added explanations which helped in explicating the interpretations and outcomes in the simplest possible forms. Considering all this sonar was one of the IoT areas where most problems were arising for future research. In addition, the implementation of AI followed a black box system and from then various attempts are being carried to convert them into a glass box with the help of explainable AI. In this research, we have developed a model that can not only classify the sonar signals but also explain the predictions given by the model. These predictions can then be used for making business decisions and further help the development of industrial IoT. Most of the research and the PhDs are fully fledged towards explainable AI and extend its use in all relevant domains.

Keywords AI in IoT · Sonar radiation · Explainable AI · Prediction · SONAR

T. Yadav (✉) · P. Anerao
Department of Mechanical Engineering, Vishwakarma Institute of Information Technology, Kondhwa-Budruk, Pune, Maharashtra 411048, India
e-mail: tanisshk.21810892@viit.ac.in

P. Anerao
e-mail: prashant.anerao@viit.ac.in

P. Mahalle
Department of Artificial Intelligence and Data Science, Vishwakarma Institute of Information Technology Kondhwa-Budruk, Pune, Maharashtra 411048, India
e-mail: parikshit.mahalle@viit.ac.in

S. Sathe
Department of Software Engineering, San Jose State University, San Jose, CA 95126, USA
e-mail: saurabh.sathe@sjsu.edu

H. O. Bansal et al. (eds.), *Next Generation Systems and Networks*, Lecture Notes in Networks and Systems 641, https://doi.org/10.1007/978-981-99-0483-9_19

1 Introduction

In the past couple of years, one can observe the increasing ubiquity of sensors in IoT platforms for the betterment of industrial solutions. Nowadays, introduction of sonar into sensors for predicting movements, patterns, sea levels, type of material, etc., is rapidly becoming prevalent. Hence, the implementation of AI for detection is becoming a great deal in the IoT sector. This paper has a predictive analysis approach covering the part of XAI, where transparent machine learning models used for predicting sonar signals consolidate new explanation styles and also the strategies to explain black box systems.

Discussions on the applicable areas of XAI and its power in adding explanations and integrating them have been stated by AI practitioners [1]. Likewise, successful use of XAI on specific use cases like solar photovoltaic power [2] and remote mobile robot [3] is achieved, and hence target explanatory observations lead to the elaboration of AI models and forecasts with the help of LIME, SHAP, etc., is demonstrated.

Although, studies have been made on using AI/ML algorithms such as neural networks, target recognition, deep learning, and classifiers for sonar radiation prediction [4, 5], applying XAI techniques is still in undersaturation, to date, it's still a novel approach even for computer sectors. Moreover, classification and neural networks have been immensely practiced for predicting the nature of signals [6, 7]. Hence, the study gives a brief outlook on the execution of explainable AI models like LIME, ELI5, SHAP, and SHAPASH which are then compared to conventional machine learning models used like KNN, logistic regression, random forest classifier, etc.

Glass box terminology w.r.t black box systems [8] and the fairness in deployment, development in explanation of accuracy, and prediction with the help of model-specific and agnostic systems is the main area of interest covered by XAI.

Dataset is typically based on sonar signals of mines and rocks. The first section covers detailed information about the chosen sonar dataset and the preprocessing that generally accompanies it. The next section covers comprehensive information on XAI models and their functionality. The successive sections discuss the approaches and experimental methods incorporated into the research, and the discussions are based on the outcomes of the different algorithms used for comparison. The final section addresses the potential outcome of all the studies and how the study differentiates from the results of previous studies.

2 Motivation

As we can state, IoT implemented in industrial as well as the household has its vast flexibility in accepting multi-disciplinary projects to obtain the wishful result, and all this is carried out by the integration of data analysis, software computing, cloud computing, sensors, and actuators. Even the intractable system can be resolved by proper channeling of IoT in the respective system. Considering the software part coding generally follows it and that's the point of the area where machine learning and artificial intelligence encompassing diverse algorithms and models are put to exercise for simplifying the arduous processes and predictions. Judged by this criterion to achieve more delineation for the gained interpretation more clarification is vital which is carried out by explainable AI methods. As of today, performing XAI is still rare and, therefore, I intend to devote my study to such application of explainable AI process ensemble in the IoT domain.

3 Related Work

A detailed explanation of related work is given below.

Paul Gorman and Sejnowski [9] proposed an analysis of hidden units in a multilayered network. The proposed work classified their experimentation as Aspect angle-dependent and -independent series. A brief comparison was made using the nearest neighbor classifier along with their study outcome and hidden units were discovered. Their research concluded with a result that the performance classifier is good than the nearest neighbor classifier and that it can provide more alternative techniques to existing ML models.

In Özkan et al. [10], classification of sonar signals in reduced sparse forms. They implemented CVANN and CVSC algorithms on the same dataset we chose and compared. The proposed work is a novel method of using a complex-valued wavelet neural network and transferring signals to the DFT domain, hence proving that CVSC and CVWANN used together is a successful method for classifying sonar echo signals.

Kummert [3] gave detailed work on the application of fuzzy algorithms in sonar. Various applications of sonar like propeller shaft blades and target tracking systems were used for applying fuzzy technology. He proposed various analytical and mechanical works like Demon analysis, fuzzification, defuzzification, etc. The overall research was more inclined toward the application of ruled-based fuzzy algorithms on different sonar applications and this study was buoyantly implemented in ATLAS sonar equipment.

Prokopovich and Shafranyuk [11] worked on a model for detecting signals of a passive sonar system. Hydroacoustic systems (HAS) were the core area of focus of this research w.r.t data pre-processing. The study proves that it is easy to stimulate signal marks to debug tracking analysis algorithms and to solve primary data display but they weren't able to cover other possible issues of close angular arrangement, the distribution density function of a stream of false marks, estimation of the consistency due to their wide scope.

Padmaja et al. [12] demonstrated research by applying data mining and machine learning techniques on metal mine detection on underwater sonar images. Implementation of ML techniques like SVM, KNN classifier, SOM, and gradient booster for image processing and ML prediction and also give an in-depth knowledge of the same AI techniques. Finally, they have proved that the gradient booster classifier results in better accuracy in the discrimination and prediction of metal mined from rock.

Arrieta et al. [1] gave a brief insight into the concepts, state of the art, future implications, and all pros and cons of XAI models. Along with that, they demonstrated practically oriented XAI explanations using various techniques and processes.

4 Gap Analysis

Studies have unambiguously shown the usage of ML and AI models in a myriad of segments using proficient tools like natural language processing, deep learning, neural networking, etc. But as they are considered black boxes, their magnitude of interpretability is doubted. To obtain high interpretability, one needs to understand the mathematical significance of underlying respective algorithms as well as the mechanism of the models.

"Explainability" plays a major role, for AI has to deal with the accuracy of the interpretability, whereas explainable AI has to do with justifying the output and hidden nets. Further comes the second crucial aspect of explainable AI which is "Transparency" which measures the degree of transparency of conventional AI which expresses different levels of features and hence the capacity of understanding it may it be algorithmic transparency, decomposability, and visual and local explanations. If to take a deep inside look, there are other esoteric terminologies in XAI like post hoc explainability, model agnostic techniques (LIME, G-REX), and feature and visual explainable techniques (SHAP, SA, ASTRID). To simply sum up, explainable AI is the most sophisticated and advanced method to find succinct explanations of AI systems.

In this paper, we have tried to achieve the same terminology where first conventional models were applied to dataset and then XAI models, to prove how XAI depicts the predictions in terms of transparency, explainability, and interpretability.

5 Proposed Work

5.1 Methodology

The perspective followed in this paper first is the collection of the appropriate data congruous with IoT domains. Accordingly, as sonar radiation is prevailing in applications of IoT and IIoT discipline; therefore, it was subsequently preferred, and the dataset was finalized. Additionally, to achieve some dexterity for analytical and experimental purposes the data was prepared, accompanied by training and feature scaling. The further approach was the building of relevant machine learning models

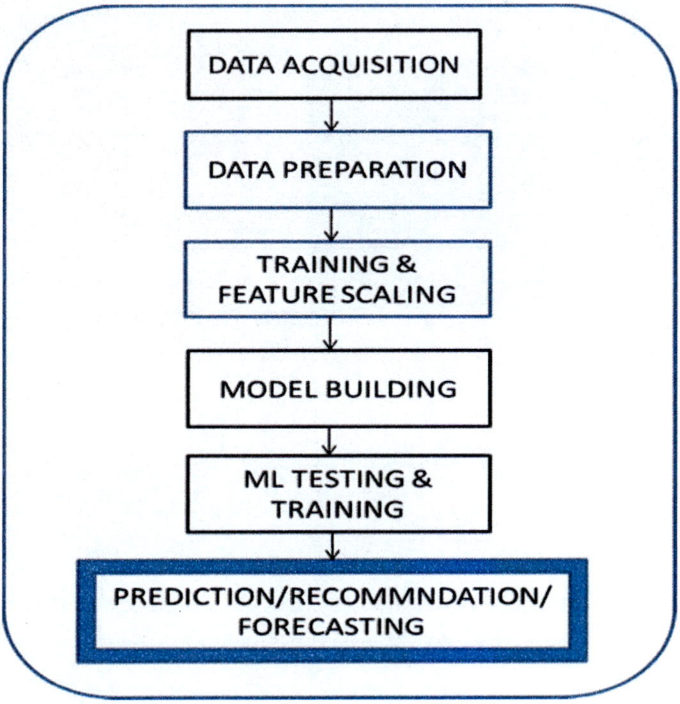

Fig. 1 Flowchart methodology

covering all the essential parts of ML testing and training of data. Finally, perceiving the mandatory outcome can be carried out in the form of prediction, recommendation, or forecasting (Fig. 1).

5.2 *Dataset*

The dataset is referred from the UCI repository. Dataset is a mixture of sonar signals gathered from bouncing signals from surfaces of rocks and mines. A total of 60 columns and 208 indexes are formed from the rebound signals of rocks and mines. Mine's signal contains 111 patterns obtained by bouncing sonar signals off a metal cylinder at various angles and under various conditions. On the other hand, rocks contain 97 patterns obtained from rocks under similar conditions. The imparted sonar signals are frequency-modulated chirps, peaking in frequency. The data contains signals of a diverse of different aspect angles, spanning 90° for the cylinder and 180° for the rock. The pattern is found to be of 60 numbers ranging from 0.0 to 1.0. Each number denotes the energy within a particular frequency, integrated over a certain period (Fig. 2).

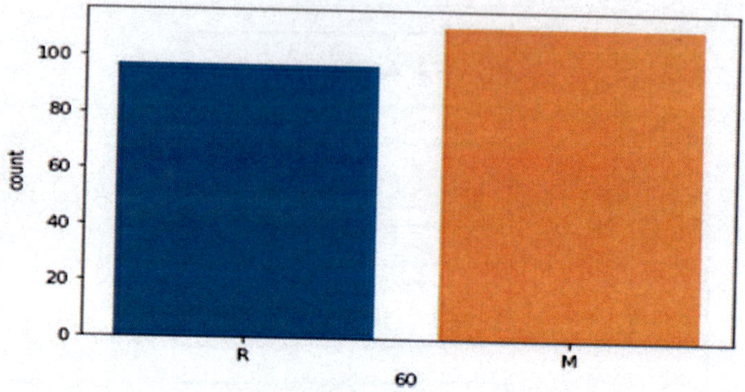

Fig. 2 Graphical representation of dataset

Fig. 3 Correlation heatmap

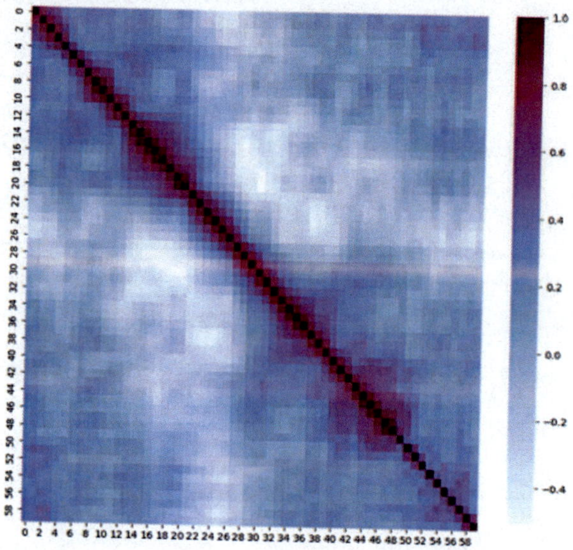

Analysis was done using a histogram count plot and to discover an appropriate heatmap of correlation was generated covering about 58 points on the X- and Y-axis (Fig. 3).

5.3 Implementation of ML Models

Initially, data preprocessing and analysis were done on the chosen dataset. Label encoding was performed on the last column as the last column indicated whether the signal reflected from rock or mine and consisted of "M" and "R". Machine learning models K-nearest neighbor classifier and random forest classifier were then implemented for further research on prediction, and comparison as well as to analyze various merits and drawbacks that occur in the process. Both the models are classification-based models which were chosen to check better efficiency by comparison.

KNN algorithm was preferred because of its easy application and because it is widely used in data classification models. The simplicity and the versatility of KNN to form different clusters of the same features are the main significance of this model. Accordingly, after the data prepossessing, the dataset was split into 90% train and 10% test set. Further, it was scaled for the features of the pipeline and then created the model for the pipeline. The next step was finding the best K parameter with the help of grid search and the respective results were generated of which the mean test score column was plotted denoting the desired result.

Results of the KNN classifier depend on distance metrics which are measured by the Minkowski distance (Fig. 4).

$$Dist(x, z) = \left(\sum_{r=1}^{d} |x_r - z_r|^p\right)^{1/p} \tag{1}$$

Correspondingly, random forest classifier, as mentioned before, was the second classification algorithm chosen to classify this dataset other than K-nearest neighbor. What is remarkable about a random forest classifier is that it can sustain big dimensional spaces along with many training samples. Widely known for its high-level

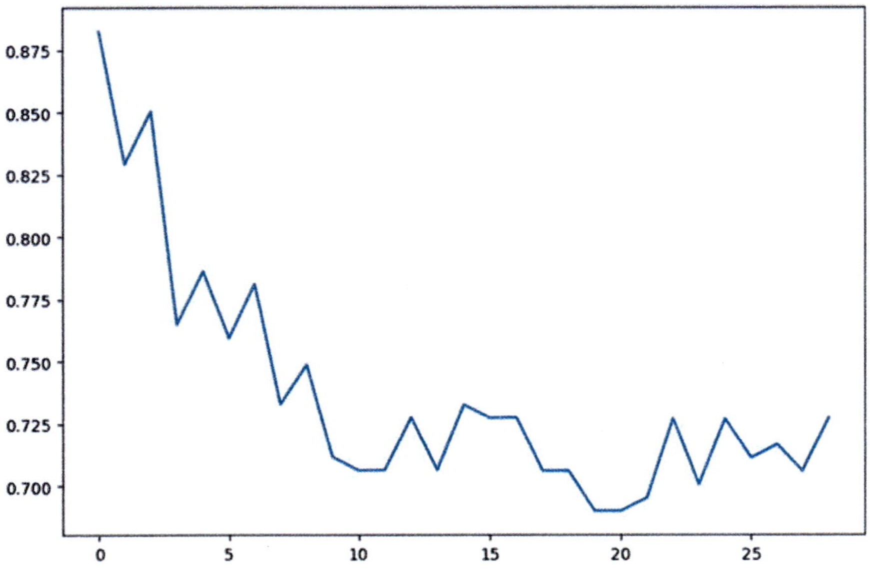

Fig. 4 The plot of mean test score (KNN grid_search_parameter)

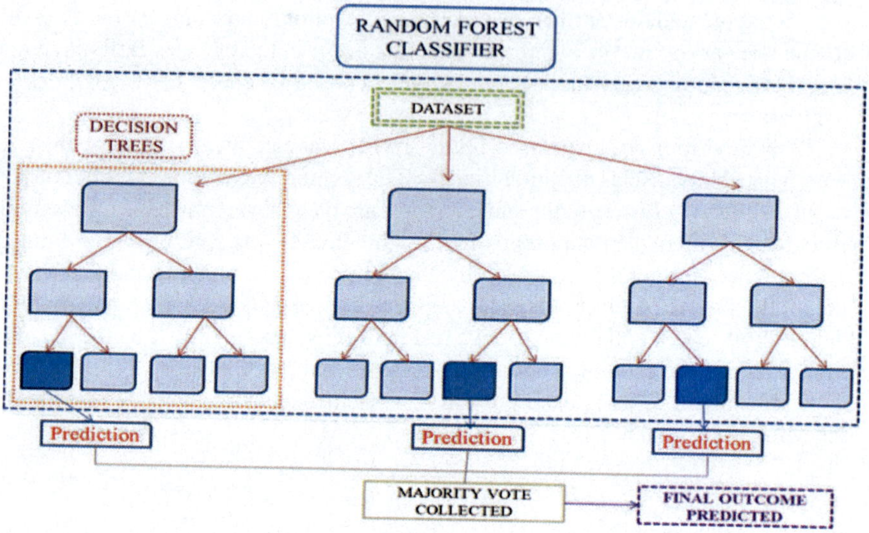

Fig. 5 Basic flowchart of random forest classifier

accuracy predictions than most classification algorithms, the key point here to note is that it yields more rational predictions without hyper-parameter tuning.

Mention should also be made of its working principle and outcomes based on many decision trees and bagging (Meta algorithm).

The main algorithm of random forest is the Gini Index which helps in the decision of decision trees' nodes (Fig. 5).

$$Gini = 1 - \sum_{i=1}^{c} (p_i)^2 \tag{2}$$

Second comes the entropy that fixes the node branching

$$Entropy = \sum_{i=1}^{c} -p_i * \log_2(p_i) \tag{3}$$

So, its use in the code was no doubt propitious, this also takes to the heart of the matter that the number of meta estimators (estimators) given was 50 which plays a major role in building the decision trees and fitting them on subsets of the main dataset before even making a prediction.

5.4 Incorporation of Explainable Model

In order to solve the black box systems adhered to artificial intelligence, the use of explainable AI was introduced to give added and also to corroborate unprecedented predictions and results in coding. So, models like LIME which is an acronym for local interpretable model-agnostic explanations, SHAP acronym for Shapley Additive explanations, ELI5, SHAPASH, etc., are some of the top-notch explainable AI models. It also should be emphasized that the XAI models not only endorse interpretable and transparent ML algorithms but also provides unique strategies for explaining black box systems. The use of random forest is explicitly applied in this research because there is no denying that it is a black box system due to a large number of decision trees.

5.5 ELI5

ELI5 is used for debugging various ML models and algorithms and thus gives elucidation for their prediction. The first and foremost thing here to initiate any explainable models is to install them from Python packages. Judging by the ELI5 library and its working, it's mainly used for unraveling XGBoost, regressors and classifiers, Keras, CatBoost, etc. Therefore, the last section of the study is the implementation of the ELI5 model which gives an absolute explanation of our outcome predictions.

5.6 SHAPASH

SHAPASH has been found to be the most effective XAI model for the best explanation of interpretations proving the superfluity of modeling algorithms like RF, SVM, etc. [9]. Expressing its outcomes in graphical representation gives the best visual explanations over resulted outcomes making explanations more comprehensive. Furthermore, the SHAPASH toolbox helps in detecting the main traits of the trained model and is considered one of the user-friendly XAI models [11].

6 Result and Discussion

Now that all the intended machine learning models, i.e., K-nearsest neighbor and random forest are applied to the dataset and are used to predict whether the signal is from rock or mine, appropriate results are obtained.

The results are represented in tabular form (Tables 1 and 2).

Table 1 Tabular representation of KNN report

	Precision	Recall	f1-score	Support
0	0.9	0.82	0.86	11
1	0.82	0.9	0.86	10
Accuracy			0.86	21
Macro avg	0.86	0.86	0.86	21
Weighted avg	0.86	0.86	0.86	21

Table 2 Tabular representation of random forest classification report

	Precision	Recall	f1-score
0	1	0.91	0.95
1	0.91	1	0.95
Accuracy			0.95
Macro avg	0.95	0.95	0.95
Weighted avg	0.95	0.95	0.95

The first thing we notice is that the precision of the random forest classifier is greater than KNN, this can be considered even if we compare the precision of macro and the weighted average for both models. Mention should also be made of the accuracy which gives the best explanation between the competencies for these two models, that the accuracy of random forest is greater by 0.09 units than KNN. All things are considered then. Random forest proves a better prediction model for this dataset and further explanation is supported by the ELI5 method. Hence, the two tables give a clear picture also on the efficiency of the application of these two models.

For interpreting our results, we have used the ELI5 and SHAPASH library.

For ELI5 the resulting tables denote that the 10th column has the highest weightage for prediction giving an outcome of value 1.

The following figures show the results of ELI5 (Figs. 6 and 7).

For SHAPASH, there are two explanations that we have provided for the model we have developed. The global importance explains overall, which were the columns that our model considered the most in making contributions or which were the columns that contributed most to the predictions.

The following graph shows the explanations provided by our explainable AI module (Fig. 8).

From the graph, it can be observed that the factor that contributed most to the predictions was column 10 followed by columns 48, 11 and so on.

The following graph shows the factors that contributed most to a specific row/instance (Fig. 9).

We observed that the final response obtained was 1. The factors those contributed most to this outcome were 10, 11, 48, 47, and 9. From this, we can understand that to get an outcome of 1 having values as the columns 10, 11, 9 for this instance, we

Fig. 6 Weightage outcome of ELI5

Weight	Feature
0.0663 ± 0.2036	x10
0.0586 ± 0.1958	x11
0.0413 ± 0.1468	x48
0.0331 ± 0.1338	x47
0.0329 ± 0.1182	x8
0.0299 ± 0.1207	x9
0.0284 ± 0.0976	x35
0.0262 ± 0.1114	x12
0.0247 ± 0.0913	x51
0.0236 ± 0.0924	x50
0.0226 ± 0.0902	x44
0.0224 ± 0.0790	x27
0.0209 ± 0.0810	x36
0.0189 ± 0.0776	x3
0.0187 ± 0.0809	x46
0.0183 ± 0.0809	x45
0.0174 ± 0.0704	x42

Fig. 7 Highest weightage outcome

y=0 (probability **0.690**) top features

Contribution?	Feature	Value
+0.535	<BIAS>	1.000
+0.042	x10	0.280
+0.036	x11	0.332
+0.020	x3	0.102
+0.019	x22	0.909
+0.019	x12	0.401
+0.019	x51	0.012
+0.018	x42	0.298
+0.018	x30	0.314
+0.015	x53	0.024
+0.014	x50	0.014
+0.011	x19	0.917
+0.011	x21	0.997
+0.009	x0	0.045
+0.009	x5	0.149
+0.008	x38	0.322
+0.008	x4	0.140
+0.007	x18	0.919
+0.006	x29	0.378

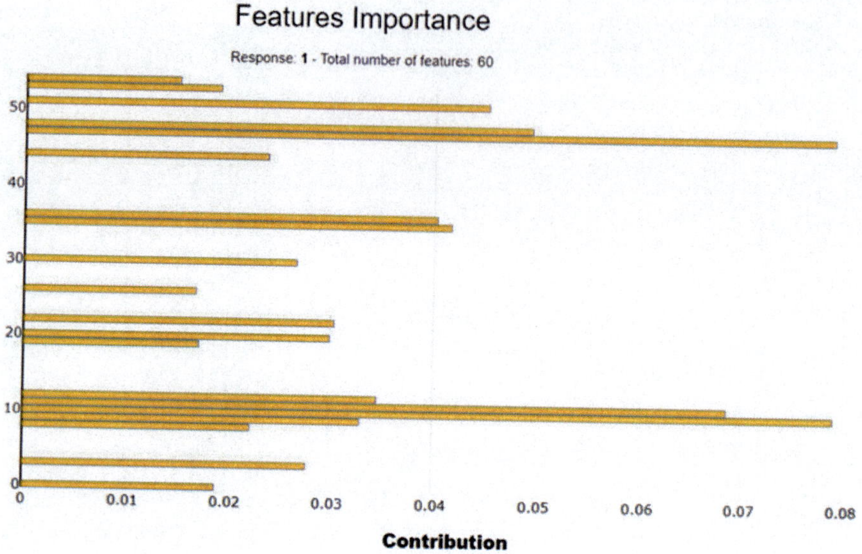

Fig. 8 SHAPASH global explanation: feature importance

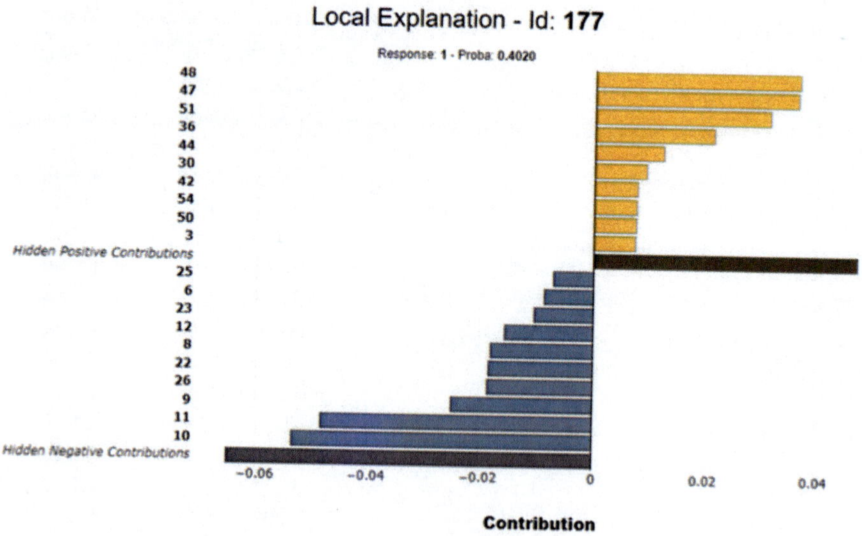

Fig. 9 SHAPASH local explanation graphical representation

can have values similar to values in column 9 and not like columns 48, 47, or 51. This is extremely important as it can help to distinguish between 2 similar instances. Strong rules and filters can also be derived to develop high-level business decisions.

7 Conclusions and Future Work

The study gives a detailed demonstration of the application of K-nearest neighbor, random forest classifier, and ELI5 to predict and explain sonar radiation signals. The final analysis proves that random forest is a better model than KNN and ELI5 was successfully implemented on the sonar dataset. Considering the results, it seems clear that it has thoroughly delineated the predictions of the random forest model and hence, eased its understanding of interpretation. Therefore, now it does not seem hard to question its explainability and transparency and makes it difficult to escape the conclusion that it provides ample information on the characteristics and attributes of the random forest model. The study is totally in favor of researchers who are working on Sonar IoT platforms.

The research provides a very productive approach to implementing XAI for sonar datasets and hence can be a further motivation toward the use of XAI in different IoT domains creating a great level of advanced explainability along with pertinent research. More studies already give a brief picture of its extrapolation in medical fields specifically in cancer [12]. This is not to say that they are restricted only to science and technology but they can yield influence in nonscientific domains such as musicology, statistics, marketing, etc. We can evidence its gradual motivation in image processing, visual computing, and data analytics. To sum up, XAI has great potential to explain diverse black box systems.

References

1. Arrieta AB, Díaz-Rodríguez N, Del Ser J, Bennetot A, Tabik S, Barbado A, Garcia S, Gil-Lopez S, Molina D, Benjamins R, Chatila R, Herrera F (2019) Explainable artificial intelligence (XAI): concepts, taxonomies, opportunities and challenges toward responsible AI, information fusion. https://doi.org/10.1016/j.inffus.2019.12.012
2. Saboni A, Ridha Ouamane M, Bennis O, Kratz F (2022) Model reports, a supervision tool for machine learning engineers and users. Int J Educ Inf Technol 16
3. Kummert A (1993) Fuzzy technology implemented in sonar systems. IEEE J Ocean Eng 18(4)
4. Rai A (2020) Explainable AI: from black box to glass box. J Acad Mark Sci 48:137–141
5. Neupane D, Seok J (2020) A review on deep learning-based approaches for automatic sonar target recognition. Electronics 9(11). https://doi.org/10.3390/electronics9111972
6. Tjoa E, Guan C, (2015) Fellow. a survey on explainable artificial intelligence (XAI): towards Medical XAI. J Latex Cl Files 14(8)
7. Emmert-Streib F, Yli-Harja O, Dehmer M (2020) Explainable artificial intelligence and machine learning: a reality rooted perspective. WIREs Data Mining Knowl Discov e1368
8. Ghosh I, Sanyal MK (2021) Introspecting predictability of market fear in Indian context during COVID-19 pandemic: an integrated approach of applied predictive modelling and explainable AI. Int J Inf Manag Data Insights 1(2):100039
9. Paul Gorman R, Sejnowski TJ: Analysis of hidden units in a layered network trained to classify sonar targets. Neural Netw 1:75–89
10. Özkan Bakbak P, Peker M (2018) Classification of sonar echo signals in their reduced sparse form using complex-valued wavelet neural network. Neural Comput Appl https://doi.org/10.1007/s00521-018-3920-4 (0123456789)

11. Prokopovich VV, Shafranyuk AV (2021) Model of signal mark detection in a passive sonar system. IOP Conf Ser : Mater Sci Eng 1215:012009
12. Padmaja V, Rajedran V, VijayaLakshmi P (2020) Study on metal mine detection from underwater sonar images using data mining and machine learning techniques. J Ambient Intell HumIzed Comput. https://doi.org/10.1007/s12652-020-01958-4

Necessity of Self-explainability in AI: Insights from the Existing Social Projects

Mayur K. Jadhav, Shilpa S. Laddha, and Sanchit M. Kabra

Abstract Self-explanatory artificial intelligence has lately gained prominence as researchers and practitioners strive to make their methodologies more understandable. It is not contentious to assert that researching how individuals explain to one another may be a useful beginning point for explanation in artificial intelligence, given that the majority of this research has focused on publicly explaining decisions or actions to a human observer. However, it is reasonable to say that the vast bulk of self-explanatory artificial intelligence research is based solely on the researchers' subjective opinions of what constitutes a "good" explanation. Extensive and essential study in science and technology has been conducted on the definition, creation, selection, assessment, and presentation of explanations by humans. This study contends that people use certain cognitive biases and social expectations in the explanation process. This paper covers relevant studies from several fields and makes the case that self-explainable artificial intelligence should expand on some current research. It highlights several key results and explores how they may be combined with research on self-explainable artificial intelligence.

Keywords Explanation · Explainability · Interpretability · Self-éxplainable AI · Transparency

M. K. Jadhav (✉)
Ph.D. Student, Department of Computer Science Engineering, Government College of Engineering, Chh. Sambhajinagar, Maharashtra, India
e-mail: mj8055x@gmail.com

S. S. Laddha
Head of Department, Information Technology, Government College of Engineering, Chh. Sambhajinagar, Maharashtra, India

S. M. Kabra
Computer Science Engineering, BITs, Pilani, India

H. O. Bansal et al. (eds.), *Next Generation Systems and Networks*, Lecture Notes in Networks and Systems 641, https://doi.org/10.1007/978-981-99-0483-9_20

1 Introduction

The phrase "Self-Éxplainable AI" in this article broadly refers to an explanatory agent that identifies underlying factors in its own or another agent's decision-making. It's crucial to say that adding more AI is not the answer to self-explainable AI. The issue is ultimately one of human–agent interaction. A definition of human–agent interaction as the nexus of social behavior system, artificial intelligence, and human–computer interaction (HCI). One issue in human–agent interaction is self-explainable AI. The top circle in Fig. 1—representing philosophy, social and cognitive psychology, and cognitive science—and its relationship to the bottom two circles—their influence on the creation of artificial intelligence and our interactions with it—are the subjects of this article.

This article's scope is divided into three areas in light of the range of self-explainable AI:

- Survey: To gather information and examine pertinent explanation-related publications, with a focus on "everyday" explanation.
- Everyday explanation: Unlike scientific explanations, which focus on more broad relationships, "everyday" explanations explain why specific facts (events, qualities, decisions, etc.) occurred. We support this perspective by noting that, rather than attempting to comprehend and develop broad theories, confidence is lost when users are unable to understand the traces of observed behaviour or decisions [3, 4].
- Relationship to Self-Éxplainable AI: To highlight key ideas from papers that address some of the several self-explainable AI subfields

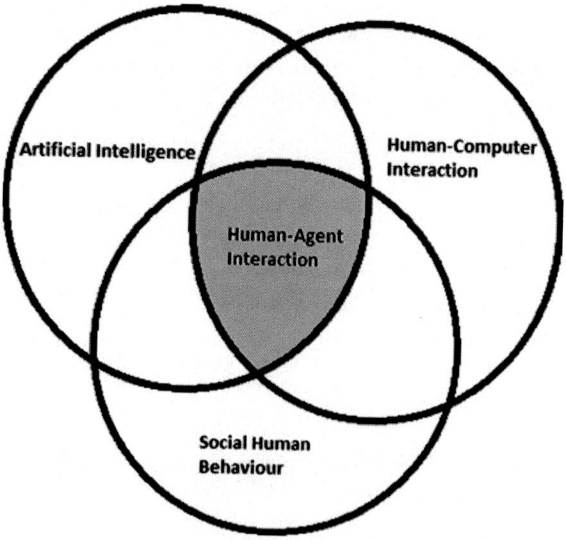

Fig. 1 Applicability of explicit self-éxplainable AI

In this study, I focus on four key results from the reviewed literature that, in my opinion, are crucial for self-explainable AI but of which the majority of researchers and practitioners are now ignorant:

1. Explanations are contrastive in nature; they are sought in response to certain counterfactual instances, known in this work as foils. That is, instead of asking why event P occurred, people tend to question why it occurred as opposed to another event Q. For self-explainable AI, this has significant societal and computational ramifications.
2. People hardly, if ever, expect an explanation to include a real and comprehensive cause of an occurrence; instead, explanations are chosen (biasedly). Humans are skilled at picking one or two reasons out of an occasionally endless number of causes to serve as the explanation. However, some cognitive biases have an impact on this choice.
3. While probabilities do matter because truth and likelihood are crucial in explanation, appealing to probabilities or statistical correlations in explanation is less successful than referring to causes. The best answer for a person is not always the most likely one, and it's also crucial to note that adopting statistical generalisation to explain why things happen is ineffective unless they are supported by a causal justification for the generalisations altogether.
4. The presentation of explanations is based on the explainer's assumptions about the explainee's ideas since they are social in nature. They are knowledge transfers that take place during conversations or interactions.

These four ideas come together to form a unified idea: explanations are contextual and include more than merely listing relationships and causes (causal attribution). While there may be numerous causes for an event, the explainee frequently only cares about a small subset of them that are pertinent to the context, the explainer chooses a subset of these causes based on a variety of criteria, and the explainer and explainee may discuss and disagree about this explanation. These three criteria, in my opinion, are crucial in many applications if we are to create really self-explainable AI, particularly intelligent systems that are able to provide explanations.

2 What is Explanation?

There are numerous related ideas in explanation that frequently show some term conflation and appear to be used interchangeably by writers and authors within articles. This section explains the differences between causal attribution and causal explanation in specific. Additionally, the distinction between explanation and interpretability will be briefly discussed.

2.1 *Defining <w.R.T.>*

2.1.1 Causality

There are several hypotheses for what causes an event or characteristic, and the idea of causality has received much attention. The two basic categories into which the various definitions of causality can be grouped are dependence theories and transference theories. Causality and different circumstances The regularity theory of causation is said to have been developed by Hume [5, Section VII]. According to this theory, if one type of occurrence is always followed by another, there is a link between the two. According to Lewis [6], the Humeian idea is about counterfactuals rather than dependence. According to Hume, the co-occurrence of events C and E as perceived does not convey important causal information. Instead, the cause should be construed in connection to a hypothetical, counterfactual situation: event C is said to have caused event E if, in some imaginary counterfactual scenario, E would not have happened if C had not occurred. Many definitions of causality are based on this notion in some form; for arguments in favor of and refinements of this definition, see Lewis [6] and Hilton [7].

2.1.2 Causal Chains

In the study of cause and explanation, the concept of causal chains is critical. A causal chain is a series of events connected by a chain of causes, where an association between an event C and an event E means that C must happen before. Root causes are any unconnected events. Hilton et al. [8] outline the five different types of causal chains in the table below, emphasizing that each type of causal chain relates to a certain sort of explanation (Table 1).

People can provide an adequate explanation even if they are unaware of the complete causal chain. Even while it is undeniably true that the reasons for physical occurrences may be traced back to events that occurred during the Big Bang, most adults can explain to a small child why a bouncing ball finally ceases.

2.2 *Explanation*

An explanation, according to Josephson and Josephson [9], is an attribution of causal responsibility. Lombrozo [10] defines explanation as both a process and a final product. However, I suggest that the explanation and outcome are actually the results of two processes:

Table 1 Types of causal chains according to Hilton et al. [8]

Type	Description	Example
Temporal	Proximal events are not constrained by distal occurrences. Events in time can be swapped without affecting the result.	A and B together cause C; the sequence of A and B is immaterial; for example, two persons flipping a coin each win if both coins are heads; it makes no difference who flips first.
Coincidental	Proximal events are not constrained by distal occurrences. In a specific situation, the causal links remain true, but not in general.	This time, A causes B, but the general link does not hold; for example, a person smoking a cigarette causes a house fire, but this does not happen very often.
Unfolding	Proximal events are tightly constrained by distal occurrences. The causal links are constant in general and in this specific situation, and they cannot be altered.	A causes B, and B causes C; for example, turning on a light switch causes an electric current to flow to the light, which causes the light to illuminate.
Opportunity Chains	The proximal event is enabled by the distal event.	A enables B, and B causes C; for example, installing a light switch allows it to be flicked, causing the light to turn on.
Pre-emptive	Distal comes before proximal, preventing the proximal from triggering an event.	If B did not occur, A would have caused C; for example, opening the car with my remote lock would have unlocked the door if my wife had not previously unlocked it with the key.

2.2.1 Cognitive Process

The process of "filling the gaps" [11] in order to select an explanandum, or explanation, for a certain occurrence. The reasons for the occurrence are discovered in this procedure, sometimes in reference to a specific counterfactual circumstance, and a subset of these causes is selected as the explanation (or explanans). The act of identifying the causes of a given phenomenon is known as attribution in human social behavior, and it is simply one stage in the whole explanation process.

2.2.2 Product

The output of the cognitive explanation process is the explanation that comes about as a result of the cognitive process.

2.2.3 Social Process

The process of information transmission between explanation and explainee, often including a group of individuals, with the purpose of supplying the explainee with enough knowledge to grasp the reasons of the event, however, other sorts of goals can be set, as we shall describe later.

2.3 *Explanation as a Product*

An explanation, according to the definition given in [12–15], is an answer to a why-question. According to Bromberger [16], a why-question combines a whether-question with the word "why" before it. A whether-question is an interrogative enquiry that requires a 'yes' or "no" response. The presupposition in a why-question is a fact referenced in the question being discussed and phrased as if it were true (or false if the question is a negative sentence). The query "why did they do that?" is an example of a why-question, with the assumption "they did that" and the inner whether-question "did they do that?" Why-questions are interesting, but they are more difficult to answer structurally.

Explanations, on the other hand, can give solutions to other types of questions. Table 2 I presents a simple model for explanatory questions based on Pearl and Mackenzie's Causation Ladder [17]. In this concept, explanatory questions are classified into three types: What-questions (such as "What happened? "), how-questions (such as "How did that happen? "), and why-questions (such as "Why did that happen?") are the three categories of inquiries. Why-questions are the most difficult to answer since they need the most intricate types of reasoning. What-questions necessitate factual explanations, and associative reasoning may be used to determine which unseen happenings also happened in light of the seen events. Questions concerning how something happened are also true, but they need interventionist thinking to discover the variables that, if removed, may prevent the occurrence from recurring. Associative thinking may also be required in this case.

Table 2 Classes of explanatory question and the reasoning required to answer

Question	Reasoning	Description
What?	Associative	Determine which unseen occurrences may have occurred in light of the observed events
How?	Interventionist	Simulate a change in the situation to check if the occurrence continues to occur
Why?	Counterfactual	Simulating different reasons to determine whether the event still occurs

2.4 Abductive Reasoning as Explanation

As a cognitive process, explanation is inextricably linked to abductive reasoning. Peirce [17] first presented the notion of abduction as a unique method of thinking, distinct from induction and deduction yet proceeding from effect to cause like induction. His studies focused on the contrast between developing a theory to explain an observed event and inducing it via scientific experiments (abduction). The abduction-related inference is defined by him as follows:

The surprising fact, C, is observed;
But if A were true, C would be a matter of course,
Hence, there is reason to suspect that A is true.

This is definitely an inference, not a deduction or induction, used to explain reality C from premise A. This, however, does not take contradictory hypotheses into account. This more forceful type of kidnapping looks like this, according to Josephson and Josephson [9]:

D is a collection of data (facts, observations, givens).
H explains D (would, if true, explain D).
No other hypothesis can explain D as well as H does.
Therefore, H is probably true.

Explanation and abductive reasoning, according to research in philosophy and cognitive science, go hand in hand. People use abductive inference to choose what they think to be the "best" explanation, especially when attempting to understand the causes of events.

2.5 Justification and Interpretability

In this section, we will quickly go over the distinctions between justification, explanation, explainability, and interpretability as they are used in this article and appear to be employed in artificial intelligence. Lipton [18] provides a taxonomy of the needs and approaches for interpretable AI. This study accepts Lipton's contention that the explanation is post hoc interpretability. According to Biran and Cotton's [9] definition, interpretability of a model is the amount to which an observer can understand the reasons behind a choice. Thus, explanation is one method by which an observer may gain information, but there are surely other methods, such as making judgements that are already easy to understand or introspection. In my perspective, interpretability and explainability are equivalent.

A justification explains why a conclusion is sound but does not always seek to illustrate the decision-making process [9]. It is critical to understand the similarities and differences between these concepts as one reads this page because certain related research is only useful for explanation.

2.6 *Why Do People Demand Explanations?*

For a number of reasons, people may seek answers. Curiosity is one of the most important criteria for humans, but there are also more practical ones, such as examination—for example, when a teacher asks her students to explain something on an exam to assess their knowledge of a particular subject—and scientific explanation—asking why we observe a particular environmental phenomenon. Because we are interested in explanations in AI, our focus in this work is on how intelligent beings could rationalize their actions. As a result, this section concentrates on why people seek "everyday" explanations for why particular events occur rather than explanations for larger scientific phenomena, however, this work is frequently applicable.

It is apparent that the primary purpose of explanation is to facilitate learning [7, 10]. We can benefit from these enhanced models because they help us understand how certain occurrences or features occur. Individuals seek explanations to enhance their understanding of someone or something, according to Heider [21], in order to construct a credible model that can be utilized for prediction and control. This argument is reinforced by studies indicating that people prefer to question about happenings or observations that are new or surprising to them [22–24].

3 Social Attribution: What Explanations Are Used for Behaviour?

We assume that analogous linkages exist between the elements of the social environment, just as precise legal relationships, causal or otherwise, connect the contents of the nonsocial environment, determining what may or will happen. Page 21 of Heider's Chap. 2 (21) This section discusses the study of social attribution, which analyses how individuals assign and (partially) explain the behavior of others. Such study is undeniably significant in a variety of AI domains.

Perception is important in social attribution. While the neuro-physical level and maybe lower levels can be used to establish the reasons for behavior, social attribution is primarily concerned with how others attribute or explain the behavior of others rather than the real causes of human behavior. Heider [21] defines social attribution as "person perception."

4 Cognitive Processes—How Do People Choose and Evaluate Explanations?

The number of explanations and causes for x is equal. Consider how various parties could have described the cause of death: the doctor could have said "multiple haemorrhages," the lawyer could have said "driver negligence," the carriage-builder could have said "a defect in the brake lock construction," and a civic planner could have said "the presence of tall shrubbery at that turning." While all of the explanations are legitimate, some are more relevant than others given the topic's details. [25, p. 54] Hanson.

Mill [26] developed the Method of Difference, which he labelled the use of "statistical" correlations to infer cause in one of the earliest examinations of cause and explanation. He did, however, provide a list of cognitive biases that individuals appear to exploit, such as unexpected situations, precipitating reasons, and unpredictability. He maintained that causal linkage and explanation selection are ultimately arbitrary and that choosing one explanation over another is "wrong" from a scientific and philosophical standpoint. In causal attribution, such covariation model principles predominated, particularly in Kelley's work [27]. However, some studies discovered that covariation models could not account for a wide range of events, despite the fact that humans may deduce the reasons of several occurrences from a single piece of evidence [28, 29].

We investigate three types of cognitive processes used in explanation: (1) causal connection, which is the process by which people determine the causes of events; (2) explanation selection, which is the process by which people select a specific subset of the identified causes as the explanation; and (3) explanation evaluation, which is the process by which an explainee determines the value of an explanation. The majority of the research shows that people employ cognitive biases in the generation, selection, and evaluation of explanations.

5 Sufficiency, Necessity, and Robustness

According to the authors [15, 30, 31], sufficiency and necessity are crucial criteria for selecting explanatory explanations. Required justifications, according to Lipton [15], are preferable to acceptable ones. Consider a beetle with DNA defects that, when kept at appropriate temperatures, allow its wings to grow longer than normal. Consider that there are two such mutations, M1 and M2, and either one can result in the mutation. In contrast to a beetle whose wings would not change, the temperature explanation is better since neither M1 nor M2 are separately necessary for the observed occurrence; just that either M1 or M2. In contrast, even when we are aware that M1 was the cause, the temperature is still required and desirable.

Sufficiency, according to Woodward [31], is another key criterion since individuals pick reasons that have an influence on their own without the need for additional causes. This differs from sufficiency in the prior situation, when any mutation M1 or M2 together with temperature is sufficient. In contrast to circumstances in which there are several adequate causes, Woodward's assertion only applies to unilaterally sufficient causes. For example, if it was discovered that the third mutation, M3, could result in larger wings independent of temperature, it would be desirable over temperature plus another mutation. This is related to the ease of notation.

According to many authors [30, 31], the degree to which a cause C is considered resilient is evaluated by whether the effect E would still have occurred if variables other than C were slightly adjusted. As a result, a cause C1 that only applies in particular situations has less explanatory value than a cause C2 that applies in a wide range of situations.

A shared meaning must be created before AI can explain something. This is frequently the goal of an explanation provided by an intelligent agent: to at least partially develop a shared understanding of the choice made by the agent and the observer.

According to Lombrozo [10] and Wilkenfeld and Lombrozo [32], explanations can serve a range of reasons in addition to delivering knowledge, such as assigning blame or persuading, and the intentions of the explainer and explainee may differ in some social explanation scenarios. Persuasion is undeniably intriguing in terms of AI explanation: Persuasion that a choice is correct may be judged more important than presenting the genuine cause in some situations if the goal of an explanation from an intelligent entity is to acquire the observer's trust. If we want someone to do something, it may be better to present a less reasonable explanation that is more persuasive to the explainee. In this instance, the explainer's objectives (to build confidence) diverge with those of the explainee (to understand a decision).

6 Open Questions

Can this method be used to simplify all topics that require explanation? Can we give a theoretical underpinning for it, if so? Can we quantify and minimise the explanation-performance trade-off? How explicit should the reasons be? What are the best methods to describe things?

The area professionals who created the AI algorithms will need to collaborate with us to discover solutions to these problems. One thing is certain: the "one model fits all concerns" approach to AI will no longer work in the future, as consumers and governments demand greater transparency.

7 Conclusion

We have studied and argued that self-explainable AI can benefit from existing models of human definition, creation, selection, presentation, and assessment of explanations. We have examined some of the most relevant and noteworthy sociological studies on human explanation, and we have provided some insight into how this work may be implemented in self-explanatory AI. New self-explainable AI models should take into account the four major conclusions presented in this review paper: Why-questions are contrastive in character, explanations are biassed in their selection, explanations are social in nature, and probabilities do not carry the same weight as causal linkages. We recognize that these notions will not be applicable in all applications, but in many cases, they have the potential to improve explanatory agents.

References

1. Miller T Explanation in artificial intelligence: insights from the social sciences. School of computing and information systems, University of Melbourne, Australia. J Artif Intell Print ISSN: 0004-3702. https://doi.org/10.1016/j.artint.2018.07.007
2. Jadhav M, Chakkarwar V, Tamane S (2021) Automatic riots detection system using deep learning. In: Recent trends in image processing and pattern recognition. Springer
3. Stubbs K, Hinds P, Wettergreen D (2007) Autonomy and common ground in human-robot interaction: a field study. IEEE Intell Syst 22(2):42–50
4. Mercado JE, Rupp MA, Chen JY, Barnes MJ, Barber D, Procci K (2016) Intelligent agent transparency in human–agent teaming for Multi-UxV management. Hum Factors 58(3):401–415
5. Hume D (2000) An enquiry concerning human understanding: a critical edition, vol 3. Oxford University Press
6. Lewis D (1974) Causation. J Philosophy 70(17):556–567
7. Hilton DJ Logic and causal attribution. In: Contemporary science and natural explanation: commonsense conceptions of causality. New York University Press, pp 33–65, 168
8. Hilton DJ, McClure JL, Slugoski Ben R (2005) The course of events: counterfactuals, causal sequences and explanation. In: Mandel DR, Hilton DJ, Catellani P (eds)c The psychology of counterfactual thinking, Routledge
9. Josephson JR, Josephson SG (1996) Abductive inference: computation, philosophy, technology. Cambridge University Press
10. Lombrozo T (2006) The structure and function of explanations. Trends Cognitive Sci 10(10):464–470
11. Chin-Parker S, Bradner A (2010) Background shifts affect explanatory style: how a pragmatic theory of explanation accounts for background effects in the generation of explanations. Cognitive Process 11(3):227–249
12. Dennett DC The intentional stance. MIT Press, p 169
13. Overton J (2011) Scientific explanation and computation. In: Thomas Roth-Berghofer DBL, Tintarev N (eds) Proceedings of the 6th international explanation-aware computing (ExaCt) workshop, pp 41–50
14. Lewis D Causal explanation. Philosophical Papers 2(166):214–240
15. Lipton P (1990) Contrastive explanation. Royal Inst Philosophy Suppl 27:247–266
16. Bromberger S (1966) Why–questions. In: Colodny RG (ed) Mind and cosmos: essays in contemporary science and philosophy. Pittsburgh University Press, Pittsburgh, pp 68–111
17. Pearl J, Mackenzie D (2018) The book of why: the new science of cause and effect, Hachette UK
18. Peirce CS (1903) Harvard lectures on pragmatism. Collected Papers 5
19. Lipton ZC The mythos of model interpretability. arXiv:1606.03490
20. Biran O, Cotton C (2017) Explanation and justification in machine learning: a survey. In: IJCAI 2017 workshop on explainable artificial intelligence (XAI), pp 8–13
21. Williams JJ, Lombrozo T, Rehder B (2013) The hazards of explanation: overgeneralization in the face of exceptions. J Exp Psychology: Gen 142(4):1006
22. Heider F (1958) The psychology of interpersonal relations. Wiley, New York
23. Hilton DJ, Slugoski BR Knowledge-based causal attribution: the abnormal conditions focus model. Psychological Rev 93(1, 166):75
24. Hilton DJ (1996) Mental models and causal explanation: judgements of probable cause and explanatory relevance. Think Reasoning 2(4):273–308

25. Hesslow G The problem of causal selection. Contemporary Sci Nat Explan: Commonsense Conceptions Causality 168:11–32
26. Hanson NR (1965) Patterns of discovery: an inquiry into the conceptual foundations of science. CUP Archive
27. Mill JS (1973) A system of logic: the collected works of John Stuart Mill, vol. III
28. Kelley HH (1967) Attribution theory in social psychology. In: Nebraska symposium on motivation. University of Nebraska Press, pp 192–238
29. McGill AL, Klein JG (1993) Contrastive and counterfactual reasoning in causal judgment. J Personal Soc Psychol 64(6):897
30. Hilton DJ, McClure J, Sutton RM (2010) Selecting explanations from causal chains: do statistical principles explain preferences for voluntary causes? Eur J Soc Psychol 40(3):383–400
31. Lombrozo T (2010) Causal–explanatory pluralism: How intentions, functions, and mechanisms influence causal ascriptions. Cognitive Psychology 61(4):303–332
32. Woodward J (2006) Sensitive and insensitive causation. Philosophical Rev 115(1)
33. Wilkenfeld DA, Lombrozo T (2015) Inference to the best explanation (IBE) versus explaining for the best inference (EBI). Sci Educ 24(9–10):1059–1077

Comparison of Blockchain Ecosystem for Identity Management

K. S. Suganya and R. Nedunchezhian

Abstract In recent years, blockchain technology is found to be the most significant discovery in this digital era, after the discovery of the Internet and Cloud Computing. Blockchain is a simple, distributed public ledger that contains all the user's transaction details in a block. The global copy of the block is then shared among all its peer-to-peer network users after validation by the blockchain miners. Once a block is validated and accepted, it cannot be altered by any users making it a trust-free transaction. It also resolves the problem of double-spending by using traditional cryptographic methods. Since the advent of Bitcoin, blockchain is the backbone for all its transactions. But in recent years, it has found its roots and uses in many fields like smart contracts, smart city management, healthcare, etc. Identity management against digital identity theft had become a major concern among financial and other organizations. To solve this digital identity theft, blockchain technology can be employed with existing identity management systems which maintain a distributed public ledger containing details of an individual's identity containing information such as digital birth certificates, citizenship number, bank details, voter details, and driving license in the form of blocks verified on the blockchain becomes time-stamped, unforgeable and publicly visible for any legitimate users. The main challenge in using blockchain technology for preventing digital identity theft is ensuring the pseudo-anonymity and privacy of the users. This survey paper will exert to study the blockchain concepts, consensus protocols, and various blockchain-based digital identity management systems with their research scope. This paper also discusses the role of blockchain in COVID-19 pandemic management through self-sovereign identity and supply chain management.

Keywords Blockchain · Consensus protocols · Bitcoin · Identity theft · Digital identity management · Pandemic · COVID-19 · Self-sovereign identity

K. S. Suganya (✉)
Department of Computer Science and Engineering, KCG College of Technology, Chennai, India
e-mail: suganyaselvaraj06@gmail.com

R. Nedunchezhian
Department of Computer Science and Engineering, Coimbatore Institute of Technology, Coimbatore, India
e-mail: nedunchezhian@cit.edu.in

H. O. Bansal et al. (eds.), *Next Generation Systems and Networks*, Lecture Notes in Networks and Systems 641, https://doi.org/10.1007/978-981-99-0483-9_21

1 Introduction

Blockchain is one of the disruptive technologies of this digital era. Blockchain has gained this momentum of importance, since the popularity of Bitcoin Cryptocurrency [1]. Bitcoin is based on the principle of blockchain, a distributed immutable ledger technology. The main reason for the success of this technology is its ability to address the lack of trust issues in a centralized environment. Blockchain intuition and innovation revolve around the following concepts, namely cryptographic hash functions, immutable ledger, distributed peer-to-peer network, proof of work, and consensus protocol. According to Gartner forecasts, the blockchain's business value can grow to reach $176 Billion by 2025 and $3.1 trillion by 2030. The blockchain digital disruption scale predicted according to Gartner reports is shown in Fig. 1. According to Fig. 1, blockchain technology was at the initial stage of disruption DD1 in 2018 and is expected to enter the DD3 by 2023 and cab expected to emerge as the revolutionizing technology in the future. A blockchain is a continuously growing list of records called blocks that are cryptographically and securely linked together and shared by the users in a distributed peer-to-peer network. The first block in the Blockchain is popularly termed as "Genesis Block". Each block contains data, previous block hash, and current block hash. The hash values in a blockchain act like a "Fingerprint" ensuring uniqueness. Some amateurs and experts around the world are arguing the potential uses of blockchain may be disregarded owing to the disruptive pattern of Bitcoin [1], a public Blockchain Cryptocurrency system, and the newly emerging Libra coins, a permissioned Blockchain Cryptocurrency system proposed by the social media giant Facebook [2]. Despite the misconception about blockchain usage, many organizations are adopting and spending huge amounts to adopt and experiment with blockchain technology.

Figure 1 this image shows the digital disruption scale predicted by Gartner in 2018. Image downloaded from https://t.co/anKGT32ByN in February 2018.

2 Evolution of Blockchain

History of Blockchain

Table 1 describes the brief history of Blockchain since the inception of the double bookkeeping system by Luca Pacioli and Leonardo Da Vinci since the beginning of 3000 BC. Later, David Chaum proposed an untraceable digital currency system named the "Blind signature system". But the discovery of timestamping a digital document [3] paved the way for today's Bitcoin [1]. Further down the timeline, many innovations paved way for the current Blockchain system and the recent launch of Ethereum hard fork blockchain technology named Byzantium and Constantinople has also been discussed.

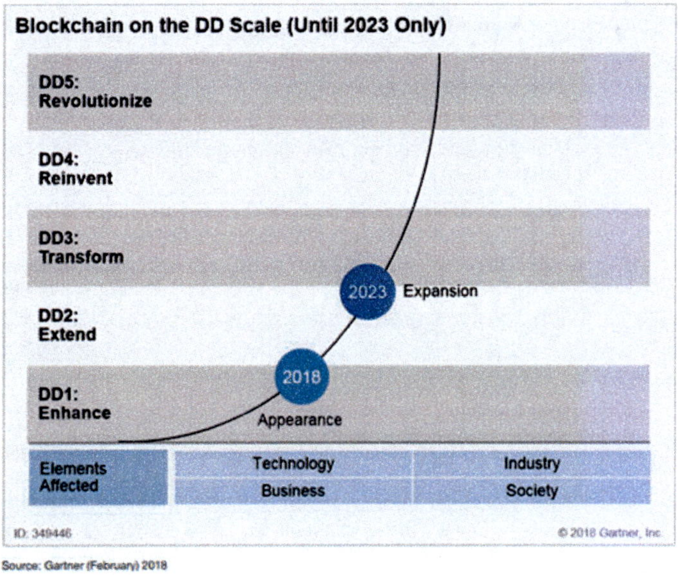

Fig. 1 Blockchain on the digital disruption scale (Until 2023)

3 Identity Management

A. *Digital Identity*

An identity is a representation of a living or non-living entity that may be a human, organization, device, or application. In today's digital era, data is digitalized due to the growth of the Internet. A digital identity is information associated with electronic information available about a person. It is commonly indicated by some of our attributes listed below.

- Who you say you are (name, citizenship, Social Security number, Aadhar number, birth certificates)?
- What do you represent yourself as (Facebook id, LinkedIn Profile, Google login, Instagram login, Snapchat login, Whatsapp login, and so on) in the digital presence?
- What you possess (bank credentials, credit card/debit card details, insurance details)?
- What you are allowed to do (eBay seller/customer profile, Amazon shopping seller/customer profile, share market trade profile, and so on)?

With the rise in digital presence, there is an alarming increase in security issues such as identity theft, phishing, online scams, etc., which poses threat to identities as often the digital identity and real-world identities vary in some cases. According to the identity theft fraud study report by Javelin Strategy [4], the number of data breaches and identity theft has fallen in the year 2019 but still raises concern.

Table 1 Comparison of various blockchain-based identity management systems

Blockchain-based IDMS	Type of blockchain	Type of identity management	Blockchain technology used	Functionalities
Abacus	Public/Private permissioned	Self-sovereign identity	Ethereum	It is suitable for financial and cryptocurrency applications and is an open-source product for streaming compliance workflows
Aries	Public/Private permissioned	Self-sovereign identity	Hyperledger Fabric	It provides a toolkit for creating, transmitting, and storing digital identities
BanQu	Public/Private permissioned	Self-sovereign identity	Not disclosed	It provides software as a service that helps farmers to track their products and helps in supply chain management by providing economic identities
BitID	Public permissionless	Self-sovereign identity	Bitcoin	It is an open authentication protocol that uses public key authentication by Bitcoin address
Bitnation	Public permissionless	Self-sovereign identity	Ethereum smart contract	It provides an open governance platform that records the identities, and other legal events using the Blockchain platform
Blockchain Helix	Public/Private permissionless	Self-sovereign identity	Not disclosed	It provides a digital identity for individuals and organizations and allows the user to choose control over the data shared during KYC and AML

(continued)

Table 1 (continued)

Blockchain-based IDMS	Type of blockchain	Type of identity management	Blockchain technology used	Functionalities
Blockpass IDN	Public/Private Permissioned	Self-sovereign identity	Ethereum network PASS tokens	It provides a digital identity verification of humans(KYC), objects(KYO), and connected devices(KYD) for businesses in regulated industries
Bloom	Public/Private Permissioned	Self-sovereign identity	Ethereum	It provides decentralized credit scoring with a reusable digital identity
Cambridge blockchain	Private	Self-sovereign identity	Cambridge blockchain	It provides a Blockchain-based identity management system and compliance for KYC and AML for financial institutions through "Personal Data Services(PDS)"
CertCoin	Public	Self-sovereign identity	NameCoin	It is a NameCoin-based decentralized authentication system
Civic	Public	Self-sovereign identity/Federated	Bitcoin	It provides decentralized identity-based solutions for US citizens and is one of the well-known platforms for cost-effective verifying a user's citizenship
Colendi	Public/Private	Self-sovereign identity	Colendi API	It is a credit-based financial service that makes use of Blockchain for managing user identities in a decentralized ledger

(continued)

Table 1 (continued)

Blockchain-based IDMS	Type of blockchain	Type of identity management	Blockchain technology used	Functionalities
Cordo R3	Public/Private Permissioned	Self-sovereign identity	Java	It provides a decentralized identity management system with high availability and governance features
CredyCo	Public	Self-sovereign identity	Bitcoin	It provides document verification as a SAAS based on Blockchain
Cryptid	Public	Self-sovereign identity	Factom blockchain	It is an open-source identification system to detect vulnerabilities in software and hardware
Datum	Public	Self-sovereign identity	Ethereum	It is a proprietary Blockchain based identity solution for providing self-sovereign identity
Decentralized Identity Foundation (DIF)	Public	Self-sovereign identity	Ethereum	It establishes an open ecosystem for Blockchain-based identity for people, organizations, apps, and devices
Dock.io	Public/Private	Self-sovereign identity	Ethereum	It is a data exchange protocol that requires a proprietary token(DOCK)
Dominode	Permissioned	Self-sovereign identity	Not disclosed	It provides a regulated Blockchain-based identity solution for license management

(continued)

Table 1 (continued)

Blockchain-based IDMS	Type of blockchain	Type of identity management	Blockchain technology used	Functionalities
e-Residency	Private	Self-sovereign identity	KSI Blockchain	It is the first implemented e-governance Blockchain-based identity solution for the Estonian government
EDDITS	Public/Private	Self-sovereign identity	Ethereum	It means "Ethereum Decentralized Digital Identity & Trust Services," which provides identity management solutions for ERC-725 identities
Engine	Permissionless	Self-sovereign identity	Ethereum	It is a permissionless P2P network that provides a secret contract for identity management
Evernym	Public/Private	Self-sovereign identity	Hyperledger	It is an open-source decentralized identity management solution developed by Hyperledger Indy and Sovrin Foundation
FinID.me	Public/Private	Self-sovereign identity	Ethereum	It provided e-ID with compliance for European nations
Guardtime's BLT	Private	Self-sovereign identity	KSI Blockchain	It provides a replacement for RSA authentication for both data in motion and data at rest

(continued)

B. *Identity management*

Identity management is the method of identifying, verifying, and validating a participant to access confidential or secret information. To explain the users and roles of Identity management, let us consider the Aadhar Unique Identification system

Table 1 (continued)

Blockchain-based IDMS	Type of blockchain	Type of identity management	Blockchain technology used	Functionalities
Gemalto Trust ID Network	Public/Private	Self-sovereign /Federated identity	Corda R3	It provides ID as a Service and authentication services and digital identity services for mobile subscriber enrollment
Hyperledger Indy	Public permissioned	Self-sovereign identity	Hyperledger	It provides a scalable and interoperable digital identity solution that works well with other Blockchain platforms
HYPR	Private	Self-sovereign identity	HYPR biometric platform	It provides Blockchain based biometric authentication solutions using Mobile phones and IoT devices
IBM Trusted Identity	Public/private permissioned	Self-sovereign identity	Hyperledger	It provides a permissioned decentralized digital identity platform using Hyperledger Indy and Sovrin
KYCLegal	Public/Private	Self-sovereign identity	Not disclosed	It is a QR code-based Blockchain-identity authentication system using Mobile phones
KYC Chain	Private	Self-sovereign identity	Ethereum	It is a Blockchain-based digital identity solution for banking and finance compliances

(continued)

Table 1 (continued)

Blockchain-based IDMS	Type of blockchain	Type of identity management	Blockchain technology used	Functionalities
Jolocom	Private	Self-sovereign identity	Ethereum	It is a Blockchain-based digital identity solution for providing self-sovereign identities
Netki	Private	Self-sovereign identity	Hyperledger	It provided a digital identity verification solution for KYC and AML
NameID	Public/Private	Self-sovereign identity	NameCoin	It is a Blockchain-based standard that turns NameCoin-based Identity into OpenID identity
NameCoin	Public/Private	Self-sovereign identity	Bitcoin	It provides an asset management Blockchain-based digital identities using Bitcoin fork
Nuggets	Public/Private	Self-sovereign identity	Not disclosed	It is an e-commerce platform for storing digital identities and banking identities in Blockchain
OIX	Public	Federated identity	Collaborative	It is a part of the Open Identity Exchange group that allows cross-platform communication for establishing federated identities
Passport ID	Private	Self-sovereign identity	Apple	It is an iOS application that provides Blockchain-based digital identity notary for Apple users

(continued)

Table 1 (continued)

Blockchain-based IDMS	Type of blockchain	Type of identity management	Blockchain technology used	Functionalities
PeerMountain	Private	Self-sovereign identity	Ethereum	It provides Blockchain-based digital identity solutions as App as a Service for Android and iOS users
Persona.im	Private	Self-sovereign identity	Ethereum	It provides Blockchain-based digital identity solutions with zero knowledge
Pillar	Public/private	Self-sovereign identity	Not Disclosed	It is a developing project proposed by pillar for Blockchain-based identity management
Rate3	Public	Self-sovereign identity	Ethereum or Stellar	It provided RTE tokens for Blockchain-based identity management
Regis	Private	Self-sovereign identity	Ethereum	It provides Blockchain-based digital identity solutions with a DNS registry
ScuttleBot	Public	Identity Provider	Scuttlebutt API	It provided a "Blockchain-based identity" protocol that embraces subjectivity over consensus
SelfKey	Private	Self-Sovereign Identity	Ethereum	It provides a "Blockchain-based identity" for managing digital assets and identities

(continued)

Table 1 (continued)

Blockchain-based IDMS	Type of blockchain	Type of identity management	Blockchain technology used	Functionalities
SecureKey	Private permissioned	Self-Sovereign/Federated identity	Hyperledger Fabric	It allows the user to authenticate themselves using smartphones and allows them to share the necessary documents. It has been used to access 80 percent of Canadian government services
ShoCard	Private	Self-sovereign identity	Bitcoin	It provides "Blockchain-based Mobile identity" for financial services with proof of signing on the Blockchain
ShoBadge	Private	Self-sovereign identity	Bitcoin	It is an enterprise-level Blockchain-based identity management by ShoCard
Sovrin	Public permissioned	Self-sovereign identity	Hyperledger	It provides a scalable and interoperable digital identity solution with Hyperledger Indy
SpidChain	Private/Public	Self-sovereign identity	Not disclosed	It is a Blockchain-based identity management For managing KYC
Taqanu	Private/Public	Self-sovereign identity	Not disclosed	It is a Blockchain-based identity management that provides GlobalID with an attestation framework

(continued)

Table 1 (continued)

Blockchain-based IDMS	Type of blockchain	Type of identity management	Blockchain technology used	Functionalities
Tierion	Private/Public	Self-sovereign identity	Bitcoin	It is a Blockchain Proof Engine that provides data verification based on the data written to the Bitcoin Blockchain
UniquID	Private	Self-sovereign identity	Ethereum	It provides Identity as a service by leveraging Blockchain for IoT platforms
UPort	Public/Private	Self-sovereign identity	Bitcoin	It provides the user with a seamless Blockchain-based digital identity management system with reusable components
Verity	Public/Private	Self-sovereign identity	Hyperledger Fabric	It is a framework for achieving a Blockchain-based digital identity management system
XAIN	Public/Private	Federated identity	Python	It is an integrated machine learning AI-based Blockchain identity management system

proposed with biometric and demographic data as a reference to the identity system proposed in [5].

Users and Roles

- Users: A person who uses the services. E.g., an Indian citizen possessing an Aadhar Unique Identification Number.
- Identity Provider: An organization or governing body issuing the identity. E.g., UIDAI – Unique Identification Authority of India.
- Service Provider: An organization or governing body verifying the user's identity. E.g., A bank verifying its account holder against Aadhar Unique Identification Number.
- Identity: User's proof of identity. E.g., Aadhar Card.

Identity management is governed by three A's namely Authentication, Authorization, Administration, Audit, and Analytics.

Authentication

Authentication is the process of verifying who you say you are with user details and a secret passphrase. It is the process of confirming a user about his/her identity. The secret passphrase may be a PIN, Password, One-Time Password, biometric, smart card, location information, and many more. In simple words, authentication is nothing more than proving the ownership of any account or access. There are two types of authentication namely single-factor authentication and multi-factor authentication where the latter is gaining more importance because of the high security.

Authorization

Authorization is the process of providing access or privilege control to a particular data, device, and application. In simple words, authorization is nothing but what you are allowed to do. In some cases, authentication and authorization go in hand with each other, e.g., Apple Id to login to an Apple device and authorized to use iTunes and other Apple services.

Administration

The administration is the process of managing authentication and authorization access to a user's resource using the user's digital identity. These are carried out in a centralized or decentralized manner.

Audit

An audit is a process of monitoring the correct administration of user identity authentication and authorization. These are often carried out by third-party auditors or in some cases federated audit management to enforce the detection and implementation of preventive policies.

Analytics

Analytics is the process by which an identity management system gains insights and future predictions by collecting and analyzing user identity data.

Let us discuss some of the identity management systems that have evolved from Identity 1.0 to Identity 3.0.

C. *Simple Identity Management:*

This is a simple identity management system where the user is authenticated with a simple request–response, i.e., username and passphrase.

D. *Centralized Identity management:*

This identity management has three roles, namely users, the single identity provider (IDP), and many service providers (SP). In a centralized identity management system, users can access all the service providers with the same identities (username and passphrase). This kind of identity management system suffers from a single entity that has complete authority and control. This centralized identity management system is often implanted using a centralized Authority (CA). These systems are prone to high complexity in terms of administration, maintenance, and audit and are not scalable across organizations of the world, and also suffer from the single point of failure. In some case, these single-entity identity providers can be compromised and thus reduces the trust among the service providers which rely on identity providers for user identities. For instance, in the year 2018, nearly 87 million Facebook users' information has been shared with a data analytic firm "Cambridge Analytica" due to a data breach by a psychology app quiz. [6].

E. *Federated Identity Management:*

In federated identity management, the users use a shared single-user identity to access multiple service providers [7]. A federated identity management system is implemented using Single Sign On mechanism (SSO) which requests the user to log in once and be authenticated to access all the services. SSO is a subset of federated identity and they vary from federated identity. The former mechanism enables users to use single identities to access multiple systems (authentication) and the latter enables users to access multiple systems (authorization). Microsoft Passport, Google, and Facebook Single Sign on are some of the popular SSO identity management systems. This system is made possible by implementing a common standard, rules, procedures, and protocols to manage the trust among the participating service providers. Federated identity is often implemented based on the open standard Security Assertion Markup Language (SAML), OpenID, and many more. The federated identity is based on the mutual trust by service providers in the identity providers and provides wide scalability across multiple systems, and cost efficiency. The major disadvantage of this system is special care needs to be taken to establish standards and rules agreed upon by the users of federated identity.

F. *Self-sovereign Identity Management:*

The term "Self-sovereign Identity" means the sole ownership and control of the person or organization's identity lie with the individual and they decide how the identity is shared. This type of identity management also prevents the use of a third-party intermediary or government organization from data breaches. There are ten guiding principles for Self-sovereign Identity management proposed by Christopher Allen [8] which are more extensive compared to other principles proposed in [9].

- **Existence**: *Users must have an independent existence.*
- **Control**: *Users must control their identities.*
- **Access**: *Users must have access to their data.*
- **Transparency**: *Systems and algorithms must be transparent.*
- **Persistence**: *Identities must be long-lived.*
- **Portability**: *Information and services about identity must be transportable.*
- **Interoperability**: *Identities should be as widely usable as possible.*
- **Consent**: *Users must agree to the use of their identity.*
- **Minimization**: *Disclosure of claims must be minimized.*

- **Protection**: *The rights of users must be protected.*

Self-sovereign identity management gained popularity because of its user-centric nature. Self-sovereign identity management is made plausible by blockchain sistributed ledger technology. This kind of digital identity management system has been implemented in countries like Estonia [10] and Dutch [11] where the former is based on the partnership with Bitnation and the latter is a truly self-reliable blockchain-based self-sovereign identity management system.

4 Blockchain in Identity Management

Blockchain identity management system provides a decentralized ledger technology for accessing user identities through self-sovereign identity management. They provide better protection against security threats such as identity theft, KYC onboarding, and privacy of personal data. The use of blockchain identity management has gained popularity since its adoption by various industries like finance and banking, healthcare, government sectors, supply chain management, telecommunication, e-commerce, food management, IT, and many more.

Lim discussed some of the identity management systems based on blockchain but some important products were missing [12]. Gruner discussed the three major offerings of blockchain-based identity management systems, namely uPort, Sovrin, and Shocard [13]. Jacobovitz provided a brief survey of blockchain-based identity management systems that were active till 2016 and failed to discuss Hyperledger Indy [14]. Michael [15] discussed the blockchain-based identity system from enterprise and ecosystem perspectives but some of the self-sovereign identity management systems like OIX, and Jolocum were not discussed. In this section, we are going to discuss some of the common blockchain-based identity management systems that were compared based on the factors such as types of blockchain, types of identity management, the blockchain technology used, and functionalities in Table 1.

5 Blockchain in Pandemic Management

The recent pandemic rise due to COVID-19 has given rise to many speculations, devastations, and changes in people's approach toward behavior, economy, and societal expectations. The economic devastation due to COVID-19 will fail the leaders and governments in this pandemic rise. At this stage, human costs are unfathomable and the need for better technologies like blockchain, artificial intelligence, and the Internet of Things (IoT) has paved the way for providing a better framework for future pandemic management instead of traditional solutions.

In March 2020, the Blockchain Research Institute convened a meeting to discuss the challenges of COVID-19 and the possibility of using blockchain technology to resolve those challenges. The main focus areas discussed in the meeting were summarized below.

G. *Self-sovereign Identity:*

For fighting a pandemic, data is the most important asset. Effective pandemic management needs data about what, where, how, who, and when. (how many people were infected? where are they located? when are they infected and recovered? how were they infected? who have they contacted?). Countries like China, Singapore, and South Korea have better control over the COVID-19 pandemic rather than Western countries like the USA, the UK, Italy, Germany, and Spain because the former countries have access to these data as they have already experienced SARS and many other pandemics rather than the latter countries. Chinese special camera surveillance helped the authorities to identify their people and their whereabouts enforcing effective lockdown but at the cost of the liberated personal identity of their citizens. Blockchain can prevent this liberated personal identity issue with a shift in control to individuals using the self-sovereign identity. Self-sovereign identity with decentralized identity management using Blockchain can enable the citizens of any country to disclose necessary information to their government during the pandemic and erase the data after the pandemic has run its course. Access Pass and VestChain are the two popular self-sovereign blockchain-based identity management which have helped the Chinese government and citizens in curbing this pandemic to a good extent. How could these self-sovereign identity apps help government officials to curb this pandemic? Let us discuss the role of blockchain in this scenario:

- By enforcing the law, the government can request its citizen to make anonymized data such as body temperature and their location for tracking the COVID-19 infected and non-infected persons with limited breach to the privacy of the citizens.
- If a suspected COVID-19 victim is being hospitalized, the victim can share the necessary anonymized medical data with medical professionals but withhold some personal medical data at their convenience.
- An incentive reward system by the government to share the anonymized medical data of a citizen can go a long way. Some citizens may share the data out of community and social responsibilities to their nation and government.
- The government can also issue "health certificates" to their citizens so that they can be verified by their organization to return to work.

Thus, with this approach to self-sovereign identity management using blockchain, citizens can own their private data and use it anonymously. But this transition will take time.

H. *Supply chain management:*

The COVID-19 pandemic has also raised many challenges in the supply chain management of ventilators, medicines, sanitizers, and personal protective equipment for medical professionals. The first challenge is tracking them. The next challenge is verifying the authenticity of the product and the supplier. The third challenge is moving them to the necessary location where they are needed. The final challenge will be increasing the production of these products ahead of its demands to prevent unnecessary hoarding. The US government is facing a major problem in this area because of their government's announcement to state governors to acquire these products on their own which resulted in greater competition among its states. A proper solution to curb these challenges will be providing transparent and collaborative supply chain management using blockchain. There are already many applications such as blockchain-based supply chain management. For example, IDlogiq [16] is a blockchain-based anti-counterfeit to protect and identify validation of medicines. Remedichain [17] is another blockchain-based supply chain management framework designed by Dr. Baker to track, measure, and donate oncology drugs to those who need them. Chinese blockchain entrepreneurs have formed a collaborative platform for matching the donation of medical supplies with the healthcare professional who need them thereby preventing fraudsters and counterfeiting. Let us discuss how to curb the challenges in supply chain management using blockchain:

- By vetting the suppliers and notarising the process using blockchain to improve the efficacy of the suppliers.
- Create a self-sovereign identity for the pharmaceutical products.
- Manage the logistics of pharmaceutical products using blockchain. IBM has launched Tradelens [18] to streamline these logistic activities using the blockchain platform.

Furthermore, blockchain provides privacy, and data integrity in the medical supply chain management to a greater extent.

I. *Incentive Model for Pandemic Management*

The COVID-19 pandemic has raised the need for collective responsibility by both individuals and institutions. An incentive model to enable the individual to share the data is a powerful tool that the world is in much need of as of the current situation. Blockchain-based digital currency or loyalty tokens can be rewarded to individuals who share the data. This has been implemented by the Heart and Stroke Foundation of the Canadian government for monitoring their citizen's hypertension levels for people over 40 age. Those who meet the criteria of no hypertension are rewarded with reward points at groceries, pharmacies, and gym membership. Similarly, patients with rare diseases share their data for the advancement of research to cure those diseases and act as an incentive. A blockchain platform can be used to collect patient data (who tested positive and negative) all over the world for economic or social incentives. Such blockchain platforms must ensure that the patient data is shared after receiving the patient's consent and anonymously.

6 Conclusion

Despite the rising or declining popularity of Bitcoin, blockchain has its roots firmly in today's technology trends because of its public, distributed, and informatics-sharing platform. Over the years, blockchain has found its applications in various non-financial applications, namely Healthcare, Digital Copyrights management, Voting, Supply chain management, digital identity management, Smart contracts, Blockchain in the Internet of Things, Intrusion Detection Prevention(IDP), and financial applications namely E-commerce, crowdfunding, Trading and Gaming, cryptocurrency, microfinance and many more. The recent COVID-19 Pandemic has profoundly changed our behavior and expectations in society and Blockchain Technology can change and help our approach toward this expectation in various ways. Self-sovereign identity management in healthcare to share the virus-host data via health records can be of significant use. Supply chain management of Pharmaceutical medicines and digital asset management of these medicines using blockchain can be of significant use. The incentive rewards model using Blockchain for sharing the patient's data in this pandemic situation can also be of significant use. This pandemic has shown us the need for better incorporation of technologies for the common good to improve the infrastructure, mitigate the crisis and as well as demonstrate the need for Blockchain in self-sovereign identity management, supply chain management, etc.

References

1. Nakamota S (2008) Bitcoin: a peer-to-peer electronic cash system, pp 1–9. www.bitcoin.org. https://doi.org/10.1007/s10838-008-9062-0.
2. Amsden Z et al (2019) An introduction to libra. Libra association Facebook Inc, pp 1–29. https://libra.org/en-us/whitepaper
3. Haber S, Stornetta S (1991) How to time-stamp a digital document. J Cryptol 3(2):99–111
4. Marchini K, Pascual A (2019) Identity fraud study: fraudsters seek new targets and victims bear the brunt, March 6, 2019. https://www.javelinstrategy.com/coverage-area/2019-identity-fraud-report-fraudsters-seek-new-targets-and-victims-bear-brunt
5. Maliki TE, Seigneur JM (2014) Online identity and user management services. In: Managing information security, 2nd ed. Syngress, Rockland, MA, USA, pp 75–118
6. Salinas S, Newberg M (2020) Facebook has been worried about data leaks like this since it went public in 2012, cnbc.com. https://www.cnbc.com/2018/04/12/facebook-warned-of-data-breaches-years-ago-when-it-went-public-in-2012.html. Accessed 12 Feb 2020
7. Chadwick DW (2009) Federated identity management. In: Aldini A, Barthe G, Gorrieri R (eds) Foundations of security analysis and design V. FOSAD 2009, FOSAD 2007, FOSAD 2008. Lecture notes in computer science, vol 5705. Springer, Berlin
8. Allen C (2016) The path to self-sovereign identity. http://www.lifewithalacrity.com/2016/04/the-path-to-self-soverereignidentity.html
9. Zyskind G, Nathan O et al (2015) Decentralizing privacy: using blockchain to protect personal data. In: Security and privacy workshops (SPW), 2015 IEEE, pp 180–184. IEEE
10. Bitnation (2015) Estonia e-residency program & Bitnation Dao public notary partnership. https://bitnation.co/blog/pressrelease-estonia-bitnation-public-notary-partnership/
11. Stokkink Q, Pouwelse J(2018) Deployment of a blockchain-based self-sovereign identity. In: 2018 ieee international conference on internet of things (iThings) and IEEE green computing and communications (GreenCom) and IEEE cyber, physical and social computing (CPSCom) and IEEE smart data (SmartData), Halifax, NS, Canada, 2018, pp 1336–1342
12. Lim SY et al (2018) Blockchain technology the identity management and authentication service disruptor: a survey. Int J Adv Sci Eng Inf Technol 8(4–2):1735–1745

13. Grüner A, Mühle A, Meinel C (2018) On the relevance of blockchain in identity management, CoRR. http://arxiv.org/abs/1807.08136
14. Jacobovitz O (2016) Blockchain for identity management. Department computer science, Ben-Gurion University, Beer Sheva, Israel, Technical Report 16–02, 2016. https://www.cs.bgu.ac.il/frankel/Technical Reports/2016/16-02.pdf
15. Kuperberg M Blockchain-based identity management: a survey from the enterprise and ecosystem perspective. IEEE Trans Eng Manag
16. IDLogiq selected as finalist in the GS1 disruptive innovation start-up connect, Business Wire, Berkshire Hathaway (2019). www.businesswire.com/news/home/20190326005427/en/IDLogiq-Selected-Finalist-GS1-Disruptive-Innovation-Startup
17. RemediChain (2020) www.remedichain.org
18. Maersk and IBM to Form Joint Venture Applying Blockchain to Improve Global Trade and Digitize Supply Chains. Press Release, IBM Corporation, 2018. www-03.ibm.com/press/us/en/pressrelease/53602.wss

Review on Deep Learning in Wireless Communication Networks

Shewangi and Roopali Garg

Abstract Internet-connected mobile devices are increasing day by day in almost every domain which created a surge in mobile data traffic so an advanced system is highly needed that can handle a large amount of data. In this paper, Deep Learning (DL) is incorporated into wireless technology. Deep learning is an emerging technology that has a diversity of applications. A large amount of heterogeneous data is generated by wireless systems which is difficult to compute in a short time. DL makes wireless communication intelligent and more robust. In this paper, a review of the DL platform and different mechanisms to deploy DL in wireless communications is summarized. Further different research areas for deep learning in a wireless system are discussed to motivate researchers in this domain.

Keywords Deep learning · Reinforcement learning · Wireless network · Fog computing · Mobile network

1 Introduction

According to a survey by Compound Annual Growth Rate (CAGR), mobile traffic increases by 20% every year. The wireless technology uses electromagnetic waves to achieve signals. All wireless communication system uses an electromagnetic spectrum that contains a range of possible frequencies. The lowest frequency (30–300 Hz) is used for power line frequencies and the highest range frequency (30–300 GHz) is used for satellite communication. Federal Communications Commission (FCC) allocates bands of frequencies according to requirements and monitors the broadcast to detect unlicensed operations and any kind of technical violations. Spectrum is a non-renewable source in nature whose demand is increasing day by day. So, there

Shewangi (✉) · R. Garg
UIET, Panjab University, Chandigarh, India
e-mail: Shivangikochhar93@gmail.com

R. Garg
e-mail: roopali.garg@pu.ac.in

H. O. Bansal et al. (eds.), *Next Generation Systems and Networks*, Lecture Notes in Networks and Systems 641, https://doi.org/10.1007/978-981-99-0483-9_22

is an urgent need for an intelligent wireless network that is motivated to meet end-user requirements and future 5G and 6G mobile-generated devices. Deep learning is a tremendous technology for various operations of future wireless communication networks. Embedding deep learning in the wireless network proves to be advantageous for growing diversity as it can handle a large amount of data. Machine Learning (ML) is an ever-growing research area that combines all the different areas of science, namely big data, wireless communication, and information theory. It is an approach to make a model which depends on prediction. ML is mainly categorized into three categories—supervised, unsupervised, and reinforcement learning. Supervised learning is called supervised because it acts as a teacher who is supervising the whole learning process. In this type of learning, users get an inference from classification or regression from trained data. The wireless system is generating a lot of unlabeled and semi-labeled data and supervised learning is a traditional way of learning which is useful only when the data is labeled and sufficient. It can also be used for routing, and mobile activity recognition [1, 2]. Unsupervised learning is called unsupervised because there is no supervision exists, unlike supervised learning. It decides the whole learning process on its own and it is left to the device to extract the information [2]. The inference is made from datasets consisting of input data but there is no labeled response. It can perform more complex tasks as compared to supervised learning because; values are taken in real time. Unsupervised algorithms are further divided into clustering, k-means, etc. Reinforcement Learning (RL) is an intermediate between supervised and unsupervised learning in which a sequence of actions is decided by a user. The most important reason for using this learning method is a trial and error method [3] in which the agent does not have full information about the external environment, the agent collects information directly by interacting with the environment.

2 Enabling Deep Learning in Mobile Networking

It is costly to deploy deep learning technology in a mobile network because super powerful hardware like GPU (Graphics Processing Unit) and software are needed to support training. Many tools exist in the literature that is used to implement deep learning [4]; some of them are illustrated in Fig. 1. By employing these tools, conclusions can be made within a few milliseconds. A summary of all the techniques and tools that enable the deployment of deep learning in wireless communication is discussed in Table 1.

(1) Advanced parallel computing: Traditional machine learning algorithms lack computation power when a large number of matrix multiplications is required so to overcome this issue GPU [8] is introduced to achieve high performance and parallel computing. It can be used for high-performance video games which consist of thousands of core chips installed in a multiplication matrix. Hashing technique is introduced in [7] that reduces computational requirement and makes deployment easy. *(2) Fog computing*: Fog computing is the same as edge computing that offloads the

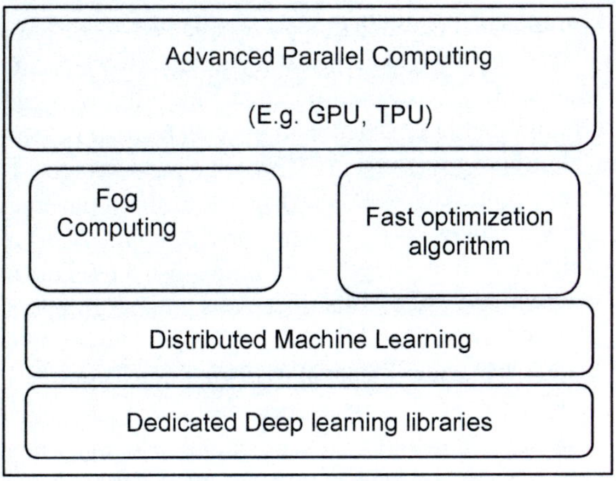

Fig. 1 Hierarchical view of deep learning enablers

Table 1 Summary of all the techniques and tools that enables the deployment of deep learning in wireless communication

Techniques	Platform	Issues	Functions	Challenges
Advanced parallel computing	GPU, TPU [5]	Mobile server, a base station	Enable parallel and fast training	Fault tolerance is generally low
Fog computing	Nm-X [6] core ML [7]	Reduce communication overhead	To offload the burden from cloud network	Its performance is medium to low
Fast convergence algorithm	Adam [8] Nadam	Training deep neural architecture	For fast convergence	Energy consumption is high
Distributed machine learning	ML base, GeePs	Allows distributed training	Distributed data centre in all area	More expensive hardware is required to install at all distributed
Dedicated libraries	Tensor flow [9], caffe [10]	High energy consumption	Deep learning implementation in mobile operating system	Medium to high performance

data traffic and reduces the burden of servers [11, 12]. There are different kinds of hardware that exist in literature to move DL computing from cloud networks to mobile devices. The author in [13] introduces a low-power and programmable DL processor. It consumes only 288 μw power and achieves 374 GOPS/w efficiency. An Intel company called Movidius develops an embedded framework that allows DL according to user requirements on ultralow power devices. Fog computing mainly focuses on smartphone and wearable devices like smartwatch, fitness band, etc. It stores, analyzes, and exchanges the data so that burden from the cloud can be offloaded. A mobile processor called NN-X (Neural Network Next) uses low energy consumption and accelerates neural network execution. *(3) Fast Optimization Algorithm:* In DL, fast optimization algorithm is needed to optimize objective function because of large data functions which are difficult to handle. Adam and Nadam is an algorithm proposed in [8] which shows fast convergence that is enabled by strong constrain. Adadelta is an algorithm that converges function and does not rely on global values [14]. *(4) Distributed Machine Learning:* Mobile data is collected from different sources and stored in the distributed data center. It is nearly impossible to store in one place so distributed machine learning helps in maintaining different interfaces. Hsieh et al. [15] develop a system called Gaia, which is distributed geographically by using Tensor flow which consists of a library to make graphs on GPU and CPU. Several issues occurred while implementing ML-like Storage: Incorporating Deep learning libraries require a lot of storage for I/O and data processing. *(5) Programming Model*: There are many programming models that exist in literature like Auto Encoder (AE), Restricted Boltzmann Machine (RBM), and Convolution Neural Network (CNN); however, it would be challenging to choose the best model. Resource Management: Resource management is important while giving workload to each node because some nodes may be left due to poor resource management. ML-base [11] is a different distributed technique to parallelize ML algorithms. *(6) Dedicated deep learning libraries*: With the growing popularity of deep learning, it becomes very important to build dedicated libraries to facilitate any technology in deep learning. TensorFlow [9] is the most common library which helps in deploying graphs on CPU and mobile devices for fast implementation on cloud and fog service. This library contains a visualization tool called tensor board which helps in debugging and helps users to understand the flow of data. It supports almost all languages, e.g., python C++ and Java. The biggest advantage of TensorFlow is that it has proper documentation which makes it easy for users to understand the model and its complete functionality. The second most common library is Caffe [10] which is famous for its fast runtime and flexible framework that enables users to make their models. It can run on MATLAB and Python. It also allows users to train the neural network on both distributed and single systems. Caffe is also famous for its implementation on mobile operating systems whether it is Android or iOS. It will be very beneficial to future mobile edge computing due to easy implementation in mobile computing. Pytorch is another famous library that is released in the Python language. It is a very lightweight toolbox that helps in building neural models that are easily reusable. It is initially developed in the Lua language and officially maintained under Facebook. The only disadvantage of this library is the lack of documentation and resources.

3 Research Areas in Wireless Network using Deep Learning

There are diverse applications of deep learning in wireless networks. The major research contributions across different wireless networking areas are discussed below and all the applications of these areas are presented in Table 2.

(1) DL-Driven Network Level—Mobile data analysis focuses on logs recorded by the Internet service provider. It improves the network state prediction and traffic classification. Network state prediction is used to increase the throughput and channel quality by identifying the number of active users in that particular cell. Traffic prediction [16, 17] is used to predict the type of traffic and identify the protocol and application used in the network. It uses the CNN model to identify any malware in the data. Call Detail Record (CDR) [18] mines the information such as cell number and total session time, of telecommunication transactions among users. *(2) DL-Driven App-Level*—The data is collected by mobile phones, Global Positioning System (GPS), and video recorders. Mobile devices act as central device to gather all the relevant data which is used for pre-processing lately. Enabling DL in app-level data helps to extract important information from large data which can be beneficial to users. Mobile data analysis is another future research direction. Mobile data changes very quickly

Table 2 Research areas for deep learning as applied to wireless communication

Research areas	Model	Functions	Application
DL-Driven network-level mobile data	MLP, AE, CNN, LSTM	Investigate network prediction and traffic	Cellular traffic forecasting
DL-driven app-level data	RBM, CNN, LSTM	Analyze health, Mobile object recognition	Mobile app classification, Heart rate prediction, Mobile Speech synthesis
DL-driven user-mobility	MLP, DBN RNN, CNN	Mobile trajectory analysis, risk prediction	Mobile mobility and location estimation
DL-driven user-localization	RBM, MLP, CNN	Accuracy is more than cellular localization, the main focus is on vehicle application	Prediction of the position of users in Indoor and outdoor location
DL in wireless sensor networks	MLP, RBM, LSTM	Improves the efficiency of energy by doing data analysis	WSN node and underwater localization
DL in network security	Stacked AE, RBM, MLP	It protects the whole network from any malicious attack	Malware detection
DL in signal processing	MLP, AE	Process the signal with the help of OFDM and MIMO	MIMO detection, modulation classification

because the user changes its position and cell constantly. As a result, analysis of data becomes very difficult so embedding deep learning proves to be more robust and high precision. There are two approaches to app-level data edge-based computing and cloud-based computing. Natural Language Processing (NLP) and Automatic Speech Recognition (ASR) are remarkable achievements in deep learning which are also included in app-level data. *(3) DL-Driven User Mobility*—Predicting the moving pattern of the mobile device has become a research area in wireless domain [19]. Mobile data is generally contained loss, fluctuation, and noise; therefore, precise modeling is important. As DL is expertise in performing automatic feature extraction, it becomes easy for person mobility modeling. Among all models, CNN and RNN are the most successful models for such types of applications. Research is still going on in predicting mobile user trajectories by using the spatio-temporal correlation of mobile events. *(4) DL-Driven User Localization* is not limited to the outdoor environment; it is now applicable to the indoor environment also. Deep neural networks examine the location of users in an indoor or outdoor environment which is based on the signal strength received from wireless nodes [20]. There are also many other models which are used for localization of users but the most common models are RBM, MLP, and CNN because the architecture of all these models is robust to multipath fading. *(5) DL in Wireless Sensor Networks (WSN)* uses DL applications for many purposes like sensing, communication, and analysis [20, 21]. It consists of several distributed sensors which collect information like temperature, pollution, and pressure through a wireless channel and send it to the centralized server. Wireless sensor applications is gaining popularity for the last many years. WSN works on the MLP model whose work is the same as fog computing that offloads the data traffic and reduces the burden of servers placed at the distribution center. *(6) DL in Network Security*—When it comes to security DL has proved to be the best candidate for network security in wireless communication. DL guards the system by employing a firewall, anti-virus software [22, 23], and Intrusion Detection Systems (IDS).

Advanced cyber system learn the signature from past experience and identify any intrusive behavior. Nowadays, mobile devices always carry a huge amount of private information. It is highly important to save this information from any attack or by installing malicious apps on smartphones for ill-conceived purposes. For this purpose, DL is also incorporated into the wireless network for detecting such threats and gets a lot of benefits from DL since they can automatically learn the signature from past experience and identify any intrusive behavior *(7) DL in Signal Processing* is used for detection and estimation of operations like channel estimation, MIMO signal, localization, and decoding with the help of the Artificial Neural Network (ANN) model [24, 25, 15]. OmniSIG software development kit is based on DL and used to train signal processing in real time. It can detect Wi-Fi and Bluetooth signals very quickly as compared to traditional wireless sensor nodes (Table 2).

4 Recent Advancement in Wireless Communication

The author in [6] implements a system that can communicate through encoders that replace the transmitter and receiver tasks. The system is built on a physical layer in which there is no requirement of individual optimization of each block. It increases the accuracy of traffic flow prediction while maintaining complexity. Experimental results show that the proposed model can realize a better performance of traffic flow prediction. It is difficult and challenging to identify Wi-Fi and Long-Term Evolution (LTE) signal over unlicensed band so Hao Gu in [26] proposed deep learning-based intelligent recognition method for identifying LTE and Wi-Fi signals by training the deep learning algorithm called CNN. The simulations show that it can efficiently recognize LTE and Wi-Fi signals. Further, it is also robust to the intelligent recognition method. Wireless Sensor Network (WSN) is a promising technology that provides an opportunity to a diversity of communication applications. To fabricate these sensors, different material is used which is discussed in [27]. The properties of this material like Hetero-Junction Bipolar Transistor (HBT) and Bipolar Field Effect Transistor (BiFET) [27] are discussed in the paper for designing wireless communication systems.

5 Conclusion

Deep learning is a subset of artificial intelligence and a key technology not explicitly to wireless communication but also many emerging technologies in wireless. In this paper, different wireless technologies are discussed which can be used with the help of deep learning, as it makes wireless communication more intelligent and versatile. Different applications and platform of deep learning in wireless have been discussed. The paper is concluded with several research areas and future challenges in wireless communication where deep learning can be deployed. [28–40]

References

1. Chakrabort S, Tomsett R (2017) Interpretability of deep learning models: a survey of results. In: IEEE smart world congress workshop DAIS, ,USA, pp 1–6
2. Jiang C, Zhang H, Ran Y, Hen Z, Cheng K (2017) Machine learning paradigms for next-generation wireless networks. IEEE Wireless Commun 24(2):98–105.
3. Liu W, Wang Z, Liu X, Zeng N, Liu Y (2017) A survey of deep neural network architectures and their applications. Neurocomputing 234:11–26
4. Pouyanfar S, Sadiq S, Yan Y, Tian H, Tao Y, Reyes MP (2018) A survey on deep learning: algorithms, techniques, and applications. ACM Comput Surv 52(5):1–92. https://doi.org/10.1145/3234150
5. Cui H, Zhang H, Ganger GR, Gibbons PB, Xing EP (2016) GeePS: scalable deep learning on distributed GPUs with a GPU-specialized parameter server. In: 11th ACM Europe conference computer system, p 4. https://doi.org/10.1145/2901318.2901323

6. Deep S, Zheng X (2019) Leveraging CNN and transfer learning for vision-based human activity recognition. In: 29th international telecommunication networks and applications conference, pp 1–4
7. Kingma D, Ba J (2015) Adam: A method for stochastic optimization. In: International conference learning representations. San Diego, pp 1–15
8. Zhang C, Straus P (2019) Deep learning in mobile and wireless networking: a survey. IEEE Commun Surv Tutor 12(3):2224–2287. https://doi.org/10.1109/COMST.2019.2904897
9. Zeiler MD (2012) ADADELTA: an adaptive learning rate method. arXiv:1212.5701
10. Dean J, Corrado G, Monga R, Chen K, Devin M, Tucker P (2012) Large scale distributed deep networks. Adv Neural Inf Process Syst 1223–1231
11. Kraska T, Talawaker T, Duchi J, Griffith R, Franklin MJ, Jordan M (2013) MLbase: a distributed machine-learning system. In: Conference on innovative data systems research, vol 1, pp 1–7
12. Vaquero LM, Rodero ML (2014) Finding your way in the fog: towards a comprehensive definition of fog computing. ACM SIGCOMM Comput Commun Rev 44(5):27–32. https://doi.org/10.1145/2677046.2677052
13. Dozat T (2016) Incorporating Nesterov momentum into Adam. In: Workshop Track, pp 1–4
14. Hsieh K, Harlap A, Vijaykumar N, Konomis D (2017) Gaia: geodistributed machine learning approaching LAN speeds. In: USENIX symposium network system design implement, pp 629–647
15. Samuel N, Diskin T, Wiesel A (2017) Deep MIMO detection. In: IEEE 18th international workshop signal process. Advance wireless communication, Japan, pp 1–5. https://doi.org/10.1109/SPAWC.2017.8227772
16. Aazam M, Zeadally S, Harras KA (2018) Offloading in fog computing for IoT: review, enabling technologies, and research opportunities. Futur Gener Comput Syst 87:278–289
17. Zhang C, Patras P (2018) Long-term mobile traffic forecasting using deep spatio-temporal neural networks. In: 18th ACM international symposium mobile Ad Hoc network computer, pp 231–240
18. Moyo V, Sibanda K (2015) The generalization ability of artificial neural networks in forecasting TCP/IP traffic trends: how much does the size of learning rate matter? Intell J Comput Sci 4(1):9–17
19. Ezema LS, Ani CI (2017) Artificial neural network approach to mobile location estimation in GSM network. Int J Electron Telecommun 63(1):39–44
20. Nowicki M, Wietrzykowski J (2017) Low-effort place recognition with WiFi fingerprints using deep learning. In: International conference automobile, pp 575–584
21. Li J, Serpen G (2016) Adaptive and intelligent wireless sensor networks through neural networks: an illustration for infrastructure adaptation through Hopfield network. Appl Intell 45(2):343–362
22. Azar MY, Varadharajan V, Hamey L, Tupakula U (2017) Autoencoder-based feature learning for cyber security applications. In: IEEE International Joint Conference Neural Network, USA, pp 3854–3861. https://doi.org/10.1109/IJCNN.2017.7966342
23. Diro AA, Chilamkurti N (2018) Distributed attack detection scheme using deep learning approach for Internet of Things. Futur Gener Comput Syst 82:761–768
24. Kumar DP, Amgoth T, Annavarapu CSR (2019) Machine learning algorithms for wireless sensor networks: a survey. Inf Fusion 49:1–25
25. Assaf AE, Zaidi S, Affes S, Kandil N (2016) Robust ANNs-based WSN localization in the presence of anisotropic signal attenuation.IEEE Wireless Commun Lett 5(5):504–507
26. Eugenio M, Nkechinyere CJ (2020) Learning to communicate with autoencoders: rethinking wireless systems with deep learning. In: International conference on artificial intelligence in information and communication (ICAIIC), pp 308–311
27. Chen M, Gao Y, wei X (2020) Deep-broad learning system for traffic flow prediction toward 5G cellular wireless network. In: IEEE international wireless communication and mobile computing, pp 940–945
28. Nguyen DD, Nguyen HX, White LB (2017) Reinforcement learning with network- assisted feedback for heterogeneous RAT selection. IEEE Trans Wireless Commun 16(9):6062–6076

29. Xiao W, Xue J, Li Z, Li W, Zhou L (2017) Tux2: distributed graph computation for machine learning. In: USENIX symposium network system design implement, USA, pp 669–682
30. Gokhale V, Jin J, Dundar A, Martini B, Culurciello E (2014) A 240 G-Ops/s mobile coprocessor for deep neural networks. In: IEEE conference computer vision pattern recognition workshops, pp 682–687
31. YayehY, Lin HP, Berie G, Adege AB, Yen L, Jeng SS (2018) Mobility prediction in mobile ad-hoc network using deep learning. In: IEEE international conference application system invention, Japan, pp 1203–1206. https://doi.org/10.1109/ICASI.2018.8394504
32. Wang X, Gao L, Mao S (2016) CSI phase fingerprinting for indoor localization with a deep learning approach. IEEE Internet Things 3(6):1113–1123. https://doi.org/10.1109/JIOT.2016.2558659
33. Liu L, Cheng Y, Cai L, Zhou S, Niu Z (2017) Deep learning based optimization in wireless network. In: IEEE International Conference Communication, pp 1–6, France. https://doi.org/10.1109/ICC.2017.7996587
34. Xu Z, Wang Y, Tang J, Wang J, Gursoy MC (2017) A deep reinforcement learning based framework for power-efficient resource allocation in cloud RANs. In: IEEE international conference communication, France, pp 1–6. https://doi.org/10.1109/ICC.2017.7997286
35. Yan X, Wang J, Fu N, Ou W, Liu B (2017) Signal detection of MIMO-OFDM system based on auto encoder and extreme learning machine. In: IEEE international joint conference neural network, USA, pp 1602–1606. https://doi.org/10.1109/IJCNN.2017.7966042
36. Alzantot M, Wang Y, Ren Z, Srivastava MB (2017) RSTensorFlow: GPU enabled TensorFlow for deep learning on commodity Android devices. In: Proceeding 1st ACM international workshop deep learning mobile system application, pp 7–12
37. Jia Y et al (2014) Caffe: convolutional architecture for fast feature embedding. arXiv:1408.5093
38. Xu X, Li D, Dai Z, Li S, Chen X (2019) A heuristic offloading method for deep learning edge services in 5G networks. Access IEEE 7:67734–67744
39. Gaun H, Wang Y (2020) Deep learning based intelligent recognition method in heterogeneous communication networks. In: IEEE international conference on communications in China (ICCC), pp 478–482
40 Gupta N, Hasim Z (2017) Recent trends and applications of materials used in fabrication of devices for wireless sensor network. Elsevier, Materials Today Proceedings, pp 10332–10335

Novel Image Processing Technique Using Thresholding and Deep Learning Model to Identify Plant Diseases in Complex Background

Saiqa Khan and Meera Narvekar

Abstract Every year tomato crops suffer huge losses due to the occurrence of disease outbreaks. Therefore, the identification of diseases has its importance in the research fraternity. However, this whole process is influenced by the presence of a complex natural environment and similarity in diseases in terms of disease symptom patterns, size of disease region, and similar color variations in leaves with its background. To counter all these issues, in this work we propose an Integrated model that helps to distinguish between various tomato diseases. The Region of Interest ROI identification process along with CNN architecture can capture discriminative features. To ensure the effectiveness of the proposed work, the model has been evaluated on a real-world dataset and Plant Village standard dataset and it is compared with several state-of-the-art CNN architectures. It is shown that the proposed model has achieved higher accuracy compared to other state-of-the-art models.

Keywords Image processing · Otsu thresholding · CNN · Recognition accuracy

1 Introduction

The main source of earning in India is agriculture. Many problems surfaced nowadays such as water scarcity, untimely rain, and plant diseases. Plant diseases are considered as one of the most important when not controlled it will have disastrous effects on the quality of food products. To successfully conduct crop diagnosis and control, crop disease identification plays a crucial role [1].

To detect the disease of the plants, human involvement becomes essential either for visual inspection or for laboratory assessment. However, with so many diseases and many crops, the farmer in a country like India cannot contact agriculture experts

S. Khan (✉)
Computer Engineering Department, M. H. Saboo Siddik College of Engineering, Mumbai, India
e-mail: saiqa.comp@gmail.com

M. Narvekar
Computer Engineering Department, D. J. Sanghvi College of Engineering, Mumbai, India

H. O. Bansal et al. (eds.), *Next Generation Systems and Networks*, Lecture Notes in Networks and Systems 641, https://doi.org/10.1007/978-981-99-0483-9_23

for timely assessment and treatment. Therefore, with mobile phone cameras and other handheld devices at their disposal, farmers can make their own decision by using application designed for this purpose. The primary benefit is to know the disease early and to apply the necessary treatments. It may lead to better productivity and cultivation. Tomato belongs to the Solanaceae family. Other vegetables in this family include eggplant, chilies, potatoes, etc. They all share similar characteristics. Therefore, their disease symptoms and identification process are almost the same. This research has worked mainly focused on biotic and abiotic disorders because it is a tedious process to locate the exact cause of the disease by visual inspection.

Crop disease recognition is divided into two types: Handcrafted feature representation and deep representation. Even though handcrafted features help in the recognition process. The most limiting factor for their usage is exhaustive pre-processing and selection of the most relevant features. These limitations motivated the usage of the deep feature representation method.

2 Literature Survey

Several authors have worked in the direction of disease identification. The leaf is the most observed part of the plant and identification of disease symptoms on leaves aid in the detection and classification process. Since the Inception of the first convolution neural network, researchers have overseen various real-world applications including image classification. With the emergence of powerful neural networks such as Alex Net [2], in the field of image processing CNN has achieved an unexampled quantum leap. The literature reported its use in the field of plant disease diagnosis and detection.

Zhong et al. in their work proposed a model based on the Residual block attention mechanism [3]. In recent times, the attention mechanism has gained a lot of importance because it assists in focusing on discriminating areas of disease. They have concluded that attention-based residual units can achieve 98% accuracy. However, the accuracy of the baseline CNN is 84%. To justify its effectiveness, the model was compared with several previous research works. Although the accuracy magnitude is quite high, the system worked under the constraint of a simple background and uniform lighting dataset.

In another work, Zong et al. explored the apple disease detection process using the DenseNet-121 deep convolution network. In their work, regression, multi-label classification, and focus loss function is investigated on six apple leaf diseases to achieve 93.71% accuracy on the test set, respectively [3]. Albert Cruz et al. have achieved an accuracy of 98.96% after applying augmentation on pre-processed images of grapevine yellows disease. They conducted several experiments ranging from applying transfer learning models to using a Support Vector Machine (SVM) for classification [4]. They have also included human subjective evaluation for assessment purposes.

Shanwen Zhang et al. have designed a CNN with the dilated kernel, and it is evaluated by considering segmented images. Their model GPDCNN is trained by mini-batch Stochastic Gradient Descent (SGD) with momentum factor [5]. They have utilized a technique of aggregating the multiscale contextual information, that tends to alleviate small disease area problems. They have demonstrated its effectiveness on the cucumber vegetable dataset.

Next, Ramesh et al. collected rice leaf images from the farm for different diseases including blast, sheath rot, brown spot, bacterial blight, and healthy [6]. They conducted pre-processing using Hue, Saturation, and Value (HSV) color space to separate diseased and non-diseased parts. A deep neural network is used for classification with the Jaya Optimization algorithm (DNN_JOA). To evaluate its overall effectiveness, the method is compared with past techniques such as ANN, DAE, and DNN.

For severity estimation, very few researchers have come forward with the intended solution. Data augmentation techniques are applied to reduce generalization errors. Authors have shown an accuracy of 95.245% for biotic disease detection and of 86.51% for severity estimation using ResNet-50 architecture [7]. Additionally, they presented results with a segmented dataset and were shown to have 97% accuracy for Coffee plantations.

Chen et al. have presented three related architectures on Rice and Maize crops for disease identification, including VGG+Inception, DenseNet+Inception and Segmentation+CNN architectures [8–10]. By considering transfer learning strategies alternately than randomly assigning weights, authors have used ImageNet dataset weights. Validation accuracy was around 91.83% on the public Plant Village dataset and 92% on the complex background dataset [11]. Table 1 represents recent work in the field of disorder detection by deep learning.

2.1 Research Gaps and Motivation

Following are the important research gaps identified:

1. It is difficult to identify tomato leaf characteristics using handcrafted features. There is a need to capture its intricacies in detail.
2. Machine learning models' outcomes depend on the selection of an optimal set of features. Therefore, automatic feature set derivation is of utmost importance.
3. At the time of designing models, it is necessary to identify parameters usage and hence, memory requirement. For real-world deployment, model accuracy vs computational efficiency must be monitored.
4. To ensure generalization, model training in real-world settings is required.

2.2 Research Contributions

Our main contributions to this research work are summarized as follows:

Table 1 Recent literature survey work

Sr. No	Work Cited	Plant species	Stress type	Pre-processing	DL Architecture	Sample type	Issues not addressed
1	Thangaraj et al. (2020)	Tomato	Biotic	Resize	Modified X-caption	Plant Village	Severity of disease Absence of Real-world images
2	Atila et al. (2020)	38 species	Biotic	Resize	EfficientNet	Plant Village	Memory constraints Absence of Real-world images
3	Mohit et al. (2020)	Cucumber	Biotic	Resize	From scratch	Farm images	Occlusion
4	Pardede et al. (2020)	Apple, corn and potato Tea leaves	Biotic	Resize	Modified architecture	Plant Village and own	Not evaluated on different conditions
5	Jiang et al. (2020)	Rice	Biotic	Mean shift segment	CNN + SVM	Own dataset	It is not mentioned how many layers and how many neurons are optimal for the architecture

5. We have collected tomato leaf images for biotic and abiotic disorders by visiting different farms across Maharashtra. Many abiotic disorders such as Drought stress or chemical injury resemble symptoms of the biotic disorder.
6. We have addressed the issue of removing cluttered backgrounds by focusing only on the Region of Interest (ROI) and designing our own ANN to train on real-world images.
7. We conducted extensive experiments to assess the validity of the proposed methodology. Visualization techniques are explored to show an impetus for improving network performance using visualization.

The research article is organized as (1) Section 2 gives details for data collection and proposed methodology (2) Section 3 presents the experimental setting and Section 4 shows comparative analysis. Finally, the conclusion is presented.

3 Materials and Methods

Different sources are used for data collection. The primary was the visit to various ecological farms where images are captured during different seasons and at different time scales to capture real-world operating settings. Occlusion is the main element in all the images and the main research gap was to have high-accuracy results on a real-world dataset. Many authors have demonstrated tremendous performance in the past. However, the ablation study shows that the efficacy of all results is on the standard Plant Village dataset where laboratory settings are maintained. We have captured images using different mobile phones. Another data source is the use of Internet-downloaded tomato leaf images. Figure 1 shows an example of a tomato leaf image.

(a) Alternaria+leaf miner (b) Bacterial leaf spot (c) Bacterial wilt (d) Early blight

(e) Healthy (f) Insect attack (g) Late blight (h) Leaf miner

(i) Mosaic (j) Physiological disorder (k) Powdery mildew (l) Septoria leaf spot

Fig. 1 Sample dataset images

3.1 Proposed Methodology

This research aims to design a novel CNN network that assists in the identification of tomato biotic and abiotic disorders. The architecture designed here identifies the region of interest and only ROI images would be given as input to the proposed network. ROI identification resembles human visual system functioning where the brain focuses on primary features and suppresses secondary details. So, we are restricting our CNN to focus on relevant areas by ignoring occlusion details.

(1) RGB to HSV Color Space conversion

HSV is a representation of human perception of colors in digital images and it is device dependent in nature. HSV colors are represented in Hue, Saturation, and Value elements [12]. In the proposed work, RGB color model is converted to its HSV counterpart.

(2) Thresholding

Thresholding is a technique used to perform image segmentation. Global and Local thresholding are the two main types of thresholding where global thresholding is further divided into traditional (Otsu), iterative (triclass), and multistage. Technically, general thresholded image $g(x,y)$ can be defined as

$$g(x, y) = \begin{cases} 1 \text{ if } f(x, y) > 1 \\ 0 \text{ if } f(x, y) \leq 1 \end{cases} \tag{1}$$

where $f(x,y)$ denotes the input image. In particular, Otsu's thresholding accesses all the possible threshold values and calculates for each pixel level, measure of spread that lies on each side of the threshold. In other words, it captures the pixels that fall in the background or foreground. Another popular technique of thresholding is the k-means clustering algorithm. The main drawback of k-means is to give the number of clusters as input.

(3) Convolution Neural Network (CNN)

With the advent of deep learning, CNN has become extremely popular. It encourages automatic extraction of features through convolution, pooling, and different activations. CNN can be designed in two ways either from scratch or a with transfer learning approach. Models such as VGG16, VGG19, ResNet, Inception, and DenseNet have proven effective in image classification. However, as they have been trained on the ImageNet database. Therefore, they are inefficient in capturing the intricacies of leaf image symptoms.

(4) Proposed work

In our proposed work, we focused on designing our own network. The input image is resized and converted to the HSV color model. Channel S is optimal for the image segmentation process. In HSV space, saturation is a component that assesses colorfulness. Therefore, it is selected for further processing. The Otsu thresholding is applied to find a mask and different contours are calculated using

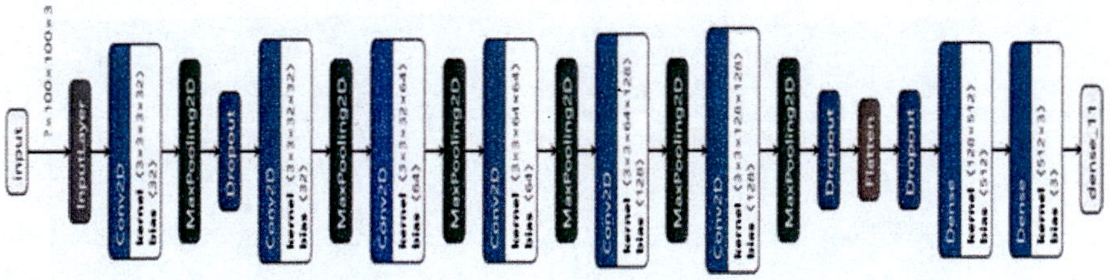

Fig. 2 Proposed deep learning model architecture

the chain approximation process. The largest contour that represents the most dominating region of the image, i.e., the leaf, is chosen from all the contours discovered in the previous steps. This selected part is given as input to the deep learning model (Fig. 2). The proposed deep learning model's architecture is described in Fig 2. It consists of 19 layers. The combination of convolution, pooling, and dropout to reduce overfitting is used. Initially, 32 filters are used to capture minute details. Then, 64 have been used to capture high-level details as we move further down in the model. As the infected region is small for some leaf diseases, therefore a smaller number of filters followed by more are used. The architecture of the deep neural network is linear here where the output of one block is used as an input for the other and eventually the last layer output acts as a classifier to predict the correct disease.

4 Experimental Results and Discussion

4.1 Experimental Platform

All experiments are executed on a machine with Windows 10,64-bit operating system with Intel® Core™ i5-7200U CPU @2.5 GHZ having 12.0 GB and NVIDIA Geforce 920 MX. Python is the programming language used. The training is performed on Google Colab. The deep learning framework used is Tensorflow 1.14 and Keras 2.2.5.

a. **Training parameters**

Here, the end-to-end training is conducted in multiple batches to improve performance on given hardware settings. The Batch size considered is 50. 100 is the number of iterations set for training. However, an early stopping mode is selected to reduce the overfitting effect. Cross-entropy loss is selected as the loss function. The main classification is done using softmax activation. The weight initialization is set using the Xavier model.

The Adam model with 0.001 as the starting learning rate is utilized for optimization. If the validation accuracy does not decline after three iterations, the model training is considered complete, and the process ends. The input image size for the model is set at 100 by 100 pixels.

Fig. 3 ROI for the input image **a**. Input image **b**. Leaf dominant image

b. **Model evaluation index**

Recognition accuracy is the main evaluation index, and it is given by the following.

$$\text{Recognition Accuracy} = (\text{Number of correctly identified diseased images}/\text{Total Number of diseased images tested}) \times 100 \quad (2)$$

Here, multiclass classification occurs.

c. **Comparative Evaluation**

ROI is selected using the process demonstrated in Section 3.1 and results are shown in the following Fig. 3 where background complexities are removed, and only the dominant leaf part is selected as the region of interest.

4.2 Evaluation of Our Own Constructed Dataset

Real-world images are divided into a ratio of 55%, 25%, and 20%, respectively, for training, validation, and testing datasets. Our constructed data set evaluation is carried out by analyzing validation accuracy and loss compared to training accuracy and loss. Results (Fig. 4) show that our model performs well on the training and validation dataset. Its testing accuracy is 86 %. The model is run for 50 epochs where an epoch is a word used in machine learning that refers to the number of passes the machine learning algorithm has made across the whole training dataset.

It is proven that its recognition accuracy is higher. The average training recognition accuracy of our model reached 97.54%. Figure 5 shows t-sne visualization for image data and Fig. 6 shows the confusion matrix. t-sne visualization is used to show a high-dimensional representation of features of training image data. The visualization shows clearly that inter-class variation is more and intra-class differences are less.

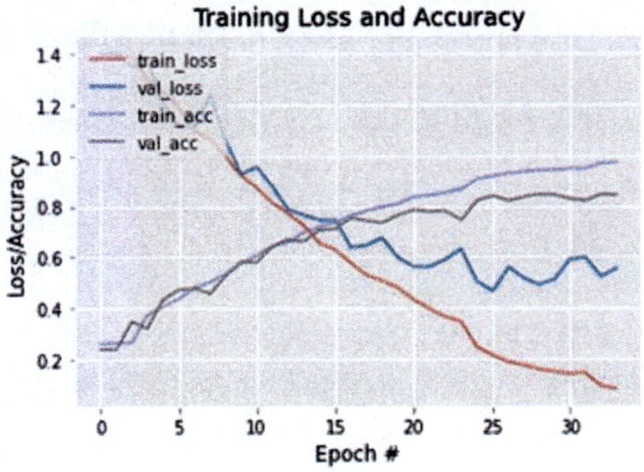

Fig. 4 Proposed model performance

This proves that features exhibited by the proposed CNN model can represent the diversity of the dataset well and this is clearly observable in the results represented in the confusion matrix where early blight, late blight, Septoria leaf spot, and healthy classes are shown as 0, 1, 2, and 3. True positives are shown diagonally in the confusion matrix and they are used to calculate the accuracy of the dataset.

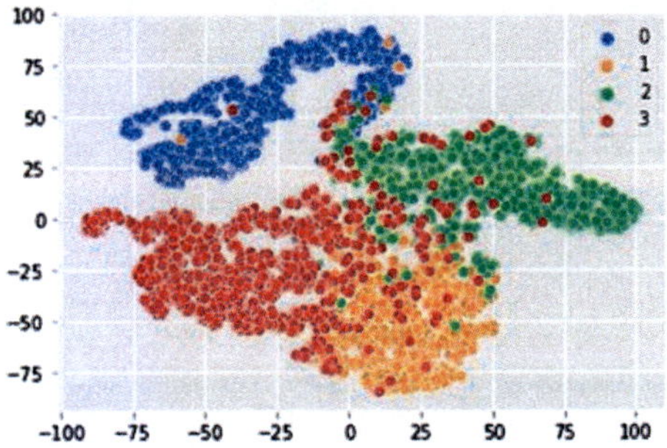

Fig. 5 t-sne visualization for input images

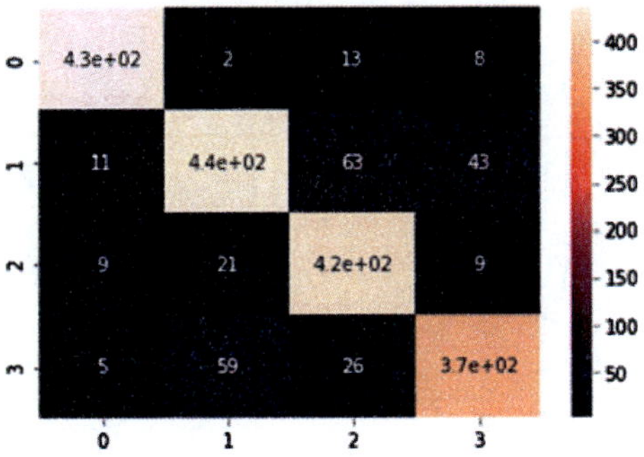

Fig. 6 Confusion matrix for test data

4.2.1 Evaluation with Plant Village Dataset

(https://github.com/spMohanty/PlantVillage-Dataset)

On a plant Village dataset, augmentation is applied to increase the number of images and to increase variations in the dataset. The training accuracy after 24 epochs is 96.29% and the validation accuracy is 91.03%.

4.2.2 Comparative Analysis with Other Models

Comparative analysis is carried out by training other models on the same Internet-downloaded dataset. Compared to the Plant Village dataset, Internet-downloaded images have many variations. Table 2 shows the performance of the proposed model compared to Karlekar et al. model [13], VGG+GAP model where VGG model is combined with the global average pooling method [14].

Visualization is conducted for hidden layers and results on the single-leaf image are presented in Fig. 7 [15, 16].

Table 2 Comparative results on Internet Downloaded dataset

Model	Training accuracy (%)	Training loss	Validation accuracy (%)	Validation loss
Karlekar et al. (without background removal)	29.19	1.3814	26.56	1.3868
VGG+GAP	68.79	0.8584	64.47	0.8961
Proposed model	94.33	0.1909	89.17	0.2494

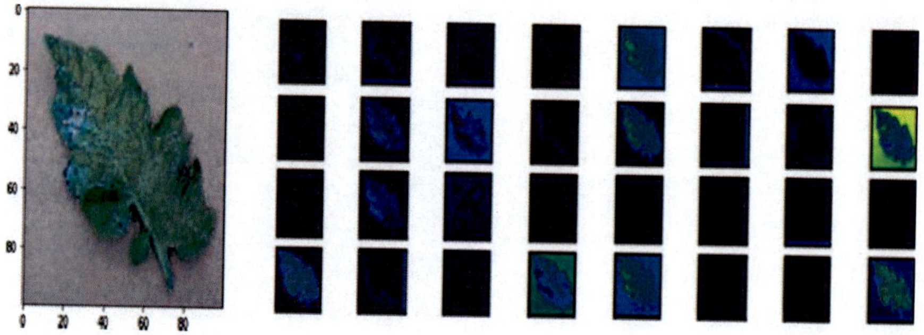

Fig. 7 Post-hoc Visualization **a** input leaf image, **b** feature maps for convolution layer

5 Conclusion

This research presents a work where it is proved that instead of directly giving input images to CNN. It is important to remove the surrounding background using thresholding and to identify the region of Interest using the contour method. The modified image is then given as input to CNN and tested on different datasets including standard Plant Village, real-world, and Internet datasets. The major challenge addressed in this study is determining the optimal input to the CNN, and result analysis shows that providing the optimized image consisting of leaf portion to the CNN network provides better accuracy than giving the original image.

Acknowledgements We appreciate the help received from Mr. Swapnil Dekhane, Research Officer at Tansa Farm, Mumbai, India. He facilitated us in data collection and classification. We would also like to thank Dr. Joshi, head of plant pathology and his team from Dapoli for briefing us on different diseases identification. On Internet downloaded dataset, system achieved 89.17% validation accuracy. In the future, we would like to extend the proposed work by including visualization techniques to improve accuracy and to optimize computational efficiency.

References

1. Li Z, Yang Y, Li Y, Guo R, Yang J, Yue J (2020) A solanaceae disease recognition model based on SE-Inception. Comput Electron Agric 178:105792
2. Krizhevsky A, Sutskever I, Hinton GE (2017) ImageNet classification with deep convolutional neural networks. Commun ACM 60(6):84–90
3. Zhong Y, Zhao M (2020) Research on deep learning in apple leaf disease recognition. Comput Electron Agric 168:105146
4. Cruz A, Ampatzidis Y, Pierro R, Materazzi A, Panattoni A, Bellis LD, Luvisi A (2019) Detection of grapevine yellows symptoms in Vitis vinifera L. with artificial intelligence. Comput Electron Agric 157:63–76
5. Zhang S, Zhang S, Zhang C, Wang X, Shi Y (2019) Cucumber leaf disease identification with global pooling dilated convolutional neural network. Comput Electron Agric 162:422–430
6. Ramesh S, Vydeki D (2020) Recognition and classification of paddy leaf diseases using Optimized Deep Neural network with Jaya algorithm. Inf Process Agric 7(2):249–260
7. Esgario JG, Castro PBD, Tassis LM, Krohling RA (2021) "An app to assist farmers in the identification of diseases and pests of coffee leaves using deep learning". Inf Process Agric
8. Chen J, Chen J, Zhang D, Sun Y, Nanehkaran YA (2020) "Using deep transfer learning for image-based plant disease identification". Comput Electron Agric 173: 105393. doi:https://doi.org/10.1016/j.compag.2020.105393.
9. Chen J, Zhang D, Nanehkaran YA, Li D (2020) Detection of rice plant diseases based on deep transfer learning. J Sci Food Agric 100(7):3246–3256. https://doi.org/10.1002/jsfa.10365
10. Chen J, Zhang D, Nanehkaran YA (2020) "Identifying plant diseases using deep transfer learning and enhanced lightweight network." Multimed Tools Appl 79(41–42): 31497–1515. doi:https://doi.org/10.1007/s11042-020-09669-w.
11. SpMohanty. "SpMohanty/PlantVillage-Dataset." GitHub. Accessed March 21, 2021. https://github.com/spMohanty/PlantVillage-Dataset.
12. Verma T, Dubey S (2019) Impact of color spaces and feature sets in automated plant diseases classifier: A comprehensive review based on rice plant images. Arch Comput Methods Eng 27(5):1611–1632. https://doi.org/10.1007/s11831-019-09364-6

13. Karlekar A, Seal A (2020) SoyNet: Soybean leaf diseases classification. Comput Electron Agric 172:105342. https://doi.org/10.1016/j.compag.2020.105342
14. Yan Q, Yang B, Wang W, Wang B, Chen P, Zhang J (2020) Apple leaf diseases recognition based on an improved convolutional neural network. Sensors 20(12):3535
15. Brahimi M, Boukhalfa K, Moussaoui A (2017) Deep learning for tomato diseases: classification and symptoms visualization. Appl Artif Intell 31(4):299–315
16. CNN Visualization: Methods Of Visualization. (2020, May 15). Retrieved September 04, 2021, from https://www.analyticsvidhya.com/blog/2018/03/essentials-of-deep-learning-visualizing-convolutional-neural-network.

Super Resolution-Based Channel Estimation

Sachin Agrawal, Pratham Agrasen, Praveen Kumar Shah, and Madhuri Bendi

Abstract For channel estimation of OFDM communication system, a deep learning (DL) technique is discussed in this paper. A two-dimensional picture is used to investigate the time–frequency response of a quickly fading communication channel. The goal of this work is to utilize certain known values at the pilot sites to identify the unknown values of the channel response matrix. A universal pipeline based on deep image processing approaches, image super-resolution (ISR), and image restoration (IRR) is presented to achieve this goal. In this technique, the pilot values are treated like a low-resolution picture and the channel estimation is performed by cascading the SR network along with a denoising IRR network. In addition, a working example of the suggested pipeline is presented. The estimated error demonstrates that the proposed approach is better than the previously used channel net approach which is comparable to the MMSE approach when all channel characteristics are known.

1 Introduction

To deal with frequency fading in wireless communication systems, OFDM (Orthogonal Frequency Division Multiplexing) is used. It is the most common modulation technique that has been used in this sector. When the signal is received at the receiver's end, it is distorted due to inherent features of communication channel. In order to get the original signal at the receiver's end, it is essential to compensate for the channel effect.

S. Agrawal (✉) · P. Agrasen · P. Kumar Shah · M. Bendi
National Institute of Technology, Delhi, India
e-mail: sachinagrawal@nitdelhi.ac.in

P. Agrasen
e-mail: 181220043@nitdelhi.ac.in

P. Kumar Shah
e-mail: 181220044@nitdelhi.ac.in

M. Bendi
e-mail: 202220007@nitdelhi.ac.in

H. O. Bansal et al. (eds.), *Next Generation Systems and Networks*, Lecture Notes in Networks and Systems 641, https://doi.org/10.1007/978-981-99-0483-9_24

In most cases, the receiver guesses channel matrix utilizing certain symbols known as pilots. Both transmitter (T_x) and the receiver (R_x), are aware about the location and value of these pilots. Three different structures can be explored based on these pilot arrangements: 1. block, 2. comb, and 3. lattice type [1]. In the first configuration, the pilots are sent recurrently at the commencing of an OFDM block at all sub-carriers, whereas, in the comb type, the pilots are present in some sub-carriers of a few OFDM symbols. In the lattice-type configuration, the pilots are placed along both the time and frequency axes with set times in a diamond-shaped structure.

Least Square (LS) and Minimum Mean Square Error (MMSE) are the most common pilot-based estimate approaches that use pilot values in time–frequency matrices to identify unknown channel matrix responses. These algorithms have already been tweaked to work in a variety of situations [2].

In contrast with the LS estimation method, which does not require knowledge of the channel's statistics, the MMSE estimation improves performance by including channel statistics and noise variance. In order to employ the MMSE in real-world applications, certain ways are described that simplify the complexity of the scheme by estimating channel statistics rather than using precise data.

The approximated linear variant of the MMSE (ALMMSE) which is implemented on fast-fading channels has low complexity than MMSE due to the smaller correlation and filtering matrices [3]. Recently, deep learning (DL) attracted a huge attention in communication system. The signal detection [4], modulation recognition [5], channel state information (CSI) feedback [6], channel equalization [7], and channel estimation [8, 9] have all been presented as ways to improve the performance of different traditional algorithms in DL-based communication systems. In this approach, communication systems are treated as a black box [8], and signal transmission/reception is accomplished via an end-to-end DL architecture.

The DL block inherently comprises channel estimation, decoding, encoding, and all other operations of any $T_x - R_x$ communication path. This approach is unable to obtain the channel time-frequency response explicitly, making it ineffective for applications that require the whole channel response.

The channel response matrix is treated as an image [9], and channel estimation is performed using a denoising network. In [10], the channel response is taken as an image followed by a cascaded deep learning approach of SR and IR. The channel between the transmitting (T_X) and receiving antenna in a multiple antenna space is the subject of the above study, which does not address the time–frequency response of each transmitter–receiver connection.

In this paper, an optimized deep learning-based framework is proposed for the estimation of channel response matrix in an OFDM communication system. In this technique, the time–frequency matrix of the channel is considered as a two-dimensional image with known pilot points. Here, the channel which will be estimated is regarded as a high-resolution (HR) image, whereas the channel response with known pilots is taken as a low-resolution (LR) image.

To estimate the channel matrix, a two-phase technique is proposed. The LR resolution is first improved using an image super-resolution (ISR) method. Afterward, the noise effects are removed using an image restoration (IRR) process. This paper's contributions are summarized as follows:

- Create a picture of the channel's time–frequency response.
- Develop the LR picture of the channel grid in the pilot locations and the predicted HR image of the estimated channel response.
- Estimate the channel response using deep learning-based picture using ISR and IRR networks.

The remaining paper is arranged as follows. Section 2 gives a quick overview of channel estimation using traditional approaches. The suggested DL-based channel estimator's structure is shown in Sect. 3. The simulation findings are reported in Sect. 4. Lastly, Sect. 5 concludes the paper.

2 Channel Estimation

The input–output relationship for the jth time period and ith sub-carrier in an OFDM system is describes as follows [12]:

$$Y_{i,j} = H_{i,j}X_{i,j} + Z_{i,j}. \tag{1}$$

Here, $Y_{i,j}$ represents the received signal, $X_{i,j}$ and $Z_{i,j}$ show the transmitted OFDM symbol and white Gaussian noise, respectively. In the case of OFDM subframe of size $N_S * N_D$, the range of the time period index j is [0, N_D-1], while the range of the sub-carrier index i is [0, N_S-1]. The (i, j) element of $H \in C^{N_S*N_D}$ is $H_{i,j}$.

For all sub-carriers and time periods, H reflects the channel's time–frequency response. H = h[1], h[2], ..., h[N_D], where each channel frequency response at jth time period is represented by h[j] is used for channel estimation, especially in fading channels.

At pilot sites, the LS technique calculates the channel response. If the LS estimated channel is represented as a diagonal grid $H_p^{LS} \in C^{N_P*N_P}$, then it can be calculated by solving [12]:

$$\hat{H}_P^{LS} = \arg_{H_p} min||y_p - H_pX_p||_2^2 \tag{2}$$

$||.||_2$ is the l2 distance and $H_p^{LS} \in C^{N_P*N_P}$ is the evaluated diagonal grid. The known pilot values are stored in x_p, while the associated observations are stored in y_p. $h_{LS} = y_p/x_p = diag(H_p^{LS})$ is the outcome of optimizing Eq. 2.

We must use a 2D interpolation approach to get the channel response value at coordinates other than the pilot positions.

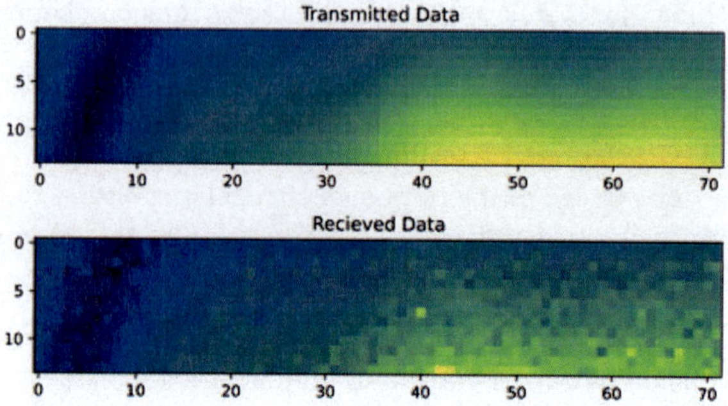

Fig. 1 T_X and R_X data as an image

3 Deep Learning-Based Channel Estimator (DLCE)

Considering the channel response matrix as a low-resolution and highly noisy two-dimensional image as shown in Fig. 1, various techniques can be implemented to produce a higher and better resolution image with low noise contents. The image super-resolution (ISR), a sub-method of deep learning could be used for resolution amplification in the images. It is observed that deep learning-based techniques, especially with fully and deeply convoluted neural networks (CNN), show better performance in retrieving high-resolution (HR) pictures from the low resolution (LR) pictures. Besides image super-resolution (ISR), image restoration (IRR) techniques can be used to reduce and remove the noise from the image. For instance, a feed-forward Denoising Convolutional Neural Network (DNCNN) is reported in [11]. It utilizes the residual learning and batch normalization to boost the training mechanism.

In the this work, SISO communication link, i.e., single input/single output link is considered between the pair of transmitter (T_x)–receiver (R_x) antennas. In this connection, the channel matrix "H" between a T_x and a R_x can be represented as a two-dimensional image which will act as an input to our model pipeline. Figure 2 depicts the proposed deep learning-based channel estimation (DLCE) pipeline. Here,

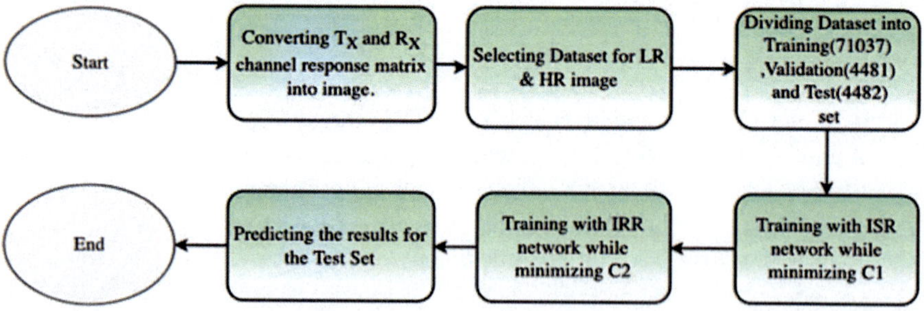

Fig. 2 The proposed DLCE pipeline

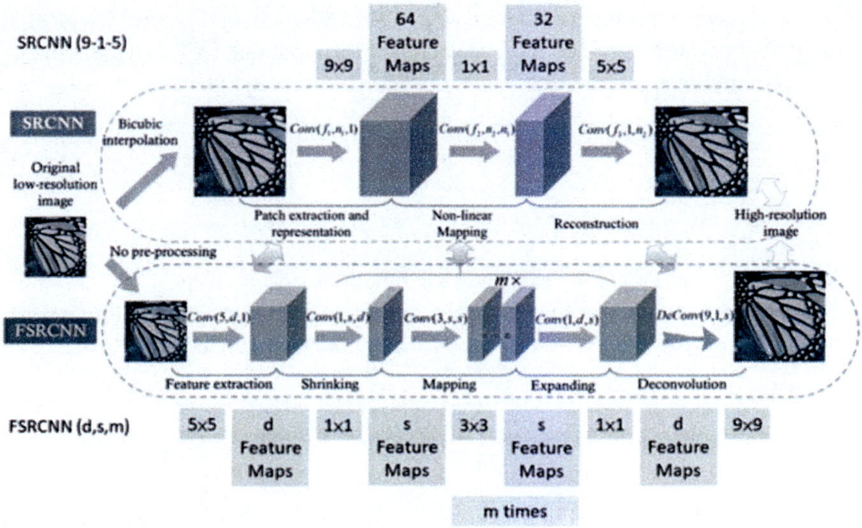

Fig. 3 FSRCNN [13] versus SRCNN model [10, 12]

the aim is to ascertain the complete time–frequency response of the channel utilizing the transmitted pilots. To get the whole channel response, a two-step training method can be implemented:

- Firstly, an ISR network is applied which takes the vectorized low-resolution 2D image as an input to the pipeline.
- Secondly, to reduce the additional noise, a denoising IRR network is implemented over the ISR network outputs to give the final estimated channel matrix.

For the ISR and IRR, we have used FSRCNN [13] and DNCNN [11], respectively. As shown in Fig. 1. The FSRCNN uses the following steps:

- Feature extraction—Firstly, it does feature extraction with 5×5 convolutions.
- Shrinking—Reduction in feature maps for optimization.
- Non-Linear Mapping—Multiple 3×3 layers are applied.
- Expanding—The feature map is then increased by a 1×1 convolutions.
- Deconvolution—HR image is then reconstructed using 9×9 filter.

The above model is better for channel estimation than the previously used SRCNN model [12] as compared in Fig. 3. At the beginning, a layer of 64 filters (3×3 x 1) is implemented and cascaded with a ReLU. Afterward, 18 convolutional layers of 64 filters of size $3 \times 3 \times 64$ are implemented. At last, a filter of size $3 \times 3 \times 64$ is realized in order to reconstruct the output. The network parameters set is denoted as

$$\theta = \theta_{ISR}, \theta_{IRR} \quad (3)$$

where θ_{ISR} represents the parameter values set of ISR network, whereas θ_{IRR} signifies the IRR parameter values set. The vector h_p^{LS} is input to DLCE and the estimated channel output response matrix is given by

$$\hat{H} = f(\theta; h_p^{LS}) \tag{4}$$

$$\hat{H} = f_{IRR}(f_{ISR}(\theta_{ISR}; h_p^{LS}); \theta_{IRR}) \tag{5}$$

where f_{ISR} and f_{IRR} are the ISR and IRR functions, respectively.

The total loss function which has to be minimized to achieve the optimized result from the training of the network is the MSE. It can be calculated as follows:

$$C = \frac{1}{||N||} \sum_{h_p \in N} ||f(\theta; h_p^{LS}) - H||_2^2 \tag{6}$$

where N is the size of training set. As we follow a two-step approach, we first minimize the error C1 of the ISR network, then we fixed the weights of the ISR network and minimize the error C2 for the IRR network to get the optimized result from the model pipeline. Here, C1 and C2 are calculated as follows:

$$C1 = \frac{1}{||N||} \sum_{h_p \in N} ||Z - H||_2^2 \tag{7}$$

$$C2 = \frac{1}{||N||} \sum_{h_p \in N} ||\hat{H} - H||_2^2 \tag{8}$$

where Z is the result of the ISR network.

4 Simulated Result

In this section, first of all, the network is trained and then MSE is calculated for two different values of SNRs. The obtained results are compared with the conventional baseline algorithms and shown in Figs. 4 and 5. It can be observed that in both cases the DLCE performs better than the Channel Net [12], which already outperforms the existing MMSE and ideal ALMMSE methods. For our deep learning-based channel estimation, Tensorflow and Keras with a GPU backend are employed to implement the proposed pipeline. The ISR and IRR networks are implemented with the training rate of 0.001 and batch size of 128 with at most 400 iterations. The training, validation, and testing sets comprise of 71037, 4482, and 4481 channels, respectively.

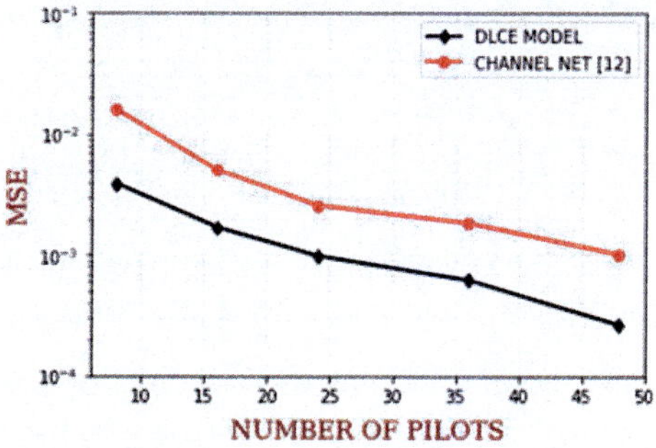

Fig. 4 Comparison of DLCE and Channel Net [12] at SNR:22 dB

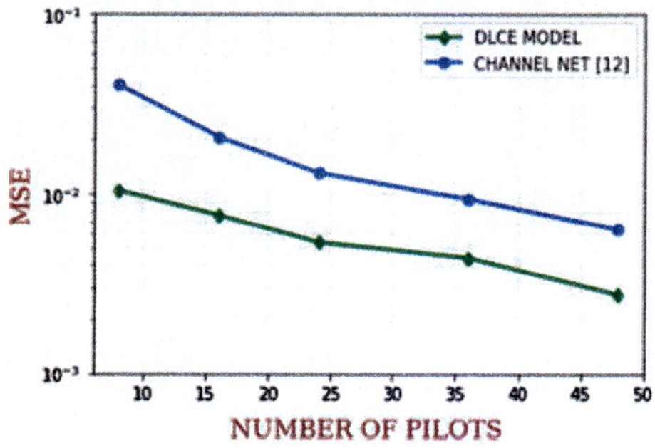

Fig. 5 Comparison of DLCE and Channel Net [12] at SNR:12 dB

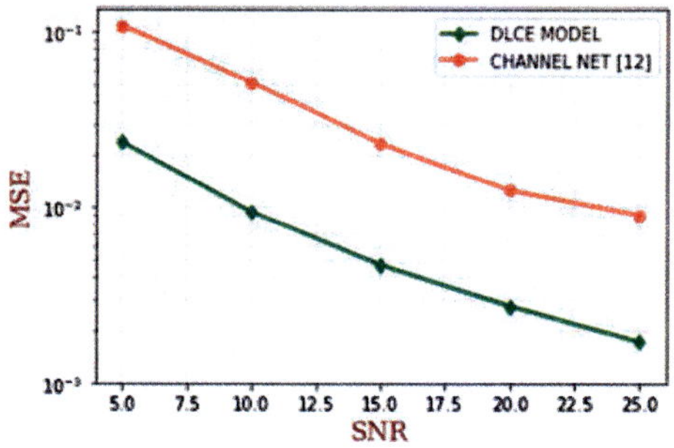

Fig. 6 Comparison of DLCE and Channel Net [12] over different SNRs

Figure 6 shows the MSE as a function of SNR when 48 pilots are used in each frame. As seen, the proposed method shows better MSE versus SNR compared to Channel Net [12].

5 Conclusion

In this paper, two algorithms ISR and IRR are utilized to find out the channel response. The MSE is calculated for different values of SNR and pilots values. The simulation results show that the DLCE performs better than the earlier reported method ChannelNet.

References

1. Coleri S, Ergen M, Puri A, Bahai A (2002) Channel estimation techniques based on pilot arrangement in OFDM systems. IEEE Trans Broadcast 48:223–229
2. Li Y, Cimini LJ, Sollenberger NR (1998) Robust channel estimation for OFDM systems with rapid dispersive fading channels. IEEE Trans Commun 46:902–915
3. imko M, Mehlfhrer C, Wrulich M, Rupp M (2010) Doubly dispersive channel estimation with scalable complexity. In: 2010 international ITG workshop on smart antennas (WSA), pp 251–256
4. Samuel N, Diskin T, Wiesel A (2017) Deep MIMO detection. In: 2017 IEEE 18th international workshop on signal processing advances in wireless communications (SPAWC), pp 1–5
5. OShea T, Hoydis J (2017) An introduction to deep learning for the physical layer. IEEE Trans Cogn Commun Netw 3:563–575
6. Wen C, Shih W, Jin S (2018) Deep learning for massive MIMO CSI feedback. IEEE Wirel Commun Lett 1–1
7. Erdogmus D, Rende D, Principe JC, Wong TF (2001) Nonlinear channel equalization using multilayer perceptrons with information theoretic criterion. In: Neural networks for signal processing XI: proceedings of the 2001 IEEE signal processing society workshop (IEEE Cat. No.01TH8584), pp 443–451
8. Ye H, Li GY, Juang B-H (2018) Power of deep learning for channel estimation and signal detection in OFDM systems. IEEE Wirel Commun Lett 7(1):114–117
9. He H, Wen C, Jin S, Li GY (2018) Deep learning-based channel estimation for beamspace mmWave massive MIMO systems. IEEE Wirel Commun Lett 7:852–855
10. Dong C, Loy CC, He K, Tang X (2016) Image super-resolution using deep convolutional networks. IEEE Trans Pattern Anal Mach Intell 38:295–307
11. Zhang DK, Zuo W, Chen Y, Meng D, Zhang L (2017) Beyond a gaussian denoiser: residual learning of deep CNN for image denoising. IEEE Trans Image Process 26:3142–3155
12. Soltani M, Pourahmadi V, Mirzaei A, Sheikhzadeh H (2019) Deep learning-based channel estimation. IEEE Commun Lett 23(4):652–655
13. Dong C, Loy CC, He K, Tang X (2015) Image super-resolution using deep convolutional networks. TPAMI 38(2):295–307

Robust Multi-Spectral Palm-Print Recognition

Poonam Poonia and Pawan K. Ajmera

Abstract In recent years, biometric has emerged as the most trustworthy and reliable technique for verifying or identifying humans. The distinctive and consistent qualities of palm prints have led to a rise in their use as a biometric attribute for user access and authentication. A palm-print template is proposed that stores the relative geometric information of the minutiae points. It is impossible to determine the orientation of the palm-print from the template since it lacks information on the orientation of the minutiae and their position. For template matching, an internal angle based on Delaunay triangulation is used, which then produces matching scores for various spectral bands. The multispectral data is finally integrated using a score-level fusion approach that minimizes the overlapping effects. The technique was evaluated on the widely used PolyU multispectral palm-print database, which provides more accurate results than a mono-spectral band. The total Correct Recognition Rate (CRR) is 96.28%, while the Equal Error Rate (EER) is 0.16%.

Keywords Fusion · Template matching · Delaunay triangulation · Minutiae · EER

1 Introduction

In today's era of advanced technology, no key is more secure than a biometric key. It is an automated way for person authentication based on unique physical or behavioral attributes [1–4]. As a more promising biometric feature, palm-print-based recognition is an active research area. Palm-print image consists of unique attributes which can be characterized as texture features and can be computed in the presence of wrinkles, ridges, and creases [5]. Among different palm-print features, minutiae points are used as significant features for person authentication [6]. Minutiae points such as core, island, termination, and bifurcation are extensively used in the area of finger and palm biometrics. Delaunay triangulation is widely opted in the area of

P. Poonia · P. K. Ajmera (✉)
Department Electrical and Electronics Engineering, Birla Institute of Technology and Science, Pilani, India
e-mail: pawan.ajmera@pilani.bits-pilani.ac.in

H. O. Bansal et al. (eds.), *Next Generation Systems and Networks*, Lecture Notes in Networks and Systems 641, https://doi.org/10.1007/978-981-99-0483-9_25

face and fingerprint recognition. In order to extract various geometries and quantitative characteristics from the generated triangles, Khan et al. [7] combined the Niblack binarization algorithm with the Delaunay triangulation method. In the presence of random position disruption, Delaunay triangle structures [8] cannot change. In [9] a net-based Delaunay triangulation method was proposed. The method extracts fingerprint features that are invariant to translation, rotation, and scaling of the matching template and the query image. Soleymani and Amirani [10] designed a hybrid matching algorithm by combining the Delaunay triangulation and Voronoi diagram.

The paper mainly focuses on making of a non-invertible template. The method of Delaunay triangulation is used to extract minutiae features (ridge bifurcation and ridge termination) from different spectral bands. Finally, score-level fusion is applied to fuse multispectral information.

The rest of the paper is organized as follows: Section 2 describes the proposed palm-print recognition system. Section 3 summarizes the experimental results and discussion of the proposed methodology on the standard database. Section 4 concludes the proposed work.

2 Proposed Method

2.1 Pre-Processing

Data collection of palm prints serves as the first phase. The next phase is pre-processing, which comes after gathering images of high quality. Then, image enhancement module is applied by using the Gabor filter that enhances perceptibility of ridge structures of the palm-print images as discussed in [11]. Block diagram of the proposed methodology is shown in Fig. 1. The images obtained are shown in Fig. 2(a)–(e). After the enhancement module, next step is binarization which converts the image of type double to an image of type logical, i.e., zeros and ones. Further, a thinning algorithm is used to eliminate the redundant pixels as discussed in [12]. It is based on the intensity of neighboring pixels as well as the pixel under consideration. Figure 2(c) shows the images obtained after thinning algorithm.

2.2 Feature Extraction

Minutiae extraction: Minutiae are the key elements of the proposed experimentations. These points consist of various features such as core, island, termination, and bifurcation. In the present approach two minutiae points, ridge bifurcation and ridge termination are used. According to the general observations, type and location of each minutiae point are obtained from the thinned images.

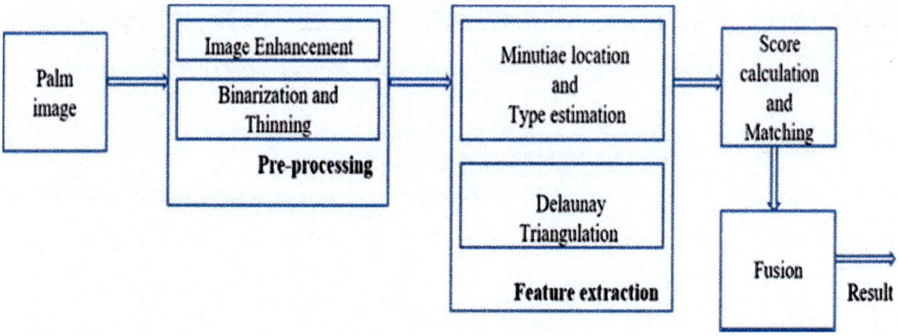

Fig. 1 Block diagram representation of the proposed approach

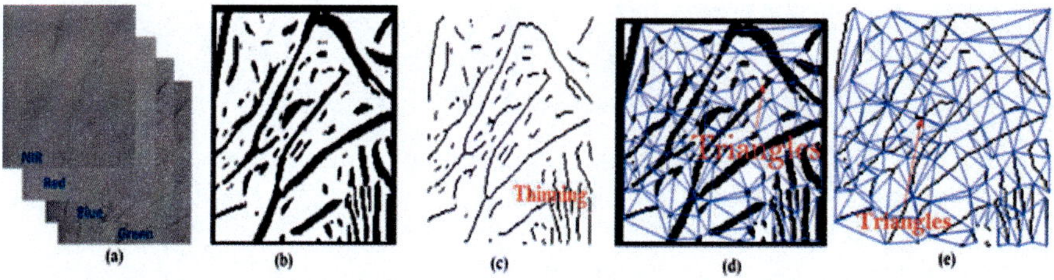

Fig. 2 **a** Multi-spectral palm-print images with different bands (Green, Blue, Red, NIR) **b** image after enhancement **c** Thinned image **d** Delaunay triangles on enhanced image **e** Delaunay triangles on thinned image

At a point *P (i, j)* the high-intensity pixel is given in Eq. (1):

$$= \sum_{i=1}^{i+1} \sum_{j=1}^{j+1} P(i, j) \tag{1}$$

A high-intensity pixel at the center of a 3 × 3 window, ridge ending occurs when the number of high-intensity pixels is 2. In the case of ridge bifurcation, the number will be 4.

Delaunay triangulation: A technique for creating original triangles in 2D or 3D space is called Delaunay triangulation [13]. The method has the ability to provide high robustness to local disturbances. In the testing phase, some of the minutiae points are not sensed. This happens due to numerous reasons such as faulty device or improper placement of palm on the measuring device.

The majority of the other triangles would remain intact, and the suggested template would only be altered locally. In the case of zooming in or out, template maintains the triangle internal angles intact. Hence, template is robust to zooming as well. The images obtained after Delaunay triangulation are shown in Fig. 2(d)–(e). Before calculating the match score, the type of minutiae and the angles in the triangles are computed. Representation of triangle features is shown in Fig. 3.

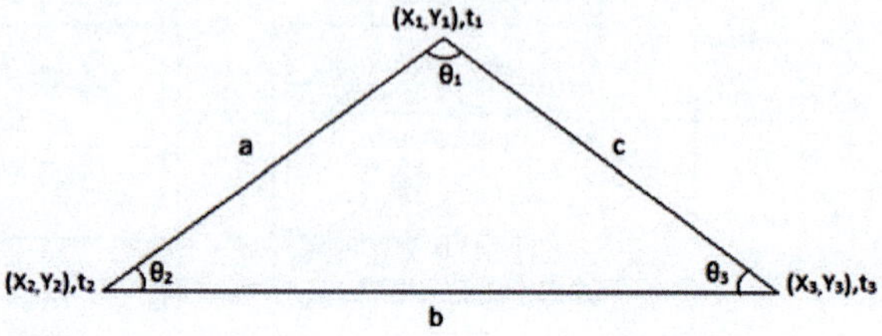

Fig. 3 Representation of triangle features

The sides and angle of a triangle are calculated as given in Eq. (2) and (3):

$$a = \sqrt{(X_1 - X_2)^2 + (Y_1 - Y_2)^2} \tag{2}$$

$$\phi_2 = \left(\frac{180}{pi}\right)\cos^{-1}\left(-\frac{a^2 - \left(b^2 + c^2\right)}{2bc}\right) \tag{3}$$

where *(xi, yi)* are 2D coordinates of the location of *i*th minutiae point and *(a, b, c)* are the lengths of the sides of a triangle. θ*i*, (*i* = 1, 2, 3) are the three internal angles as shown in Fig. (3).

Finally, feature vector is calculated as given in Eq. (4):

$$FV_i = (\phi_1, \phi_2, \phi_3, t_1, t_2, t_3) \tag{4}$$

where *(i = 0)* to number of triangles in the template and the t_i the minutiae type for the respective vertices.

Score calculation and matching: For each combination of the given test template with each of the other test templates in the database, used one at a time, a distinct score is determined.

The score formula is computed as given in Eq. (5):

$$score_f = \frac{\left(count_m + \frac{count_{nm}}{length(T_{db})}\right)}{\left(\frac{length(T_{db}) + lenth(T_{in})}{2}\right)} \tag{5}$$

where T_{in} is the template of the input image and template T_{db} of database image. $count_m$ and $count_{nm}$ number of matches and not matches attempts, respectively.

The match score of the given test image and an individual image from the database, considering all the combinations of the '*n*' feature vectors FV_{in} and '*m*' feature vector FV_{db}, respectively.

The algorithm for score calculation is given below:

1. procedure
2. for $a \leftarrow$ length (test template T_{in})
3. $count_m \leftarrow 0$

$count_{nm} \leftarrow 0$
$match_{flag} \leftarrow$ false
for $b \leftarrow length\ (T_{db})$
$\Delta\theta_{bc} \leftarrow \min(T_{db}[b,c] - T_{in}[a,c])$
$\Delta\theta_{b,avg} \leftarrow \text{avg}(T_{db}[b,c] - T_{in}[a,c])$
if $(\Delta\theta_{bc} < \Delta\theta_{tol})\Lambda(\Delta\theta_{b,avg} < \Delta\theta_{b,avg})\Lambda$(minutiae type)
$\Delta\theta_{bc} \leftarrow \Delta\theta_{tol}$
$match_{flag} \leftarrow$ true
end if
if $match_{flag} = =$ true then
$count_m \leftarrow count_m + 1$
delete $T_{db}[b]$ from T_{db}
else
$count_{nm} \leftarrow count_{nm} + 1$
end if
end for
end for
Mat_Sc$(T_{in}, T_{db}) \leftarrow score_f$
end.

2.3 *Score Level Fusion*

Based on the general observations, more discriminative information yields better performance [14, 15]. For the fusion of different spectral bands, a simple sum rule is applied. Due to the overlapping of different spectral information, the simple summing rule may not provide as much the final accuracy. Therefore, score-level fusion scheme is investigated to reduce the overlapping effect that further improves the performance [16].

The weighted sum is calculated as given in Eq. (6):

$$d_{ws} = \sum_{k=1}^{n} w_k d_k \tag{6}$$

where W_k is the weight on the distance of kth band d_k and n is number of bands and d_{ws} is distance using weighted sum.

3 Experimental Results and Discussion

The proposed framework is validated on PolyU multispectral palm-print database [17] having four spectral bands: (a) Blue, (b) Green, (c) Red, (d) NIR (near infrared).

Table 1 EER and CRR without fusion

Spectral bands	Blue	Green	Red	NIR
EER (%)	0.393	0.402	0.375	0.412
CRR (%)	78.5	78.41	78.41	77.08

The database contains 6000 palms from 250 volunteers for one illumination. A total of 24 palm images are captured from each volunteer for a single illumination.

The samples are acquired in two different sessions with an average interval between 2 sessions was about nine days. Equal Error Rate (EER) and Correct Recognition Rate (CRR) are used to calculate the performance of the proposed approach. The EER values for mono-spectral bands are given in Table 1. The minimum value of EER is 0.375% and 0.393% for red and blue mono-spectra bands, respectively. The maximum CRR value is 78.5% on Blue spectral as given in Table 1. Blue and Red spectra bands give better performance. The results of the fusion approach are tabulated in terms of EER and CRR in Table 2. The fusion of spectra bands gives the better results in comparison to mono-spectra. Table 2. shows the overall EER value of 0.16% for (Blue and Red) and (Blue, Red, Green, and NIR). The EER value of 0.17% for (Blue, Red, and NIR) and (Blue, Red, and Green) bands.

The highest CRR value is 96.28% for fused spectra bands of (Blue, Red, and NIR) and (Blue, Red, Green, and NIR) as given in Table 2. The peak CRR for all spectral bands is achieved using 100 feature vectors. Figure 4 shows the ROC curve for fused spectra bands.

4 Conclusion

In this paper, a non-invertible template, Delaunay triangle internal angle was proposed that based on the local geometry of minutiae features for palm-print recognition. The proposed method uses minutiae transform algorithm that is proven to be robust to the local disturbances. The template is robust against the variations due to rotation, translation, and distortion. The fusion scheme of spectral bands provides more discriminative information which yields better results in comparison to mono-spectral information. The proposed recognition system yields CRR and EER of 96.28% and 0.16%, respectively.

For future scope, explore some image enhancement techniques in the preprocessing stage to improve the recognition performance. Aim of getting better results we can try to fuse fingerprint and palm-print features using the proposed system.

Table 2 EER and CRR with fusion

Spectral bands	Fusion of spectral bands										
	Blue and Green	Blue and Red	Blue and NIR	Green and Red	Green and NIR	Red and NIR	Blue Green Red	Blue Green NIR	Blue Red NIR	Green Red NIR	Blue Green Red NIR
EER (%)	0.25	0.16	0.22	0.24	0.245	0.19	0.17	0.18	0.17	0.18	0.16
CRR (%)	89.17	96.28	88.34	92.01	87.19	94.11	96.22	90.56	96.22	90.14	96.28

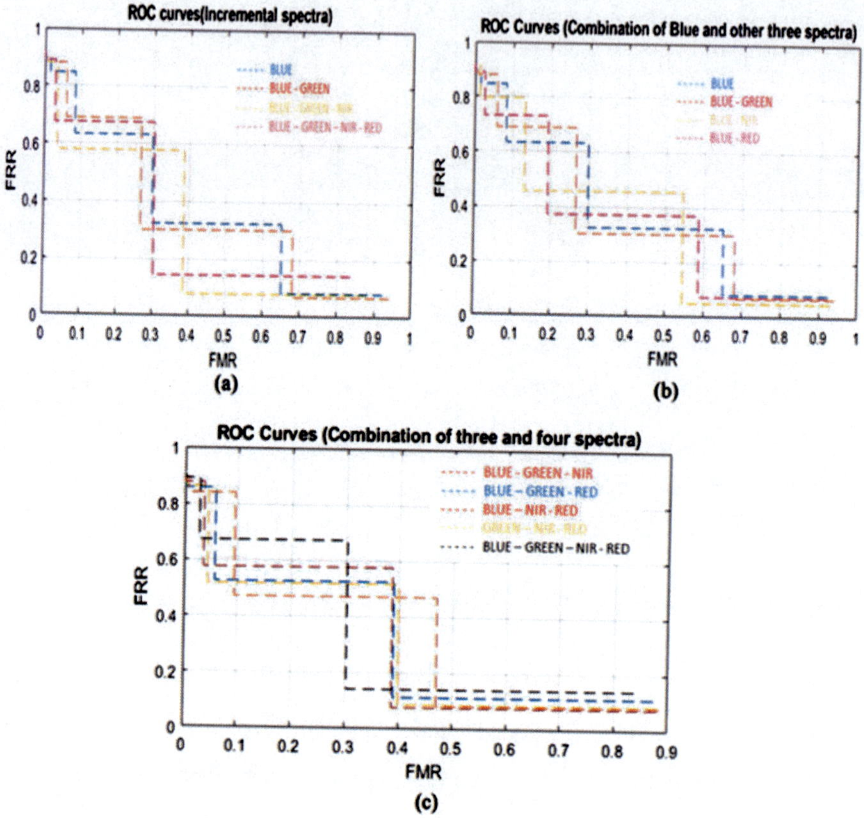

Fig. 4 ROC curve for **a** Fused incremental spectral bands **b** Fused spectral bands with Blue illumination **c** Fusion of three and four spectral bands

References

1. Ross J, Prabhakar S (2004) An introduction to biometric recognition. IEEE Trans Circuits Syst Video Technology 14(1):4–20
2. Ajmera PK, Jadhav DV, Holambe RS (2011) Text-independent speaker identification using Radon and discrete cosine transforms based features from speech spectrogram. Pattern Recogn 44(10–11):2749–2759
3. Zhang D, Kong WK, You J, Wong M (2003) On-line palm print identification. IEEE Trans Pattern Anal Mach Intell 25(9):1041–1050
4. Han CC, Cheng HS, Lin CL, Fan K (2003) Personal authentication using palm-print features. Pattern Recogn 36(2):371–381
5. Nigam A, Gupta P (2015) Designing an accurate hand biometric based authentication system fusing finger knuckleprint and palmprint. Neurocomputing 151:1120–1132
6. Hu D, Feng G, Zhou Z (2007) Two-dimensional locality preserving projections (2dlpp) with its application to palmprint recognition. Pattern Recogn 40(1):339–342
7. Khan ZK, Qureshi RJ, Ahmad J (2015) On feature based Delaunay triangulation for palmprint recognition. J Platf Technology 3(4):9–18

8. Bebis G, Deaconu T, Georgiopoulos M (1999) Fingerprint identification using Delaunay triangulation. In: IEEE Conf. on Information Intelligence and System. Springer pp. 452–459
9. Parziale G, Niel A (2004) A fingerprint matching using minutiae triangulation. Int Conf Biom Authentication 3072:241–248
10. Soleymani R, Amirani MC (2012) A hybrid fingerprint matching algorithm using Delaunay triangulation and Voronoi diagram. 20th Iran Conf Electr Eng (ICEE) 2: 752–757
11. Hong L, Wan Y, Jain A (1998) Fingerprint image enhancement: algorithm and performance evaluation. IEEE Trans Pattern Anal Mach Intell 20(8): 777 789
12. Lam L, Lee SW, Suen CY (1992) Thinning methodologies—a comprehensive survey. IEEE Trans Pattern Anal Mach Intell 14(9):869–885
13. Abellanas M, Hurtado F, Ramos PA (1999) Structural tolerance and Delauny triangulation Inform. Process Lett 71:221–227
14. Srivastava GS, Bhardwaj S, Bhargava S (2016) Fusion of palm-phalanges print with palmprint and dorsal hand vein. Appl Soft Computing 47: 12–20
15. Jaswal G, Kaul A, Nath R (2018) Multiple feature fusion for unconstrained palm print authentication. Comput Electr Eng 72:53–78
16. Zhang D, Guo Z, Lu G, Zhang L (2010) An online system of multispectral palmprint verification. IEEE Trans Inst Measurement 59(2):480–490
17. PolyU multispectral palmprint database. Available at http://www4.comp.polyu.edu.hk/biometrics/MultispectralPalmprint/MSP.htm.

Lightweight Authentication Scheme for Fog-Based Smart Grid Network

Priya Deokar, Sandhya Arora, and Aarti Amod Agarkar

Abstract Smart grid is the future for electricity generation and utilization. Two-way communication between the participating nodes requires a secure environment for message transfer. Individual smart meter's data must be securely handled in the network. In this paper, we have proposed the fog-based lightweight authentication scheme which preserves the privacy of smart meter's data and also satisfies the security requirements. The proposed scheme uses hash and exclusive-OR operations for pairwise key generation and is lightweight in terms of communication and computation overhead. The scheme is simulated using the AVISPA tool and is secure against the simulated attack.

Keywords Authentication · Privacy preservation · Smart grid network · Security · Fog computing

1 Introduction

Smart grids are recognized as the future of power grids and are currently being deployed across the world, in developed as well as developing countries. The number of smart electricity meters is forecasted to reach 142.8 million in North America by 2024. The Asia-Pacific is believed to be the largest market for smart meters and instaled smart meter base is forecasted to reach 975 million units by 2024 [1].

P. Deokar (✉)
Smt. Kashibai Navale College of Engineering, Cummins College of Engineering for Women, Pune, India
e-mail: priyadeokar23@gmail.com

S. Arora
Cummins College of Engineering for Women, Pune, India
e-mail: sandhya.arora@cumminscollege.in

A. Amod Agarkar
Marathwada Mitra Mandal's College of Engineering, Pune, India
e-mail: pratibha26@gmail.com; aartiagarkar@mmcoe.edu.in

H. O. Bansal et al. (eds.), *Next Generation Systems and Networks*, Lecture Notes in Networks and Systems 641, https://doi.org/10.1007/978-981-99-0483-9_26

The National Institute of Standards and Technology (NIST) framework considers the smart grid system as consisting of seven important domains, viz. Service Provider, Generation, Transmission, Distribution, Operations, Markets, and Customer [2, 3]. For smooth functioning of smart grids, actors in these domains need to interact with each other. Data collection from the smart meters and processing is the responsibility of Advanced Metering Infrastructure (AMI) which is one of the important components of a smart grid system. Two-way communication between electricity providers and consumers is facilitated by a smart grid. They also have self-healing capability. Smart meters periodically collect consumer data from household appliances and send that to suppliers.

Fog computing paradigm fits well into the hierarchical structure of smart grids. Use of fog nodes in smart grids can offload computing and data processing tasks to the edge of the network thereby reducing latency of communication. Fog nodes can also be used to perform billing operations.

Fog computing can definitely be a contributing factor to smart grid but it brings security challenges along with it. As fog nodes, smart meters are located at consumer's premises and use public channels for communication, they are vulnerable to tampering and other active attacks by the adversaries. Authentication protocol for smart grid communications should be designed in such a manner that unauthorized users do not get access to entities involved in the communication. Authorized entities should be given access based on the role they have in the system. This is considered as the user authentication part. Other than this, based on the communicating entities, pairwise keys should be generated between them. This forms the key management part. As entities such as smart meters are resource-constrained devices, techniques used for addressing this problem should not contain computationally heavy cryptographic operations.

Our research work is based on symmetric key cryptography and light cryptographic techniques using hashing. The goal of our proposed technique is to provide data confidentiality and availability and, at the same time, maintain the integrity of data. Additionally, message exchanges are minimized between communicating entities and used light cryptographic techniques thereby making the operations more efficient considering the communication and computational costs.

Below are the specific contributions of this paper:

1. The proposed scheme incorporates a fog paradigm which reduces latency of communication.
2. The paper presents a scheme which is based on symmetric key cryptography and hashing technique. As both of these techniques are light and efficient, it reflects in the communications of this scheme.
3. Communicating entities are authenticated and key generation, renewal using lightweight techniques is part of the scheme.

The paper is organized as follows. Section 2 represents the literature survey, Sect. 3 defines the system model. The proposed fog-based authentication scheme is described in Sect. 4. Simulation of the scheme using AVISPA is presented in

Sect. 5. Section 6 describes the security analysis followed by performance evaluation in Sect. 7. Section 8 presents the conclusion.

2 Literature Survey

Authentication techniques are very important for maintaining security and privacy of smart grid systems. Identity-based authentication, ECC-based authentication, authentication schemes using lattice-based cryptography, schemes based on hash–XOR operations are majorly used approaches in the literature.

Hu et al. [4] **used face identification for authentication which satisfies integrity, confidentiality, and availability of the system. Badar et al.** [5] proposed a scheme which makes use of an identity-based authentication technique for smart grid communications. In their scheme, physical unclonable function is used. Tsai et al. [6] employed identity-based encryption (IBE) and identity-based signature (IBS) for anonymous key distribution. Their scheme is claimed to be secure under a random oracle model. Later on, Odelu et al. [7] found out that scheme [6] doesn't provide security and privacy of smart meter credentials needed for shared key generation.

Schemes [7–10] used elliptic curve cryptography for their work. The advantage of ECC is, it is lightweight in terms of computational overhead. The lattice-based cryptography techniques are also becoming popular. Schemes [11–13] proposed lattice-based authentication and key agreement protocol which are resilient to quantum attacks and are lightweight.

Prosanta Gope [14] proposed a scheme based on multi-factor authentication. Physically unclonable functions, reverse fuzzy extractor, and one-way hash function are employed in this scheme. He suggested the use of PUF to protect against physical attacks. Wazid et al. [15] proposed the SAKA-FC scheme which is based on hash function and bitwise XOR computations for IoT devices and biometric fuzzy extractors for fog servers. Lu et al. [16] protocol is quantum-resistant and assumes three-layer network structure. Their scheme is focused on microgrids and security of the scheme relies on the learning parity with noise (LPN) problem which is an NP-complete problem.

Authentication in smart grids is a widely studied problem and many schemes are proposed in the literature to address it. Considering the previous work and security loopholes, our aim is to propose a system that is lightweight and suitable for fog environments.

3 System Model

This section presents the details of network model, adversary model, and security goals.

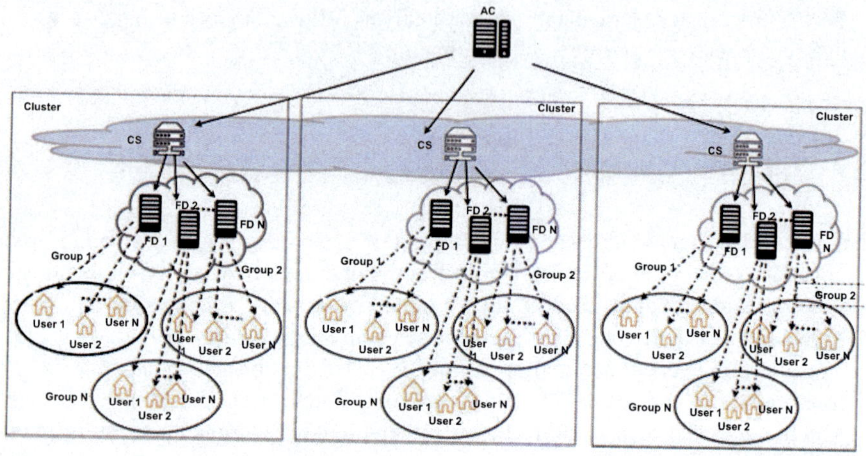

Fig. 1 A Fog-based smart grid network

3.1 *Network Model*

The goal of our scheme is to provide security, data integrity, and availability. The network model shown in Fig. 1 consists of Smart Meters (SM), Fog Nodes(FN), Cloud Servers (CS), and Analysis Centre (AC). Devices in these layers are logically grouped together to form several clusters. Each cluster consists of a Cloud Server, multiple fog nodes, and smart meters in the NAN of fog nodes.

Following are the responsibilities of entities involved in the network.

- Smart Meter periodically collects data from appliances and sends encrypted data to Fog nodes.
- Fog Nodes decrypt data received from smart meters and aggregate the data. Fog Nodes may also store this data for billing and analysis purposes. Aggregated data is then encrypted with the shared symmetric key between fog node and cloud server and sent to the cloud server.
- Cloud Server collects data from the fog nodes, aggregates data received from all the fog nodes from the cluster, and sends the data to the Analysis center.
- Analysis Center is the topmost layer which receives consumption data of the smart grid. All this data can be analyzed further.

3.2 *Adversary Model*

- AC, FD, and CS are considered honest-but-curious entities. They don't launch active attacks and are part of the system.

- $\mathcal{A}$ is an adversary who can eavesdrop on the communication between entities that are part of the system model. $\mathcal{A}$ is even capable of launching other active attacks.

4 Proposed Authentication Scheme for Fog Based Smart Grid

The proposed scheme works in two phases: Initialization phase and data transfer phase. Figure 2 shows the sequence diagram, and Table 1 shows the symbols and its description used in the proposed scheme. Operations performed during these phases are explained with respect to a single cluster.

Initialization Phase Key generation process is initiated by AC. Symmetric keys are generated between communicating parties at each hierarchical level.

1. Shared symmetric key generation between AC and CS: AC and CS have a shared password which is PSW_c. Nonce is generated at AC, n_A which is used to compute $A_{T1} = h\,(ID_A||PSW_c||ID_C) \oplus n_A$ and $A_{T2} = h\,(ID_A||n_A) \oplus TS_{ac}$. TS_{ac} is the transaction sequence number maintained at AC. Message $M1 : (A_{T1}, A_{T2})$ is sent to CS. AC also computes $K_{AC} = h\,(A_{T1}||n_A||TS_{ac})$ which is a shared symmetric key between AC and CS. CS finds values of n_A and TS_{ac} from the received message. Nonce is generated using $n_A = h\,(ID_A||PSW_c|ID_C) \oplus A_{T1}$ and transaction sequence number is generated using $TS_{ac} = h\,(ID_c||n_A) \oplus A_{T2}$. CS also computes shared symmetric key using $K_{AC} = h\,(A_{T1}||n_A||TS_{ac})$.

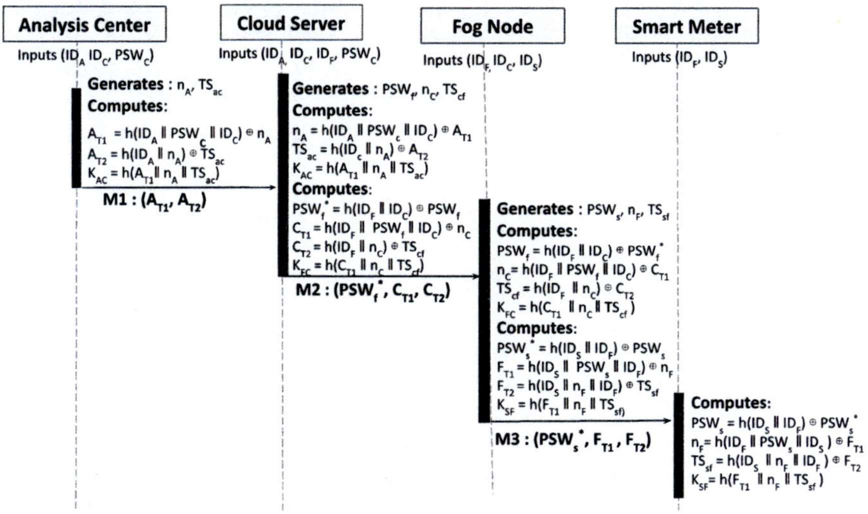

Fig. 2 Symmetric key generation

Table 1 List of symbols

Symbol	Description
H(.)	Hash Function
$\oplus$	Exclusive-OR operation
$\|\|$	Concatenation operation
SM	Smart Meter
FN	Fog Node
CS	Cloud Server
AC	Analysis Center
ID_S	Identity of Smart Meter
ID_F	Identity of Fog Node
ID_C	Identity of Cloud Server
ID_A	Identity of Analysis Center
PSW_c	Shared password between Analysis Center and Cloud Server
PSW_f	Shared password between Fog Node and Cloud Server
PSW_s	Shared password between Fog Node and Smart Meter
TS_{ac}	Transaction sequence number between Analysis Center and Cloud Server
TS_{cf}	Transaction sequence number between Fog Node and Cloud Server
TS_{sf}	Transaction sequence number between Fog Node and Smart Meter
K_{AC}	Shared symmetric key between Analysis Center and Cloud Server
K_{FC}	Shared symmetric key between Fog Node and Cloud Server
K_{SF}	Shared symmetric key between Fog Node and Smart Meter

2. Shared symmetric key generation between CS and FN: CS generates password PSW_f. Nonce n_C is generated at CS which is used to compute $C_{T1} = h\left(ID_F||PSW_f||ID_C\right) \oplus n_C$ and $C_{T2} = h\left(ID_F||n_C\right) \oplus TS_{cf}$. TS_{cf} is the transaction sequence number maintained at CS. Message $M2 : \left(PSW_f*, C_{T1}, C_{T2}\right)$ is sent to FN. CS also computes $K_{FC} = h\left(C_{T1}||n_C||TS_{cf}\right)$ which is a shared symmetric key between CS and FN. FN finds values of n_C and TS_{cf} from the received message. Nonce is generated using $n_C = h\left(ID_F||PSW_f||ID_C\right) \oplus C_{T1}$ and transaction sequence number is generated using $TS_{cf} = h\left(ID_F||n_C\right) \oplus C_{T2}$. FN also computes shared symmetric key using $K_FC = h\left(C_{T1}||n_C||TS_{cf}\right)$.
3. Shared symmetric key generation between FN and SM: FN generates password PSW_s. Nonce n_F is generated at FN which is used to compute $F_{T1} = h\left(ID_S||PSW_s||ID_F\right) \oplus n_F$ and $F_{T2} = h\left(ID_S||n_F||ID_F\right) \oplus TS_{sf}$. TS_{sf} is the transaction sequence number maintained at FN. Message $M3 : (PSW_s*, F_{T1}, F_{T2})$ is sent to SM. FN also computes $K_{SF} = h\left(F_{T1}||n_F||TS_{sf}\right)$ which is a shared symmetric key between FN and SM. SM finds values of n_F and TS_sf from the received message. Nonce is generated using $n_F{=}h\left(ID_F||PSW_s||ID_S\right) \oplus F_{T1}$ and transaction sequence number is generated using $TS_{sf}{=}h\left(ID_S||n_F||ID_F\right) \oplus F_{T2}$. SM also computes shared symmetric key using $K_{SF} = h\left(F_{T1}||n_F||TS_{sf}\right)$.

Data Transfer Phase Once the symmetric keys are generated between the pair of controllers, the data transfer phase is introduced. During data transfer phase, smart meters send the electricity consumption data to fog nodes in a secured manner. Smart meter encrypts the electricity consumption data, M with the shared secret key with fog node as $C = E_{K_{SF}}(M)$. Fog nodes receive the message and decrypt the data as $M = D_{K_{SF}}(C)$. Fog node aggregates the data received from smart meters under its cluster and encrypts the aggregated data with K_{FC}. Cloud server receives the encrypted data from fog nodes, decrypts it, and saves the aggregated value of electricity consumption at its end. This electricity consumption data is used by the analysis center for defining the prediction about the electricity requirement in future. The communication between analysis center and cloud server happens with the help of shared symmetric key.

Key Renewal Phase The symmetric keys shared between the controllers are refreshed after a particular duration. Smart meter generates the nonce N_S, encrypts it with K_{SF}, and sends to fog node. Smart meter and fog nodes compute the new symmetric key as $K'_{SF} = K_{SF} \oplus N_s$
$K_{SF} = K'_{SF}$
where K'_{SF} is the new symmetric key computed based on the nonce N_S. In the same fashion, the symmetric keys between fog node and cloud server, cloud server, and analytical center are refreshed.

5 Simulation of the Scheme Using AVISPA

Simulation of initialization and symmetric key generation is done in the Automated Validation of Internet Security Protocols and Applications (AVISPA) tool. Pairwise symmetric keys are generated between Analysis Center and Cloud Server, Cloud Server and Fog Node, Fog Node, and Smart Meter. Figure 3 shows the simulation

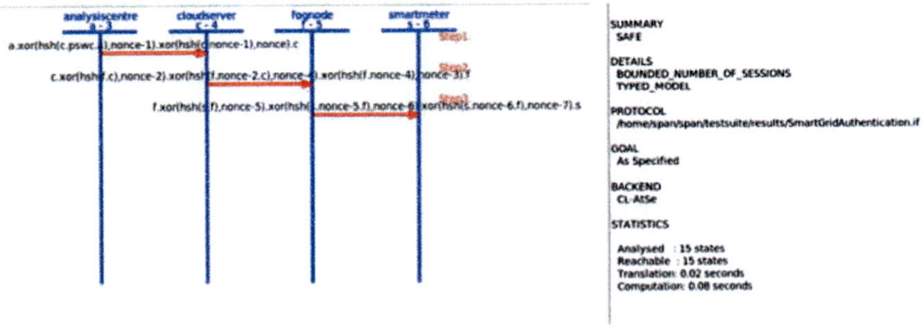

Fig. 3 Simulation of symmetric key generation

of messages sent during symmetric key encryption and the summary of simulation results using the CL-AtSe model. It indicates that the scheme is secure.

6 Security Analysis

Security analysis of the proposed system shows that the system is secure and preserves the privacy of individual smart meter readings while being transferred.

1. Privacy Preservation: In the proposed scheme, SM aggregates data from individual appliances and sends the aggregated value to FN. FN aggregates data from all the SMs and sends the aggregated value to CS. Thus, consumption data of individual appliance is not available to FN, CS, and AC thereby preserving privacy of data.
2. Authentication: In the proposed scheme, third party is not involved in the key generation and management process. Pairwise symmetric keys are generated mutually within the communicating parties with the help of ID, password, and Transaction Sequence Numbers at the respective parties. Data is encrypted using these respective symmetric keys.
3. Confidentiality: Encrypted data is transferred between the communicating parties. Pairwise Symmetric keys used for encryption are known only to those parties, viz SM-FN, FN-CS, and CS-AC. This ensures confidentiality.
4. Integrity: During key generation process, data which includes nonce, transaction sequence number, etc., is transferred in encrypted format. Along with this, hash values of the data are also sent to ensure integrity.
5. Data Freshness: During the key generation process, transaction sequence number is used which keeps track of the received messages. It also helps prevent replay attacks.

7 Performance Evaluation

The aim of our scheme is to provide a secure authentication and key management technique based on hierarchical data aggregation. It uses hash functions and exclusive-OR operations which require minimal execution time which makes the scheme lightweight. Performance of the system is evaluated based on its communication and computation overhead. Communication overhead is determined by the number of messages exchanged over the network among the communicating parties for key agreement. Computation overhead is determined by the total execution time required by each operation that is performed during the key generation process.

7.1 Communication Overhead

During the symmetric key generation phase, analysis center sends one message to each cloud server. Each cloud server sends one message to each fog node under its cluster. The fog nodes generate messages for the smart meters under their control. For example, consider that the scenario contains one analysis center, c number of cloud servers. Each cloud server communicates with f fog nodes, and every fog node handles with s smart meters. Thus, the number of messages generated by the analysis center is c, the number of messages generated by all cloud server is $c \times f$, and the number of messages generated by all fog nodes are $c \times f \times s$. Thus, the total number of messages generated is $c + c \times f + c \times f \times s$. During data transfer phase, after either 15 min or 30 min smart meters send the electricity consumption data to fog nodes. Thus, the number of message transfers from smart meters is s. Each fog node sends one message to cloud server containing the aggregated electricity consumption value. The data of electricity consumption is saved at cloud server. Analysis server fetches data from cloud server when needed for prediction purpose. When the analysis center sends the query to cloud server about the electricity consumption data, cloud server sends a single message to analysis center.

7.2 Computation Overhead

The proposed scheme makes use of hash and exclusive-OR operations for pairwise symmetric key generation. In data transfer phase, the data is encrypted using shared symmetric key at each level in the hierarchy. Consider T_{HSH} is the time required for computation of hash function, T_{XOR} is the time required for computation of exclusive-OR operation, T_{DR} is the time required for decryption operation, T_{Aggr} is the time required for computation of aggregation operation. Table 2 shows the number of operations performed at each node.

We have compared our protocol with [8]. To compare the protocol, we have considered the scenario which contains one SM, one FN, and one CS. The [8] protocol requires $4T_{pm} + 25T_h$ where T_{pm} is the ECC point multiplication and requires 20.23 ms. T_h is the hash function and needs 0.03 ms. Thus, the total time requirement is 81.67 ms. For our proposed scheme, the total time requirement is $22 \times T_{HSH} + 16 \times$

Table 2 Computation overhead during symmetric key generation

Message	Analysis center	Cloud server	FN	SM
M1	$3T_{HSH} + 2T_{XOR}$	$3T_{HSH} + 2T_{XOR}$	–	–
M2	–	$4T_{HSH} + 3T_{XOR}$	$4T_{HSH} + 3T_{XOR}$	–
M3	–	–	$4T_{HSH} + 3T_{XOR}$	$4T_{HSH} + 3T_{XOR}$
TOTAL	$3T_{HSH} + 2T_{XOR}$	$7T_{HSH} + 5T_{XOR}$	$8T_{HSH} + 6T_{XOR}$	$4T_{HSH} + 3T_{XOR}$

T_{XOR}. The time requirement of exclusive-OR operation is negligible. Hence, total time requirement for our protocol is 0.66 ms which is very less compared to [8]. Thus, our scheme is lightweight in terms of computation and communication overhead.

8 Conclusion

In smart grid, a large amount of personal electricity consumption data is shared on the network. Privacy preservation of the smart meter readings is the need for smart grid network. In this paper, we have presented the fog-based authentication scheme which preserves privacy of the smart meters' data and at the same time satisfies the security requirements including authentication, confidentiality, and integrity. The scheme is efficient in terms of communication overhead and computation overhead. The scheme is simulated using AVISPA and results showed that the scheme is secured against MITM attack.

References

1. Global Smart Metering Markets. https://www.globenewswire.com/news-release/2019/07/22/1885589/0/en/Global-Smart-Metering-Markets-2018-2019-Forecast-to-2024.html
2. Guidelines for Smart Grid Cybersecurity. https://nvlpubs.nist.gov/nistpubs/ir/2014/NIST.IR.7628r1.pdf
3. Gopstein A, Nguyen C, O'Fallon C, Hastings N, Wollman D (2021) NIST framework and roadmap for smart grid interoperability standards, release 4.0: special publication (NIST SP), national institute of standards and technology: Gaithersburg, MD. https://doi.org/10.6028/NIST.SP.1108r4
4. Hu P, Ning H, Qiu T, Song H, Wang Y, Yao X (2017) Security and privacy preservation scheme of face identification and resolution framework using fog computing in internet of things. IEEE Internet Things J 4(5):1143–1155
5. Badar HMS, Qadri S, Shamshad S, Ayub MF, Mahmood K, Kumar N (2021) An identity based authentication protocol for smart grid environment using physical uncloneable function. IEEE Trans Smart Grid 12(5):4426–4434. https://doi.org/10.6028/NIST.SP.1108r4
6. Tsai J, Lo N (2016) Secure anonymous key distribution scheme for smart grid. IEEE Trans Smart Grid 7(2):906–914. https://doi.org/10.1109/TSG.2015.2440658
7. Odelu V, Das AK, Wazid M, Conti M (2018) Provably secure authenticated key agreement scheme for smart grid. IEEE Trans Smart Grid 9(3):1900–1910. https://doi.org/10.1109/TSG.2016.2602282
8. Chen CM, Chen L, Huang Y et al (2021) Lightweight authentication protocol in edge-based smart grid environment. J Wirel Commun Netw 68. https://doi.org/10.1186/s13638-021-01930-6
9. Zhang L, Tang S, Luo H (2016) Elliptic curve cryptography-based authentication with identity protection for smart grids. PLoS ONE 11(3): e0151253 . https://doi.org/10.1371/journal.pone.0151253
10. Abbasinezhad-Mood D, Nikooghadam M (2018) An anonymous ECC-based self-certified key distribution scheme for the smart grid. IEEE Trans Ind Electron 65(10):7996–8004. https://doi.org/10.1109/TIE.2018.2807383

11. Yang Y, Huang J, Chen J, Han X, Zhao Y, Wei Z (2020) INAKA: improved authenticated key agreement protocol based on newhope. IEEE Access 8:41764–41773. https://doi.org/10.1109/ACCESS.2020.2977137
12. Bos J et al (2018) CRYSTALS - kyber: a CCA-secure module-lattice-based KEM. In: IEEE European Symposium on Security and Privacy, pp 353–367. https://doi.org/10.1109/EuroSP.2018.00032
13. Feng Q, He D, Zeadally S, Kumar N, Liang K (2019) Ideal lattice-based anonymous authentication protocol for mobile devices. IEEE Syst Jo 13(3):2775–2785. https://doi.org/10.1109/JSYST.2018.2851295
14. Gope P (2020) PMAKE: privacy-aware multi-factor authenticated key establishment scheme for advance metering infrastructure in smart grid. Comput Commun 152:338–344. ISSN 0140-3664. https://doi.org/10.1016/j.comcom.2019.12.042
15. Wazid M, et al (2018) Design of secure key management and user authentication scheme for fog computing services. Futur Gener Comput Syst. https://doi.org/10.1016/j.future.2018.09.017
16. Lu S, Li X (2021) Quantum-resistant lightweight authentication and key agreement protocol for Fog-based microgrids. IEEE Access 9:27588–27600. https://doi.org/10.1109/ACCESS.2021.3058180
17. Kilinc HH, Yanik T (2014) A survey of SIP authentication and key agreement schemes. IEEE Commun Surv Tutor 16:1005–1023. https://doi.org/10.1109/SURV.2013.091513.00050

Human Emotion Recognition Using Facial Characteristic Points and Discrete Cosine Transform with Support Vector Machine

Kanchan S. Vaidya, P. M. Patil, and Mukil Alagirisamy

Abstract This paper introduces a method for human emotion recognition using facial characteristic points (FCPs) and discrete cosine transform (DCT) coefficients as the feature vectors, extracted from the human emotion facial images. In order to classify the emotions based on the feature vectors, Support Vector Machine is used as a classifier. For experimentation purpose linear function, polynomial, non-homogenous polynomial, sigmoid, Gaussian, and radial basis functions are used as kernels. In this work two types of features as facial shape features and appearance features are used for facial emotion recognition. In the first type thirty FCPs are extracted as the facial shape features, whereas in the second type first hundred coefficients of DCT are used as the feature vectors. In order to extract the FCPs, the facial image is localized and divided into various regions. Performance of the proposed algorithm has been evaluated with the help of confusion matrices, average accuracy, and timing analysis on seven human emotion facial images from Japanese Female Facial Expression (JAFFE), and in-house generated databases independently. The databases is divided in two parts as train and test databases in the ratio of 70% and 30%, respectively. The performance is compared with the amount of confusion for the same number of the class created on the same database. The highest average accuracy obtained with the proposed algorithm for emotion recognition is 92.86%. The results show that the radial basis function of SVM is more suitable for person-associated situation and the linear function describes person-irrelevant problems better.

Keywords Facial emotion recognition · Facial characteristic points · Discrete cosine transform · Support vector machine · Confusion matrix · Face localization

K. S. Vaidya (✉)
Lincoln University, Petaling Jaya, Malaysia
e-mail: vaidyaskanchan@gmail.com

P. M. Patil
SNDCoERC, Nashik, Maharashtra, India
e-mail: drpmp66@gmail.com

M. Alagirisamy
Lincoln University, Petaling Jaya, Malaysia
e-mail: mukil.a@lincoln.edu.my

H. O. Bansal et al. (eds.), *Next Generation Systems and Networks*, Lecture Notes in Networks and Systems 641, https://doi.org/10.1007/978-981-99-0483-9_27

1 Introduction

Emotions play an important role in human life because humans express emotions through daily interactions [1]. In general, seven types of facial emotions are considered, and various facial expressions such as happiness, sadness, surprise, anger, fear, disgust, and a neutral face indicate a person's inner emotions and emotional states [2]. Facial expressions can be captured by the movement of muscles with the characteristics of eyebrows, mouth, nose, eyes, cheeks, and chin, but facial expressions are not considered neutral. Recent advances in computer vision and machine learning have made it possible to detect emotions from facial images. Facial expression recognition is one of the most active research topics due to its wide range of uses in the field of human–computer interaction [3, 4]. Facial emotion recognition involves finding the regularity of the dataset to analyze. These regularities can be used to recognize emotions as well as faces. Various techniques are used to perform tasks, but they fall into two broad classes: Parameterization method and detection method [5]. It is argued that effective intelligent human–computer interaction requires computers to interact with users in a natural way, just as humans interact [6]. Face recognition of human emotional states can be an invaluable tool if you want to achieve more effective human–computer interaction (HCI). Therefore, facial sentiment analysis is the most important task in emotional computing, human–computer interaction, and computer vision [7]. Over the last two decades, significant efforts have been made by academia, industry, and government to discover more robust ways to assess truthfulness, deception, and credibility to improve HCI.

From the Literature review it is observed that most of the research work is reported on recognition of facial expressions. The researchers have used expressions including happiness, sadness, surprise, anger, fear, disgust, and a neutral face indicate a person's inner emotions and emotional states. More researchers have focused to increase the human computer interaction. Very less research work is reported on texture based appearance features and shape features.

The present paper focuses on facial emotion recognition based on facial characteristics points. For the recognition of emotions discrete cosine transform along with support vector machine classifier is used. The experimentation used linear function, polynomial, non-homogenous polynomial, sigmoid, Gaussian, and radial basis functions as kernels.

The objective of this paper is to enhance the accuracy percentage using SVM classifier with radial basis function kernel and discrete cosine transform coefficients on specific databases.

This paper is organized as follows: Section 2 describes the localization of the facial part of a monochrome and color facial image of human emotions, followed by the extraction of FCP. Section 3 describes the theoretical aspects of the DCT and SVM classifiers used with the various kernels used in SVMs. Intermediate steps, including mathematical background and corresponding diagrams, are also mentioned where appropriate. Section 4 contains a detailed description of the various steps involved in the proposed algorithm. Section 5 shows the results of the proposed algorithm for

JAFFE and the internally created database, along with performance comparisons for confusion matrix, average accuracy, and time analysis, and Section 6 concludes.

2 Face Localization and FCPs Extraction

The human face is a highly correlated object because there is no change in skin color. Facial features (i.e., mouth and eyes) are different in color from skin color, but areas of hairless skin dominate the surface of the face and stand out against most backgrounds. Therefore, it is expected that there will be a border around the face. In addition, the scalp and hair usually trigger the contours of the edges that lengthen the tap on the face. As a result, both face and head structures can be expected to act as blobs or symmetrical enclosures around the center point. Detecting and identifying faces in a single image are the most important part of almost any facial recognition system. Face localization aims to determine the image position of the face for the purpose of verifying documents such as passports, driver's licenses, and ID cards.

In this section, we have developed two face localization algorithms. The first algorithm was developed to extract the facial part from a monochrome image, and the second algorithm was developed to extract the facial part from a color image. Localization algorithms for extracting facial parts of human emotional images consist of two main stages. In the first stage, the column variance of the facial image is calculated. In the second stage, the facial part (excluding hair) of the facial image was extracted through the skin filter.

2.1 Face Localization for Monochrome Facial Image

The ultimate goal of face localization is to find objects in the image that resemble the shape of the face. Many algorithms have been proposed for face position and recognition based on the use of shape [8–10], color information [11], and movement [12]. These methods can be divided into four categories: Knowledge-based methods, functionally immutable approaches, template matching methods, and appearance-based methods [13].

2.1.1 Phase 1: Face Localization

The face localization algorithm is based on the fact that the variance has peaks at the edges of the face region of the image (see Fig. 1). It extracts the face part from the background and reduces the image size. The various steps involved in the algorithm are shown below:

***Step* 1**: Calculate the variance for each column of each image.

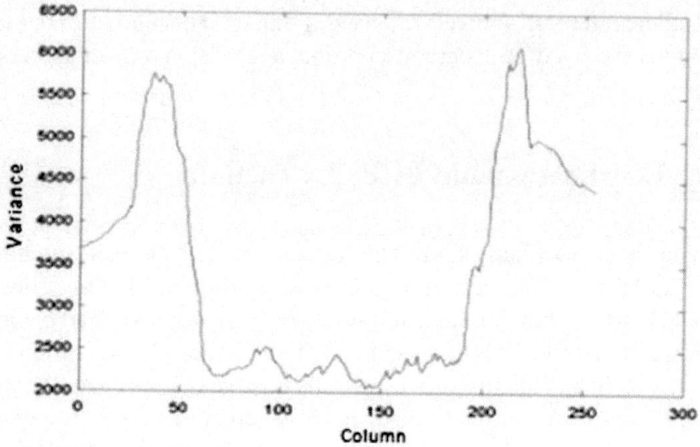

Fig. 1 Column variance for cropping of the facial portion of an image

***Step* 2**: Select the column *m*, with maximum value ν. Check if this column is on the right or left side of the image. If this column is close to the edge of the image, select the highest value column on the other side.

***Step* 3**: Select the column *k* (empirical), with $0.2 \times \nu$ near the margin as first edge of new cropped image. Crop a new image that is half the size of the original image based on *k*.

***Step* 4**: Resize the cropped image and vectorize it as the desired normalized image.

2.2 *Face Localization for Color Facial Image*

The color face image algorithm works in two phases to extract the part of the face. The first phase is the same as that used to crop facial parts from a monochrome facial image using column distribution. In the second stage, fine adjustments are made through the skin filter to extract only the facial area (excluding hair). In general, it has been observed that the background of all images in the available database is flat and the width of most faces of each image is half the width '*w*'.

2.2.1 Phase 2: Fine tuning using skin filter

The skin color detection algorithm is used to fine-tune a colored face image after performing a crop on the face section (as described in Phase 1). Doing so will cause the trimmed face to pass through the skin filter and generate a map on the skin color pixels. This map is used to extract facial features from intensity features. The skin filter is based on the Fleck and Forsyth algorithm. Input color image should be in *RGB* format with color intensity values ranging from 0 to 255. The *RGB* matrices are "zeroed" to supposedly prevent desaturation when the image is converted from RGB color space to $I\ R_g\ B_y$ color space. A minimum intensity value greater than 10 pixels from the edge of any of the three color planes is set as the null response of the image. This value is subtracted from all three color planes.

The RGB image is transformed to log-opponent ($I\ R_g\ B_y$) values and from these values the texture amplitude, hue, and saturation are computed. The conversion from RGB to log-opponent is calculated according to a variation on the formula given by Fleck and Forsyth as,

$$I = \frac{L(R) + L(B) + L(G)}{3} \tag{1}$$

$$R_g = \frac{L(B) - [L(G) + L(R)]}{2} \tag{2}$$

$$B_y = \frac{L(B) - [L(G) + L(R)]}{2} \tag{3}$$

where

$L(x)$ operation is defined as $L(x) = 105 \times \log 10(x + 1)$.

The R_g and B_y matrices are then filtered. Color image cropping windowing median filter of with sides of length $4 \times SCALE$. The $SCALE$ value is calculated as being the closest integer value to $(height + width)/320$. Median filtering is a rate-determining step during the skin detection process and can be improved by implementing the window median filter approximation proposed by Freck's multi-ring operator. The texture amplitude map is used to find areas with little texture information. The skin in the image tends to have a very smooth texture, so one of the limitations when detecting areas of skin is to select only the areas with low texture. The texture map is generated from the matrix I by the following steps:

***Step* 5**: Median filter I with a window of length $8 \times SCALE$ on a side

***Step* 6**: Subtract the filtered image from the original I matrix

***Step* 7**: Take the absolute value of the difference and median filter the result with a window of length $12 \times SCALE$ on a side.

From log opponent to hue is $hue = a\tan^2(R_g, B_y)$, where the resulting value is in degrees. The conversion from log opponent to saturation is Hue, and saturation has been used to select those regions whose color matches that of skin. The conversion of saturation is computed as $saturation = \sqrt{(R_g^2 + B_y^2)}$. Using constraints on texture amplitude, hue, and saturation, regions of skin can be marked. The resulting cropped facial portion after applying skin filter for the sample image from the in-house generated database is as shown in Fig. 2.

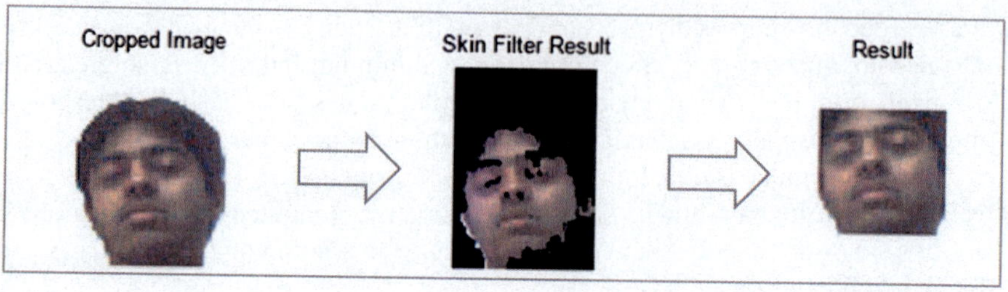

Fig. 2 Face localization for color facial image (Skin color cropping)

2.3 Facial Characteristic Points (FCPs) Extraction

Geometric features can be generated by the segments, perimeters, and regions of some shapes formed by the points in the facial image. Thirty FCPs were used as landmarks to recognize human emotions (see Fig. 3). These FCPs are features that represent important movements in emotional expression. During the human emotion recognition process, the FCP shift between blank and strong faces can be used as input to the SVM classifier.

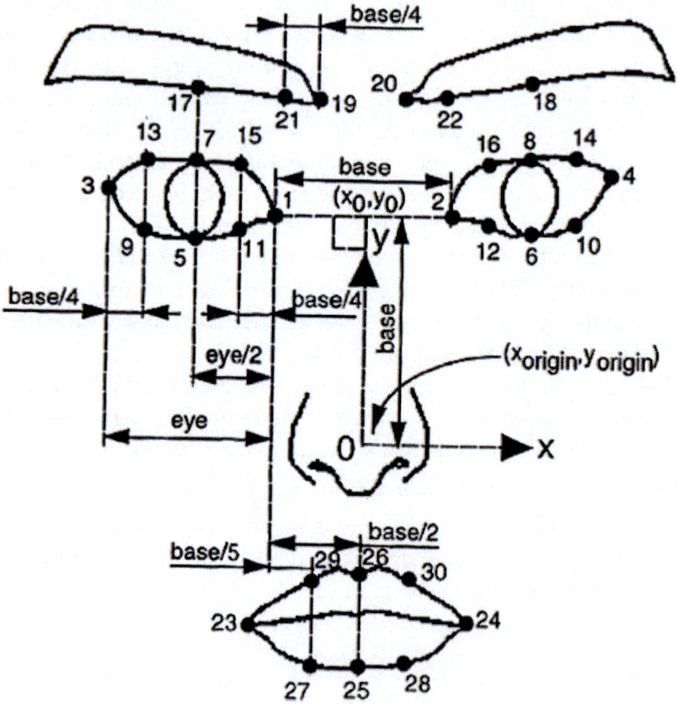

Fig. 3 Facial Characteristic Points (FCPs)

The extracted FCPs from different images are not identical/unique; therefore several pre-processing steps have been performed to achieve this task. To compensate for the variations in the size, orientation and position of the face in images as well as the variations in the size of the face components i.e. eyes, eyebrows, and mouth, the coordinates of the FCPs are transformed so that they are comparable across the set of facial images. Hence the coordinates of these FCPs undergo four transformations: Translation, rotation, normalization, and displacement calculation. They are described as follows:

1. **Translation**: The origin of the image coordinate system is in the upper left corner of the image, but the origin of the FCP coordinate system should be moved to the tip of each individual's nose.
2. **Rotation**: It is used to correct the rotation of the face so that the coordinates are expressed with respect to the z-axis (camera- to- face) in the FCP coordinate system.
3. **Normalization**: It is introduced to correct the distance effect between the camera and the face. The FCP of the rotated blank and expressive faces is divided by the value and the base. These normalized values for expressive faces are subtracted from the normal values.
4. **Displacement Calculation**: The FCP displacements are obtained from subtracting the FCPs of the expressionless face from the FCPs of an expressive face.

After performing the preprocessing steps above, the processed data is ready as an input vector for facial expression classification. Fig. 4 shows the complete flow chart for feature extraction.

3 Theoretical Aspects

The proposed human emotion detection algorithm can be divided into two parts. First part extracts FCPs or DCT features and in second part as SVM that classify the human emotions using these feature vectors. Theoretical aspects of Discrete Cosine Transform and Support Vector Machine classifier are as follows:

3.1 Generalized Discrete Cosine Transform

Due to its excellent properties and simplicity, the DCT is widely used in image processing and texture analysis. DCT is widely used in image and signal processing for lossy data compression. The DCT has strong energy compression characteristics. The generalized form of the DCT was originally proposed by N. Ahmed et al. [14]. It is used by many researchers in the field of digital imaging as a framework for understanding field characteristics through mathematical refinement. With proper

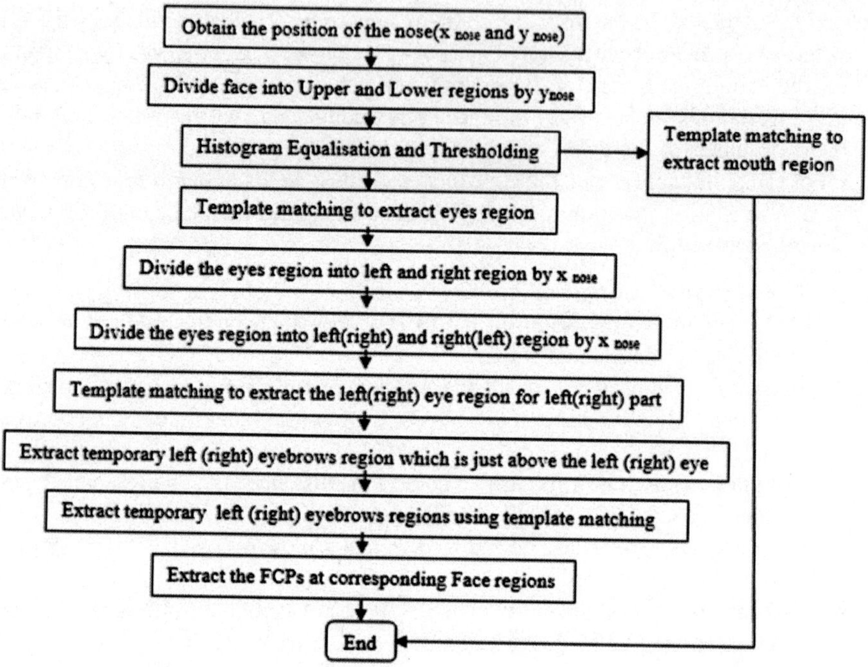

Fig. 4 Flowchart for the FCPs feature extraction algorithm

selection of DCT parameters, texture features of human emotional images can be extracted. DCT is a mathematical formula for generating frequency components. An important feature of the DCT is that it is very easy to group the coefficients of the DCT matrix into two groups based on their values. The upper left corner is for high value coefficients, and the low value coefficients are in the lower right corner of the $M \times N$ size image array. This helps reduce the dimensions of the extracted features without using reduction techniques such as principal component analysis. This reduces the computational complexity of the proposed algorithm. The DCT helps extract local and global texture features of human emotional facial images. For every human emotion facial image $f(x, y)$ of dimension $M \times N$, the DCT coefficient matrix $f(u, v)$ is calculated using,

$$f(u, v) = C(u)\ C(v) \sum_{x=0}^{M-1} \sum_{y=0}^{N-1} f(x, y)\ \cos\left[\frac{(2x+1)\pi\ u}{2M}\right] \cos\left[\frac{(2y+1)\pi\ v}{2N}\right] \tag{4}$$

where

$f(u, v)$ is the DCT coefficient matrix of the human emotion facial image $f(x, y)$,

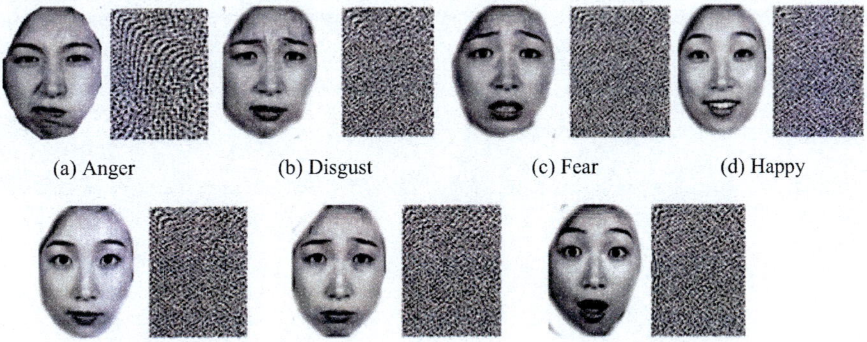

(a) Anger (b) Disgust (c) Fear (d) Happy

(e) Neutral (f) Sad (g) Surprise

Fig. 5 Original images along with DCT coefficient matrix of seven human emotion facial images

$$C(u) = \begin{cases} \frac{1}{\sqrt{M}} & for\ u = 0, \\ \sqrt{\frac{2}{M}} & otherwise. \end{cases}$$

$$C(v) = \begin{cases} \frac{1}{\sqrt{N}} & for\ v = 0, \\ \sqrt{\frac{2}{N}} & otherwise. \end{cases}$$

The DCT coefficient matrix extracted from the database by applying (4) to the emotional images of all the faces of the subject is stored in the form of a vector matrix. Fig. 5 shows the DCT coefficient matrix of the original image and the six facial images of human emotions modeled in the JAFEE database. The proposed approach presented in this paper uses a higher (empirically selected) 100 coefficient attribute. This number was selected after running some fields of the attribute. The highest accuracy rate was found using the first 100 attributes of the DCT. Fig. 6 shows the ability to discriminate the DCT coefficient attribute matrix for the first 100 coefficients plotted against the seven human emotional facial images of the modeled object from the JAFEE database.

3.2 Support Vector Machine Classifier

The problem of facial image recognition of human emotions is actually a multi-class problem. A support vector machine is a classifier that uses the kernel method to calculate the hyper-plane of the feature space and then separates the classes from different classes of samples. The hyper-plane of the feature space corresponds to the non-linear surface of the input space. For human emotional faces, the goal is to train and test a classifier that can be recognized based on several images of the face from the training dataset, and the emotions from the test dataset are identified by

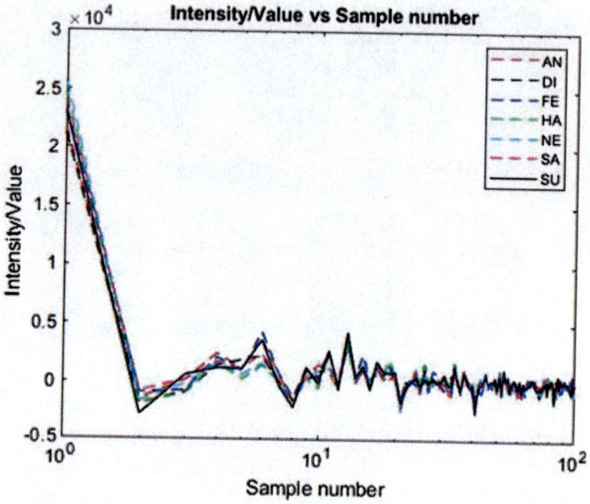

Fig. 6 DCT coefficient attribute plot of first hundred coefficients for seven human emotion facial images

the classifier. In this application, each image should be considered as having a point in N-dimensional space. Where N is the number of pixels in the image. Therefore, SVM is a kind of supervised learning machine.

The role of the classifier is to determine which class in the set of different classes the unknown object belongs to [15]. The classifier needs to find the best boundaries between the different classes. An optimal classifier, called a Bayesian classifier, minimizes overall costs. This cost depends on the previously selected penalty function. The most common penalty function for classification is the Kronecker operator. A good classification costs nothing, but a bad classification costs one. Thus, the optimal classifier computes for an object X:

$$X \in \omega_i \iff P(\omega_i|X) \geq P\left(\omega_j|X\right) \forall j \neq i \tag{5}$$

where

(ω_i) is the set of classes and

$P(\omega_i|X)$ is the probability of the class ω_i for an object X.

There are different types of classifiers, and there are also classification methods. SVM is one of them. Developed by V. Vapnik in the 90's, SVMs are based on the separation hyper-plane. If the dimensions of the data are high enough, the different classes that form different clusters are linearly separable by the hyper-plane. Kernel tricks allow you to handle non-linear split planes because the data is first transposed into a higher dimensional space (called feature space) where the split hyper-plane is found in the input space. Fig. 7 shows the principle.

This hyper-plane in high dimensional space corresponds to the non-linear surface of the input space. This hyper-plane is chosen to maximize the margin between itself and the closest data. It's very intuitive that this is a way to minimize post-training errors, as test objects (unknown objects) can be "close" to training objects. The goal

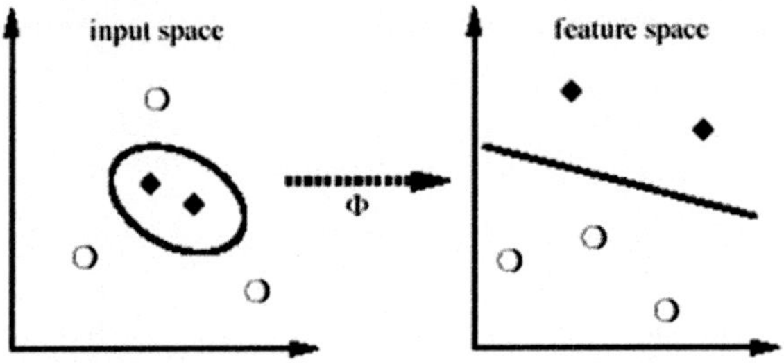

Fig. 7 Basis of SVMs (nonlinear bound in the input space is replaced by a linear separation in the feature space)

of SVM training is to find a hyper-plane (if the classes are linearly separable) or a surface that separates seven different classes.

SVM is a useful data classification technique. Algorithm learning is based on risk (or test risk) minimization. Consider a training set of instance-label pairs $(x_i, y_i), i = 1, 2, \ldots, l$, where $x_i \in R^n$ and $y \in \{1, -1\}$. The SVM requires the solution of the following optimization problem:

$$\min_{\omega,b,\xi} \frac{1}{2}\omega^T \omega + C\sum_{i=1}^{l} \xi_i, \tag{6}$$

Subject to the constraints: $y_i\left[\omega^T \varphi(x_i) + b\right] \geq 1 - \xi_i, \xi_i \geq 0$.

Training vectors are mapped into a higher dimensional space by the function φ. Furthermore, $K(x_i, x_j) \equiv \varphi(x_i)^T \varphi(x_j)$ is called the kernel function. The choice of kernel function is also significant when designing SVM based classifiers. The basic kernels that are generally used are

$$\text{Linearfunction} : K\left(x_i, x_j\right) = {x_i}^T x_j \tag{7}$$

$$\text{Polynomial} : K\left(x_i, x_j\right) = \langle {x_i}^T x_j \rangle^d \tag{8}$$

$$\text{Non}-\text{homogenous Polynomial} : K\left(x_i, x_j\right) = \langle {x_i}^T x_j + c\rangle^d \tag{9}$$

$$\text{Sigmoid} : K\left(x_i, x_j\right) = tanh(K\langle {x_i}^T x_j\rangle + \vartheta) \tag{10}$$

$$\text{Gaussian} : K\left(x_i, x_j\right) = exp\left[\frac{\|{x_i}^T - x_j\|^2}{2\sigma^2}\right] \tag{11}$$

$$\text{Radialbasisfunction(RBF)} : K\left(x_i, x_j\right) = (\gamma {x_i}^T x_j + r)^d, \gamma > 0. \tag{12}$$

All kernels have uniform, invariant properties. The RBF kernel has translation-invariant properties. The kernel allows you to calculate nonlinear variants of algorithms that are usually represented by the inner product. For SVMs, the nonlinear limit is calculated using the inner product of high dimensional space without knowing the transposed function φ.

4 Proposed Algorithm

The proposed human emotion detection algorithm, which uses FCP and DCT coefficients as feature vectors along with SVM as a classifier, is performed in four steps. The first two steps extract the FCP as a feature vector. The third step uses the DCT coefficient as a feature vector to extract local and global texture features of human emotional facial images and the fourth step trains an SVM classifier. Details of the various steps involved in the proposed algorithm are given below:

4.1 Step 1: Face localization

Phase1: The column variance of the image is calculated first and is used to crop the face portion of the image by selecting the column with the maximum variance value, as described in Section 2.1 of the monochrome image.

Phase2: This phase is only performed if the human emotional face image is a color image. In this phase, the skin filter is used to make fine adjustments to extract only the facial area (excluding hair). The skin color detection algorithm is used to trim a portion of the face and then fine-tune the colored facial image (Phase 1). By doing so, the trimmed face passes through a skin filter based on the Fleck and Forsyth algorithm to generate a map on the skin color pixels. It is used to extract facial features through strength features. The color image entered in the format is converted to the color space opposite the logo, and these values are used to calculate the amplitude, hue, and saturation of the texture. Hue and saturation values were used to select areas where the color matches the skin color.

4.2 Step 2: Extraction of FCPs from human emotion facial image

Several pre-processing steps are performed to make the FCPs extracted from different images of the same emotion unique. To compensate for the variations in the size, orientation and position of the face in images as well as the variations in the size of the face components i.e. eyes, eyebrows, and mouth, we have transformed the coordinates of the FCPs so that they are comparable across the set of facial images. Therefore, the coordinates of these FCPs are subject to four transformations: Translation, rotation, normalization, and translation calculation. After pretreatment, we extracted 30 FCPs as landmarks of the geometric features generated by the segments, perimeters, and regions of several shapes formed by the points in the facial image. These FCPs are features that represent important movements in emotional expression. During the human emotion detection process, the FCP shift between expressionless and expressive faces was used as input to the SVM classifier.

4.3 Step 3: Extraction of DCT coefficients from human emotion facial image

The DCT coefficient matrix is extracted by applying (4) to the emotional images of all the faces of the subject and stored in the form of a vector matrix. The DCT coefficient is a frequency component that can be very easily grouped into two groups based on its value, which helps reduce the dimensionality of the extracted features and reduces the computational complexity of the proposed algorithm. Since very few DCT coefficients retain the energy of the whole image and reconstruct the original image with minor loss of information, for experimentation purpose we utilizes the higher hundred coefficient attributes as the best accuracy rate yielded using the first hundred attributes of DCT.

4.4 Step 4: Training of SVM classifier and testing of human emotion facial image

SVM is considered the standard method of supervised classification, which finds the optimal hyper-plane that can classify test database images based on input training patterns (supervised approach). It encloses a hyper-plane for classifying input data into multidimensional space and separating different classes. Create the optimal hyper-plane to minimize errors. Extracted texture features such as FCP and DCT coefficients from the training dataset are sent to the SVM classifier for training. During the test, the features (FCP and DCT coefficients) extracted from the test image are sent to the SVM. We also used the various kernels described in Section 3.2. The RBF kernel used in SVM classifiers has been observed to provide superior accuracy and perform faster predictions.

5 Results and Discussions

The algorithm was implemented and tested on a Core i7 (6th generation) with a 3.4 GHz processor and 8 GB RAM in a MATLAB environment. Performance of the proposed algorithm has been evaluated with the help of confusion matrices, average accuracy, and timing analysis on facial images from Japanese Female Facial Expression (JAFFE), and in-house generated databases independently. The database was divided into two parts, a train database and a test database, at 70% and 30%, respectively. Performance is compared to the amount of confusion for the same class number created in the same database. After selecting thirty FCPs as feature vectors and hundred DCT coefficients by convolving a human emotion facial image from the train databases with DCT, we have generated the feature vectors and applied it to the input layer SVM classifier. For experimental purposes, we used different kernels for SVMs. The RBF kernel used in SVM classifiers has been observed to provide superior accuracy and perform faster predictions. Therefore, it is recommended to use the RBF kernel with SVM. All the features from the train databases are applied one by one till the training of SVM is complete. Once the SVM is trained, when a query human emotion facial test image from the same test database is applied, the same feature vector is generated and applied to the input layer of SVM. As the SVM is fully trained, the output of a particular neuron gives a high value to the class to which the sample test human emotional facial image belongs.

The performance of the proposed algorithm using the DCT coefficient was evaluated using the confusion matrix of the JAFFE database with the RBF kernel and SVM. As shown in Table 1, the performance of the algorithm is compared to the level of confusion for the same number of classes created. The average accuracy obtained for classifying the entire facial images from test datasets are observed to be 92.86%. The performance of the proposed algorithm using FCPs has been evaluated with the help of confusion matrices on the in-house generated database with RBF kernel and SVM. The performance of the algorithm is compared with the amount of confusion for the same number of the class created as shown in Table 2. The average accuracy obtained for classifying the entire facial images from test datasets are observed to be 85.43%. It's easy to see that the results of Sleepy and Surprise expressions are more likely to occur due to the large FCP shifts. Observing the above confusion matrix, it is clear that the confusion decreases as the distance between FCPs increases. Table 3 shows the time analysis using the same database. The RBF kernel used in SVM classifiers has been observed to provide superior accuracy and perform faster predictions. Recall time is more important than the training time because network training is required only once.

6 Conclusions

Emotion recognition is a difficult task due to the inherent ambiguity of emotions in the perception of the human mind. The main purpose of this paper is to introduce an emotion detection system by training an SVM classifier with texture-based appearance features using DCT coefficients and shape features using FCP. The RBF kernel used in the SVM classifier provides excellent accuracy and performs faster predictions. Experimental results prove that SVM classifier with RBF kernel achieve an

Table 1 Confusion matrix of proposed algorithm of DCT coefficients with RBF kernel in SVM on JAFFE dataset

True class	Assigned class						
	Anger	Fear	Happy	Sad	Surprise	Disgust	Neutral
Anger	**94%**	3%	0	0	0	2%	1%
Fear	3%	**92%**	0	0	0	5%	0
Happy	0	0	**94%**	0	4%	0	2%
Sad	0	0	0	**93%**	0	6%	1%
Surprise	0	0	4%	0	**96%**	0	0
Disgust	2%	5%	0	6%	0	**86%**	1%
Neutral	1%	0	2%	1%	0	1%	**95%**

Table 2 Confusion matrix of proposed algorithm of FCPs with RBF kernel in SVM on in-house generated dataset

True Class	Assigned Class						
	Anger	Fear	Happy	Sad	Surprise	Disgust	Neutral
Anger	**82%**	5%	0	5%	4%	3%	1%
Fear	5%	**86%**	0	3%	0	6%	0
Happy	0	0	**85%**	0	13%	0	2%
Sad	5%	3%	0	**84%**	0	7%	1%
Surprise	4%	0	13%	0	**83%**	0	0
Disgust	3%	6%	0	7%	0	**83%**	1%
Neutral	1%	0	2%	1%	0	1%	**95%**

Table 3 Timing analysis of the proposed algorithm

Database	Training time (sec)	Recall time/pattern (sec)
JAFFE	63.58	0.1
In-house	96.11	0.14

accuracy of 92.86% on JAFFE dataset using DCT coefficients while 85.43% accuracy on in-house generated database using FCPs as feature vector. Further research should include efficient texture features with good classifier to increase the system performance.

References

1. Varghese AA, Cherian JP, Kizhakkethottam JJ (2015) "Overview on emotion recognition system." In: 2015 International conference on soft-computing and networks security (ICSNS). pp 1–5 IEEE
2. Dubey M, Singh L (2016) Automatic emotion recognition using facial expression: A review. Int Res J Eng Technol (IRJET) 3(02):2395–72
3. Azcarate A, Hageloh F, Van de Sande K, Valenti R (2005) "Automatic facial emotion recognition." Universiteit van Amsterdam pp 1–6.
4. Ferreira PM, Marques F, Cardoso JS, Rebelo A (2018) Physiological inspired deep neural networks for emotion recognition. IEEE Access 6:53930–53943
5. Mehta D, Siddiqui MFH, Javaid AY (2018) Facial emotion recognition: A survey and real-world user experiences in mixed reality. Sensors 18(2):416
6. Martinez B, Valstar MF, Jiang B, Pantic M (2017) Automatic analysis of facial actions: A survey. IEEE Trans Affect Comput 10(3):325–347
7. Kaviya P, Arumugaprakash T (2020) "Group facial emotion analysis system using convolutional neural network." In: 2020 4th International conference on trends in electronics and informatics (ICOEI)(48184). pp 643–647 IEEE.
8. Ahuja N (1996) "A transform for multiscale image segmentation by integrated edge and region detection". IEEE Trans Pattern Anal Mach Intell 18(9):1211–1235
9. Amit Y, Geman D, Wilder K (1997) "Joint induction of shape features and tree classifiers",. IEEE Trans Pattern Anal Mach Intell 19(11):1300–1305
10. Cai J, Goshtasby A, Yu C (1998) "Detecting human faces in color images." In: Proc. 1998 Int'l workshop multi-media database management systems. pp 124–131
11. Amit Y, Geman D, Jedynak B (1998) "Efficient focusing and face detection," In: Wechsler H, Phillips PJ, Bruce V, Fogelman-Soulie F, Huang TS (eds) Face recognition: From theory to applications. vol 163, pp 124–156
12. Chellappa R, Wilson CL, Sirohey S (1995) Human and machine recognition of faces: A survey. Proc IEEE 83(5):705–740
13. Yang M, Kriegman DJ, Ahuja N (2002) "Detecting faces in images: A survey." IEEE Trans Pattern Anal Mach Intell 24(1)
14. Ahmed N, Natarajan T, Rao KR (1974) Discrete cosine transform. IEEE Trans Comput 100(1):90–93
15. Pal M, Mather PM (2005) Support vector machines for classification in remote sensing. Int J Remote Sens 26(5):1007–1011

Infrastructure and Algorithm for Intelligent Bed Systems

Jainil Patel and Swati Jain

Abstract This paper contains the algorithms and the software which the intelligent bed manufacturing company uses to make the sleep of the user much more better. The smart bed, equipped with sophisticated hardware, is powered by some intelligent algorithms for different purposes. The work done during this project revolves around those algorithms and makes some further improvements to them, which directly impacts the end users. People tend to use Fitbit or OURA rings to track their sleep but during the sleep also our body moves, the bed should adjust the firmness to make the sleep more comfortable. The intelligent bed captures the data like pressure, humidity, temperature and much more to make the sleep more and more user's personal sleep assistance. The bed learns the user preferences of the sleeping pattern and adjusts itself to it. The algorithms are the bed detection algorithm, bed sleep-wake algorithm, the sleep-stage algorithm, the wake-up alarm algorithm, the surface motion, the wave algorithm, etc. With the advancement in the Internet of Things world and due to the acceptance of cloud computing platforms like AWS, the bed is becoming more user-friendly.

Keywords Bed algorithm · BCG heart rate measurement · Sleep-stage predictions · Snoring detection · Air beds

1 Introduction

People tend to neglect the importance of the sleep. In this modern era where everyone rushes behind the money, and the stress comes along with it [14]. People are getting less quality sleep as they age [10]. People tell that their day is not good but the reason for that is their night was not good. As the importance of personal data is increasing

J. Patel (✉) · S. Jain
Institute of Technology, Nirma University, Ahmedabad, Gujarat, India
e-mail: 20mced07@nirmauni.ac.in

S. Jain
e-mail: swati.jain@nirmauni.ac.in
URL: https://nirmauni.ac.in

H. O. Bansal et al. (eds.), *Next Generation Systems and Networks*, Lecture Notes in Networks and Systems 641, https://doi.org/10.1007/978-981-99-0483-9_28

and the cost of saving the data is decreasing, the Internet of Things is booming up and making products that can gather the data and use it for the betterment of the user. People are working toward creating infrastructure which would provide personalized and better sleep experience to the users using sleep science and data science. A part of the infrastructure is the embedded code in the bed's Raspberry Pi, the application running on the Android and IOS, which gives users an ability to control the bed from their mobile phone, the database and infrastructure of AWS services.

The human spend average of 36% of life in sleeping. During that portion of time, the information processing, memory storage from volatile fast memory to non-volatile slow memory (memory consolidation), toxin clearance, tissue repairs, rebuilding of metabolic activities, energy replenishment is done [14].

1.1 What is a Smart Bed?

The bed which uses sensors to adjust the bed firmness to make the sleep easier, and control the IoT devices without the need of the phone at night is called the smart bed. The bed uses the information from the sensors to self-adjust to improve the sleep of the user. The smart bed also delivers the sleep information to the smartphone, where it can report that how well the person is sleeping.

The features of the smart bed are like [17].

Sleep Tracking: The tracking of the sleep and identifying the stages of the sleep is done in real time so that when the person tries to get the information on the phone, he gets it as soon as he wakes up.

Temperature Control: Beds have temperature control system built into it which can warm the bed and even cool the bed. The bed adjusts the temperature accordingly the sleep stages so that it is easy to sleep and easy to wake up.

Air Chambers: The air chamber can inflate and deflate and can help the body to be in certain posture. The bed can adjust the firmness using these chambers. It can avoid the situation where the nerves of the people get stuck in the middle of night due to some bad posture in the night, by identifying the posture of the person. The small pumps which do not make much noise are used to inflate and deflate the chambers.

App Integration: Beds can seamlessly connect to the virtual assistant like Alexa and Google Assistant or Apple's Siri to get the command and to send the command to other devices like the lights, air conditioner, coffee machine, speakers, music system, TV and switches so that person has not to move out of bed in the morning.

Light control of the room: The bed has the sensors to calculate the light inside the room. The lights are turned on when there is night and someone is found on the bed. The lights are turned off when the person is found asleep automatically. The lights are turned on for certain time when the person wakes in the middle of the night to go for washroom.

1.2 What is Sleep?

Sleep is a recurring state of mind and body, which has characteristics of altered consciousness [16]. Human needs good quality sleep but in this stressful world the quality of sleep is decreasing. Good quality of sleep is needed for the brain to charge itself again. If person sleeps longer than normal than too they feel less energetic. People spent approximately one-third of their life on bed but they spent very less on it and change it very rarely. People don't understand the benefit of spending money on a bed that can give better quality of sleep (Fig. 1).

1.3 What is Restorative Sleep?

Restorative sleep is the sleep that produces improved subjective alertness, mood, cognitive function, energy and well-being relative to the immediate pre-sleep period. Restorative sleep improves the subjective ratings of alertness, cognitive function, mood, energy, physical symptoms or function, well-being [14].

After you wake up, you feel less tired, less sleepy, in a good mood, rested, refreshed or restored, ready to start the day, energetic, mentally alert than that sleep is normally called as restorative sleep that restores your brain energy. The sleep that is adequately high enough in both quality and quantity such that the sleeper feels restored when they wake up is called restorative sleep. When a person has had restorative sleep, they wake feeling refreshed, energized and empowered to live their best possible life each day.

The key to get restorative sleep is to make the mind empty while sleeping by doing meditation, and removing the stress of tomorrow.

Decreasing amount of sleep quantity: Sleep pattern has a tendency to change as we age. People find that the cause of them to have harder time falling asleep is their age. They wake up often during the night and earlier in the morning. Especially after

Fig. 1 Good quality of sleep [18]

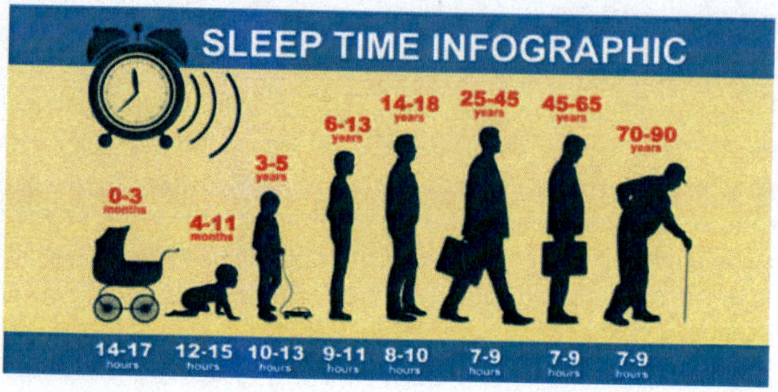

Fig. 2 How the sleep quantity is decreasing as we age [20]

having kids the wake ups in the nights get increased so much for 1 year that their sleep is affected a lot. Total sleep time usually slightly decreases from 8 to 6.5 h or stays the same. It may be harder to fall asleep and we may feel that we spend more time on bed in awake condition. Older people feel like they are light sleeper than when they were young [19]. People spend less time in deep, dreamless sleep. Number of wake ups for older people are average 3–4 each night. They are also more aware of awake and feel it harder to sleep again. The reason is anxiety, discomfort, pain in the body, stress in the mind, low income, abnormally high income, chronic illness and many more. Due to less sleep they get, they feel even less energetic as they grow older [4] (Fig. 2).

1.4 Why People Tend to Use Airbeds?

Normally a male partner tend to have a more firm bed to get better sleep, and a female partner tend to have a less firm bed. The double beds available in the markets are not suitable for this purpose. As the age changes the preference of the firmness of the bed also changes. But the air bed can provide the variable firmness.

People when do a travel and sleep at hotel or at other house, they feel that the bed is not as same as their home bed. Due to the change in the firmness and quality of the bed, the sleep takes longer time to come. The quality of sleep also degrades if you are not frequent traveler and you are habituated to one bed.

So normally companies that make airbeds tend to sell their beds at cheaper rate to customers and at higher rate to hotels. The customers will feel the same firmness of the bed and feel more comfortable in the hotel. This benefits both hotels and customer. The ecosystem of the bed is so intelligent and they use the past profile of the customer to set the hotel beds automatically.

This type of ecosystem is possible due to the availability of the smartphones to almost all the people, Internet going so cheaper, the cloud services like AWS offering such low prices, and the Internet of Things booming up in the market.

1.5 How the Intelligent Bed Affects the Sleep?

The body of the human needs different temperature in the different stages of sleep [4]. The bed can adjust the temperature of the surface after detecting the sleep stage of the person. The bed can also wake a person gently by having optimal wake-up temperature.

The humidity also plays an important role in determining what temperature the body will need in a particular sleep stage. So the beds have humidity sensors to calculate the humidity of the air. People tend to move when they are in dreams so the bed has to adjust the surface so the pressure points can be eliminated to have a constant pressure. The adjustments can be done by the air beds by using the pumps and the valves.

The air beds also provide a lull, a wave type of feeling that can help a person to sleep, but if the person is habituated to that, it is nearly sure that he will buy the same company's bed again. The only reason of people who are using iPhone are still using iPhone and the people who are using Android are still using Android is because of the data that they have in that ecosystem and the hardness of copying personal data from one ecosystem to another ecosystem.

Another advantage is that person don't have to move out of the bed to control the switches in the room. The connection of the appliances can be directly made to the smart bed and the bed can detect the gestures, actions like going at washroom at night, and other commands by recognizing the speech in various languages to activate the appliance like the light, fans, air conditioner, television, etc. The person can set the action that when the bed detects the sleep the television and the music system should be turned off automatically, then the bed would perform that action.

1.6 Technologies Used for the Project

Python programming language, PyCharm, NumPy and Pandas for computations; Matplotlib and Bokeh for visualization; AWS services like event bridge, S3, lambda, cloudwatcher, simple notification service, MQTT, IoT, Raspberry Pi and apps like LightBlue to give Bluetooth commands.

The mobile apps to control the bed, APIs using flask which interacts with the database, slack application for the AWS error notifications and kinesis for the streaming data analysis.

2 Literature Summary

In 2015, a PhD thesis named On the analysis and classification of sleep stages from cardiorespiratory activity [1] by Xi Long gives information about the sleep-stage identification and using the features of heart and respiration to calculate the sleep stages. It is much useful to get initial understanding of the sleep stages and sleep classification. The thesis includes variety number of ways to track the sleep by using different hardwares. Mostly the hardware that are used to track the sleep of small babies in the hospital are demonstrated in the thesis. It shows the general framework of the sleep-stage classification about the data acquisition, signal preprocessing, feature extraction, feature post-processing, feature selection and sleep-stage classification. The paper includes the HRV estimations from the pressure data. The thesis also describes about using the heart rate and the respiration rate-based feature for the bed sleep-stage detection. The problem in the paper is that it describes about the bed that doesn't adjust itself or has the movements of the pumps and valves affecting the data that is collected. When the bed is adjusting, the sleep-stage detection needs to be different as there is more noise in the data collected by the device.

In 2008, a paper named Strobe Lights, Pillow Shakers and Bed Shakers as Smoke Alarm Signals by Thomas [2] which explains a testing standard of a 100 Hz vibration, presented for 4 min, not considering sleep stage: woke 95% of 20 legally deaf adults, woke 77–100% 10–19 year old depending on age group. Commercial motion alarms woke the subject up 80% of the time from deep sleep. The paper demonstrates the possibility of waking up without the sound just by shaking the beds. But it doesn't include the effect on the circadian rhythms over the time usage of the bed shakers. Demographic and Environmental Factor Subjects greater than or equal to 60 years old are less likely to wake up than those less than 60 years old, from motion alarms. Commercial bed-shaker alarms do not wake up 42% of moderately alcohol impaired young adults from deep sleep Audible Alarms: the most effective wake-up alarm is an audible 520 Hz square wave T-3 sound at 75 dBA at the pillow. The paper suggests that no one can wake the alcohol drank person without shaking him up thoroughly. Bed shakers at intensity 6 have a chance of waking 96% of the people. The aim of the paper was to make sure that the deaf person also wakes up in case of some emergency like fire.

In 2020, a paper named Automated Sleep Stage Analysis and Classification Based on Different Age-Specified Subjects From a Dual-Channel of EEG Signal [3] by Santosh Kumar Satapathy which uses the techniques like support vector machines, decision tree, and K-nearest neighbors to the dual-channel heart rate signals to make the predictions of the sleep stage was published. It also focuses on classifying some sleep disorders using the heart rate data. It proposes to use SVM classifier, decision tree classifier, K-nearest neighbor classifier, and ensemble methods to get accuracy of 98.4, 97.2, 96.7, 97.5%, respectively. It follows acquisition of EEG signals, feature extraction, and feature selection followed by sleep disorder detection through classification algorithm to get the predictions of the sleep stages.

In 2019, a review paper named A Review of Approaches for Sleep Quality Analysis [4] by Sheikh Shahnawaz Mostafa was released which tells about the techniques used for the sleep scoring which includes features like duration, intensity, continuity, stability, frequency, and sleep episodes. It is review of the techniques that other companies uses to give a sleep score. The percentage weightage that is given to individual features is mentioned in the paper.

As per the dataset at https://data.world/makeovermonday/ 2019w23 we can see that the Americans are sleeping more as the year passes. They are getting less quality of sleep so they require more time in the bed. This is due to multiple factors like stress, or more income, or less work to do daily, and many more. If this trend continues, then people will need sleep more than 15 h (approx.) in 2050, which will make us more lazy and less energetic.

In Sleep Disorders and Sleep Deprivation: An Unmet Public Health Problem [14], author describes about the sleep architecture, four stages of NREM, circadian rhythms controlling the sleep wake, sleep pattern changing with age, and many other useful information. It describes about what is the difference between the sleep stage. It describes about the sleep healing.

In Kirjavainen T., Cooper D., Polo O., Sullivan C. E. Respiratory and body movements as indicators of sleep stage and wakefulness in infants and young children [13], author uses the bed with the static charge, similar to what we have in the touch screen displays on the bed for the babies to detect the sleep stages using the variations in the static charge of the bed. The paper suggests to have a sleep monitor for the babies because they don't have a fix schedule of sleep as their circadian rhythms are not formed properly. But there were just 21 subjects in total for the analysis.

In the paper Agent-Based Simulation of Smart Beds With Internet of Things for Exploring Big Data Analytics, [12] the author describes about the simulation for the bed and data generated by it. But the truth value of the data cannot be got because there is no guarantee about the sleep stages.

The paper sleep-stage prediction with raw acceleration and photoplethysmography heart rate data derived from a consumer wearable device [15] describes about using various machine learning techniques like random forest, logistic regression, k-nearest neighbors, neural networks to predict the sleep stages. But nowadays the device like Fitbit band itself can predict the sleep stages by their own algorithms. But the problem is that people don't like wearing such devices while sleeping.

The research paper sleep/wake state prediction and sleep parameter estimation using unsupervised classification via clustering [11] explains about how the unsupervised method can predict the sleep wake by using activity counts. But the algorithm for getting the activity counts is not mentioned in the paper. Author uses k-means clustering to predict the sleep wake from the activity counts.

3 Bed Algorithms

This chapter includes the algorithms that an intelligent bed uses to keep track of the sleep of the user.

3.1 Hardware and Data Generated By It

It is observed that when the person is sleeping on the bed, the heart rate, respiration rate, and the movements signals captured inside the pressure sensors have higher standard deviation than the bed without person. This causes the standard deviations of the pressure to be higher when the person is on the bed [5]. This algorithm uses the standard deviation of pressure values over a 30 s window, per pressure sensor. The bed consists of some pressure sensors, therefore at every 30 s we have some values which are the standard deviations of the pressure values. We define a threshold of at a value and if any some values of the pressure deviation is above this threshold then it is called as the person is in the bed. There is also a smoother which smooths the bed detection values by using forward–backward smoother. We need certain numbers of standard deviation values above the threshold so that the pets are not detected in place of humans. This means if a baby is sleeping we are not detecting that at all too. A part of the project was to reiterate the bed occupancy algorithm because the old version was not predicting the correct result when the person is sleeping in another direction. A part of the project was about going over all the available historical data of the beta testers, plotting the data, generating the stats, calculating error, and choosing the new threshold.

The raw pressures are stored in the array. The standard deviation of the array is taken at certain intervals after applying the bandpass filter on the pressure values. Those standard deviations are stored in the array. At every epoch the mean of the standard deviation array is taken after removing the outliers. The standard deviations of certain number of chambers should be higher to tell that the person is moving, in bed or out bed.

When the bed is adjusting itself or when the wave is running, the above described bed detection algorithm always predicts that the person is inside the bed. But those fake predictions are removed by the smoother, as we know that the bed adjusts itself quickly. A part of the project was to make the bed detection algorithm agnostic of the chamber of the bed that the head of person lies. Also the part of the project was to find the threshold for different models of the bed.

Thus by measuring the small pressure changes of the heart rate, respiration rate, blood pressure rate, movement of the person, and other signals, the bed can predict the person is on the bed by 96% of accuracy. But the algorithm can be fooled if two bigger pets lie on the bed at the same time, then the algorithm will predict as human is on the bed.

The bed detection system is completely working separately than the sleep-wake algorithm and the sleep-stage algorithm. This is to make the complexity of one model small and to distribute the workload on other models that can work parallel to compute the sleep stage in real time (Fig. 3).

For evaluating the correctness of the prediction made by the various sleep algorithms deployed in the Raspberry Pi, a group of people have signed up as the beta testers, who wear the Fitbit while sleeping every day on the smart beds. A part of the code fetches the data from the Fitbit servers for all such users, which is then

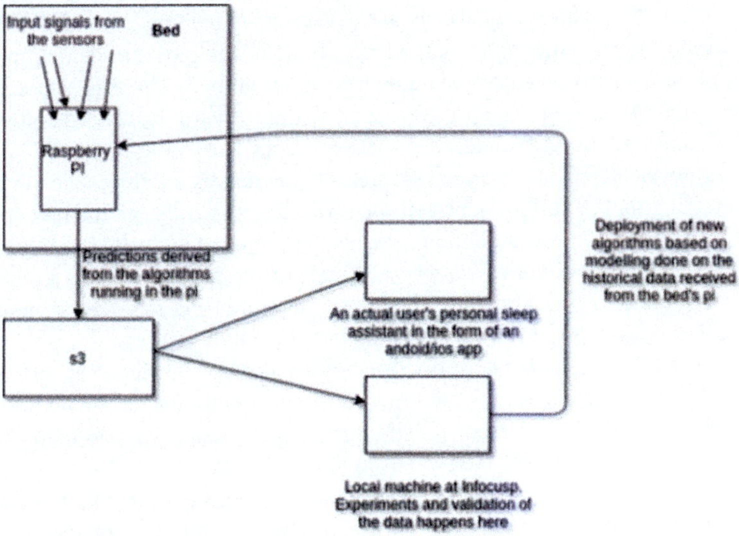

Fig. 3 High-level overview of the system

aligned with the data from the bed according to the local time. Fitbit provides many functionalities but functionalities of our interest are heart rate, sleep-wake detection, and sleep-stage predictions. Various statistical measures are used to compare the 24 h sequential data coming from the bed and Fitbit. Even though Fitbit is not perfectly accurate, its ease of use and the ease of availability make it good for the validation purposes. Sometime withings data and oura ring data are also used but they are not much accurate. For more accurate ground truth data and developing features, the Polysomnography (PSG) machines at sleep lab were used. Any inconsistencies in the prediction by the algorithm are identified by comparing those data with the Fitbit/PSG data. As the sleep pattern of each and every user is somewhat different so the output pattern of the prediction algorithm for each and every user can be different. The end goal is to reduce the average error for the sleeping patterns of all the users. (Example for two users 85% accuracy is more preferred than the 99% accuracy for one user and 60% accuracy for another user.)

3.2 Bed Occupancy Detection Algorithm

It is observed that when the person is on the bed the bed has some pressure deviations from the target that it has to achieve because of the weight of the person. This causes the standard deviations of the pressure to be higher when the person is on the bed [5]. This algorithm uses the standard deviation of pressure values over a 30 s window, per pressure sensor. The bed consists of some pressure sensors, therefore at every 30 s we have some values which are the standard deviations of the pressure values. We define a threshold of at a value and if any some values of the pressure deviation is above this threshold then it is called as the person is in the bed. There is also a smoother which smooths the bed detection values by using forward–backward smoother. We need certain numbers of standard deviation values above the threshold so that the pets are not detected in place of humans. This means if a baby is sleeping we are not detecting that at all too. A part of the project was to reiterate the bed occupancy algorithm because the old version was not predicting the correct result when the person is sleeping in another direction. A part of the project was about going over all the available historical data of the beta testers, plotting the data, generating the stats, calculating error, and choosing the new threshold.

Here we store the raw pressures detected by the bed at high rate into the array and apply the bandpass filter to the raw pressure signals for removing noise. Then we take the standard deviations on sliding window of size 2 s and store them into the array. Take the mean of the stored standard deviation when the array size is 30 s. Apply thresholds on the mean that is calculated to determine whether the person was moving, in bed or out bed. We know that when the person is inside the bed then the standard deviations of the pressures are higher than a certain threshold. There are certain numbers of chambers, so the threshold should be high for certain number of chambers to determine that the person is in the bed.

When the bed is adjusting itself or when the wave is running, the above described bed detection algorithm always predicts that the person is inside the bed. But those fake predictions are removed by the smoother, as we know that the bed adjusts itself quickly. A part of the project was to make the bed detection algorithm agnostic of the chamber of the bed that the head of person lies. Also the part of the project was to find the threshold for different models of the bed.

Thus by measuring the small pressure changes of the heart rate, respiration rate, blood pressure rate, movement of the person, and other signals, the bed can predict the person is on the bed by 96% of accuracy. But the algorithm can be fooled if two bigger pets lie on the bed at the same time, then the algorithm will predict as human is on the bed.

The bed detection system is completely working separately than the sleep-wake algorithm and the sleep-stage algorithm. This is to make the complexity of one model small and to distribute the workload on other models that can work parallel to compute the sleep stage in real time.

3.3 *Bed Sleep-Wake Prediction Algorithm*

The bed doesn't have cameras to predict whether the user is sleeping or not. So it uses the algorithms that convert the pressure values to the sleep wake. It is observed that when someone is in wake stage, then in 30 s the body at least moves causing the change in the pressure features that the bed is recording [6]. If there is no movement in the bed than it is said that user is sleeping. The sleep-wake algorithm's result is

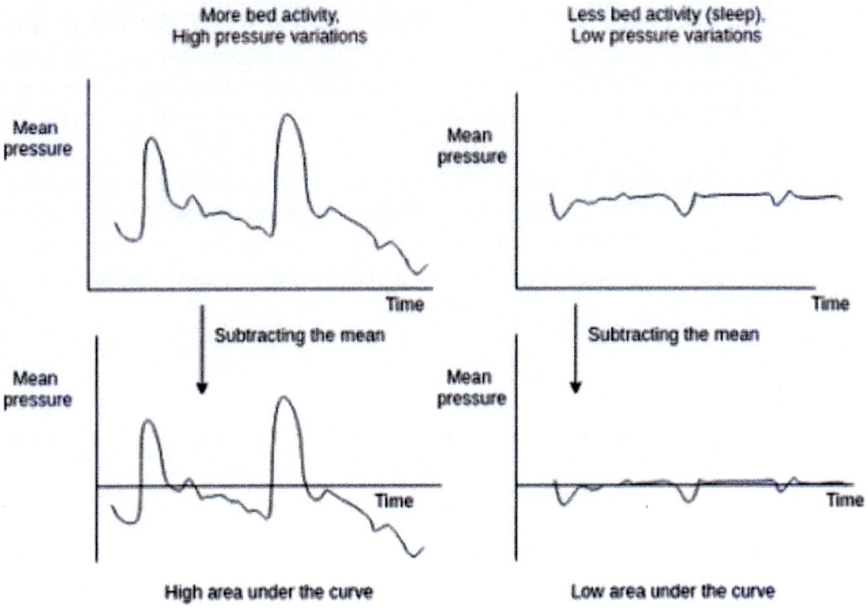

Fig. 4 Explaining the area under the curve value

post-processed by the bed detection. When the person is sleeping, the movements and physical activities are significantly low compared to when a person is lying in a bed but not sleeping [11]. Based on this concept the prediction of the sleep wake is done. It uses the variation in the sequential pressure values in all the chambers, if there is any activity the pressure values would be having sharp variations and when the person is sleeping then the pressure values would have minimal variations. We set a threshold value and define a statistical function over the pressure values of all the sensors. If the output of the function exceeds the threshold values, then we predict that the person is awake, as the activity is high enough to exceed the threshold (Fig. 4). Some functions are listed below:

(1) Area under the curve This function is applied on 30 s of sequential mean pressure values captured at 64Hz by each pressure sensor, if the output of this function is greater than the threshold for certain numbers of sensors, then it suggests higher bed activity and hence the person is awake.
    The area under the curve is obtained after shifting the data to zero line by subtracting the mean for eight chambers.
    AUC = SUM(ABSOLUTE(PRESSURE ARRAY - AVERAGE(PRESSURE ARRAY)))
(2) Median Quantile Difference: Median quantile difference is the difference between median of AUC and 15th percentile value of area under the curve function.

MQD = (MEDIAN OF AUC BUFFER) - (15 PERCENTILE OF AUC BUFFER)
It is observed that when the bed is adjusting itself or executing the surface motion, the AUC buffers get a lot of higher values which can lead to a false wake detection. The AUC buffer should be cleaned if there is more activity of the bed.

(3) Mean diff squared: First the mean pressures are maintained for all the chambers. Then the difference is taken between the current mean value and the last epoch's mean value. This difference is squared so that the difference is amplified. This gives us a total eight (one per pressure chamber) mean squared difference values. A median is taken to return a single value.
MDS = (PRESSURE MEAN AT EPOCH E - PRESSURE MEAN AT EPOCH E-1)
If the value of median quantile difference or median of mean difference squared is higher than their threshold values then the prediction is wake. Thus by capturing the movement-related data from the sensitive pressure sensors, the bed can identify whether the person is sleeping or awake.

Sleep-wake algorithm by respiration rate-based filtering: The normal sleep-wake algorithm described above works fine for the majority of times, except in the starting and ending of the sleep session when a person is about to sleep or about to wake up. A person when in sleep has a comparatively lower respiration rate compared to when he/she is awake. This understanding can be used to improve the sleep-wake prediction of the algorithm. The sleep wake is filtered from front and end till we find the respiration rate lower than a custom threshold at certain percentile value of individual user. The small sleep detected till the larger block of sleep is found is removed. The filtering of the small sleep blocks from the middle is also done.

Basically if the person is laying on the bed and reading the book, then the sleep wake will predict false sleep, but we remove the sleep until we get continuous blocks of certain epochs of sleep.

Sleep-wake algorithm when adjusting the bed: When the wave is running or the bed is adjusting, then the above method will predict the false wake. So in that case we use different model to predict the sleep wake.

Mean Absolute Difference (MAD) = AVERAGE(DIFF(PRESSURE DATA).

When there is certain movement on the chamber, the MAD gets increased. We make sure that there is no wave movement on one chamber, and take the count since MAD has achieved a certain threshold. If the count since is more than 0 than it means that the person has moved his head. The extra sleep detected is removed by the smoothing. We make sure that at least one chamber has no movement by the bed pumps during the wave. Ideally after sometime, we will predict the sleep wake using the model made by heart rate and respiration rate, but currently the normal threshold model made by movement data is giving more accuracy than the model made using the heart rate and respiration rate.

Signals captured by the bed: Any disturbance in the medium can be measured by the device. The signals of heart rate and respiration rate captured by the pressure sensor which is situated near valves. The information from the signal is separated by filtering the digital signal. Thus by standard measurement of the signal and placing the sensor at the same location, by doing the signal processing we can extract the information needed from the data by running certain algorithms.

We can notice that an earthquake is also a disturbance, which creates change in the signal. that change can be captured by multiple beds in the state. The alarms on the mobile phones of those users can be triggered at night that an earthquake is detected by multiple beds.

Fire can also be detected by checking the respiratory effort that the body has to do. If we see that more effort is needed by the body than usual then we can predict that there is fire. Also there are certain smoke sensors available which can also help in detecting smoke at night.

Handling multiple heartbeats: The pressure signals measured also contain the information of the heartbeats [9]. Multiple heartbeats from single side occur when partners are sleeping together on one side, and other possibility is that the woman is pregnant and due to that multiple heartbeat signals are captured by the bed. It is very tough to separate those signals. There are certain zones in the beds and each zone is giving us signal of HR. The heart rate with the best score is given priority. We need to make algorithm to check whether the signals received from some zones are similar or not, by checking the delta loss function of those signals.

Gesture control for the beds: The beds have certain chambers and pressure deltas from those are recorded. The bed can understand the gestures at least approximately five gestures. For example, if the alarm is running and someone in sleep makes great pressure delta by jumping on the bed, then it should stop the wave and start rebalancing. The support for IFTTT (if this then that) can be given by integrating the service into the bed, and by that the webhooks can be configured. By configuring the webhooks, if the user punches the bed at zone 2 for two times then it would mean light off, and the light off signal can be made using the IFTTT service easily.

If the pressure moves from up to down on the bed then the gesture control will decrease the bed temperature, and if the pressure moves from down to top it means that the bed would increase the bed temperature. The bed temperature is regulated by the coils in the beds.

4 Sleep-Stage Prediction

The chapter includes the information about the sleep stages and why they are important to identify. The chapter also covers the algorithm that intelligent bed uses to identify the sleep-wake prediction.

4.1 Sleep Stages

During sleep, we usually pass through three stages which are light sleep, deep sleep, and REM sleep. These stages of sleep normally progress like light, deep, light, and REM then again the cycle repeats. As the night progresses the amount of deep sleep that the body is getting decreases. Deep sleep is the best sleep that gives us more relaxation to the brain. The REM sleep is the least good sleep that doesn't give us relaxation. Infants spend almost half of their sleeping time in REM sleep. Adults and children spend almost 20% of their total time in REM sleep [7] (Fig. 5).

- Light sleep During the light sleep we drift in and out of sleep and can be awoken easily. Our eyes move very slowly and muscle activity reduces. Some people experience sudden muscle contractions often preceded by a sensation of starting to fall. If we are in the light sleep in the morning then on waking up we feel more comfortable.
- Deep sleep During deep sleep, the brain waves called delta waves begin to appear in the heart ECG. There is no eye movements in deep sleep. People awakened during deep sleep do not adjust immediately and often feel disoriented for several minutes after waking up. The mind is completely resting and charging up for the

Fig. 5 Stages of sleep

next day during this period of time. People are getting less deep sleep because of diseases or stress and the main focus of the bed is to make the environment which provides more deep sleep.

- REM sleep When we switch into REM sleep, the breathing becomes more rapid, shallow, and irregular. Eyes jerk rapidly in various directions and our limb muscles become temporarily paralyzed while in this sleep stage. The heart rate increases and blood pressure increases. People often see dreams in this stage of the sleep. As the night progresses, the duration of REM sleep cycle increases and the duration of deep sleep cycle decreases.
- Wakes during the sleep. When the person moves in bed without knowing that the body moved during the night is called as wake during the sleep. The wake stage is nearly similar to the REM sleep.
  In bed if we see any movement on the bed, than that period is marked as wake period, and if we see the disturbances in the heart rate and respiration rate without the body movement then we call that as the REM sleep. Thus the sleep-wake prediction is completely separated by the REM sleep detection.

4.2 Sleep-Stage Predicting Algorithm

The bed and infrastructure aims at increasing the deep sleep proportion of the user. Here, the objective is to predict the sleep stage of a person sleeping in the bed. It is done by offsetting certain features, making some new features by using a rolling window. The features developed are some pressure and some respiration rate-based features.

The algorithm uses two naive Bayes models to predict the sleep stages individually. First the prediction is made using the pressure-based feature. Then if we have the respiration-based features and enough respiration coverage available then the prediction is corrected by the respiration-based model. The smoothing works by taking a rolling window of certain epochs by applying majority voting. Certain illegal transition from one stage to another is which were predicted by the model by some mistake is corrected because the model is not using the previous predictions to predict the next, instead it is making predictions using probability on the current epoch only. The despiking of the sleep-stage prediction is also done. In future the heart rate model will also be implemented which will predict the sleep stage using other heart rate features like alpha, beta, gamma, and delta waves that have been identified occurring in the sleep stages [3] (Fig. 6).

Naive Bayes has the property that each feature is independent and has equal importance [8]. The formula for Gaussian naive Bayes is

$$P(x_i \mid y) = \frac{1}{\sqrt{2\pi\sigma_y^2}} \exp\left(-\frac{(x_i - \mu_y)^2}{2\sigma_y^2}\right)$$

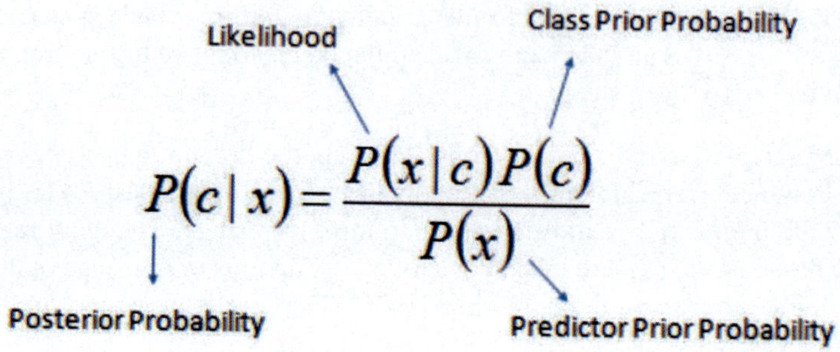

$$P(c \mid X) = P(x_1 \mid c) \times P(x_2 \mid c) \times \cdots \times P(x_n \mid c) \times P(c)$$

Fig. 6 Bayes rule of probability

4.3 *How Sleep-Stage Algorithm Works?*

When person is dreaming then there is some movements of the body at least in 30 s that is captured by the sensors, and it is called as person is in REM sleep because it has some movements which are higher. The sleep-wake detection is completely working independently so the REM stage is always overwritten by a wake stage if it is predicted as wake from the sleep-wake detection system. In the deep sleep, the person is on the bed but causing the minimal pressure variations. If we receive minimal pressure variations then the naive Bayes model will predict it as deep sleep because it has that probability function learnt. When there is not deep and nor an REM sleep prediction by the model then the only other option left is the light sleep. So the pressure-related naive Bayes works on that principle.

The respiration-based naive model works on the principle that in the deep sleep the variations in the respiration peaks are very low and even the respiration rate is also low. In the REM sleep, the variations in the respiration peaks are very high, the respiration rate also increases in the REM sleep especially when someone is dreaming. The light sleep has moderate respiration frequency changes. Using these features trained by the probabilistic model the predictions of the sleep stages are done.

4.4 *Scoring the Sleep for Every Night*

Sleep efficiency: The sleep efficiency denotes that how faster the person gets to sleep and how much faster he gets out of bed after waking up. Time to get out of bed is calculated when the bed detects the last epoch of sleep till the bed detection is person detected. Sleep latency is the time to fall asleep and it is detected by taking the difference of the time when we get the first sleep and the bed detection time when the bed was detected. Some person like to read the books or watch TV while lying on the bed which cause the sleep latency to be more which is wrong sleep latency. To avoid it the bed detection is smoothed by activity counts.

Total sleep time is the number of epochs where we get the sleep-wake detection as sleep. Here let the sleep duration is the subtraction of epoch number where we encountered the last sleep and epoch number where we encountered the first sleep.

$$\text{sleep efficiency} = \min(100,\ \text{total sleep time} * 100/\text{sleep duration})$$

Sleep duration score: To score the sleep duration that the person has took, we have to find the median of the total sleep time that the bed has been recording since long from the historical data.

The sleep duration score is the number of epochs where the sleep has detected/ median of the epochs where the sleep has detected in the past * 100.

The sleep duration score matters on how much other humans are sleeping more or less that a particular person for whom the sleep duration score is being measured, so it is a relative quantity.

Sleep continuity: The sleep continuity depends on the number of wakes detected in the middle of the night and the number of smoothed interrupt transitions that are detected. Interrupt transitions are the interrupts in the sleep where the person wakes up for water or to go to washroom.

Waso is the wake after the sleep onset which is equivalent to the number of wakes detected in the middle of the sleep.

Minimum per interrupt percentile is the waso/interrupt transition * 100/median of the (waso/interrupt transition) that we have seen in the past.

Interrupt percentile is the interrupt transition * 100/median of the interrupt transition that we have seen in the past.

Sleep continuity is 100—(the weighted average of the interrupt percentile and the minimum per interrupt percentile).

There are various other ways to calculate the sleep continuity which includes the amount of deep sleep, light sleep, REM sleep, and wakes in the consideration to calculate the sleep continuity. But here we just take the number of wakes in the consideration to calculate the sleep continuity.

Sleep score: There are various methods to give a score to sleep and there is no standardized method. The method that we follow is to give the weighted average of the sleep efficiency, sleep continuity, and sleep duration to make a sleep score for every night.

The sleep summaries are stored per day, per week, and per year. So that when the user is asking for the sleep summaries, it doesn't take time to calculate it from the data.

5 Snore Detection without Microphone

The snoring affects the other partner sleep adversely, so the bed uses intelligent algorithms to detect the snoring and to prevent it by adjusting the firmness of the bed so that the sleep of the user gets more deeper. The bed doesn't have an active microphone because of the privacy issues of the users. So the bed uses the respiratory features to determine whether the person is snoring or not. The model is trained using the ground truth obtained by the Ralph snore clock app.

5.1 Data Cleaning

The raw data was taken from the app but it is not as clean and in the format that we wanted. The bed makes 30 s interval feature files, so the ground truth has to be modified to have that frequency only, also the merging of the ground truth with the bed's feature file with respect to time was done to make a complete dataset of almost 1 month for one user. We have used the closest match method on the time for merging the ground truth with the data.

The epochs where the bed didn't found the respiration mean were removed from the dataset. The points where the bed was adjusting itself were removed from the dataset to make the dataset more clean. The categorical variables like sleep wake, bed detection, and others were changed to numerical form, so that the model could understand that.

5.2 Feature Selection

The data with the ground truth contained a lot of features and the feature selection was important. The important factor of decision tree was used to get the most important feature from the dataset. We have found that certain features like standard deviation of the respiration rate, moving average of the respiration rate with small window size, moving average of the respiration rate with a larger window size, the difference of the moving average of the smaller and faster window size. Thus by using these columns from the dataset the model was made.

5.3 Removing the Imbalance from the Dataset

The dataset was highly imbalance because the person doesn't snore most of the time. We need to remove the imbalance from the dataset to make the model better (Fig. 7).

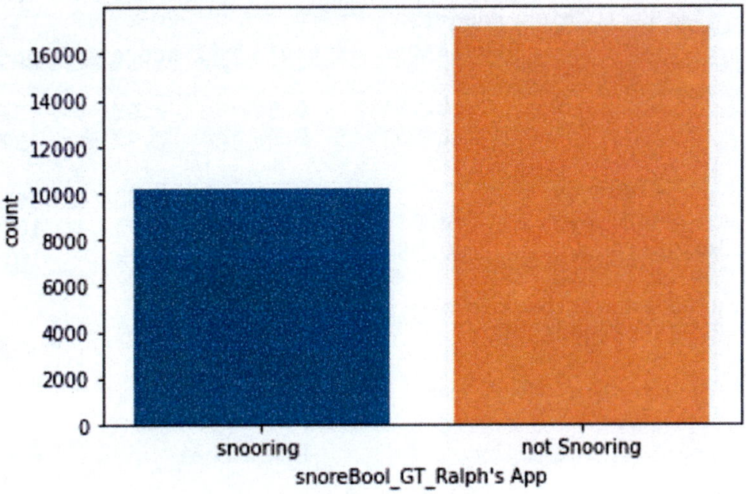

Fig. 7 Imbalance in the dataset

We have tried to use SMOTE oversampling, ADASYN oversampling, random oversampling, borderline SMOTE, and ALLKNN undersampler to make the better model. We have found that the Random oversampler worked the best to get the better recall. Recall = TP/TP+FN which we were trying to improve.

5.4 *Model Selection*

There are many classifiers out of which we have tried DecisionTreeClassifier, RandomForestClassifier, KNeighborsClassifier, KNeighborsClassifier, and svm.SVC. Out of them the Random Forest classifier worked the best and the recall value of that model was high. 70% of the data was used in training purpose and 30% of the data was used in testing purpose.

The model was

```
rfc_best=RandomForestClassifier(class_weight=None, criterion='entropy',
            max_features='auto', n_estimators=200)
```

5.5 *Results*

We achieved the recall of 0.96 with precision of 0.90. The f1-score made from it was 0.93. The accuracy of the model was 0.93. The Random Forest models tend to

```
classification report:
              precision    recall  f1-score   support

           0       0.90      0.96      0.93      5110
           1       0.95      0.90      0.92      5080

    accuracy                           0.93     10190
   macro avg       0.93      0.93      0.93     10190
weighted avg       0.93      0.93      0.93     10190

area under the curve
0.9836083909887977
```

Fig. 8 Classification report for the snore detection using the random forest

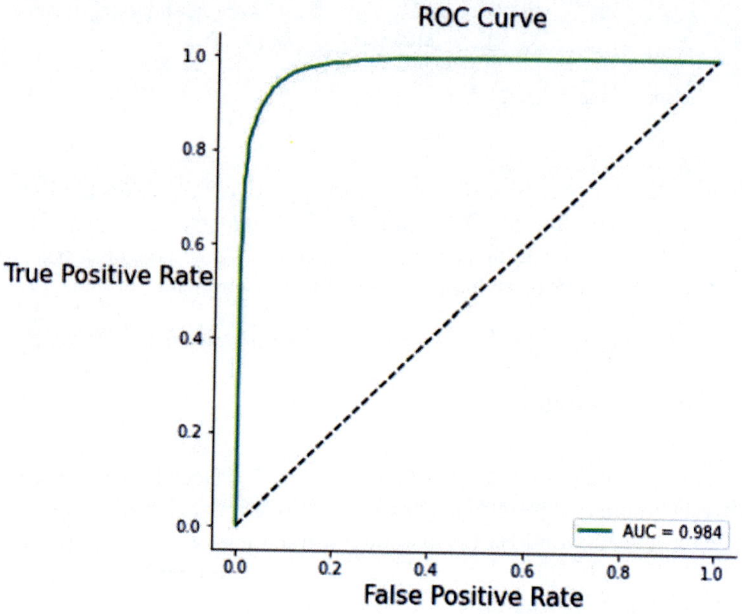

Fig. 9 ROC curve for the snore detection using the random forest

overfit more so we have tested the model using the n-fold method and the results were acceptable. The area under the curve was 0.98 (Figs. 8, 9, 10, and 11).

The random forest model was good but it tends to overfit on one person's data, because it was made by using just one person dataset. We didn't had the ground truth of different peoples. But there is a significant relationship in the respiratory features and the snoring sound (Fig. 12).

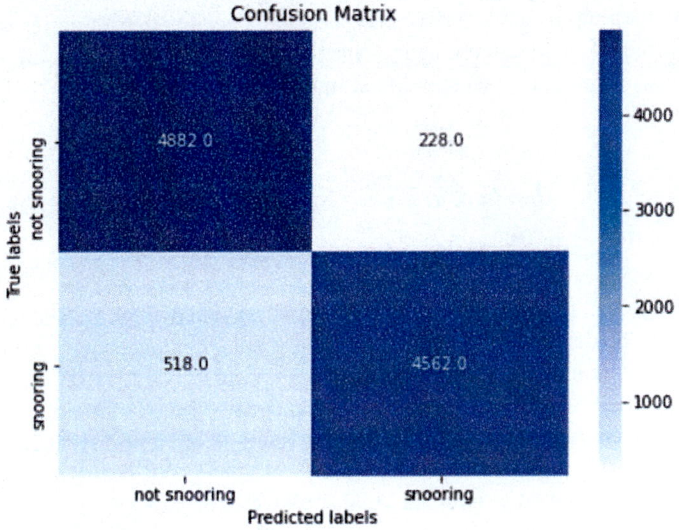

Fig. 10 Confusion matrix for the snore detection by random forest

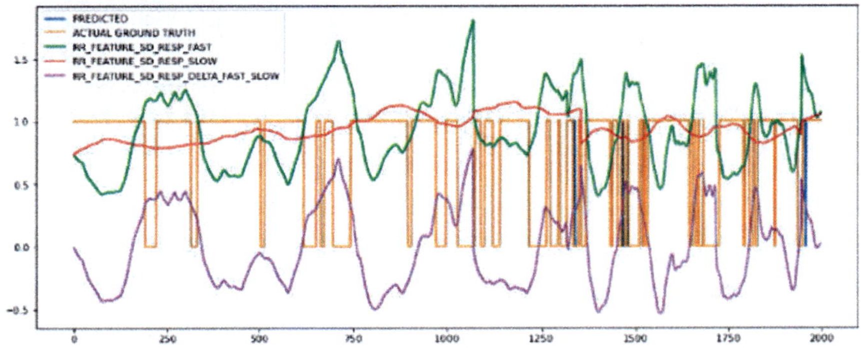

Fig. 11 The predictions made by the model

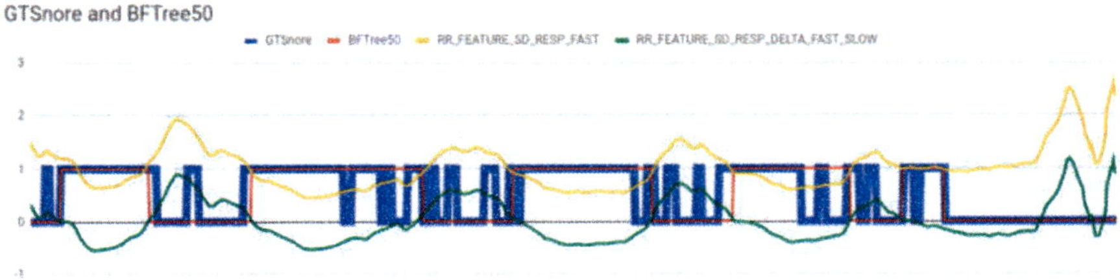

Fig. 12 The snore detection predictions made by BFTree50

Other methods tried were BFTree50 best first decision tree classifier and its results are given below. The real challenge was the pre-snore detection model that predicts snore 30 s ahead, which seems almost impossible to make.

5.6 Reducing the Feature File Size without Affecting the Model

The models are trained by using high precision values which means more decimal digits after the decimal obtained from the sensors. Companies collect the data at the full sensor potential for some time to make the model better. But after some time when the model is ready to save the space and money on S3, companies reduce the precision of the features and the sensors are giving data with less decimal values to S3.

But to determine how much decimal places we can reduce, each feature is taken and the standard deviation, minimum value, and maximum value are seen. Then the feature files with high decimal precision are reduced to see the effect of the reduction on the model's prediction. We have reduced the feature file size by around 40% by just reducing the decimal places of each feature and the prediction of the model was the same as it was predicting earlier.

The reduced decimal value features not only help to save the space, but also help in predicting faster. The space and time required by the model are also reduced.

6 Surface Motion

6.1 Surface Motion

The bed can execute a wave which is done by inflating and deflating some chambers for certain amount of time. This wave can help the user to be relaxed and helps in the sleep. The surface motion is the configurable wave using patterns written in string that can be ran on the bed.

6.2 Segments of the Surface Motion

The surface motion uses a config which has the starting states and repeated states defined for a motion. The segments are inflate, deflate, equalize, and rest. These segments are used in combination of the chamber number. The inflate segment means that the pressure of that chamber will increase for a particular amount of time defined by the segment duration. The deflate segment means that the pressure of that chamber will decrease for a particular amount of time. The equalize segment means that the pumps will be off and the valves will be opened so that the chamber can equalize the pressure of the air in them. The rest segment is for the time delay if wanted. The vibrate segment means that the valves will be opened at inflate side and the pumps will be oscillated between on and off states at a very high frequency.

The config of surface motion contains the strings of segments separated by comma, which contains the states to be executed in parallel. The custom operations like Reset, set firmness, change segment duration, and change pump numbers were also introduced for the surface motion. Example: "I1 I2, D1 D2, E1 E2". Here first segment I1 I2 means we have to inflate the zones 1 and 2 together for segment duration amount of time. D1 D2 means in the next segment we have to deflate the chambers 1 and 2, E1 E2 means we have to equalize the pressure in two chambers 1 and 2.

6.3 *Communication with the IoT Device*

The communication to the IOT devices can be done from using a ssh port if the port number and password are known. Other ways of communication are the MQTT, Bluetooth, Simple Notification service from Amazon AWS, etc. The MQTT messages contain the ID and the body of the message. The Bluetooth protocols have the bytes and position of them defined to parse that. Thus in the front end we can see the mobile app but at the backend the communication is done via the Bluetooth or MQTT or Wi-Fi. Some IoT devices use the infrared for controllable by the remote control.

The MQTT commands as well as Bluetooth commands were made to save the config file to the bed, save and get the profile properties like wake-motion-defaults, relaxation-motion-number, and wake-mode. The mobile app can communicate using SQS also but the sqs commands were not made for the surface motion.

MQTT is the Message Queue Telemetry protocol where the subscriber bed is listening to the broker that is running on the server. The broker sends the messages to each bed securely. The Bluetooth is a protocol which uses 20 bytes of packets to communicate with the bed. The Bluetooth is also secured by authentication methods in the embedded.

6.4 *Wake-Up Alarms*

The wake-up alarms are the time settings set by the users which trigger the surface motion at user-defined intensity to wake them up slowly and gently. Usually it is hard to wake up someone in the deep sleep, so we need to be sure that the duration of the wave is more than the block size of the deep sleep to make sure that they can wake up the user. The key to the good sleep is the consistent sleeping timings schedule that should be maintained.

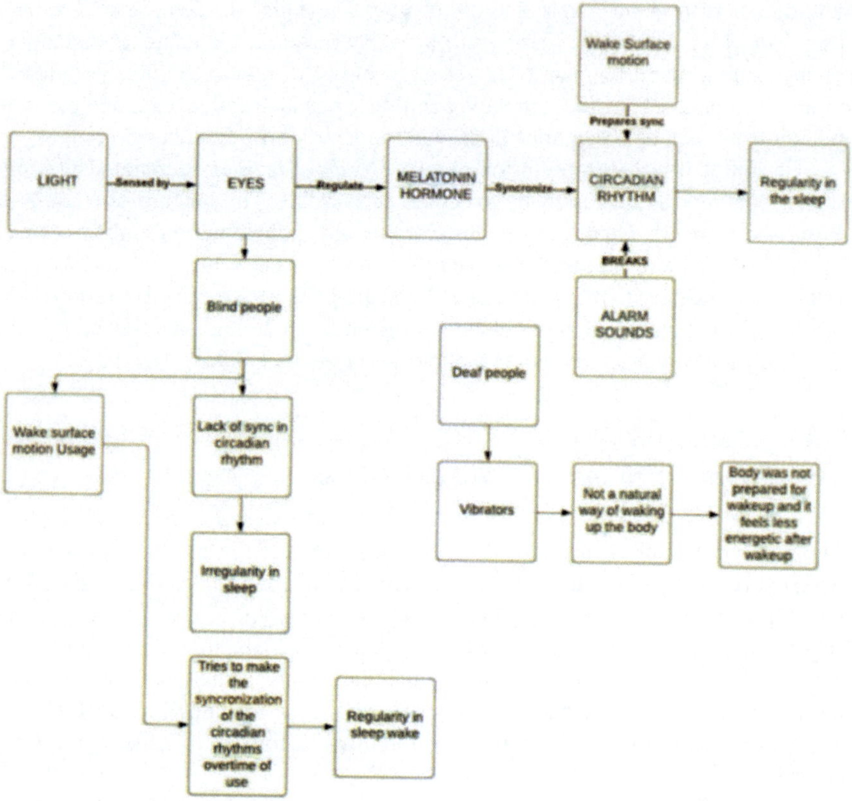

Fig. 13 High-level overview of the system

6.5 *Waking Up Naturally*

There are evidence that when other people (mother, etc.) woke up by shaking your body, the body feels more energetic than when it is woke up by the alarm clock. Body feels the natural wake up and gets to sync to wake up automatically at that time after repeated wake ups. Sleeping in the dark place where the light does not reach at morning is not recommended because the clock in the mind of human does not synchronize well in that scenario (Fig. 13).

Circadian Rhythms—The biological clock inside the mind of humans that directs the body that when to generate how much energy, and when to sleep. Melatonin—Hormone that manipulates the circadian rhythm. The secretion of the hormone is directed mostly by the light that is seen by the eyes (Fig. 14).

Light is responsible for our biological clock (circadian clock) to synchronize and prepare our body for wake up when the time comes. The melatonin can be measured to check whether the biological clock is working good and to check whether there is

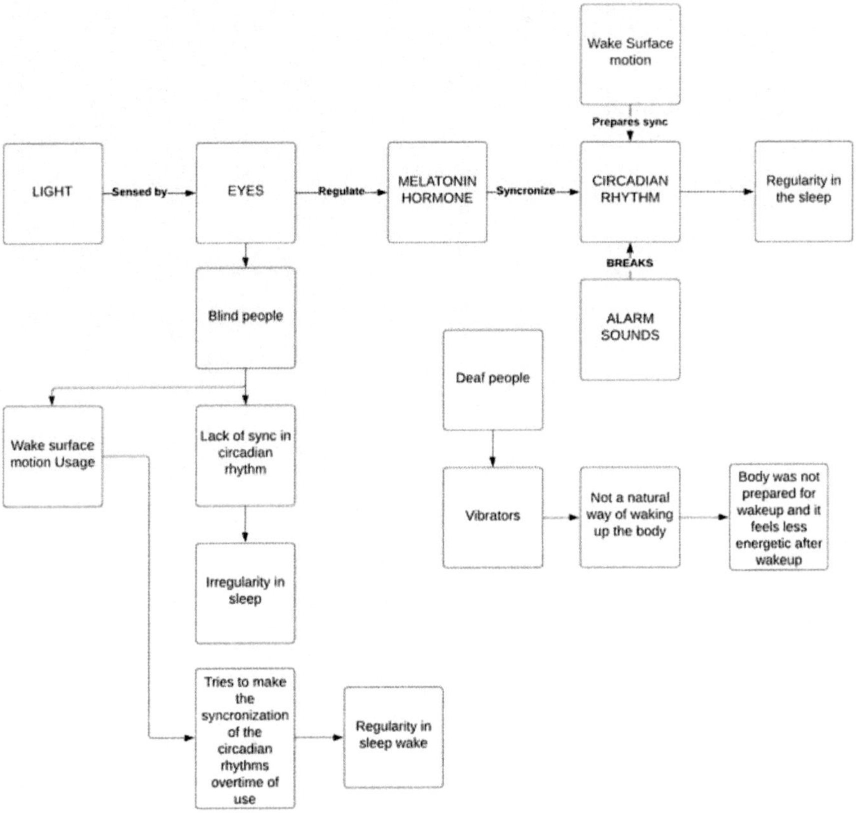

Fig. 14 High-level overview of the circadian clock of brain

any irregularity in it. When the biological clock is not in rhythm of the blind people, we see short sleeps and more naps. When the athlete flies to different countries, they experience jet lags often as they see the light at different time. The peak performance time of the body also changes due to that. They usually take pills to control the melatonin, but it is not good for health every time. Because the blind people do not have the sensation of light, the circadian clock does not get synchronized. So the melatonin release is irregular which makes the person sleep less and more often sleep during the day and night. The vibration from the watch or alarm sounds doesn't synchronize the circadian rhythm. The research was done on "Can the surface motion helps to synchronize the circadian rhythm for the people too?"

In the study, it is found that the surface motion helped 66% of users to set their circadian rhythm better, and after using the surface motion they feel easy to wake up on the alarm or before the alarm. Surface motion on the bed was changed after certain days during the study, and it can be seen that it is not effective way of waking up in the beginning, but after a certain time, when the body adapts to that, the body

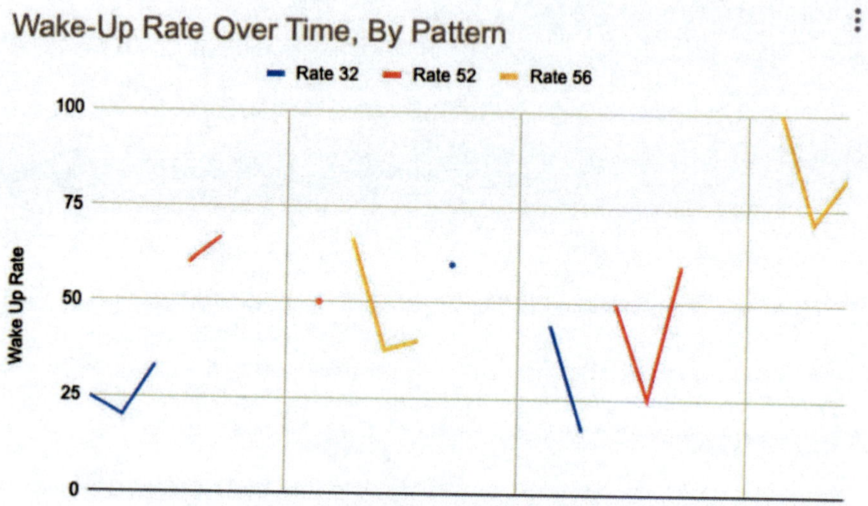

Fig. 15 How the performance of the silent alarm increases over time

knows that some motion is going to happen at certain time. And the body learns to wake up by setting the circadian rhythm before that (Fig. 15).

There are three types of sleepers in the world, 50% are those whose sleep is sensitive to the motion that happens beneath the skin. 30% of people are those who occasionally wake up to the motion. 20% are those deep sleepers who don't care about the environment while sleeping and are unworkable by the surface motion. Those 20% of users take a long time to set their circadian rhythm.

7 Analysis of the Beds

This chapter is about analyzing the bed's lifetime, algorithms, leaks, sleep stages, and lot more.

7.1 IOT Device Tracking

Any IOT device whether it has GPS or not, whether you give permission of the location or not, still the device can log the external IP address from which the approximate location of the device can be detected. The approximate location is then used to promote the device in the particular areas. There are many APIs available which convert the IP to the location. This location's zip codes can be easily visualized in the map. IOT devices are not using the VPN most of the time so the external IP addresses

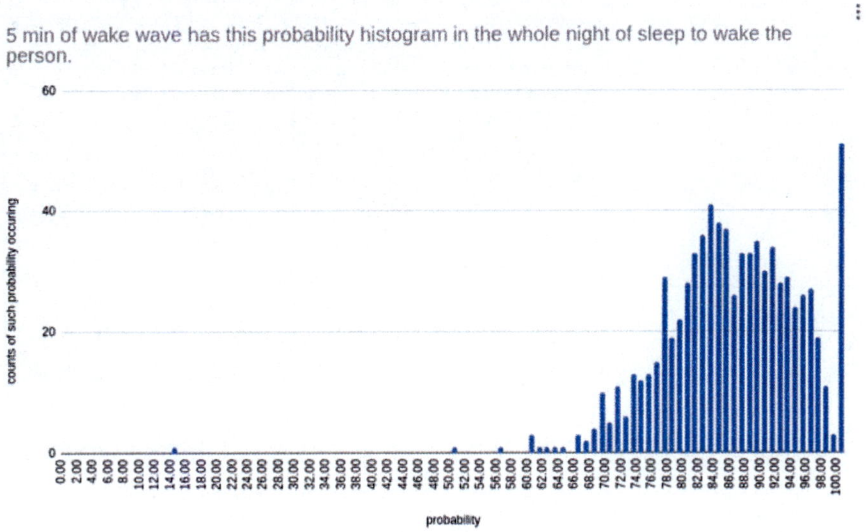

Fig. 16 Histogram of the probability of waking someone up using a 5 min alarm at any time of night

are pretty much accurate every time. The routers in the home are not configured to virtual private network. So when the device pings the website, its profile data and the external IP address can be easily recorded for further analysis.

In the modern era companies keep record of when the product was purchased, and does that product needs any repairs or not so that they can manufacture the parts that will be needed in repairing of the product. Companies are trying to figure out if it is possible to sell that product again or not. This type of targeted advertisement is happening on the internet every day.

7.2 Probability of Waking Someone in the Night

A person can be easily awake if he is in the light or wake condition during the sleep. The probability of waking someone up changes as the night progresses. The probability of waking someone in whole night is not equal to waking someone up during the last 2 h of sleep and it is not equal to waking someone up during the last 15 min of the sleep. To calculate that for what amount of time the alarm should be played so that we have enough probability to wake someone up during the last 2 h of sleep, the analysis of 1000 nights was made to make the histogram of the probability. The window of size 5 and 10 min and other minutes too was taken to analyze the probability (Figs. 16 and 17).

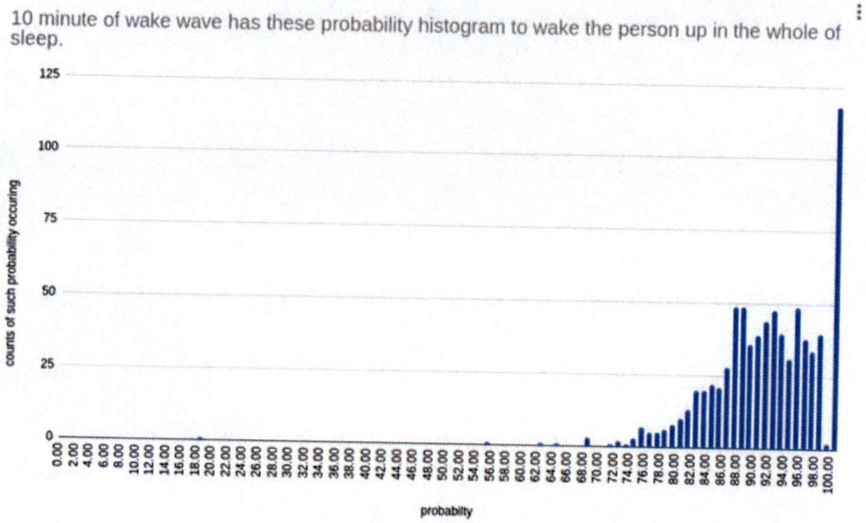

Fig. 17 Histogram of the probability of waking someone up using a 10 min alarm at any time of night

The probability of waking someone using 5 min alarm during any time at night is around 84% and by using 10 min alarm it is around 96%. So if you wanted to be sure to wake up at anytime during the night, make sure that your alarm length is 10 min to have a 96% f chance of waking a person up (Figs. 18 and 19).

The probability of a 5 min wake-up alarm to find the light sleep period to wake a person up is around 88% in the last 2 h of sleep. So if a person want to wake up at morning at usual time, the 5 min wake-up alarm can wake a person up by probability of 88%. The probability of the alarm to wake up by using a 10 min alarm is around 92–100% in the last 2 h of sleep (Figs. 20 and 21).

8 Quality of Sleep

This section discusses about the changes in sleep patterns with changes in the thoughts of mind.

8.1 Sleeping Issues Related to Fear

Every human has certain fears, for example, fear of losing someone, fear of death, fear of heights, fear of darkness, and much more. In the life most of the people go to

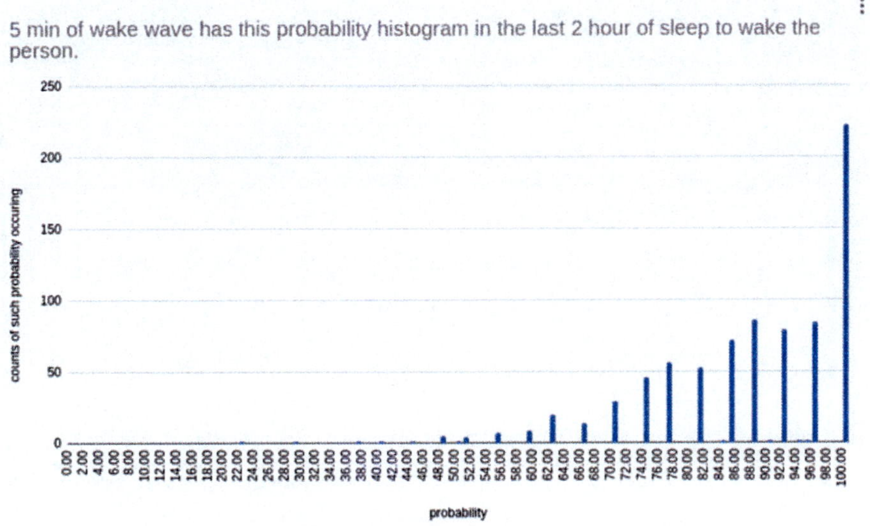

Fig. 18 Histogram of the probability of waking someone up using a 5 min alarm in last 2 h of sleep

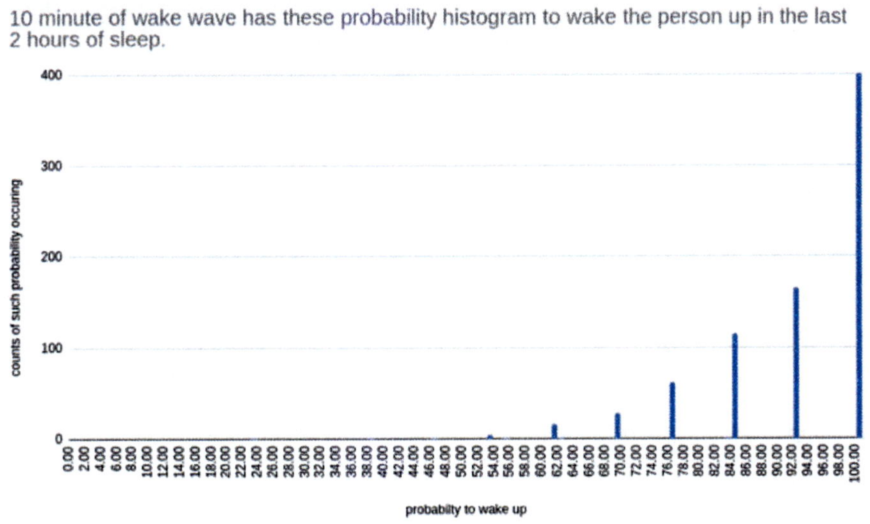

Fig. 19 Histogram of the probability of waking someone up using a 10 min alarm in last 2 h of sleep

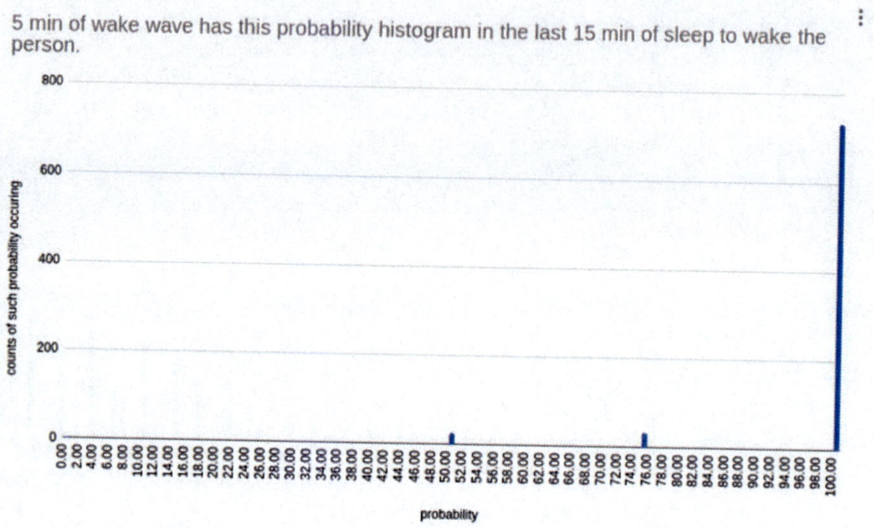

Fig. 20 Histogram of the probability of waking someone up using a 5 min alarm in last 15 min of sleep

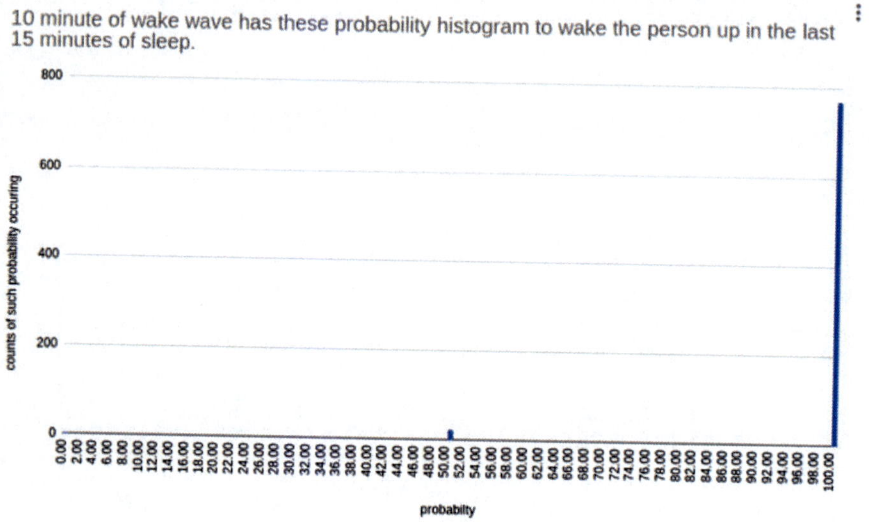

Fig. 21 Histogram of the probability of waking someone up using a 10 min alarm in last 15 min of sleep

the fear of losing someone, maybe their parents at a certain age. The fear sometimes affect the sleep timings. People feeling worried about their relatives to be in disease like corona get less sleep than normal. The hormone that is needed for the good sleep does not build up in the mind when people are in stress. There is also an impact of bad news that we see in the television. People like watching bad news so media houses are showing them but in realty the mind affects adversely if they are watched for a long amount of time.

To overcome the problem, people should understand one thing that every born human on the earth will die one day. If not today then sometime in future a human will die. And by accepting this bitter truth one should remove the fear of losing someone. Also we should watch less intense news especially which includes the crime scenes.

8.2 Sleeping Issues Related to Stress

Stress hormone cortisol secreted by adrenal glands and ACTH secreted by pituitary glands create a cycle that also affects the circadian clock somewhat. Due to having extra stress in mind, the sleep gets disturbed also. When people are in stress, they think more while sleeping and that causes the delay in the start of the sleep, it also affects the circadian clock. And because of stress, the person wakes up with less energy.

The best way to overcome this is to forget everything bad that happened in the past and meditate with empty mind, to get better sleep.

8.3 Sleeping Issues Due to Jet Lag and Work Shifts

Travel in either direction can disrupt circadian rhythms and trigger excessive daytime naps, sleep disturbance, and less energy to do work. Prompt circadian adaptation to the new time zone may be achieved via appropriately timed exposure to bright light and darkness. People should be educated in the flight that at what time they have to see the bright light compulsory for some time in the new country that they are landed for. But the issue is that they don't care about that much and due to it many people suffers. On exposing the bright light can either advance the circadian clock or delay the circadian clock depending on which time zone you are traveling from and traveling to. They should advice that at what time it is good to look out of window of airplane and at what time it is better to see dark to cheat the circadian clock while in the plane itself. This should be done avoiding taking the melatonin from outside as a pill, and use the natural sunlight to change the circadian clock naturally.

8.4 Sleeping Issues Due to Blindness

The melatonin required for the proper synchronous circadian rhythm clock is regulated by the eyes looking at the bright light. Even if they use the alarm, the internal clock of the brain doesn't synchronize much.

People with complete blind eyes have frequent naps in the day time, and less constant sleep in the night time. They eventually require more time to sleep in total in 24 h to get enough deep sleep to refresh the brain.

8.5 Sleeping Issues Caused by Diseases Like Sleep Apnea and Snoring

Sleep apnea is a serious sleep disorder that occurs when your breathing stops and begins to fall asleep. If left untreated, it can lead to serious problems such as severe snoring, daytime fatigue, or heart problems or high blood pressure.

This condition is different from normal or basic snoring. Primary snoring can be caused by nose or throat position, sleep style (especially lying on the back), being overweight or older or alcohol or other depressants. When both primary snoring and sleep apnea-related snoring occur when the tissue in the back of your throat vibrates, people with sleep apnea may have

Snoring louder than normal snoring.

Pause while breathing (more than 10 s).

Take a shallow breath, inhale or exhale.

Be restless.

Sleep apnea can be detected by calculating the distances between the peaks in the respiratory signal (Fig. 22).

To detect the sleep apnea statistically:

Step 1: Take the 30 s window of the respiration signal from the bed.

Step 2: Take the time duration between peaks.

Step 3: Calculate the IQR from that.

Step 4: If the value is more than 1.5 * IQR, then store that as the abnormality.

Step 5: If the number of abnormal counts increases in the certain amount of time than a threshold value, then we can tell that there is a defect in the breathing cycles of the person.

8.6 Quality of Sleep Helping in Curing Disease Like COVID-19

This section is based on author's personal experience only. When we get the disease like COVID-19, the body gets weaken automatically and the antibodies are being

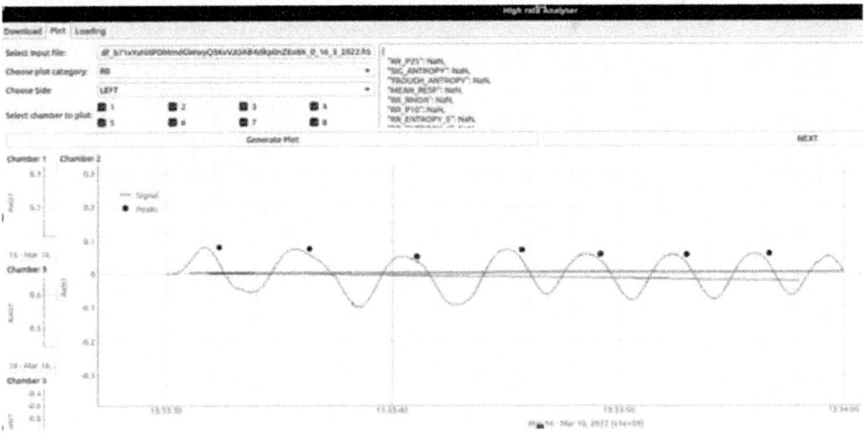

Fig. 22 High rate analyzer for the bed

made by the body. The first symptom that we can notice is the weakness followed by fever and headache in COVID. The headache can be solved only by taking rest and sleep, there is no other cure of it. The heavy dose of medicine that we take makes us drowsy and sleepy.

When the body is resting, the body can decide what to do to fight the virus. The body is active the whole time and we sometime feels that we didn't get enough sleep, even after sleeping. The body doesn't get energy when we wake up, and it feels like the body has weakened a lot.

9 Future Work

Following are the work that needs to be done in future:

- Making a heart rate-based sleep-stage detection model so that whenever we have heart rate, we get the sleep-stage prediction from that model.
- Smoothing the sleep-wake predictions.
- Moving the computation to the cloud so that the load on the Raspberry Pi decreases.
- Making better algorithms to handle the sleep stages during the surface motion is going on.
- Analyzing the anomalies that occur.
- Combining the feature files on the cloud and doing computation of the post-processing using the cloud so that the delay of the sleep summary on the mobile can be reduced.
- Making a pre-snore detection algorithm that can detect that person will snore after 30 s, so that it can be prevented by slightly moving the person.

10 Conclusion

As the Internet is available and becoming cheaper, as the storage drive's cost per GB is becoming cheaper, as the cost of cloud infrastructures is becoming more cheaper, the era of the Internet of Things is coming near. The things would try to make the ecosystem which would make sure that you purchase a certain other item of a company only if you have purchased one. The data flow from one ecosystem to other ecosystem would be stricter.

There is a need of the privacy and security if the Internet of Things are added to the life so quickly. The intelligent bed can know if you are alive or not. The bed can also know that you are in danger or not. The bed can even know that are you in house or have gone to vacation. The heart never lies like your brain does, so the bed can also know whether you have low blood pressure or high blood pressure problems. The bed can also be as intelligent to know that you are drunk or you are angry. The bed can predict that the user has some sleep disorder or not. So if the bed wants to predict it can predict a lot more things about a person. Even if you type your gender wrong in the profile, the bed in fact can also predict whether you are male or a female. The bed of course knows your size, weight, and height. It is no longer hidden from the bed.

In the near future, the bed would be so intelligent that it can even make and feel vibrations when you listen to music or watch television. It can even make the wave patterns using the music that you hear and will make you feel that you are inside the music. We are moving toward a future where a person would become more and more lazy and would like to operate everything from sitting on the bed via phone or voice.

People uses the divide-and-conquer approach to solve a bigger task, which current neural network can easily solve. But using a neural network may cause delay in the outcome of the result. So still the statistical approaches are used so that the result can be achieved quickly. The example of divide and conquer the bigger problem is that, the bed detection and the sleep-wake detection is done separately but combining their individual result gives the correct sleep wake. The sleep-wake and sleep-stage detection are done separately but combining their result gives the correct sleep stage at the end. So whenever we make models to solve the huge problem, we can make smaller models that solve a part of the huge problem and then combine the results of those models to achieve best result rather than making a huge model and waiting for it to predict after a long time.

References

1. Long X (2015) On the analysis and classification of sleep stages from cardiorespiratory activity. Technische Universiteit Eindhoven. https://pure.tue.nl/ws/files/11271404/20151230_Long.pdf
2. Ian T, Dorothy B (2009) Strobe lights, pillow shakers and bed shakers as smoke alarm signals. Fire Saf Sci 9. https://doi.org/10.3801/IAFSS.FSS.9-415

3. Satapathy SK, Ravisankar M, Logannathan D (2020) Automated sleep stage analysis and classification based on different age specified subjects from a dual-channel of EEG signal. In: 2020 IEEE international conference on electronics, computing and communication technologies (CONECCT), pp 1–6. https://doi.org/10.1109/CONECCT50063.2020.9198335.
4. Mendonça F, Mostafa SS, Morgado-Dias F, Ravelo-García AG, Penzel T (2019) A review of approaches for sleep quality analysis. IEEE Access 7:24527–24546. https://doi.org/10.1109/ACCESS.2019.2900345
5. Adami A, Hayes T, Pavel M, Singer C (2006) Detection and classification of movements in bed using load cells. Conf Proc IEEE Eng Med Biol Soc 589–592. https://doi.org/10.1109/IEMBS.2005.1616481. PMID: 17282250
6. EL-Manzalawy Y, Buxton O, Honavar V (2017) Sleep/wake state prediction and sleep parameter estimation using unsupervised classification via clustering, pp 718–723. https://doi.org/10.1109/BIBM.2017.8217742
7. https://www.verywellhealth.com/the-four-stages-of-sleep-2795920
8. Rish I (2001) An empirical study of the Naïve Bayes classifier. IJCAI 2001 Work Empir Methods Artif Intell 3
9. Shaffer F, Ginsberg JP (2017) An overview of heart rate variability metrics and norms. https://doi.org/10.3389/fpubh.2017.00258
10. Engle-Friedman M (2014) The effects of sleep loss on capacity and effort. Sleep Sci 7(4):213–224. ISSN 1984-0063, https://doi.org/10.1016/j.slsci.2014.11.001
11. El-Manzalawy Y, Buxton O, Honavar V (2017) Sleep/wake state prediction and sleep parameter estimation using unsupervised classification via clustering. IEEE Int Conf Bioinform Biomed (BIBM) 2017:718–723. https://doi.org/10.1109/BIBM.2017.8217742
12. García-Magariño I, Lacuesta R, Lloret J (2018) Agent-based simulation of smart beds with Internet-of-Things for exploring big data analytics. IEEE Access 6:366–379. https://doi.org/10.1109/ACCESS.2017.2764467
13. Kirjavainen T, Cooper D, Polo O, Sullivan CE (1996) Respiratory and body movements as indicators of sleep stage and wakefulness in infants and young children. J Sleep Res 5(3):186–194. https://doi.org/10.1046/j.1365-2869.1996.t01-1-00003.x. PMID: 8956209
14. Institute of Medicine (US) Committee on Sleep Medicine and Research; Colten HR, Altevogt BM (eds.). Sleep Disorders and Sleep Deprivation: An Unmet Public Health Problem. Washington (DC): National Academies (US); 2006. 2, Sleep Physiology. Available from: https://www.ncbi.nlm.nih.gov/books/NBK19956/
15. Walch O, Huang Y, Forger D, Goldstein C (2019) Sleep stage prediction with raw acceleration and photoplethysmography heart rate data derived from a consumer wearable device. Sleep 42(12), zsz180. https://doi.org/10.1093/sleep/zsz180
16. Ferri R, Manconi M, Plazzi G, Bruni O, Vandi S, Montagna P et al (2008). A quantitative statistical analysis of the submentalis muscle EMG amplitude during sleep in normal controls and patients with REM sleep behavior disorder. J Sleep Res 17(1): 89–100. https://doi.org/10.1111/j.1365-2869.2008.00631.x. PMID 18275559. S2CID 25129605
17. https://sleepgadgets.io/smart-mattress-smart-bed/
18. https://theoutlooknewspaper.org/2832/opinions/are-you-getting-enough-sleep/
19. mccannltc.net. N.p., Web (2022) https://mccannltc.net/news/losing-sleep-over-sleep
20. https://www.sleepassociation.org/about-sleep/what-is-sleep/

Air Quality Monitoring Platform

Prastuti Gupta, Surchi Gupta, Divya Srivastava, and Monika Malik

Abstract Nowadays, Pollution became one of our biggest concerns. Air pollution is a mixture of various harmful solid particles and hazardous chemicals and gases in the atmosphere. Various kinds of sensors are interfaced in this model to determine air quality, humidity, pressure, and the temperature of the surroundings. The NodeMCU board is being interfaced with the sensors indicated above. The MQ135, DHT11, and the HW611 e/p 280, which determine air quality, temperature, humidity, and atmospheric pressure, respectively, are all connected to the Esp8266, and the results are immediately supplied into the ThingSpeak an IoT platform. The resulting value of parameters of air can be seen just by measured readings over a threshold value email alert can also be sent on a user ID utilizing an easy mail transfer protocol which makes the user get notified and take necessary action. The Arduino IDE software was utilized to run the NodeMCU board, Embedded C is used in the coding of the board. To implement the model on ThingSpeak, Cloud computing is used. In this study, the sensors utilized are connected to their respective pin configurations on the board to determine numerous factors in real time and to acquire detection results given by them at that instant.

Keywords Air quality sensor (MQ135) · Humidity and temperature sensor (DHT11) · HW611 e/p 280 · ThingSpeak · Arduino IDE

1 Introduction

There are many platforms like ThingSpeak which help in analyzing the real-time data with the help of ThingSpeakMatlab the sensors used in IoT applications can directly send data through cloud which is further analyzed using ThingSpeak. This paper presents the study of data sensed by various sensors connected to NodeMCU. It measures parameters that are the main components of air. The main purpose of measuring the content of carbon dioxide in air, air pressure, temperature, and

P. Gupta · S. Gupta · D. Srivastava (✉) · M. Malik
JSS Academy of Technical Education, Noida, Uttar Pradesh, India
e-mail: srivastavadivya964@gmail.com

H. O. Bansal et al. (eds.), *Next Generation Systems and Networks*, Lecture Notes in Networks and Systems 641, https://doi.org/10.1007/978-981-99-0483-9_29

humidity is to know the present condition of the environment. While using these platforms, we can set up a threshold, and whenever any parameter goes up from the threshold level, safety measures can be taken to protect the environment. To monitor air quality, a few sensors are taken into consideration, namely DHT11 temperature and humidity sensor, MQ135 gas sensor, and HW611 atmospheric pressure. This setup requires two steps; one is the detection and the other is the monitoring of the environment. Cloud computing helps in Monitoring live data, in the later stage, we make use of Thinkspeak. The data so obtained is monitored in real time using Thinkspeak. Sensors are associated with NodeMCU board and readings will start coming as per the concentration of gas particles present in the air. In the last stage, Wi-Fi is used for transmitting data to the cloud. Parameters are recorded as per the resultant values.

It is very important that the system should be able to give a clear picture of the air quality so that necessary measures can be taken within the time and without much loss. Since the quality index of air is varying for urban and suburban areas, the system should be able to deliver the best result for air pollution control. In this work, we are making air quality detection systems more efficient, reliable, easy to use, and cost-effective. The system depicts the air quality in PPM using Arduino as well as on the cloud so that it could be monitored very efficiently and will also send mail alerts to user ids to alert them of the readings beyond normal range. This system can further be researched and used in various other researches as well.

1.1 Global Warming

The term "global warming" is defined in this way. Standard worldwide temperatures have increased at an alarming speed with the exception of one of the 16 warmest years since 2000; specialists are aiming to accelerate this trend: One of NASA's 134-year records. People who deny environmental change claim that rising global temperatures have caused "stagnation" or "sluggishness," yet multiple recent studies, including one published in the journal Science in 2015, refute this notion. According to scientists, if global warming continues, the average temperature in America might rise to 10 degrees Fahrenheit by the end of the century. CO_2 and other air pollutants and greenhouse gases accumulate in the atmosphere, capturing sun rays and radiation that bounce off the Earth's surface, resulting in global warming. This radiation usually escapes into space, but impurities stay in the atmosphere for generations, trapping heat and causing the earth to overheat. The greenhouse effect is the name for this phenomenon. The main source of heat-trapped pollution in the United States is the use of fossil fuels to generate electricity, which produces tons of carbon dioxide every year. By far, the most polluting industry is coal-fired power plants. The transportation industry releases 1.7 billion tons of CO_2 per year, making it the country's second-greatest source of carbon pollution.

1.2 Air Pollution

Pollution of the air results in the presence of toxic and harmful toxic waste and gases, disqualifying air from being used by plants and animals. Changes in all the components of the air can cause a decline in the quality of life in the stratosphere. Many factors contribute to air pollution. The first and sole cause of any type of environmental air pollution is human activity. Anything that burns any substance, whether it's household electrical equipment or industrially dangerous chemicals, emits harmful fumes that can pollute the air. The main contribution of hydrofluorocarbons is the huge amount of traffic on the roads. Industries are another major source of air pollution in the atmosphere. In humans, air pollution can cause respiratory problems, cancer, and cardiovascular problems. Even children can grab diseases such as pneumonia and asthma. Climate Change as well as Climate Change Global warming poses a threat to the environment. Another effect is acid rain, which results in losing the fertility of the land. Not only human beings, but also plants and animals are in danger due to the disturbance in the complete ecosystem.

1.3 The Influence of Temperature and Humidity on the Environment

Atmospheric parameters such as wind speed, wind direction, relative humidity, and temperature influence the concentration of air pollutants in the atmosphere. This study examines the effect of temperature and relative humidity on the concentrations of SO_2, NO_2, and suspended Particulate matter in the Indian coastal city of North Chennai from the year 2010–11. According to the learning findings, both SO_2 and NO_2 are interrelated in the summer (for SO_2 and NO_2) and fairly and positively after the rainy season (for SO_2 and NO_2). Except after the rainy season, RSPM and SPM have a positive relationship with temperature in all seasons. A huge change in temperature scale, suggest that the temperature in air pollutants (SO_2 and NO_2) is more effective in the summer season. In all environmental situations, the association between temperature and uncertain pollutant levels is very poor, demonstrating the effect of unstable thermal variability in the coastal arena. There exits significant negative coorelation between moistture and particulate matter RSPM .Experimental ststistics on SPM shows that during all four seasons, the coorelation was found to be very less as compared to other three seasons, also no major association was found between dampness, humidity, and sulfur dioxide. All season long, knuckles. According to this research, the moisture effect may have an impact on the survival of particulate matter near the coast.

1.4 Impact of Air Pressure in Air Pollution

Pollution levels are also influenced by air pressure. Air is normally permitted to increase pollution levels during high-pressure systems, while during low-pressure systems, the weather is often damp and windy, leading pollutants to spread through rain or weather [1]. discussed real time application scenario for controlling and monitoring Air pollution. [2] discussed the novel concept of smart air in which all the parameters are observed and processed directly on LTE. Authors in [3] observed and measured morning and afternoon air profile. [4] studied the high pollution and pollution density on an year basis for metro cities. Authors in [5] highlighted the seasonal study of air pollution and evaluated various parameters their study concluded high level of pollution in winter season. [6] discussed three years of evaluation of air quality and its control. [7] discussed intelligent control system to observe air pollution. Authors in [8] discussed wireless sensor based air pollution monitoring. [9] gives the study of toxics present in air and its monitoring using web. Authors in [10] displayed the complete set up of ZigBee for finding air pollution. [11] described the collection of data from the sensors deployed in the environment.

2 Methodology

Various sensors are employed in this article to detect humidity, temperature, atmospheric pressure, and CO_2 gas in the air. The NodeMCU board is used to connect the sensors indicated above. The MQ135 air pollution gas sensor, as well as the DHT11 temperature and humidity sensor, and the environment/pressure sensor HW-280 atmospheric pressure detection sensor, are all interfaced with NodeMCU.

The result of the atmospheric pressure sensor detecting air pressure, humidity, and temperature will be automatically fed into the ThingSpeakIoT Platform. It may then read our derived value via IoT. We utilized the Arduino IDE s/w tool to run the NodeMCU board. Arduino IDE software is an open source software tool that facilitates trouble-free code authoring, executing, and uploading the code according to the designed circuit (Fig. 1–3).

2.1 Installation of Arduino IDE

- STEP-1: Download and install the version of Arduino IDE for the laptop.
- STEP-2: Now, download the required libraries and add the zipped libraries in the tools section/ manage libraries for all the sensors.
- STEP-3: Setup the IDE by selecting the port and board (NodeMCU-1.0(Esp-12E Module)).

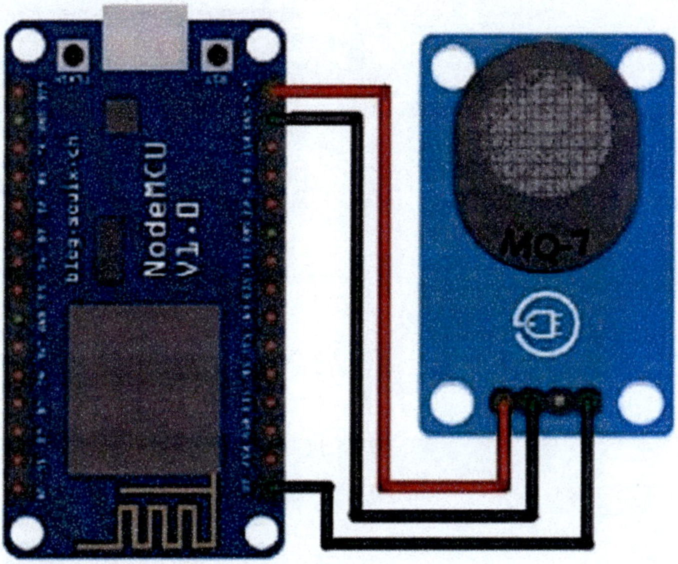

Fig. 1 Interfacing NodeMCU with MQ135

- STEP-4: Write the code and upload it and check the desired results on the Serial Monitor.

2.2 *Designing and Interfacing of Sensors with NodeMCU*

- STEP-1: Connect NodeMCU A0 pin to MQ135 A0 (Fig. 1)
- STEP-2: Connect NodeMCU G pin to MQ135 G
- STEP-3: Connect NodeMCU Vin pin to MQ135 Vcc

- STEP-4: Connect HW-166 E/P Pin VINN to NodeMCUPin 3V
- STEP-5: Connect HW-166 E/P Pin GND to NodeMCUPin GND
- STEP-6: Connect HW-166 E/P Pin SCL to NodeMCU PinGPIO3
- STEP-7: Connect HW-166 E/P Pin SDA to NodeMCU PinGPIO2

- STEP-8: Connect DHT11 Pin VCC to NodeMCU Pin VV
- STEP-9: Connect DHT11 Pin GND to NodeMCU Pin GND
- STEP-10: Connect DHT11 Pin Data to NodeMCU Pin D1.

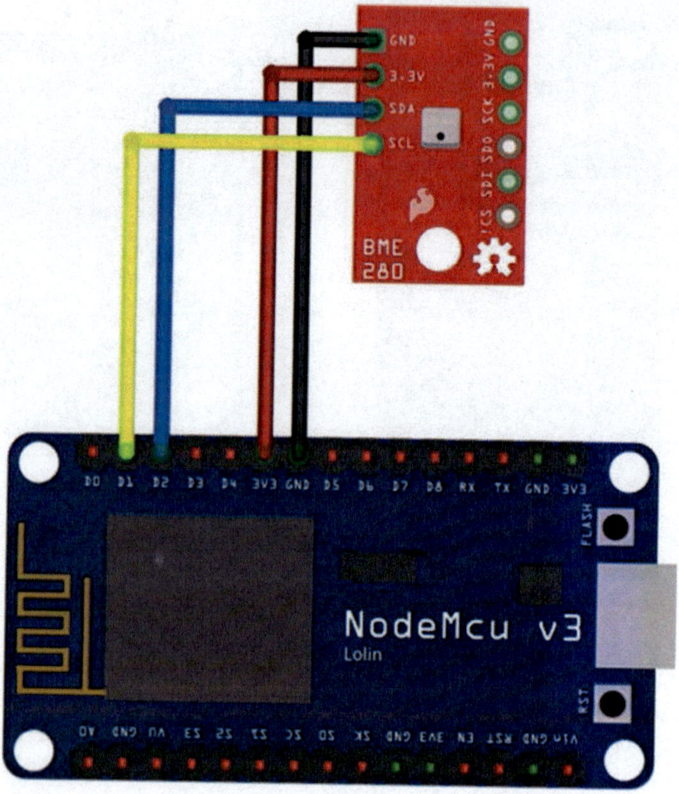

Fig. 2 Interfacing of HW-166 E/P 280 with NodeMCU

3 Results

As shown in Fig. 4, the values were obtained on the Arduino IDE Serial Monitor.

The results are obtained using Arduino IDE. The software obtained values for temperature, atmospheric pressure, humidity, and air quality values present in the atmosphere.

3.1 Notification Received on Mail

After the sensor's values are received on the Serial Monitor, the data is then transferred to the ThingSpeak Platform to aggregate and to get transferred into visual form, i.e., graphs, so that it can be easily analyzed and studied. If any reading of any sensor is above a preset range of threshold values, then through Simple Mail Transfer

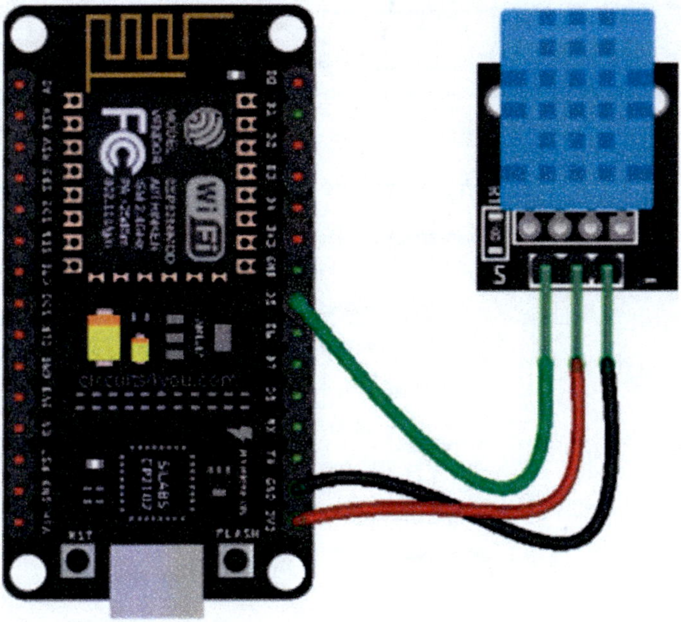

Fig. 3 DHT11 interfacing with NodeMCU

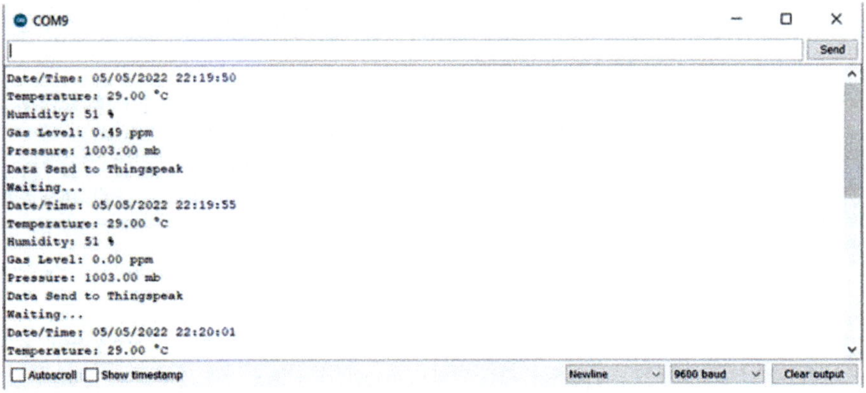

Fig. 4 Value on Arduino IDE serial Monitor

Protocols, we will get notification alerts on the user's Mail-Id, through which the user can get notified and take necessary actions (Fig. 5).

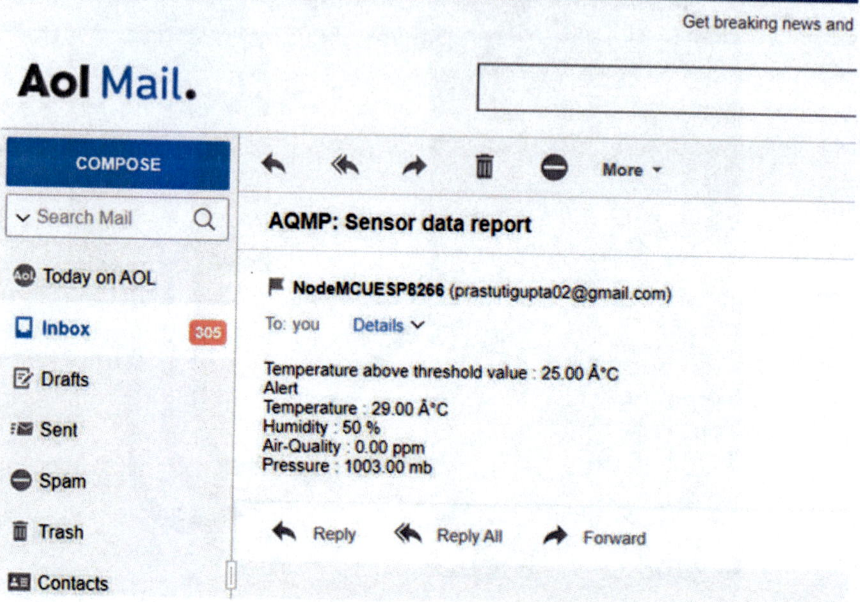

Fig. 5 Shows mail alert

3.2 Data Retrieved on ThingSpeakIoT Platform

The ThingSpeak app allows us to visualize data from the ThingSpeak cloud service. It is only necessary to input the public channel id, after which captured real-time data on mobile phones from all around the world can be viewed. The cloud application retrieved real-time data analysis on the ThingView application. ThingSpeak Humidity, Temperature, and Pressure data are represented in Fig. 6–9 showing the air quality.

4 Conclusion

Using a MQ135 and an Arduino Uno board, this project measures the temperature, humidity, and atmospheric pressure of the environment, as well as detect CO_2. The influence of temperature is also discussed, and DHT11 is used to measure the current temperature. The temperature is a key, if not the main, source of global warming. The humidity level in the environment, as well as temperature and air pressure, may be measured and monitored using the DHT11. Various attempts have been made at various levels of society to prevent the causes of air pollution. It should also be in charge of keeping the environment safe.

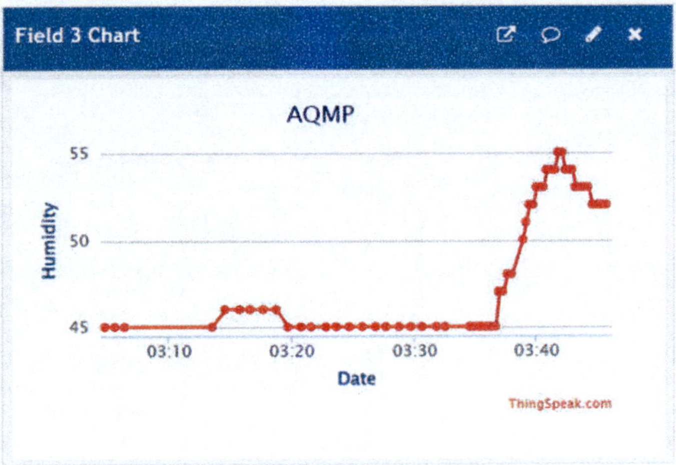

Fig. 6 ThingSpeak humidity values

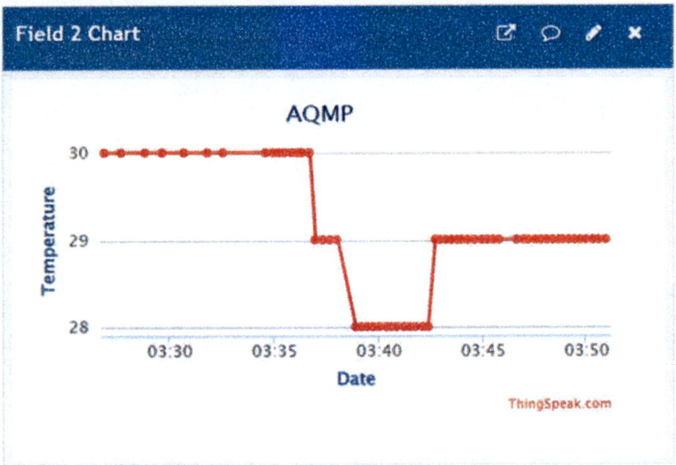

Fig. 7 ThingSpeak temperature values

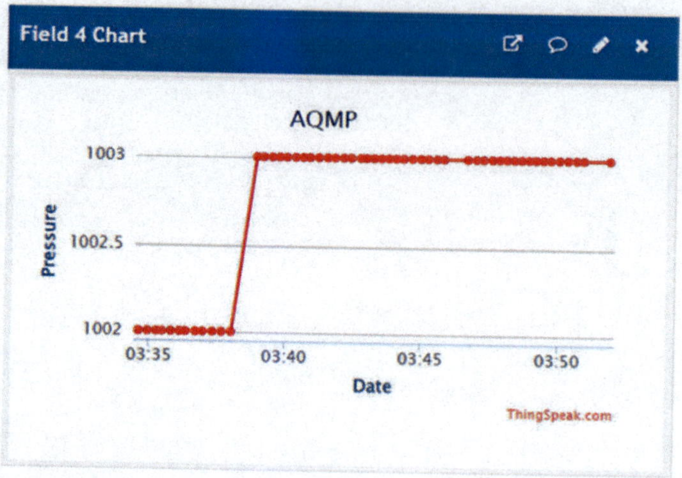

Fig. 8 ThingSpeak pressure values

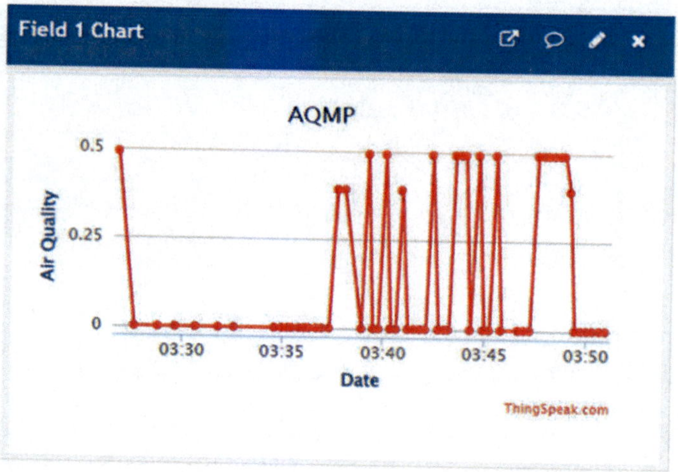

Fig. 9 ThingSpeak air quality values

References

1. Singh R, Gaur N, Bathla S (2020) IoT based air pollution monitoring device using Raspberry Pi and cloud computing. IEEE
2. Han WY (2020) Development of a IoT based indoor air quality monitoring platform. J Sens/Hindawi
3. Holzworth GGC (1967) Mixing depths wind speed and air pollution potential for selected locations in the United States. J Appl Meteorol 6:1039–1044
4. Karar K, Gupta AK, Kumar A, Biswas AK (2006) Seasonal variations of PM10 and TSP in residential and industrial sites in an urban area of Kolkata, India. Environ Monit Assess

118(1–3):369–381
5. Chauhan A, Powar M, Kumar R, Joshi PC (2010) Assessment of ambient air quality status in urbanization, industrialization, and commercial centers of Uttarakhand (India). J Am Sci 6(9):565–568
6. Dominick D, Latif MT, Juahir H, Aris AZ, Zain SM (2012) "An assessment of influence of meteorological factors on PM10 and NO_2 at selected stations in Malaysia." Sustain Environ Res 22(5): 305–315
7. Sonune AH, Hambarde SM "Monitoring and controlling of air pollution using intelligent control system". Int J Sci Eng Technol ISSN: 2277-1581 4(5): 310–313
8. Yaswanth D, Umar Dr S (2013)"A study on pollution monitoring system in wireless sensor networks". IJCSET 3(9): 324–328
9. Martinez K, Hart JK, Ong R "Environmental SensorNetworks." IEEE Comput 37(8): 50–56
10. Chourasia NA, Washimkar SP (2012) "ZigBee based wireless air pollution monitoring." Int Conf Comput Control Eng (ICCCE 2012)
11. Rajagopalan R, Varshney PK (2006) "Data aggregation techniques in sensor networks: A survey." IEEE Commun Surv Tutor 8(4): 48–63

CLAMP: Criticality Aware Coherency Protocol for Locked Multi-level Caches in Multi-core Processors

Arun Sukumaran Nair, Aboli Vijayanand Pai, Geeta Patil, Biju K. Raveendran, and Sasikumar Punnekkat

Abstract Cyber-physical systems that combine sensing, computing, control and networking with physical items and infrastructure, such as automotive, avionics and robotics, are rapidly becoming mixed criticality systems (MCS). The increasing expectations for computing ability and predictable temporal behaviour of these systems necessitate substantial enhancements in their memory subsystem architecture. The use of locked caches to have predictable execution time is one such optimization. There is no comprehensive method in order to manage coherency in locked caches in any of the current cache coherence protocols like MOESI. CLAMP—A criticality aware coherency protocol for locked multi-level caches in multi-core processors is an updated variant of MOESI and as an extension of MOESIL, to improve the data consistency of locked caches. The work CLAMP proposes an improvised locked cache coherence protocol for multiple levels of cache in multi-core MCS, whereas MOESIL is restricted to two-level cache architecture. Experiments using real-time benchmark programs on CACOSIM reveal an average cache miss rate reduction of 18% for high-criticality jobs.

Keywords Hierarchical cache architecture · Cache locking · Cache partitioning · Mixed criticality systems · Multi-core/many core architecture · Cache optimization

A. S. Nair (✉) · A. V. Pai · B. K. Raveendran
BITS Pilani, K. K. BIRLA Goa Campus, Goa 403726, India
e-mail: p20150024@goa.bits-pilani.ac.in

G. Patil
BMS Institute of Technology and Management, Doddaballapur, Main Road, Avalahalli, Yelahanka Bengaluru, Karnataka 560064, India

S. Punnekkat
Mälardalens University, Högskoleplan, Västerås, Sweden

H. O. Bansal et al. (eds.), *Next Generation Systems and Networks*, Lecture Notes in Networks and Systems 641, https://doi.org/10.1007/978-981-99-0483-9_30

1 Introduction

In mixed criticality systems (MCS), a shared platform is assigned to functionalities of varying degrees of importance [1], i.e., tasks with high criticality (HC) and low criticality (LC) coexist. The criticality is a measure for the required level of certainty to ensure task's completeness [1]. Predictive computation and HC task deadline adherence are the key design considerations of MCS to guarantee their safety certification [2]. In presence of task and core interference, providing optimum cache performance along with predictable computing of HC workload is a challenge. Locking in high level caches and partitioning in last level cache (LLC) are widely used to limit cache interference by other tasks/core and hence bound the worst-case cache behaviour. Garcioli et al. [3] and Pellizzoni et al. [4] measured the increased computation time by 296–384%, respectively, due to the impact by shared memory accesses in multi-core systems.

In multi-core MCS, data sharing leads to cache coherence issues. The snoopy cache coherence protocols [5], such as MESIF, MOESI, MOSI and MESI, invalidate a core's cache lines if another core modifies that memory block. The invalidation of cache lines belonging to HC jobs by a job executing in another core is one of the most common causes of catastrophic failures in MCS due to deadline misses of HC jobs. The coherency-related invalidations in existing cache coherence protocols abate the basic purpose of locking technique. This scenario warrants for an enhanced cache coherency protocol. The work proposes a novel cache coherence protocol CLAMP—A criticality-aware coherency protocol for locked multi-level caches in multi-core processors. CLAMP guarantees the availability of consistent locked shared data in a multi-level cache hierarchy multi-core processor, even when the data are updated by other cores. This work amends the most efficient cache coherence protocol—MOESI—and its extension MOESIL [5] to suit the bounded execution times of HC jobs in MCS, running in a system having hierarchical memory subsystem.

The remaining parts of the manuscript are structured as follows: Review of the literature is discussed in Sect. 2. Section 3 presents system and memory models considered in the work. Section 4 presents the CLAMP design overview and algorithm. Section 5 presents the results and discussions. Section 6 concludes with final thoughts and recommendations for the future.

2 Related Works

The rise of multi-core systems and anticipated increase in locked cache [6, 6] usage necessitate improved cache coherence mechanisms. Kaur et al. [8] investigated the effectiveness of cache coherence algorithms in a shared memory dual-processor system and discovered MOESI to be superior to MESI/MSI. The analysis of MI, MESI, MOESI and MESIF cache coherence algorithms by Patil et al. [9] indicates that MOESIF should be utilized to maximize off/on-chip bandwidth utilization. The importance of taking into account cache coherence influences on latency constraints is highlighted by the fact that arbitration delay scales linearly with the number of cores, whereas coherence latency scales quadratically. An efficient time-based cache coherence for limited execution was proposed by Sritharan et al. [10]. A hardware cache coherence approach that is criticality conscious was introduced by Kaushik et al. [11] to allow data flow between LC and HC operations safely and without interruption. The novel paradigm of locked/reserved caches has not been explored by any of the existing cache coherence methods. Since the rise of RT systems, improvements in the memory sub-system have played a crucial role in assuring optimum WCETs, with cache-locking strategies. The manuscript's objective is to offer a novel cache coherence mechanism for locked multi-level caches for multi-core processor MCS to lessen the impact of inter-task and inter-core interference caused by LC tasks while executing HC tasks.

3 System Model

The rise of multi-core system model is considered an m-core processor with a multi-level cache hierarchy and quick-path interconnection. The system is expected to follow automobile safety criteria, with criticality levels increasing from automotive safety integrity level (ASIL) A to D. The L1 cache is a split private cache for data and instructions, and the L2 cache is a single private cache for each core. The unified LLC is a shared cache connected to all cores with quick-path interconnect allowing limited access from other cores. The memory subsystem is assumed to be inclusive in nature across cache hierarchies.

4 Proposed Work

The memory subsystem in multi-core MCS needs to ensure optimistic WCETs for HC tasks. In multi-core system having shared data across cores, coherency misses because shared data invalidation takes place. Whenever the shared data gets modified/written in any other core, the corresponding block in all other cores—irrespective of whether the block is locked or not—gets invalidated. This violates the purpose of predicted WCET and locking of HC jobs in established cache coherence methods. This could cause HC jobs to miss their deadlines, which would cause MCS to fail drastically. To achieve deterministic WCET for HC workloads, the solution, CLAMP, takes care of all the locked blocks in cache. When another core modifies a locked cache line, CLAMP follows write-update rather than write-invalidate. This ensures that a locked cache line will never be invalidated in CLAMP and will always have the most recent data. As a result, HC jobs are guaranteed never to surpass their predictive WCET. CLAMP protocol follows the same states as the conventional MOESI protocol. In both protocols, the M state denotes the updated valid copy. The frequency of write-backs is decreased by using O state to distribute dirty copies among various cores. There is a fresh and exclusive copy in the E state cache line. The cache line is among the many shared valid copies, as indicated by the S state. Depending on whether the cache line of another core is in the O state, it can be a clean copy or a dirty copy. Before making any attempt at access, the data must be fetched from the cache first, because state I is an invalid cache line. Locking is possible, only after the memory block has been brought into the cache.

CLAMP follows the same state transfers and control signals as that of MOESI if none of the cores—local or remote- are locked. CLAMP works differently when cache lines are locked to ensure the validity of locked data even after modification happens in remote cores. CLAMP works with the assumption that anytime a read or write request arrives, it checks for valid data in the following order as shown in Fig. 1—(1) Local L1 cache of requestor, (2) snoop from remote L1 caches, (3) Local L2 cache of requestor, (4) L2 snoops it from remote L2 caches, (5) Local L2 cache snoops it from shared LLC and (6) Shared LLC receives data from primary DRAM memory.

If any read of a data item yields the value that has been most recently written, the memory subsystem is coherent. Read/write requests from the processor side and data snooping from the bus side are the two ways to receive requests for data. In any of the memory request scenarios, the cache coherence mechanism should perform admirably. It can work at the same cache level, such as local L1 to remote L1 transfers, local L2 to remote L2 transfers and vice versa, as well as coherence protocols that function at different cache levels, such as L1 to L2, L2 to L3 and so on.

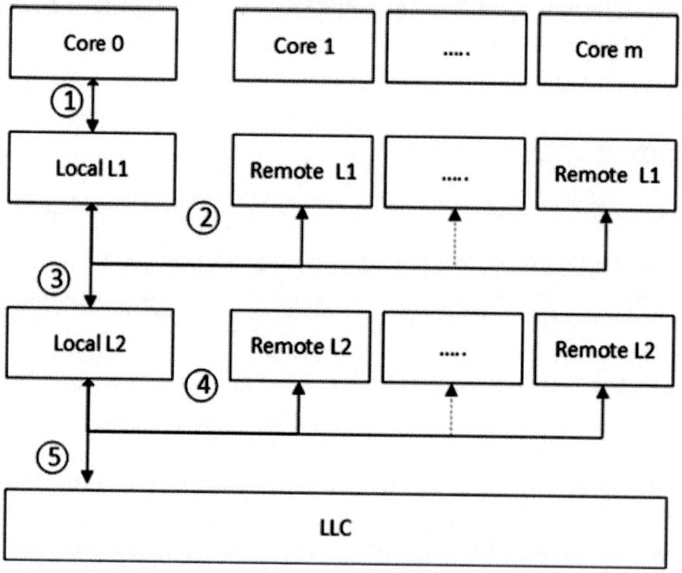

Fig. 1 Cache access sequence

4.1 Coherence Mechanism at Same Level

In comparison to MOESI, extra signals SIG_WRT_DATA, SIG_LOCK_ACK and SIG_WRT_ACK are required in both CLAMP and MOESIL [5]. These signals are necessary for snooping data transfers between nearby and distant L1 caches and ensuring that the newest data is available even if a shared memory update is performed by any core. The implementation, state transition and algorithm of MOESIL presented in Nair et al. [5] are applicable for CLAMP. CLAMP extends MOESIL with multiple levels to support hierarchical cache architecture.

4.2 Coherence Mechanism Between Different Levels

The state transition of cache line for control and data flow between upper and lower levels of a cache subsystem is detailed here. The reservation of cache line at different levels can be dependent on the architecture, design and nature of cache subsystem. It also depends on whether the content of one cache is present in subsequent cache levels, i.e., inclusive or not. If it is inclusive, whether all blocks in the locked higher-level cache are required to be locked in its subsequent lower-level cache. The maximum amount of cache space used for locking is a system-configurable value that typically ranges from 50 to 70% of total cache space.

Unlocked L2 Access by Unlocked L1. Figure 2 shows the cache state transitions for unlocked L2 cache access by unlocked L1 cache for read and write operations. For L1 read hits (cases from R1 to R20), the unlocked cache line states of both L1 and L2 stay unaltered. If both L1 and L2 are in I state (R25), L1 cache state is transitioned to E and L2 cache line state is transitioned to either S state, if the data is shared with any other L2 cache line, or E state if it is brought from the LLC. For the remaining cases, when L1 is in I state (cases from R21 to R25), as the data is available with the local L2 cache, a read request changes the L1 cache line to E state leaving L2 cache line state unchanged. When a write request arrives, if the cache line is in O or S state and the same line is locked in some remote L1 cache, the cache line changes to O state. For all the other cases, L1 cache line is transitioned to M state. L2 cache line continues to remain in the same state, if both L1 and L2 cache line state are not in I state. If both the cache lines are in I state, then L1 cache line changes to M state and L2 cache line changes to either E (if it is brought from LLC) or S (if it is shared with any other L2 cache) state.

Locked L2 Access by Unlocked L1. Figure 3 shows the cache state transitions for Locked L2 cache access by unlocked L1 cache for read and write operations. The state transitions of cache lines are same as in Sect. 4.2, except I state of locked L2 cache line states. The states L5, L10, L15, L20, L25, U5, U10, U15, U20 and U25 do not exist as there is no I state for locked cache line.

ACCESS	Read Unlocked Cache (L_1)→					Write Unlocked Cache(L_1)→				
↓Unlocked Cache (L_2)	M	O	E	S	I	M	O	E	S	I
M	R_1 M/M	R_6 O/M	R_{11} E/M	R_{16} S/M	R_{21} E/M	W_1 M/M	W_6 M/M‖O/M	W_{11} M/M	W_{16} M/M‖O/M	W_{21} M/M
O	R_2 M/O	R_7 O/O	R_{12} E/O	R_{17} S/O	R_{22} E/O	W_2 M/O	W_7 M/O‖O/O	W_{12} M/O	W_{17} M/O‖O/O	W_{22} M/O
E	R_3 M/E	R_8 O/E	R_{13} E/E	R_{18} S/E	R_{23} E/E	W_3 M/E	W_8 M/E‖O/E	W_{13} M/E	W_{18} M/E‖O/E	W_{23} M/E
S	R_4 M/S	R_9 O/S	R_{14} E/S	R_{19} S/S	R_{24} E/S	W_4 M/S	W_9 M/S‖O/S	W_{14} M/S	W_{19} M/S‖O/S	W_{24} M/S
I	R_5 M/I	R_{10} O/I	R_{15} E/I	R_{20} S/I	R_{25} E/E‖E/S	W_5 M/I	W_{10} M/I‖O/I	W_{15} M/I	W_{20} M/I‖O/I	W_{25} M/E‖M/S

Fig. 2 Unlocked L2 Access by Unlocked L1

ACCESS	Read Unlocked Cache (L_1)→					Write Unlocked Cache(L_1)→				
↓Locked Cache (L_2)	M	O	E	S	I	M	O	E	S	I
M	R_1 M/M	R_6 O/M	R_{11} E/M	R_{16} S/M	R_{21} E/M	W_1 M/M	W_6 M/M‖O/M	W_{11} M/M	W_{16} M/M‖O/M	W_{21} M/M
O	R_2 M/O	R_7 O/O	R_{12} E/O	R_{17} S/O	R_{22} E/O	W_2 M/O	W_7 M/O‖O/O	W_{12} M/O	W_{17} M/O‖O/O	W_{22} M/O
E	R_3 M/E	R_8 O/E	R_{13} E/E	R_{18} S/E	R_{23} E/E	W_3 M/E	W_8 M/E‖O/E	W_{13} M/E	W_{18} M/E‖O/E	W_{23} M/E
S	R_4 M/S	R_9 O/S	R_{14} E/S	R_{19} S/S	R_{24} E/S	W_4 M/S	W_9 M/S‖O/S	W_{14} M/S	W_{19} M/S‖O/S	W_{24} M/S
I	L_5 –	L_{10} –	L_{15} –	L_{20} –	L_{25} –	U_5 –	U_{10} –	U_{15} –	U_{20} –	U_{25} –

Fig. 3 Locked L2 access by unlocked L1

Unlocked L2 Access by Locked L1. Figure 4 shows the cache state transitions for unlocked L2 cache access by locked L1 cache for read and write operations. The state transitions of cache lines are same as in Sect. 4.2, except I state of locked L1 cache line states. Hence the states L21 to L25 and U21 to U25 do not exist.

Locked L2 Access by Locked L1. Figure 5 shows the cache state transitions for locked L2 cache access by locked L1 cache for read and write operations. The state transitions of cache lines are same as in Sect. 4.2, except I state of locked L1 and L2 cache line states. Hence the states L5, L10, L15, L20, L21 to L25, U5, U10, U15, U20 and U21 to U25 do not exist.

ACCESS	Read Locked Cache (L_1)→					Write Locked Cache(L_1)→				
↓Unlocked Cache (L_2)	M	O	E	S	I	M	O	E	S	I
M	R_1 M/M	R_6 O/M	R_{11} E/M	R_{16} S/M	L_{21} –	W_1 M/M	W_6 M/M‖O/M	W_{11} M/M	W_{16} M/M‖O/M	U_{21} –
O	R_2 M/O	R_7 O/O	R_{12} E/O	R_{17} S/O	L_{22} –	W_2 M/O	W_7 M/O‖O/O	W_{12} M/O	W_{17} M/O‖O/O	U_{22} –
E	R_3 M/E	R_8 O/E	R_{13} E/E	R_{18} S/E	L_{23} –	W_3 M/E	W_8 M/E‖O/E	W_{13} M/E	W_{18} M/E‖O/E	U_{23} –
S	R_4 M/S	R_9 O/S	R_{14} E/S	R_{19} S/S	L_{24} –	W_4 M/S	W_9 M/S‖O/S	W_{14} M/S	W_{19} M/S‖O/S	U_{24} –
I	R_5 M/I	R_{10} O/I	R_{15} E/I	R_{20} S/I	L_{25} –	W_5 M/I	W_{10} M/I‖O/I	W_{15} M/I	W_{20} M/I‖O/I	U_{25} –

Fig. 4 Unlocked L2 access by locked L1

ACCESS ←Locked Cache (L_2)		Read Locked Cache (L_1)→					Write Locked Cache(L_1)→				
		M	O	E	S	I	M	O	E	S	I
	M	R_1 M/M	R_6 O/M	R_{11} E/M	R_{16} S/M	L_{21} –	W_1 M/M	W_6 $M/M \Vert O/M$	W_{11} M/M	W_{16} $M/M \Vert O/M$	U_{21} –
	O	R_2 M/O	R_7 O/O	R_{12} E/O	R_{17} S/O	L_{22} –	W_2 M/O	W_7 $M/O \Vert O/O$	W_{12} M/O	W_{17} $M/O \Vert O/O$	U_{22} –
	E	R_3 M/E	R_8 O/E	R_{13} E/E	R_{18} S/E	L_{23} –	W_3 M/E	W_8 $M/E \Vert O/E$	W_{13} M/E	W_{18} $M/E \Vert O/E$	U_{23} –
	S	R_4 M/S	R_9 O/S	R_{14} E/S	R_{19} S/S	L_{24} –	W_4 M/S	W_9 $M/S \Vert O/S$	W_{14} M/S	W_{19} $M/S \Vert O/S$	U_{24} –
	I	L_5 –	L_{10} –	L_{15} –	L_{20} –	L_{25} –	U_5 –	U_{10} –	U_{15} –	U_{20} –	U_{25} –

Fig. 5 Locked L2 access by locked L1

4.3 CLAMP: Working

The state changes in the cache lines by following the proposed protocol for operations that read, write and snoop cache lines are represented by the algorithms StateAfterRead, StateAfterWrite and StateAfterSnoop as mentioned in Nair et al. [5]. Read/Write operations are handled by the individual cores although read/write operations trigger snoop to be invoked in remote cores. Read, write and snoop operations at highest Level (e.g. L1) follows the same strategy as described in Nair et al. [5].

Read and Write Operations at Next Level. When a read or write request occurs at a higher cache level (e.g. L1) and valid data is not available in local and remote cores of that level, the cache controller invokes the StateAfterRead [5] algorithm at subsequent lower level. When the requestor cache line state is M or O during the eviction of victim data, the cache controller invokes StateAfterWrite [5] algorithm at its subsequent lower level. Snooping data from remote cores invokes the StateAfterSnoop [5] algorithm to snoop data from remote cores at that cache level.

5 CLAMP: Results and Discussion

For cache coherence protocol evaluations, CACOSIM [9], uses SPEC2006 [12] real-time benchmark programs. The setup is as follows: shared LLC, divided L1 cache sizes from 2 to 32 KB, unified L2 cache sizes from 4 to 128 KB and a 3-tier cache architecture with 64B cache line size. The simulator computes L1/L2 miss rates, L1-to-L1 transfers, L2 fetch rates, cache access time and Read/Write/Lock/Invalidate/Acknowledgement signals.

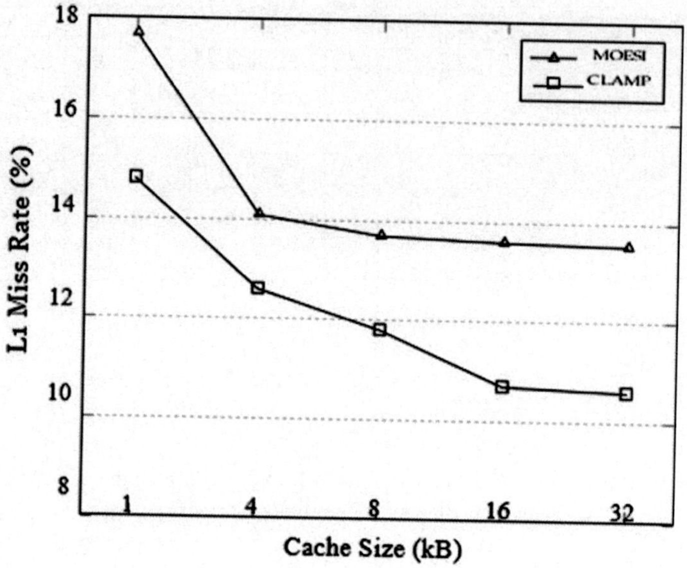

Fig. 6 L1 Miss rate (%) versus cache size

5.1 *L1 Miss Rate*

Figure 6 depicts the L1 cache miss performance as a function of cache size. Regardless of cache size, proposed cache coherence protocol provides greater cache hit performance. CLAMP, on average, has 2.89% lower miss rate than MOESI. Miss rate is decreased by an increase in L1-to-L1 data snoop and obtaining the current data in the local cache's locked cache lines.

5.2 *L2 Miss Rate*

Figure 7 depicts the L2 cache miss performance as a function of cache size. Regardless of cache size, CLAMP protocol provides better L2 cache hit performance. The improved cache hit is attributed to the snooping of data from remote cores' L2, keeping the latest data available with the locked local cache. On an average, CLAMP reduces L2 miss rate by 0.5% to 3% compared to MOESI protocol.

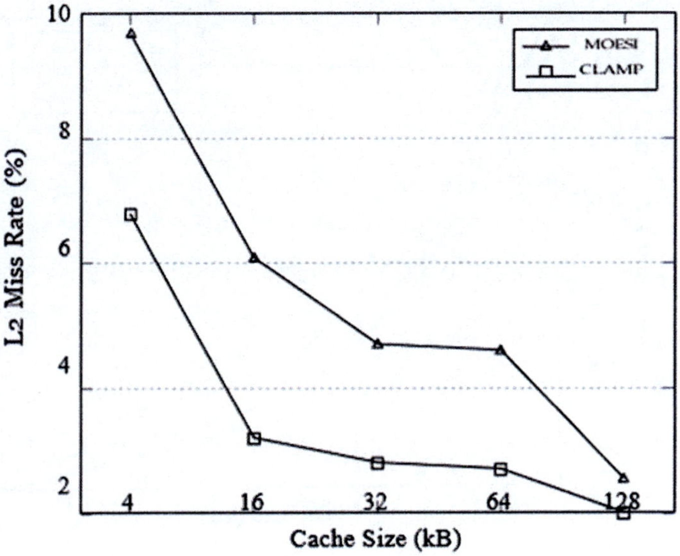

Fig. 7 L2 Miss rate (%) for varying cache sizes

5.3 5WCET of HC Tasks

Deterministic WCET is higher than WCET in any architecture as complete cache is not available exclusively for the task during the execution, while $WCET_{REF}$ measurement assumes the exclusive availability of cache for tasks. The execution time in both architectures compared by varying cache sizes 2 KB to 32 KB for 64 B cache line is as shown in Fig. 8. It was observed that CLAMP shows a reduced response time compared to MOESI and there is predictive variation of execution time of HC tasks with respect to $WCET_{REF}$. Due to its improved cache coherence, CLAMP offers high cache hit rates for HC jobs while no impact on cache misses for LC tasks. The lower miss rate makes it easier to meet HC job deadlines within the bounded duration.

6 Conclusion

CLAMP, an upgraded MOESI cache coherence protocol and an extension of MOESIL [5] suitable for locked multi-level multi-core caches, was proposed in this paper. Experiments using SPEC2006 and CACOSIM simulator show that the CLAMP improves performance. This novel coherence protocol reduces cache access time and confirms that HC jobs in multi-core MCS are executed predictably without deadline miss. We propose to enhance the work by constructing an end-to-end criticality-aware coherent memory subsystem consisting of hierarchical cache

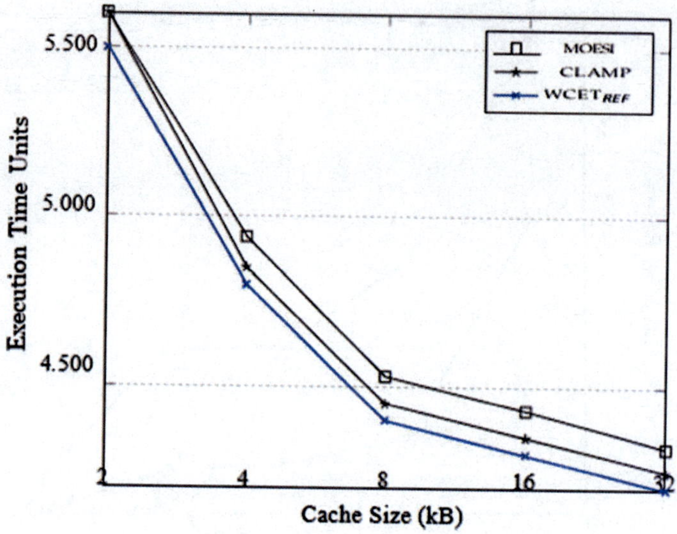

Fig. 8 (Worst-case) execution time comparison

memories supporting locking and partitioning at different levels of the hierarchy for a multi-core/many-core MCS.

References

1. Burns A, Davis RI (2017) A survey of research into mixed criticality systems. CSUR'17 50(6):1–37
2. Ward BC, Herman JL, Kenna CJ, Anderson JH (2013) Outstanding paper award: making shared caches more predictable on multicore platforms. In ECRTS'13. IEEE, pp. 157–167
3. Gracioli G, Fröhlich AA (2014) On the influence of shared memory contention. In: real-time multicore applications, SBESC'14, IEEE, pp. 25–30
4. Pellizzoni R, Schranzhofer A, Chen J, Caccamo M, Thiele L (2010) Worst case delay analysis for memory interference in multicore systems, In: DATE'10, pp. 741–746
5. Nair AS, Pai AV, Raveendran BK, Patil G (2021) $MOESI_L$: a cache coherency protocol for locked mixed criticality l1 data cache. In: IEEE/ACM 25th International Symposium on Distributed Simulation and Real Time Applications (DS-RT). IEEE, pp. 1–8
6. Chetan Kumar NG, Vyas S, Cytron RK, Gill CD, Zambrano J (2014) Jones, Cache design for mixed criticality real-time systems. IEEE 32nd International Conference on Computer Design (ICCD). IEEE, 513–516
7. Puaut I, Arnaud A (2006) Dynamic instruction cache locking in hard real-time systems. In: Proc. of the RTNS'14
8. Kaur DP, Sulochana V (2018) Design and implementation of cache coherence protocol for high-speed multiprocessor system. In: 2nd IEEE International Conference on Power Electronics, Intelligent Control and Energy Systems (ICPEICES), IEEE
9. Patil G, Mallya NB, Raveendran BK (2019) MOESIF: a MC/MP cache coherence protocol with improved bandwidth utilisation. Int J Embedded Syst 11:4

10. Sritharan N, Kaushik A, Hassan M, Patel H (2019) Enabling predictable, simultaneous and coherent data sharing in mixed criticality systems. In: IEEE Real-Time Systems Symposium (RTSS)
11. Kaushik A, Tegegn P, Wu Z, Patel H (2019) Carp: A data communication mechanism for multi-core mixed-criticality systems. In: IEEE Real-Time Systems Symposium (RTSS)
12. Henning JL (2006) SPEC CPU2006 benchmark descriptions. ACM SIGARCH Comp Archit News 34(4):1–17

Video Event Description Booster By Bi-Modal Transformer, Identity, and Emotions Capturing

Kiran P. Kamble and Vijay R. Ghorpade

Abstract In today's era, interactive media material consisting of medium text, audio, picture, and video basically has a multimedia style. With the assistance of advanced deep learning (DL) methods, some of the rare computer vision (CV) issues have been successfully resolved. Video Captioning is automatic description generation from digital video. It converts the audio tracks of people within a video into text. When the recording is played that text will be displayed in segments that are synchronized with specific words as they are spoken. It is like representing a whole activity which is happening in a video in a textual format. The captions which are generated in the process of video captioning are nothing but a transcription of dialogue and visual content in a video. They appear as a text on the bottom of the display screen [1, 2]. In this paper, we have studied and modified video captions generated using Bi-modal Transformer (BMT). We have also added facial recognition features to BMT to give more information about the expressive actions in the video. The proposed model also detects emotional expressions on a person's face in a video.

Keywords Bi-Modal Transfer (BMT) · Facial recognition · Emotion recognition

1 Introduction

Video captioning is not just making video content more accessible to viewers with hearing impairments, but that it can actually improve the effectiveness of video content as well. Both native and foreign language speakers benefit from caption. According to study by Ofcom [3], 80% of people who use video captions do not even have hearing impairments. Adding captions to a video can compensate for poor

K. P. Kamble (✉)
Department of Technology, Shivaji University Kolhapur, Kolhapur, India
e-mail: kirankamble5065@gmail.com
URL: http://apps.unishivaji.ac.in

V. R. Ghorpade
Bharati Vidyapeeth College of Engineering, Shivaji University Kolhapur,Kolhapur, India
e-mail: vijayghorpade@rediffmail.com

H. O. Bansal et al. (eds.), *Next Generation Systems and Networks*, Lecture Notes in Networks and Systems 641, https://doi.org/10.1007/978-981-99-0483-9_31

audio quality or background noise [4]. When watching video without sound or in a noisy environment, captions help to make the video easier to watch [5]. Video captions provide viewers with a way to search within video. Captioning is required to make video content accessible to viewers who are deaf or hard of hearing. Considering captions also benefits a wide variety of learners in a variety of situations, they are considered good universal design for learning [6]; globally, cable TV subscriptions are being replaced by Internet-based streaming services for better content accessibility. Over-the-top (OTT) streaming services have become increasingly popular as a new platform for media distribution. Users can access and view content on OTT platforms anytime, anywhere, and on any device connected to the Internet. Some of the top OTT platforms in the world include Netflix, Disney+Hotstar, HBO Max, ESPN+, and Amazon Prime Video, among others [7]. Video captioning has been proven as one of the primary requirements for these OTT platforms. RealGood, an online service that searches streaming services, reports Netflix has 3,781 movies [8]. There are 62 languages on Netflix in which the Netflix content can be watched. Every show or movie has subtitles or captions available in different languages [9]. This subtitle feature of Netflix has increased its revenue in huge amounts since users can watch other countries' shows, movies in their own language and enjoy a variety of movies and shows. Any user simply wants an effortless and seamless experience of watching movies and web series. Platforms like Netflix and Amazon Prime use video captioning to give an effortless experience to their user. They act as an accessibility for deaf or hard of hearing viewers. They can take help from captions generated to enjoy the content they are watching. Video captioning also increases the SEO (search engine optimization) rate of any live streaming platform. For humans to access the outside world, vision and language are two fundamental bridges. Video retrieval is one of the active topics to link these two counterparts in the research community. This measure compares a video's caption with its corresponding video in embedding space in terms of semantic similarity. Improve productivity by reducing the effort of manual annotations with semantic-based video retrieval. Captions generated from videos can be used with the visual semantic-enhanced reasoning network, where the first regions or frames of the video are captured and caption is created. Later semantic reasoning is performed on the engendered caption and an incipient more consequential caption is engendered [10]. For example, if in a video where a roadside parking is shown and if the question is like "Can a vehicle be parked here?" and if the model is giving the correct answer as "yes" then a question will arise like "Why is it legal to park a car here?" To this, the model will answer (give a visual reasoning) that there is a parking sign on the road, which means it's legal to park here. In self-driving cars also the system gives users confidence signals by generating natural language descriptions of the reasons behind its decisions. Visual Dialog requires an AI agent to engage humans in meaningful conversation and natural language on visual content. Specifically, given an picture, a dialogue history, and a tracking question on the picture, the job is to respond to the question.

Basically, an AI-Bot generates captions or questions based on the frames captured by him. For example, in the above image (Fig. 1), a bot might generate captions like "Is a dog sitting on a motorcycle?", "Has the dog worn sunglasses?", "Is the car

Fig. 1 Visual dialog scenario

parked?". This kind of visual dialogue representation helps robots to communicate with humans [11]. A model is asked a series of questions sequentially in a dialogue or conversational manner. The model tries to answer (no matter right or wrong) these questions. Sign Language Recognition is one of the most growing fields of research. Sign Language is mainly used for communication of deaf people. Sign Language is the most natural and expressive way for the hearing-impaired people. People, who are not deaf, never try to learn sign language to interact with the deaf people. This leads to isolation of the deaf people. But if the computer can be programmed in such a way that it can translate sign language to text format, the difference between the normal people and the deaf community can be minimized. To convert this sign language into text video captioning can be used [12]. Video captioning can also be used to convert actions in a video into simple instructions. For example, a simple video of cooking (making a pizza) all the steps required to cook a pizza which are presented in a video can be converted into captions. That is textual information that can be provided to the user.

2 Literature Survey

The dense video subtitling task requires a template to first localize events in a video and then to produce a text description of a sentence of what takes place during the event. The task of closed-captioning video bifurcates from the task of closed-captioning video, which is to close-caption a video without locating the event. The field of video captioning has evolved from handmade rule templates to encoder–decoder architectures inspired by advances in machine translation. Later, the captioning models were further enhanced by semantic tagging, reinforcement learning, attention, extended memory, and other modalities. The task of dense video captioning, as well as a test-bed, ActivityNet Captions dataset, was introduced by Krishna et al. [25], where they utilized the idea of the Deep Action Proposals network to generate event proposals and an LSTM network to encode the context and generate captions. The idea of context-awareness was further developed in which a bi-directional variant of Single-Stream Temporal Action proposal network was employed (SST) which makes better use of the video context, an LSTM network with attentive fusion and context gating was used to generate context-aware captions. Zhou et al. [26] adapted transformer architecture to tackle the task and used the transformer encoder's output as input to a modification of ProcNets to generate proposals. Recently, the idea of reinforcement learning was found to be beneficial for image captioning (self-critical sequence training (SCST)) and, hence, applied in dense video captioning as well. More precisely, the SCST was used in a captioning module to optimize the non-differentiable target metric, e.g., METEOR. Specifically, Li et al. [27] integrated the reward system and enriched the single-shot-detector-like structure with descriptiveness regression for proposal generation. Similarly, Xiong et al. [28] used an LSTM network trained with the sentence- and paragraph-level rewards for maintaining coherent and concise story-telling, while the event proposal module was adopted from structured segment networks. Mun et al. [29] further developed the idea of coherent captioning by observing the overall context and optimizing two-level rewards: an SST module is used for proposal generation and a Pointer Network to distill proposal candidates. Another direction of research relies on weak supervision which is designed to mitigate the problem of laborious annotation of the datasets. To this end, Duan et al. [30] proposed an auto-encoder architecture which generates proposals and, then, captions them while being supervised only with a set of non-localized captions in a cycle-consistency manner. However, the results appeared to be far from the supervised methods [13].

3 Dataset

For the dataset for ActivityNet captions we have used videos that contain a variety of human activities. The training and validation of dense video captioning with bi-modal transformers have 10 and 5 k videos, respectively, and for test data 2 k videos are used. On average a video is 2 min long and has four captions, each caption consists of 14 words. All these videos are taken from YouTube. Some of the videos are no longer available on YouTube so we managed to obtain 91% of all videos [14]. For the face recognition dataset, we have used the Kaggel dataset which contains celebrity images. We have used a total of 32 celebrities and each of them contains 140–160 images. So, there are 3945 images [15]. For emotion detection, We have used the FER2013_VGG19 dataset. Data consists of 48 by 48 pixel grayscale images of faces.

4 Methodology

To strengthen the video caption we proposed model that takes input video, audio and predict possible caption along with name identity of person object and emotion of person. Detailed architecture is depicted in Fig. 2. FER2013 dataset contains registered faces with labeling. The job is to classify each face into one of the seven categories such as Angry, Disgust, Fear, Happy, Sad, Surprise, and Neutral. Training data have two columns, one for emotion and second for pixel representation of each emotion. Statistics of training data is 28,709 samples and the testing data of 3,589 samples [16]. Previously, captions were generated using only video features. Video captioning model is required to generate one sentence for an entire video which might **not be sufficient for a full length film. Instead, a dense video event representation model should be able to first confine the vital events and then generate captions for** each of them. Bi-modal transformer generates an architecture which achieves an outstanding performance on dense video captioning. It consists of two major parts: a bi-modal encoding and a bi-modal decoding. It inputs audio and visual feature sequences. Audio and video features are encoded via VGGish and I3D along with caption tokens using GloVe and CNN architecture which are used for face recognition and emotional detection.

Before going into the bi-modal encoder, audio and visual sequences are trimmed up by proposal boundaries. Then, VGGish, I3D, and CNN features are passed through the stack of N bi-modal encoder layers where audio and visual sequences are encoded to, what we call audio-addressed visual and video-addressed audio attribute. The bi-modal encoder consists of three pairs of blocks: self-heed, bi-modal heed, and position-wise fully connected net. Self-heed allows the model to attend to all positions within one modality while bi-modal attention attends to all positions in the accompanying modality. Finally, the position-wise fully connected net is applied to every position in the input sequence. The bi-modal encoder inputs features of any length and size which can be distinct for each modality and outputs audio-addressed

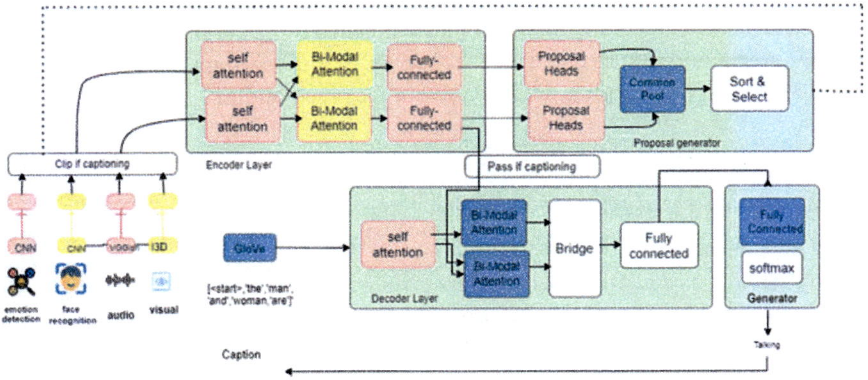

Fig. 2 Strengthen Bi-modal transformer with proposal generator

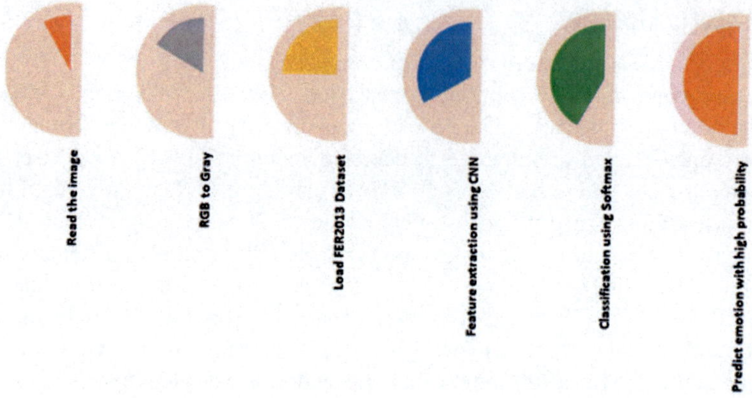

Fig. 3 Emotion detection flow diagram

visual and video-addressed audio attributes. The multi-headed proposal generator takes two streams of features coming from bi-modal encoder; both streams go into an individual stack of proposal generation heads. The architecture of each proposal head is inspired by the YOLO detection layer. We apply k-means clustering on the proposal. The bi-modal decoding inputs both streams from the encoder and previously generated caption words and consists of four blocks: self-heed, bi-modal heed, bridge layer, and position-wise, fully connected net. Finally, decoder output is used to generate a distribution for the next caption word [17].

For emotion detection, it takes the same frame and it loads the FER2013_VGG19 dataset for emotion recognition. After this it rescales the image using the softmax function in torch.nn.functional. Then, feature extraction takes place followed by CNN classification. After CNN classification, appropriate emotion gets detected, Fig. 3.

For face recognition, the LBPHFace recognizer needs to be loaded. Training data is loaded for only one time during the process. The model takes a frame from the video then that frame image is converted to a grayscale image. Next, it checks for the face in the frame. If a face is detected then it loads the Haar Cascade classifier and predicts the label of the image. If the confidence is greater than 37 then it means it has not recognized the image, so it continues to the next frame, else the face gets recognized. Figure 4 depicts detailed flow of face recognition.

5 Evaluation Metric

There are multiple factors to be considered while evaluating "Video caption strengthening generated by Bi-modal transformer", for sample: The fluidity, a scene containing multiple actions, prioritizing what is important according to bias. **SVO Accuracy** [18] SVO (subject, verb, object) accuracy is used to measure coherence of the language topology with ground truth .The purpose is to just focus on the matching

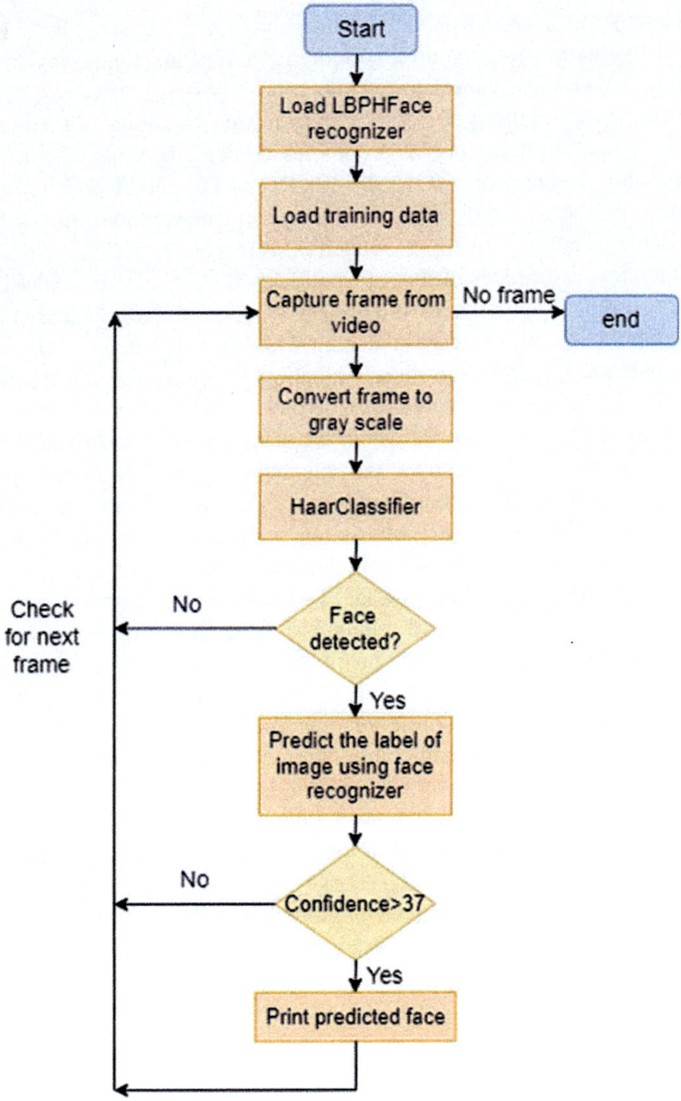

Fig. 4 Face recognition flow

of broad semantics .The visual and language details are ignored here. **BLEU** [19], Bilingual Evaluation Understudy Score in the field of machine translation, it is one of the popular metrics. Closeness of numerical translation between two sentences is measured by computing G-means of n-gram match counts, which as a result is sensitive to position matching. The complex contents become hard to adopt, as it may favor shorter sentences. **ROUGE** [20], Recall-Oriented Understudy for Gisting Evaluation, measures the overlapped n-gram sequences among predicted sentence and available reference sentences. Thus it is similar to the BLEU score. The difference is that the occurrences in sum of candidates are considered in BLEU, while the n-gram occurrences in total sum of number of references are considered in ROUGE. ROGUE favors long sentences, as this metric highly relies on recall. In **CIDEr** [21], Consensus-based Image Description Evaluation, the consensus among the reference sentences and candidate sentences provided by human labeler is measured as a part of evaluating the set of sentences which are descriptive for an image. This measure originally correlates highly with human judgements. As this captures importance and saliency, grammatical correctness and its accuracy, this evaluation metric is different from others. **METEOR** [22], Metric for Evaluation of Translation with Explicit Ordering, is computed based on alignment between a set of candidate references and given hypothesis sentences. METEOR compares exact stemmed tokens, token matches, paraphrase matches and semantic matches by using WordNet synonyms. This distinguishing aspect of semantically similar matches makes it different from others. In the literature, it is shown that METEOR is mostly better than ROUGE and BLUE, and when the number of references is small it outperforms CIDEr. METEOR modifies the precision and recall computations, replacing them with a weighted F-score. **WMD** [23], Word Mover's Distance, calculates the distance between two documents (or sentences) by using word embedding so that there is no problem if there is no common word. Here, the assumption is that vectors would be similar for the words that are similar. WMD leverages the power of word embedding to overcome the basic limitations of distance measurement. **SPICE** [24], Semantic Propositional Image Captioning Evaluation, is a principle metric that compares the semantic propositional content. It is an automatic image caption evaluation method. In terms of human evaluation agreements, SPICE outperforms metrics, BLEU, ROUGE, CIDEr, and METEOR.

6 Result and Discussion

Experiment 1: Study and implementation of video captioning using bi-modal transfer (BMT). The important events are identified by localizing and describing important events. The existing methods accomplish the task only by exploiting the visual information; it neglects the audio track in the untrimmed videos. The caption tokens are encoded with Glove, while the visual and audio features are encoded with I3D and VGGish features. A set of proposals are generated after passing the mentioned features to the bi-modal multi-headed proposal generated, by using information from both the modalities.

Experiment 2: Adding facial recognition features to above existing model (BMT) to give more information about the actions in the video. We have trained a celebrity-faces dataset which contains about 32 folders (each with almost 140–160 images) of a few Indian celebrities. Then, loaded the trained dataset for basic face recognition algorithm used, which is LBPH (local binary pattern histogram). In the previous experiment, the caption generated contained the sentences for trimmed meaningful frames with start and end time mentioned with it. In addition to this, the names of recognized faces are also detected after performing the required changes while processing the video.

Experiment 3: Adding emotion or expression recognition to the above model (BMT) which detects each face into one of the seven classes Angry, Disgust, Fear, Happy, Sad, Surprise, Neutral We have downloaded the pre-trained model, "FER2013_VGG19" from [13].This experiment uses the CNN implementation method on facial expression recognition, i.e., FER2013 and CKplus [14]. Thus, accomplished the latest improved model of 73.11% in FER2013 and 94.65% in CK+ dataset. In addition to the captions generated in the above-listed experiments, the emotions of the detected faces are also included for the given time frame.

Current modification in captions displays separate attributes: name and emotion. But the captions can be strengthened by including name and emotion in the base sentence, i.e., the sentence displayed with the help of BMT model. For example, in the last experiment (refer Table 1) base sentence is "A boy is riding a bike" , our model identifies name as separate entity, i.e., "Tiger Shroff " and the expression or emotion of the identified face as "neutral". So, the final modified sentence will be "Tiger Shroff is neutrally riding a bike" . Here the proper noun referred for human is replaced with the identified name for detected face. The adjective for its corresponding expression or emotion is also added in it to get the required sentence.

Table 1 Caption structure for scene from "War" movie (some frames of the provided video input are shown in Fig. 5)

Experiment (s)	Generated caption for specific time frame
Bi-modal transformer	{Sentence: A boy is riding a bike}
Bi-modal transformer + Facial recognizer	{Sentence: A boy is riding a bike, name: Tiger Shroff}
	{Sentence: Tiger Shroff is riding bike}
Bi-modal transformer + Facial recognizer + Expression recognizer	{Sentence: A boy is riding a bike, name: Tiger Shroff, emotion: neutral}
	{Sentence: Tiger Shroff is neutrally riding a bike}

Fig. 5 Few sample frames taken from the input video scene

7 Conclusion and Future work

In presented work, we have generated the captions for the videos. For the generation of the caption to get appropriate results along with visual data, we also have used audio data, face recognition as well as emotion recognition so that the reader can understand detailed scenarios happening in the video. So by using this model we can fulfill many related application needs. So we have combined the models.

1. BMT which uses audio and video features.
2. Face recognition.
3. Emotion recognition.

Again, to obtain optimized results we have removed the duplicate sentences from the generated results. Despite the good results, it cannot be compared with captions generated by humans. In further works, accuracy of the model still can be improved and the applications specified can be implemented.

Acknowledgements We are thankful to Shivaji University and Department of Technology for providing research facility to accomplish this task.

References

1. Video Captioning Automatic description generation from digital video. https://towardsdatascience.com/video-captioning-c514af809ec
2. Panopto to record and share videos. https://www.panopto.com/blog/frequently-asked-questions-faqs-about-video-captioning-answered
3. Ofcom communications services. https://www.ofcom.org.uk
4. 3Play Media provides captioning, transcription, and audio description services. https://www.3playmedia.com/blog/what-is-closed-captioning
5. Cielo24 quality media data solutions that help creators maximize ROI through innovative. https://cielo24.com/2016/12/captions-and-subtitles-difference
6. 3Play Media provides closed captioning, transcription, and audio description services. https://www.3playmedia.com/blog/importance-of-captioning

7. Zartek designs and develops mobile and web apps. https://www.zartek.in/what-is-an-ott-platform/
8. Reelgood streaming hub. https://blog.reelgood.com/which-streaming-service-is-the-best-bang-for-your-buck-in-2020
9. Guide to what's new and what's coming soon to Netflix 2022. https://www.whats-on-netflix.com/news/does-netflix-have-too-much-foreign-content
10. Feng Z et al (2020) Exploiting visual semantic reasoning for video-text retrieval. arXiv:2006.08889
11. Murahari V et al (2020) Large-scale pretraining for visual dialog: a simple state-of-the-art baseline. In: European conference on computer vision. Springer, Cham
12. Conversion of Sign Language into Text. https://www.ripublication.com
13. Iashin V, Rahtu, E (2020) A better use of audio-visual cues: dense video captioning with bi-modal transformer. arXiv:2005.08271
14. Iashin V (2020) Source code for bi-modal transformer for dense video captioning (BMVC 2020). https://github.com/v-iashin/BMT
15. Lyons J, Kamachi M, Gyoba J (1997) Database of digital images. https://www.kaggle.com/datasets
16. wujie1010, A CNN based pytorch implementation on facial expression recognition (FER2013 and CK+), achieving 73.112% (state-of-the-art) in FER2013 and 94.64% in CK+ dataset. https://github.com/WuJie1010/Facial-Expression-Recognition.Pytorch
17. Iashin V, Rahtu E (2020) A better use of audio-visual cues: dense video captioning with Bi-modal transformer
18. SVO: Fast semi-direct monocular visual odometry. In: 2014 IEEE international conference on robotics and automation (ICRA). IEEE
19. Papineni K et al (2002) Bleu: a method for automatic evaluation of machine translation. In: Proceedings of the 40th annual meeting of the association for computational linguistics
20. Lin C-Y (2004) Rouge: A package for automatic evaluation of summaries. Text summarization branches out
21. Vedantam R, Lawrence Zitnick C, Parikh D (2015) Cider: Consensus-based image description evaluation. In: Proceedings of the IEEE conference on computer vision and pattern recognition
22. Banerjee S, Lavie A (2005) METEOR: An automatic metric for MT evaluation with improved correlation with human judgments. In: Proceedings of the acl workshop on intrinsic and extrinsic evaluation measures for machine translation and/or summarization
23. Huang G et al (2016) Supervised word mover's distance. Adv Neural Inf Process Syst 29
24. Anderson P et al (2016) Spice: Semantic propositional image caption evaluation. In: European conference on computer vision. Springer, Cham
25. Krishna R, Hata K, Ren F, Fei-Fei L, Niebles JC (2017) Densecaptioning events in videos. In: Proceedings of the IEEE international conference on computer vision, pp 706–715
26. Wang X, Wang Y-F, Wang WY (2018) Watch, listen, and describe: Globally and locally aligned cross-modal attentions for video captioning. In: Proceedings of the 2018 conference of the north american chapter of the association for computational linguistics: human language technologies, vol 2 (Short Papers). New Orleans, Louisiana, pp 795–801. Association for Computational Linguistics. https://doi.org/10.18653/v1/N18-2125.
27. Li Y, Yao T, Pan Y, Chao H, Mei T (2018) Jointly localizing and describing events for dense video captioning. In: Proceedings of the IEEE conference on computer vision and pattern recognition, pp 7492–7500
28. Xiong Y, Dai B, Lin D (2018) Move forward and tell: a progressive generator of video descriptions. In: Proceedings of the European conference on computer vision, pp 468–483
29. Mun J, Yang L, Ren Z, Xu N, Han B (2019) Streamlined dense video captioning. In: Proceedings of the IEEE conference on computer vision and pattern recognition, pp 6588–6597
30. Duan X, Huang W, Gan C, Wang J, Zhu W, Huang J (2018) Weakly supervised dense event captioning in videos. In: Advances in neural information processing systems, pp 3059–3069

Real-Time Data-Based Optimal Power Management of a Microgrid Installed at BITS Supermarket: A Case Study

Pavitra Sharma, Devanshu Sahoo, Krishna Kumar Saini, Hitesh Datt Mathur, and Houria Siguerdidjane

Abstract The energy sector is seeing the emergence of microgrids, which mark a paradigm change away from distant central station power plants, toward more localized distributed generation. Microgrids are resilient because of their ability to operate independently of the main grid, and their capacity for flexible and parallel operations enables the supply services and boosts the grid's ability to compete. However, energy management of microgrid is very essential for their effective optimal operation. Moreover, battery energy storage (BESS) is becoming a major component of a microgrid due to uncertain nature of renewable energy sources present in a microgrid. In this regard, this study formulates an energy management algorithm that minimizes the operating cost of the microgrid considering the detailed cost model of the BESS. The developed cost objective function is solved using a famous meta heuristic optimization algorithm named, particle swarm optimization (PSO). The effectiveness of the proposed algorithm is examined with two scenarios of solar generation, i.e., full sunshine and cloudy day. The proposed algorithm is validated with the data obtained from a real-time microgrid installed at BITS PILANI supermarket. The obtained optimal operating cost of microgrid for full sunshine and cloudy day is 2.84$/Rs.227.2 and 11.43$/Rs.914.4, respectively.

Keywords Microgrid · Optimal power management · Battery energy storage systems · Solar PV · Particle swarm optimization (PSO)

P. Sharma (✉) · D. Sahoo · K. K. Saini · H. D. Mathur
Birla Institute of Technology and Science, Pilani, Rajasthan, India
e-mail: p20190027@pilani.bits-pilani.ac.in

H. Siguerdidjane
Université Paris-Saclay CNRS, CentraleSupélecLaboratoire des Signauxét Systémes, 91190 Gif-Sur-Yvette, France

H. O. Bansal et al. (eds.), *Next Generation Systems and Networks*, Lecture Notes in Networks and Systems 641, https://doi.org/10.1007/978-981-99-0483-9_32

1 Introduction

For any country's development, per capita energy consumption plays a vital role. Therefore, governing bodies are significantly focusing on providing 24 h uninterrupted electricity to everyone. In this regard, government of India has launched various schemes such as the Saubhagya Scheme, also known as the Pradhan Mantri Sahaj Bijli Har Ghar Yojana. With the help of these schemes, over 2.82 billion homes, or over 91% of rural households in India, has access to grid energy as of March 2021. However, it is a harsh fact that still 58.6% of grid energy comes from fossil fuels [1]. Likewise, there are several issues with the traditional power grids such as they are centralized, they involve interconnections among various power system devices including transformers, transmission lines, substations, distribution lines, and loads among many others [2, 3]. The renewable energy sources (RES) based distributed energy resources (DER) are emerging as a potential source for providing the power in a more practical, reliable effective, economical, and environmentally responsible way. It is an alternative to the antiquated power network where it generates power in a distributed and decentralized unit to make sure effortless coordination control of the power grid [4, 5].

Microgrids (MGs) provide an effective solution to the problems of traditional power grid system. A microgrid is the combination of groupings of DERs and energy storage systems (ESSs) including capacitors, batteries, flywheels, and hydrogen [6, 7]. Hybrid renewable energy systems (HRES), also referred to as "MGs," are a combination of two or more RES and conventional electricity sources [8]. The efficient use of both renewable and conventional energy sources is the key financial concern for MGs. By adjusting MGs' size and operation, the energy is reduced [9–13]. The authors of [13] offer a method for achieving an MG's optimum operation and sizing that ensures the lowest energy cost. For economic dispatch considering generation, storage, and responsive load offers, the authors have attempted to design a multiperiod artificial bee colony optimization algorithm. Uncertainties are taken into account when predicting non-dispatchable power generation and load needs using an artificial neural network and Markov chain (ANN-MC) method [14]. The authors manage microgrid storage considering the integration of electric vehicles and load responsiveness [15]. In [16] the authors have a threefold approach. First, energy tariffs incentivize specific customer behaviors while managing battery storage resources. Second, demand response load curtailment techniques will maximize power availability. Lastly, social communication will educate users in individual energy stewardship to engage participants in a utility collective. The authors of [17] consider the remaining battery power and solar power as the benchmark to propose a method that performs the energy management between the power generation, storage, and consumption for building microgrid. In [18] the authors have tried for an optimal solution to two conflicting objectives- cost and emission. This is solved using a modified multi-objective particle swarm optimization (MOPSO) algorithm. The results of the proposed energy management strategy obtain a trade-off between least emissions and least cost. This information can be used by the microgrid operators to take an intelligent decision about the functioning of the microgrid.

After a through literature, it is found that the detailed cost model of battery energy storage system (BESS) is not considered in the previous reported literatures. Therefore, this paper proposed an energy management algorithm that minimizes the operating cost of the microgrid considering detailed O&M cost model of BESS. The proposed algorithm aims at minimizing the operating cost of microgrid along with maximization of the use of renewable energy, effectively utilizing the BESS, and minimizing the dependency of microgrid on grid. The renowned particle swarm optimization is used to minimize the formulated objective function. The proposed algorithm is validated with the data obtained from a real-time microgrid installed at BITS PILANI supermarket. This real-time microgrid consists of 77-roof top solar panels (Longi LR5-72HBD 530–550 M) along with 30 Tall Tabular batteries each of capacity 225 Ah@C/20 which are connected in series to form a BESS.

2 Microgrid Components

The major components of the developed real-time microgrid are as mentioned in Sects. 2.1–2.3.

2.1 Solar Photovoltaic Cells (PV Cells)

A 41.2 kW of solar photovoltaic system has been installed. In total 77, solar panels (Longi-LR5-72HBD 530–550 M)) are installed having a rated peak power of 535 W at NOCT.

2.2 Battery Energy Storage Systems (BESS)

The battery energy storage system is formed using the series connection of 30 batteries each rated 12 V, 225 Ah. The battery make is Solance STT22500, Solance Tall Tabular Battery Series. The total rating of BESS becomes 360 V, 225 Ah, or 81 kWh. The BESS is mathematically modeled using Eqs. (1)–(3), and (4).

$$0 > P^{t}_{\mathrm{BESS,Ch}} > P^{m}_{\mathrm{BESS,Ch}} \tag{1}$$

$$0 < P^{t}_{\mathrm{BESS},Dch} < P^{m}_{\mathrm{BESS},Dch} \tag{2}$$

$$\mathrm{SoC}^{\mathrm{Min}}_{\mathrm{BESS}} < \mathrm{SOC}^{t}_{\mathrm{BESS}} < \mathrm{SOC}^{\mathrm{Max}}_{\mathrm{BESS}} \tag{3}$$

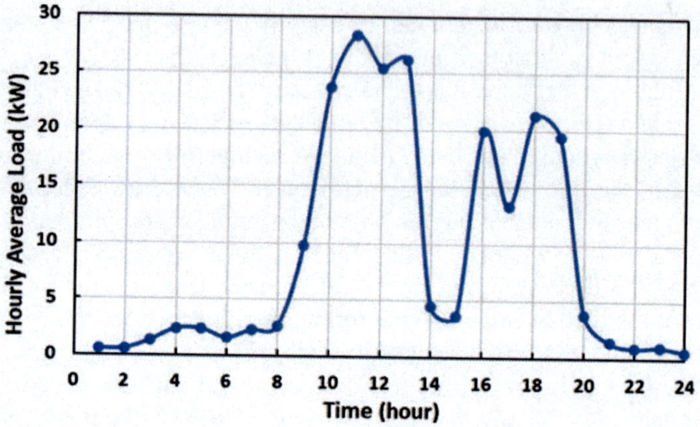

Fig. 1 Hourly average load profile at BITS Supermarket

$$SOC_{BESS}^{t+1} = SOC_{BESS}^{t}(1 - sd) + \Delta t\left(\frac{P_{BESS,Ch}^{t}\eta_{BESS}^{Ch}}{E_{BESS}^{R}} - \frac{P_{BESS,Dch}^{t}}{\eta_{BESS}^{Dch}E_{BESS}^{R}}\right) \quad (4)$$

where $P_{BESS,Ch}^{t}$ and $P_{BESS,Dch}^{t}$ are the charging and discharging power of the BESS at time instant 't', respectively. $P_{BESS,Ch}^{m}$ and $P_{BESS,Dch}^{m}$ are the maximum charging and discharging limit of the BESS. SoC_{BESS}^{Min} and SoC_{BESS}^{Max} are the minimum and maximum limits of the state of charge of the BESS, respectively. sd is the self-discharge rate of the BESS. η_{BESS}^{Ch} and η_{BESS}^{Dch} are the charging and discharging efficiencies of the BESS. E_{BESS}^{R} is the rated energy capacity of BESS.

2.3 *Load Profile at BITS Supermarket*

Figure 1 shows the average daily load profile of the BITS supermarket. The time durations having load less than 5 kW are considered as light load duration and other than that it is heavy load duration.

3 Formulation of Objective Function

The objective function which is to be minimized is given in Eqs. (5), (6), and (7)

$$C_T = \min \sum_{t=0}^{t=24} (C_{PV} + C_{BESS}^{t} + C_{G}^{t}) \quad (5)$$

$$C_{\mathrm{PV}} = K_{o\&m}^{\mathrm{PV}} \tag{6}$$

$$C_G^t = K_G^t * E_G^t \tag{7}$$

where, C_T is the total operation cost of the real-time microgrid, C_{PV} is the operation and maintenance cost of solar PV system. and C_G^t is the cost associated with the power exchanged from the grid at time 't'. E_G^t is the energy exchanged with grid at time 't', respectively. $K_{o\&m}^{\mathrm{PV}}$ is the operation and maintenance coefficient for the installed PV system in $/h. K_G^t is the energy trading price of the grid at time 't' in $/kWh.

The cost of battery (C_B^t) can be obtained from Eq. (8).

$$C_{\mathrm{BESS}}^t = C_{bf}^t * \left(C_{\mathrm{BESS}}^{O\&M}\right) * P_{\mathrm{BESS,Ch}/D\mathrm{ch}}^t * t_{\mathrm{BESS}}^{\mathrm{on}} \tag{8}$$

where $C_{\mathrm{BESS}}^{O\&M}$ is the operation and maintenance cost of BESS in $/kWh. $P_{\mathrm{BESS,Ch}/D\mathrm{ch}}^t$ is the charging/discharging power of the BESS at time instant 't'. C_{bf}^t is the arbitrary battery cost factor which depends on BESS operation (charging/discharging) and $\mathrm{SOC}_{\mathrm{BESS}}^t$ of the BESS at time 't'. Using Equations given in (9) and (10) the C_{bf}^t is obtained for charging and discharging operation of BESS, respectively.

$$C_{\mathrm{bf}}^t = \frac{\exp\left(\left(\frac{25}{\mathrm{SOC}_{\mathrm{BESS}}^t} - 1\right) * 7\right)}{5} \quad \text{when } P_{\mathrm{BESS}}^t < 0 \tag{9}$$

$$C_{\mathrm{bf}}^t = 1 + \exp\left(\left(\frac{25}{\mathrm{SOC}_{\mathrm{BESS}}^t} - 1\right) * 7\right) \quad \text{when } P_{\mathrm{BESS}}^t > 0 \tag{10}$$

The formulated objective function is minimized subject to the constraints given in Eqs. (11), (12) and (13):

$$P_L^t - P_{\mathrm{PV}}^t - P_{\mathrm{BESS,Ch}/D\mathrm{ch}}^t - P_G^t = 0 \tag{11}$$

$$0 < P_G^t < P_{G,\mathrm{Ex}}^{\mathrm{Max}} \tag{12}$$

$$P_{G,\mathrm{Im}}^{\mathrm{Max}} < P_G^t < 0 \tag{13}$$

where P_L^t, P_{PV}^t and P_G^t is the load demand on the microgrid, power generated by solar PV, and power exchanged with grid at time instant 't'. It is to be noted that, $P_{\mathrm{BESS,Ch}/D\mathrm{ch}}^t$ becomes positive when BESS discharges and is negative when it is charging. Likewise, P_G^t is positive when grid is exporting to microgrid and becomes negative when grid is importing from microgrid. $P_{G,\mathrm{Ex}}^{\mathrm{Max}}$ and $P_{G,\mathrm{Im}}^{\mathrm{Max}}$ are the maximum limits of power export and import with grid.

4 Proposed Algorithm

In order to achieve the formulated objectives, i.e., minimization of the total cost of microgrid, maximization of the use of solar energy, and effectively use the BESS, two levels of power exchange (between BESS and microgrid/grid and microgrid) are formulated such as 'slow' and 'fast'. For the BESS, it is slow charging (SC) /discharging (SD) and fast charging (FC)/discharging (FD), whereas, for grid, it is slow importing (SI)/exporting (SE) and fast importing (FI)/exporting (FE). The daily load demand is segregated into two categories such as light load and heavy load. So, status (slow/fast) of BESS and grid is dependent on load demand category and amount of load demand not fulfilled from the solar PV system ($P_{L,\text{unmet}}^{t}$). Its mathematical expression is as given in Eq. (14):

$$P_{L,\text{unmet}}^{t} = P_{L}^{t} - P_{\text{PV}}^{t} \tag{14}$$

Figure 2 shows the state flow graph to obtain the status of BESS as well as grid. The complete flow chart of proposed algorithm is shown in Fig. 3.

5 Case Study and Results

In order to examine the effectiveness of the proposed algorithm two scenarios of solar generation have been taken into account, i.e., full sunshine day and cloudy day. Figure 4 shows the hourly solar power generation at real-time microgrid during full sunshine day and cloudy day. Figure 5 shows the schematic of the installed real-time microgrid.

The values of the different parameters used in the optimization are shown in Table 1.

The hourly dynamic price of energy trading between grid and microgrid is shown in Fig. 6. The initial SOC of BESS is taken lowest to consider the worst case scenario.

Figures 7 and 8 show the obtained optimal schedule for grid and BESS power in case of full sunshine and cloudy day, respectively. It can be observed from Figs. 7 and 8 that maximum utilization of solar PV, effective utilization of BESS, and minimization of microgrid's dependency on the main grid is obtained along with minimum operating cost. Figure 9 shows the SOC of BESS for full sunshine and cloudy day.

The total operating cost of the microgrid on full sunshine and cloudy day is summarized in Table 2.

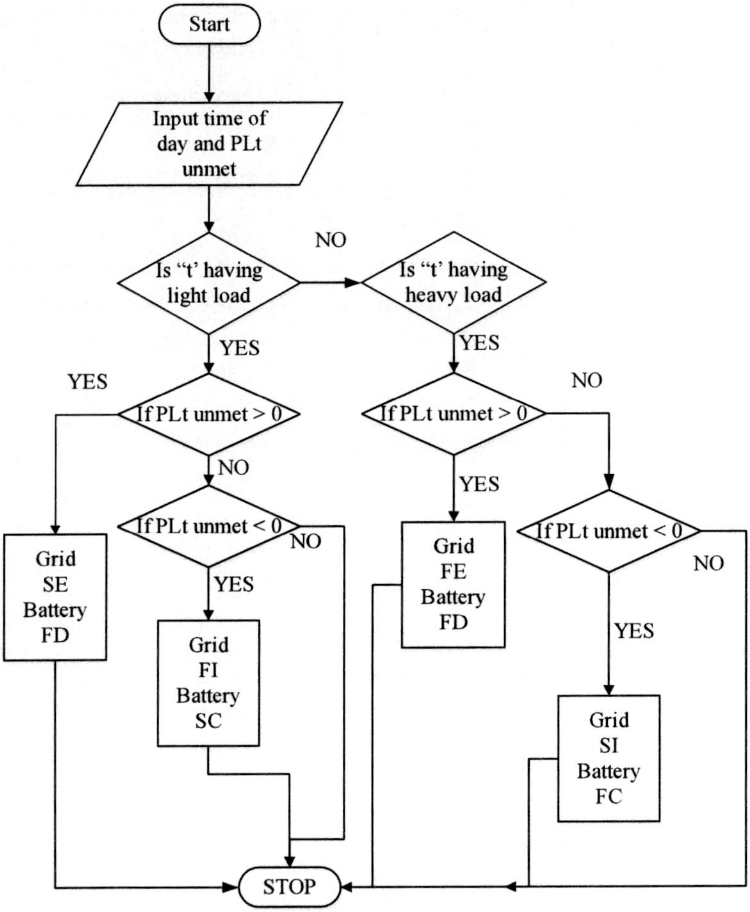

Fig. 2 State flow graph to obtain the status of BESS and grid

6 Conclusion

An energy management algorithm is proposed to minimize the operating cost of the microgrid considering a detailed cost model of battery energy storage system (BESS). The major aim of the proposed algorithm is to minimize the operating cost of microgrid along with maximizing the use of renewable energy, and effectively utilizing the BESS. The efficacy of the proposed algorithm is examined with two scenarios of solar generation, i.e., full sunshine and cloudy day. The validation of the proposed algorithm is performed with the data obtained from a real-time microgrid installed at BITS PILANI supermarket. The minimum operating cost of microgrid obtained for full sunshine and cloudy day is 2.84$/Rs.227.2 and 11.43$/Rs.914.4, respectively. These operating costs are obtained for worst case scenarios with lowest

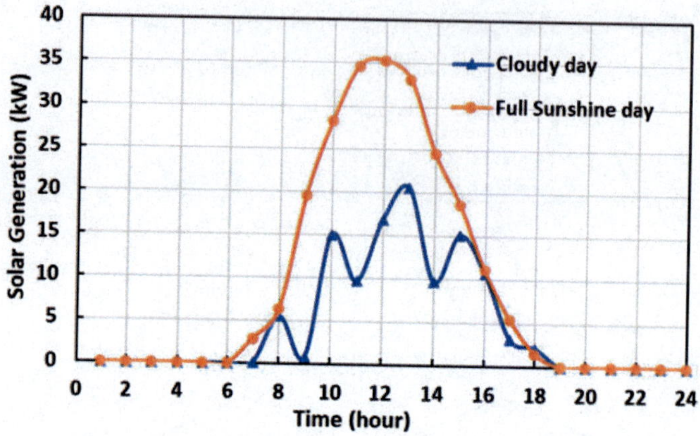

Fig. 3 Hourly solar power generation at real-time microgrid

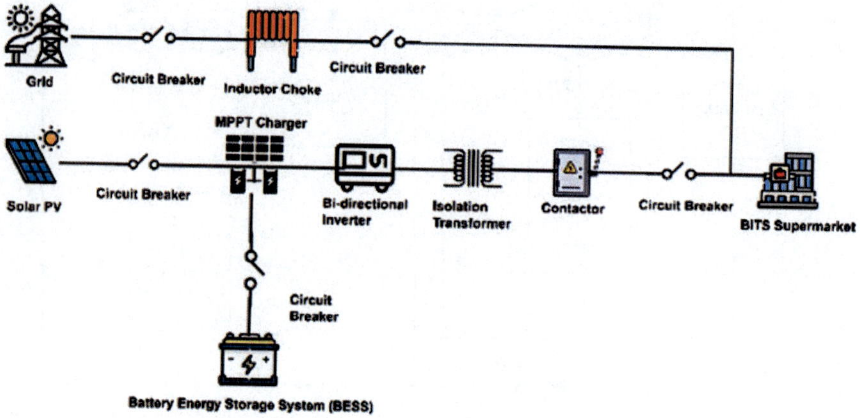

Fig. 4 Schematic of the installed real-time microgrid

initial SOC of BESS. It is found out that for full sunshine day, microgrid operator is having 84.86% profit after installing this real-time microgrid, whereas it is around 39.10% in case of cloudy day.

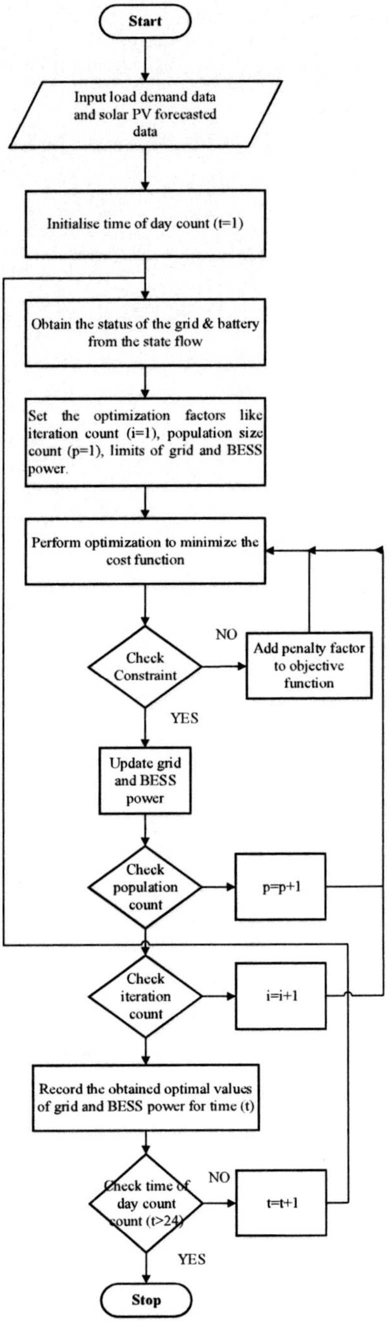

Fig. 5 Flow chart of the proposed algorithm

Table 1 The values of the different parameters used in the optimization

Parameter	Value
E_{BESS}^{R}	81 kWh
sd	0.0004%
$\text{SoC}_{\text{BESS}}^{\text{Min}}$	0.2
$\text{SoC}_{\text{BESS}}^{\text{Max}}$	1
$\text{SoC}_{\text{BESS}}^{t=0}$	0.3
$P_{\text{BESS,Ch}}^{m}$ (SC)	−5 kW
$P_{\text{BESS,Ch}}^{m}$ (FC)	−15 kW
$P_{\text{BESS},Dch}^{m}$ (SD)	5 kW
$P_{\text{BESS},Dch}^{m}$ (FD)	15 kW
$\eta_{\text{BESS}}^{\text{Ch}}/\eta_{\text{BESS}}^{Dch}$	0.8
$C_{\text{BESS}}^{O\&M}$	0.000175 $/kWh (0.0140 Rs/kWh)
$K_{o\&m}^{\text{PV}}$	0.057 $/h (4.56 Rs/h)
$P_{G,\text{Im}}^{\text{Max}}$ (SI)	−5 kW
$P_{G,\text{Im}}^{\text{Max}}$ (FI)	−15 kW
$P_{G,\text{Ex}}^{\text{Max}}$ (SE)	10 kW
$P_{G,\text{Ex}}^{\text{Max}}$ (FE)	20 kW

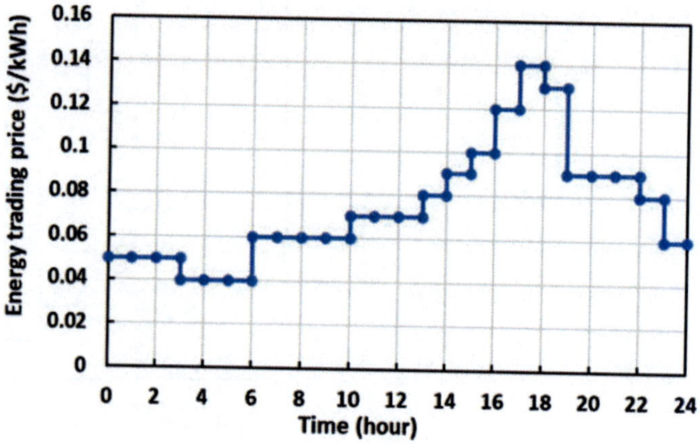

Fig. 6 Hourly price of energy trading between grid and microgrid

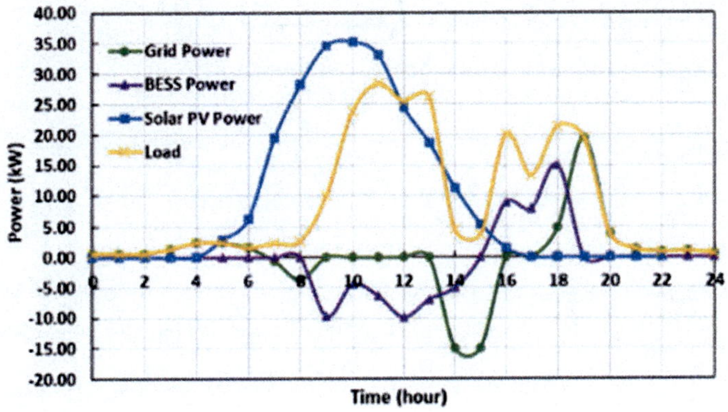

Fig. 7 Optimal schedule for grid and BESS power for full sunshine day

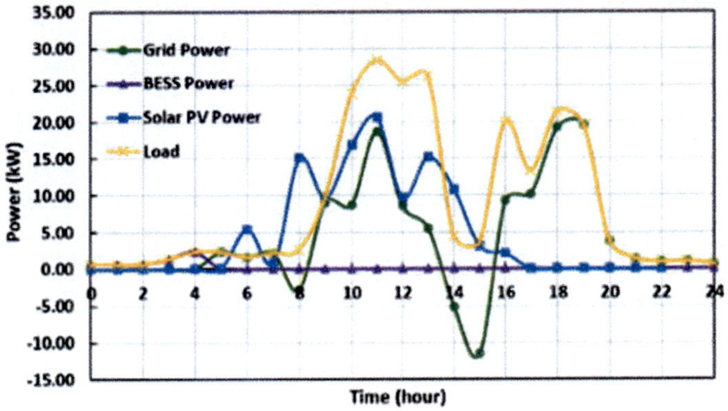

Fig. 8 Optimal schedule for grid and BESS power for cloudy day

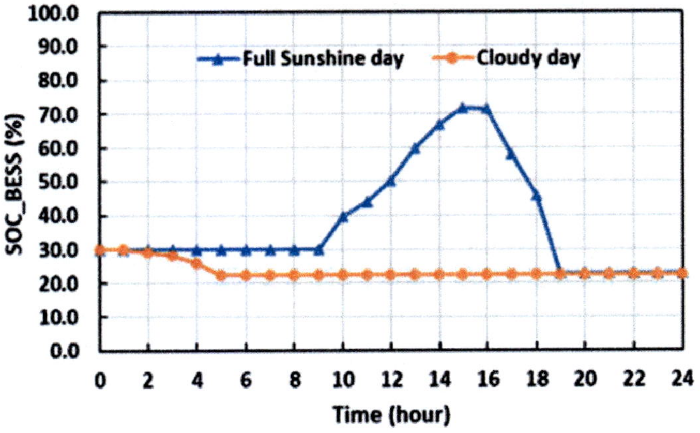

Fig. 9 State of charge of BESS for full sunshine and cloudy day

Table 2 Total operating cost of the microgrid on full sunshine and cloudy day

Day type	Total operating cost ($/Rs)	Cost paid by BITS if real-time microgrid was not installed	Profit of microgrid operator (%)
Full sunshine day	2.84$ (Rs.227.2)	18.77$ (Rs.1501)	84.86
Cloudy day	11.43$ (Rs.914.4)	18.77$ (Rs.1501)	39.10

Acknowledgements This work is supported by the Department of Science and Technology, Govt. of India, New Delhi under the "Internet of Things (IoT) Research of Interdisciplinary Cyber-Physical Systems Programme" letter no.: DST/ICPS/CLUSTER/IoT/2018/General.

References

1. Central Eectricity Authority (2021) All India installed capacity (in MW) of power stations installed capacity (in Mw) of power utilities in the states/UTS located. Cent Electr Auth Minist Power 4:1–7
2. Polprasert J, Wannakhong K, Narkvichian P, Oonsivilai A (2021) Home energy management system and optimizing energy in microgrid systems. In: 2021 international conference on power, energy and innovations, pp 126–129
3. Kumar D, Sharma P, Mathur HD, Bhanot S, Bansal RC (2020) Modified deloading strategy of wind turbine generators for primary frequency regulation in micro-grid. Technol Econ Smart Grids Sustain Energy. https://doi.org/10.1007/s40866-020-00083-7
4. Sharma P, Mishra P, Mathur HD (2021) Optimal energy management in microgrid including stationary and mobile storages based on minimum power loss and voltage deviation. Int Trans Electr Energy Syst 31:e13182
5. Thirugnanam K, El Moursi MS, Khadkikar V, Zeineldin HH, Al Hosani M (2021) Energy management of grid interconnected multi-microgrids based on P2P energy exchange: a data driven approach. IEEE Trans Power Syst 36:1546–1562
6. Sharma P, Dutt Mathur H, Mishra P, Bansal RC (2022) A critical and comparative review of energy management strategies for microgrids. Appl Energy 327:120028
7. Maulik A, Das D (2017) Optimal operation of microgrid using four different optimization techniques. Sustain Energy Technol Assess 21:100–120
8. Dawoud SM, Lin X, Okba MI (2018) Hybrid renewable microgrid optimization techniques: a review. Renew Sustain Energy Rev 82:2039–2052
9. Sharma P, Mishra AK, Mishra P, Dutt Mathur H (2021) Optimal capacity estimation and allocation of distributed generation units with suitable placement of electric vehicle charging stations. In: 2021 IEEE Region 10 Symposium, pp 1–7
10. Gabbar HA, Labbi Y, Bower L, Pandya D (2016) Performance optimization of integrated gas and power within microgrids using hybrid PSO–PS algorithm. Int J Energy Res 40:971–982
11. Fossati JP, Galarza A, Martín-Villate A, Fontán L (2015) A method for optimal sizing energy storage systems for microgrids. Renew Energy 77:539–549
12. Das CK, Bass O, Kothapalli G, Mahmoud TS, Habibi D (2018) Overview of energy storage systems in distribution networks: placement, sizing, operation, and power quality. Renew Sustain Energy Rev 91:1205–1230
13. Zenginis I, Vardakas JS, Echave C, Morató M, Abadal J, Verikoukis CV (2017) Cooperation in microgrids through power exchange: an optimal sizing and operation approach. Appl Energy 203:972–981

14. Marzband M, Azarinejadian F, Savaghebi M, Guerrero JM (2017) An optimal energy management system for islanded microgrids based on multiperiod artificial bee colony combined with Markov chain. IEEE Syst J 11:1712–1722
15. Peanjad P, Khomfoi S (2022) Electric vehicle charging station incorporating with an energy management and demand response technique. In: 2022 international electrical engineering congress, pp 1–4
16. Neal D, Ding Y, Bello I, Han R, Byamukama M, Kebir N, McCulloch M, Rogers D, Nyamongo I, Waweru K (2020) Demand side energy management and customer behavioral response in a rural islanded microgrid. In: 2020 IEEE PES/IAS PowerAfrica, pp 1–5
17. Chen C-I, Chen Y-C, Chen C-H (2020) Energy management between power generation, storage, and consumption for building microgrid. In: 2020 international symposium on computer, consumer and control, pp 335–338
18. Aghajani G, Ghadimi N (2018) Multi-objective energy management in a micro-grid. Energy Rep 4:218–225

Congestion Management in Power System Using FACTS Devices

Nazeem Shaik and J. Viswanatha Rao

Abstract Since recent increases in electrical power consumption have been quite high. The Transmission Lines, which were installed many years ago, were overloaded. Congestion is the term used to describe this overloading of transmission lines. The Power System Operator's primary difficulty is managing this transmission line congestion. On an IEEE 5-Bus test system, we're going to use this paper to clear a transmission line of congestion. This paper includes detailed explanation and tabular columns that demonstrate how congestion is decreased. In order to control the power flow through the transmission lines, FACTS devices are utilized. Transmission lines for the 5-Bus System are overloaded; however, by using a (Thyristor controlled Series Capacitor) TCSC, congestion on a line is minimized.

Keywords FACTS · Congestion management · TCSC · MATLAB simulink · Real power flow · Load flow

1 Introduction

The transmission lines are under so much stress that they are unable to transmit the power from the generating stations to the loads. Congestion is the name given to this condition of transmission lines. The following factors contribute to congestion: Line overloading, transmission line interruptions resulting from a problem or lack of maintenance, transmission line ageing, and transmission line contingency state. By installing new transmission lines, we can lessen congestion in addition to the use of FACTS devices, but because this process is extremely expensive and time-consuming, we cannot replace all the existing transmission lines with new lines. As a result, we use a variety of congestion management strategies, with FACTS being

N. Shaik (✉) · J. V. Rao
Department of Electrical and Electronics Engineering, VNR Vignana Jyothi Institute of Engineering and Technology, Hyderabad, India
e-mail: nazeemsk19@gmail.com

J. V. Rao
e-mail: viswanatharao_j@vnrvjiet.in

H. O. Bansal et al. (eds.), *Next Generation Systems and Networks*, Lecture Notes in Networks and Systems 641, https://doi.org/10.1007/978-981-99-0483-9_33

the most effective strategy when it comes to time, effort, and cost–benefit analysis for reducing transmission line congestion [1, 2].

Congestion control begins with a study of the load flow analysis of a power system, which has been done here [3] for a 5-Bus system, and observation of the real and reactive power flow in the transmission lines. Modeling transmission lines in the MATLAB Simulink software while incorporating their values and line and bus data [4]. Finding the best spot for the FACTS devices to be installed within the power system network has been looked at in [5, 6]. Designing and simulating Thyristor controlled switched capacitors (TCSC) and simulating their gate circuits allows us to control the TCSC [7]. This paper explains the details of power flow and how the TCSC relieves transmission line congestion by positioning it optimally in the transmission lines on the IEEE 5-bus test system for an overloading condition that would be observed in the "simulation and results".

2 Objective

The primary goal of this paper is to demonstrate how actual electricity may be diverted from a busy line to a transmission line that is not congested. Let's continue to examine this using the example of two transmission lines that are linked, as shown in Fig. 1, between source and load sides of two buses. Bus 1 has a line reactance of X1, while bus 2 has a line reactance of X2. $X_{Total} = (X1 \| X2)$ is the total reactance in this parallel combination of X1 and X2. The equation below gives the power flow equation between two buses.

Where $ø = (ð_1\text{-}ð_2)$, $ð_1$ & $ð_2$ are the voltage phase angles of buses 1 and 2 respectively. A FACTS device is deployed in the line 2 as shown in the Fig. 1. Wherever the power carried in the line exists the maximum limit of the line, the FACTS device kicks in and increases line 2's reactance, which allows more power to be transferred from line 1 and successfully provide the requested load without exceeding the maximum power limit. The FACTS device reduces the reactance of line 2 whenever the line 1 power limit is exceeded in a similar manner. in order to preserve system stability, the power flow is divided between the lines. In the same way whenever the line 1 power limit is exceeded, the reactance of line 2 is reduced by the FACTS device. Such that the power flow is shared between the lines maintaining the system stability.

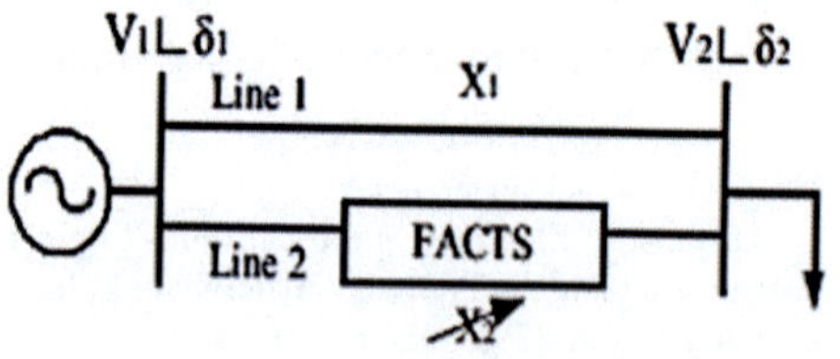

Fig. 1 Two bus system

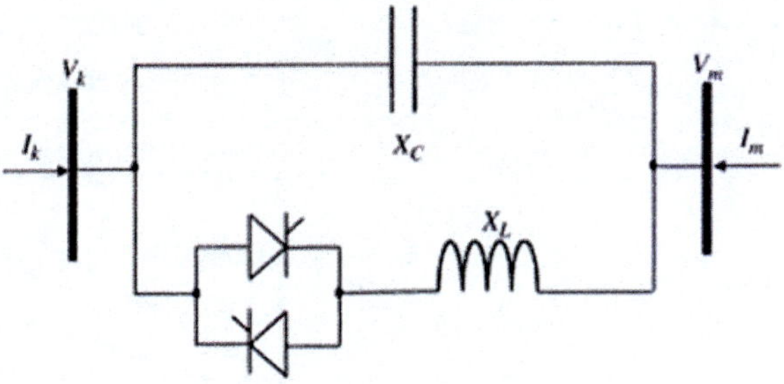

Fig. 2 Equivalent circuit of TCSC

In the same way this technique can be implemented on a 'n' bus power system with the help of FACTS devices varying the line reactance by rerouting the power to relieve from congestion.

3 Thyristor Controlled Series Capacitor (TCSC)

The above Fig. 2 shows the simplified representation of Thyristor Controlled Series Capacitor. A TCSC is made up, as shown in the Fig. 2, of a capacitor connected in parallel with a thyristor and a reactor. Using a TCSC has the benefit of providing greater control over power flow in a line than other FACTS devices. The firing angle of the thyristors may be readily adjusted in a TCSC operation to change the equivalent reactance. By altering the firing angle of the thyristors, the net reactance of the TCSC may be either capacitive or inductive [5].

4 System Model

Figure 3 depicts an IEEE 5-Bus test system, and Tables 1 and 2 (all of which were derived from Ref. [2]) display the system's bus and line data.

5 Stimulation and Discussion

A 5-bus model is simulated using the data from Tables 1 and 2 above in the MATLAB Simulink software, as illustrated in Fig. 4. An overloading of line situation is created by increasing the load in bus 4 and bus 5 by 50%.i.e. the load in bus 4 has been increased from 50 to 75 MW, and the load at bus 5 has been increased from 60 to 80 MW. we from here further continue the analysis by two cases. The first case is without TCSC in the transmission line and the second case is with TCSC in the line.

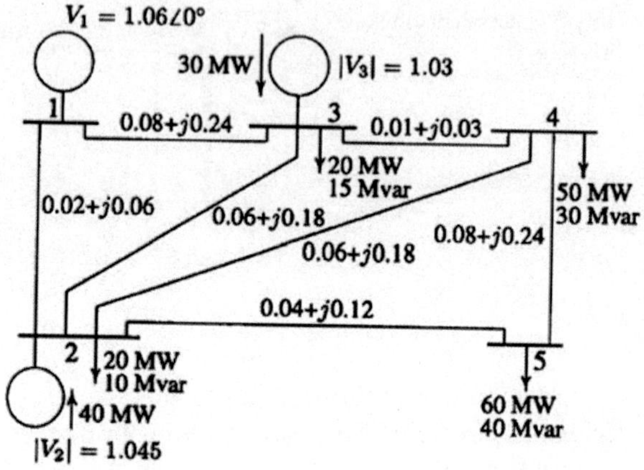

Fig. 3 IEEE. 5-Bus test system

Table 1 Bus data

Bus no	Bus type	Bus voltage		Load		Generation	
		Mag	Angle	P(MW)	Q(MVAr)	P(MW)	Q(MVAr)
1	Slack	1.06	0.0	0	0	0	0
2	PQ	1.045	0.0	20	10	40	30
3	PQ	1.03	0.0	20	15	30	10
4	PV	1.00	0.0	50	30	0	0
5	PV	1.00	0.0	60	40	0	0

Table 2 Line data

Line (bus to bus)	Resistance R(p.u.)	Reactance X(p.u.)	Line charging Y/2(p.u.)	Line MAX power limit (MW)	Tap setting value
1–2	0.02	0.06	0.030	100	1
1–3	0.08	0.24	0.025	50	1
2–3	0.06	0.18	0.020	50	1
2–4	0.06	0.18	0.020	50	1
2–5	0.04	0.12	0.015	100	1
3–4	0.01	0.03	0.010	100	1
4–5	0.08	0.24	0.025	50	1

5.1 *Load Flow Analysis Without TCSC*

The line connecting buses 1 and 2 is overloaded, as discovered by doing a load flow analysis of the overloaded situation of the transmission system; the loading of the lines is documented in Table 3. Although line 1–2 has a 100 MW maximum power

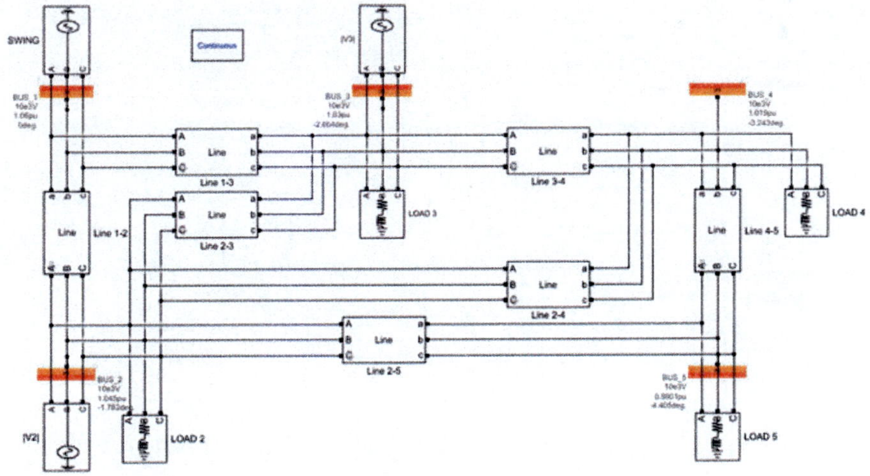

Fig. 4 Simulation diagram of 5-bus power system network

Table 3 Load flow analysis results without TCSC

S. no.	Line	PMAX	MW	MVAR
1	1–2	100	103.10	−0.53
2	1–3	50	38.84	1.91
3	2–3	50	17.26	1.71
4	2–4	50	28.05	7.19
5	2–5	100	75.91	34.13
6	3–4	100	64.85	30.63
7	4–5	50	16.95	10.64

capacity, the current power flow is 103.10 MW, which is significantly congested. Table 3 further shows that, aside from line 1–2, the other lines of the 5-bus system are much below the maximum limit. As a result, the electricity in lines 1–2 is now diverted to other lines by TCSC, which is discussed in the case after that.

5.2 *Load Flow Analysis with TCSC*

According to the simulation shown in Fig. 5, a TCSC is linked in lines 1–2. The line's reactance is raised from 0.06 p.u. to 0.1 p.u. by connecting the TCSC in series with the line. After adding 0.04 p.u. of reactance to the line, we can observe that the line 1–2 is relieved from congestion which can be seen from the results of Table 4. Where all the lines are loaded less than the maximum power limit, the line 1–2 is given more importance.

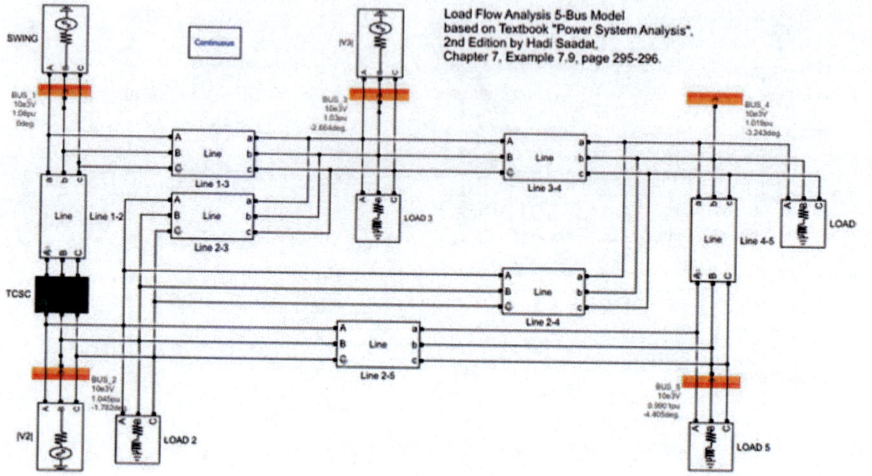

Fig. 5 Simulation diagram of 5-bus power system network with TCSC

Table 4 Load flow analysis results with TCSC

S. no.	Line	PMAX	MW	MVAR
1	1–2	100	94.22	3.89
2	1–3	50	47.82	−0.16
3	2–3	50	13.32	2.80
4	2–4	50	24.92	7.99
5	2–5	100	74.40	34.41
6	3–4	100	69.41	29.31
7	4–5	50	18.42	29.31

6 Conclusion

This paper presents the idea of using TCSC to inject reactance into a congested line, bringing the line's power flow back below its maximum point and redirecting any excess power to other lines. Two cases were the subject of a thorough examination, the results of which are tabulated in Tables 4 and 5. Table 5 compares the power flow with and without TCSC. Real power flow in transmission line 1–2 has been reduced from 103.10 to 94.22 MW in that area. The results demonstrate in the paper, where we found that in this instance, more over 8% of the additional power in the line was diverted, are more encouraging. Congestion is efficiently relieved by using FACTS devices properly and deploying them in the best locations. This work can be expanded using the same methodology to n buses.

Table 5 Comparison of power flow

5-bus system	Maximum power limit	Without TCSC	With TCSC
Real power (MW) in line 1–2	100	103.10	94.22

References

1. Kirthika N, Balamurugan S (2016) A new dynamic control strategy for power transmission congestion management using series compensation. Electr Power Energy Syst 77:271–279 (2016)
2. Choudekar P, Sinha S, Siddiqui A (2018) Congestion management of IEEE 30 bus system using thyristor controlled series compensator. In: International conference on power energy, environment and intelligent control (PEEIC), April 13–14, 2018
3. Saadat H (1999) Power system analysis, 2nd ed. Milwaukee School of Engineering. A Division of the McGraw-Hill, Chapter 7, Example 7.9
4. Kalaimani P, Mohana Sundaram K (2018) Congestion management in power transmission network under line interruption condition using TCSC. Int J Eng Res Technol (IJERT) 6(02). ISSN:2278-0181.PECTEAM-2K18
5. Siddiqui AS, Jain R, Gupta CP (Oct 2011) Congestion management in high voltage transmission using thyristor controlled series capacitors (TCSC). J Electr Electron Eng Res 3(8):151–161. ISSN-2141-2367
6. Ushasurendra S, Parathasarthy SS (2012) Congestion management in deregulated power sector using fuzzy based optimal location technique for Series FACTS devices. J Electr Electron Eng Res, 12–20
7. Bathina VR, Gundavarapu VNK (2015) Thyristor controlled series capacitor for generation reallocation using firefly algorithm to avoid voltage instability. Majl J Electr Eng 9(2)

Novel SVD-DWT Based Video Watermarking Technique

B. S. Kapre and A. M. Rajurkar

Abstract In this paper, a novel Singular Value Decomposition (SVD)-Discrete Wavelet Transform (DWT) based video watermarking technique has been presented. The main purpose of using a combination of SVD and DWT is that it improves the imperceptibility and robustness of the proposed watermarking scheme. We embed the watermark information into singular values of SVD with entropy of singular value using additive method in low frequency sub-band of DWT. A blind detection technique is used in the extraction procedure. Several attacks are used, and various performance metrics are generated, to analyze the robustness and imperceptibility of the proposed watermarking system. The results show that the proposed technique is resistant to a variety of attacks and achieves a high level of invisibility.

Keywords SVD · DWT · Entropy · Embedding · Extraction

1 Introduction

The advancement in computer technology is causing several issues for the multimedia content in everyday life. Many types of digital data, including video files, are easily being copied, moved and altered without degrading the visual quality. Consequently, one of the main concerns with multimedia technology is the security of digital media. As a result, a variety of ways for protecting video content have already been presented. Data is encrypted using cryptographic methods based on a secret key to protect the video contents [1, 2]. In fact, one of their main weaknesses is that the original formats of the videos are not maintained. To handle this problem, video watermarking techniques provide a promising security solution [3]. It is the process of hiding a watermark signal in the video frames. An image, a logo, or any

B. S. Kapre (✉) · A. M. Rajurkar
Department of Computer Science and Engineering, MGM's College of Engineering, Nanded, India
e-mail: kapre_bs@mgmcen.ac.in

A. M. Rajurkar
e-mail: rajurkar_am@mgmcen.ac.in

H. O. Bansal et al. (eds.), *Next Generation Systems and Networks*, Lecture Notes in Networks and Systems 641, https://doi.org/10.1007/978-981-99-0483-9_34

other type of information content is used as watermark signal. A video watermarking system consists of two operations. The first is embedding operation, which refers to the combination of a watermark and the host video. It can also be created by utilizing the properties of video frames. The second operation in a video watermarking system is watermark extraction. It is the procedure of recovering hidden information from a watermarked video that has been tampered and used to determine the validity of video content.

In general, the most practical video watermarking systems consider three fundamental needs [4–6]. The first is imperceptibility. It refers to the visual quality of watermarked video, which should be as close to the original as feasible. The second criterion is robustness. It refers to an ability of watermark to withstand both incidental and intentional attacks. The third requirement is capacity, which is defined as the quantity of secret information hidden in the host video. Invisibility, robustness, and capability are all inversely proportional. As a result, it is critical to strike a fair balance between all of the watermarking qualities in one technique [7]. According to the embedding domain criterion, watermarking techniques are split into two categories: spatial domain techniques [8–10] and frequency domain techniques [11–13]. The watermark is inserted in the spatial domain based watermarking by directly changing the pixel values of the host video. Whereas, in the transform domain watermarking, the watermark is inserted into the selected coefficients from the host video frames after they have been transformed using transform domain approach. The computational complexity of spatial domain-based approaches is minimal. However, they are unreliable in the case of various image processing technologies and have a low bit capacity. On the other hand, the frequency domain watermarking approaches are more resistant to common distortions such as noise addition, compression, and rotation. Furthermore, they are effectively achieving the compromise between imperceptibility and robustness requirements of digital watermarking techniques. In this paper, we present a blind and robust video watermarking system in the frequency domain, based on two transformations: Discrete Wavelet Transform (DWT) and Singular Value Decomposition (SVD).

The remainder of this paper will be organized as follows. A survey of existing frequency domain video watermarking techniques is presented in Sect. 2. Section 3 describes the proposed watermarking scheme. Section 4 contains the experimental data as well as a comparison to other methodologies. This study is described and concluded in the final section.

2 Related Work

A variety of video watermarking techniques have been presented in the literature, and they can be classified into several classes depending on characteristics like working domain, watermark visibility, and watermark resilience, etc. As previously stated, the embedding procedure in frequency domain is preceded by several transformation methods applied to the cover video frames. The most commonly used methods in this domain are Fast Fourier Transform (FFT), Discrete Cosine Transform (DCT), Discrete Wavelet Transform (DWT), and Singular Value Decomposition (SVD) [13–25]. We will focus at digital video watermarking in the transform domain in this section. With much interest to this, we propose a video watermarking scheme in the frequency domain based on dual transformations; DWT and SVD.

A single transform and combination of transform domain techniques were used to develop video watermarking system to improve imperceptibility and robustness. In [14] a non-subsampled contour let transform and SVD were used to demonstrate a hybrid video watermarking technique. The non-subsampled contour let transform was used to decompose both the original video frame and the watermark. The watermark's singular values were then combined with those of the original video using an additive non-blind method.

The authors of [15] presented four techniques for video watermarking; all were based on DWT and SVD. The amount of frames to be used for the embedding process differs between these four methods. In order to choose the embedding medium, the authors have used a scene change detector. The insertion takes place in the singular value corresponding to the low frequency sub band taken from just identified scene change frames. A robust video watermarking system using four distinct domain transformation algorithms based on DWT, SVD, FFT, and DCT is proposed in [16]. DWT and SVD domains based video watermarking scheme was presented in [17, 18]. In which, additive approach was used to insert watermark into the mid-frequency subbands of DWT. The extraction of the watermark was done using a blind extraction algorithm. The presented video watermarking technique provides high robustness against several attacks. However, provides low quality of watermarked video.

A single transformation technique was used in some video watermarking schemes. A DWT-based watermarking algorithm has been presented in [19]. This video watermarking method takes advantage of the features of the human visual system to meet the imperceptibility requirement. In fact, after being transformed using the DWT, blocks with the highest motion vector magnitude were chosen to embed watermark. The DWT-based watermarking method has been proposed in [20]. This approach only embeds the watermark on scene-changed frames that are determined by a scene-change detection algorithm. These frames were then decomposed using the three levels DWT and transformed to grey scale. The low frequency components were then added with an additive technique to insert the watermark bits. Sethuraman et al. [21] have introduced a key-frame based watermarking technique wherein the structural similarity index metric—absolute difference metric (SSIM-AMD) techniques were adopted for identification of non-redundant frames. Then, entropy–AMD method was used to select key- frame. Further, DWT was applied to decompose key-frame into sub-bands. To avoid false-positive attacks the principal component of watermark image block was computed and embedded into middle band of DWT. The strength of the watermark was decided by calculating scaling factor using the ant colony optimization (ACO) technique. It was observed that, this scheme was robust against video processing and false-positive attacks. It provides high performance in terms of imperceptibility and robustness. A blind technique based on DCT was proposed by the authors in [22]. The luminance component Y of input video frame was retrieved and divided into DCT 8 × 8 blocks. The even–odd quantization approach was then used to hide the watermark bits on a number of randomly selected DCT coefficients.

To detect and localize tampered area locations in [23] authors have developed a chromatic DCT based video watermarking approach. In this scheme, tamper detection was done by using different features of H.264/AVC coding standards. An experimental result shows that the developed technique was used to detect spatial attacks as well as helps to localize tempered regions. Although, DWT-based approaches are offer very good imperceptibility and provide low robustness. In DWT, embedding is typically carried out in the LL sub-band, which contains the majority of the image's information and offers more robustness. False Positive Rate, often known as FPR, refers to the detection of a watermark from a watermarked image even though no watermark was actually embedded in it. Using only SVD for embedding watermark is not able to solve the problem of FPR therefore other transform techniques are used with SVD.

The summary of current video watermarking techniques shown above makes it evident that using a combination of SVD and DWT transformation domain techniques offers more resistance to various attacks than using a single transform. As a result, the watermark embedding in the proposed work is done in the multi-frequency domain.

3 Proposed Scheme

The proposed scheme is a DWT and SVD based frequency domain video watermarking scheme. By utilizing their complementing qualities, these two domain transformation techniques enhance the watermark's resilience and imperceptibility. The embedding process and the extraction procedure are the two stages of the suggested system. Figures 1 and 2 represent the block diagrams for these processes, which are explained separately in the two next sections.

3.1 Watermark Embedding Process

The video is divided into sequences of n-consecutive frames. To increase the resilience against frame dropping attacks, the same watermark is put into every frame of each video sequence. Every video sequence a binary watermark is used as a watermark. The following technique is applied to all frames in each sequence. Since pixel values in RGB color space are more correlated than in color space, each video frame is transformed from RGB to YCbCr. Only the luminance component Y is chosen for the watermark insertion because the Human Visual System cannot detect changes in regions of high luminance. After that, the Y-component of the frame is split into 4*4 non-overlapping blocks. Following that, the DWT transform is used to create four sub-bands: LL1, HL1, LH1, and HH1. Where, LL1 low frequency sub-band represents approximation sub-image while HL, LH, and HH are high frequency

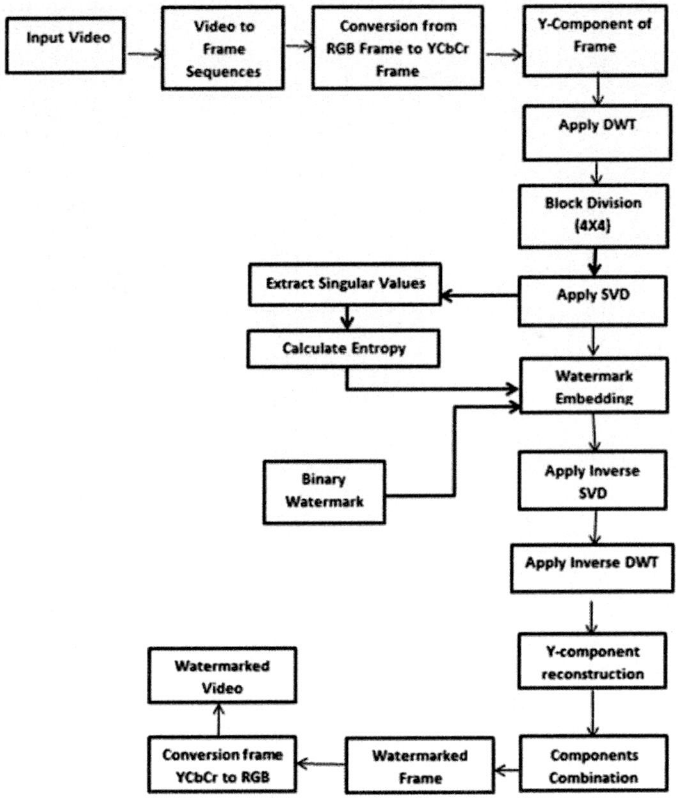

Fig. 1 Block diagram of the proposed watermarking embedding technique

sub-bands represent detail sub-images. While the approximation sub-image is the convergence of strength of original image, the detail sub-images contain the fringe information. Since the majority of the image energy concentrates here, the approximation sub-image is far more stable than the detail sub-images. In order to increase the imperceptibility, the watermark information is embedded within approximate sub-image. The LL sub-band is then divided into 4×4 sub-blocks to increase the payload capacity. Then, SVD is employed on each sub-block to get three separate matrices, U, S, and V. Where, S is a diagonal matrix, U and V are orthogonal matrices, and the embedding method just takes into account the S matrix. Because, a small change in the singular values of S matrix does not result in a significant visual change in the host video. Additionally, it displays desirable properties including translation invariance, transposition invariance, and rotation invariance. Singular values of the S matrix corresponding to the LL1-subband contain watermark information. At first entropy value of singular value is calculated and used for embedding watermark information using scaling factor α and β. The following formulas determine how the embedding process for each block b works:

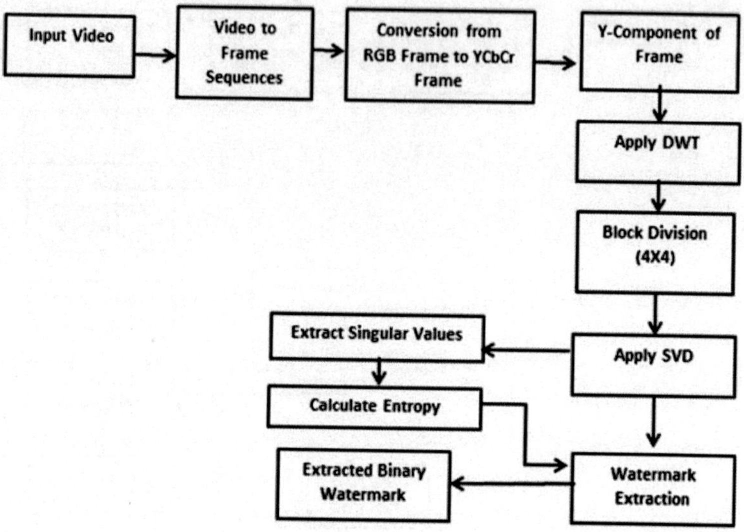

Fig. 2 Block diagram of the proposed watermarking extraction technique

```
IF W(i,j) ==1
    Sw(1,1)= S(1,1) + Ent *α
    Sw(2,2)= S(2,2) + Ent *α
ELSE
    Sw(1,1)= S(1,1) + Ent *β
    Sw(2,2)= S(2,2) - Ent *β
End
```

where, original singular values, Sw watermarked singular values, Ent is an entropy value of a block, α and β are two factors allowing balancing imperceptibility and robustness.

In order to obtain the watermarked luminance component Y, the inverse SVD operation is employed on the modified matrix *Sw*. In the same way, inverse SVD is applied on every block of the LL1 sub-band. Then, the resulting watermarked LL1 sub-band is combined with non- watermarked sub-bands (LH1, HL1 and HH1) by applying inverse DWT. To create the final watermarked video frame, this watermarked Y is combined with the unchangeable Cb and Cr elements and changed to RGB color. To get the watermarked video frame, the above procedure is applied to every frame in each sequence to generate watermarked video.

3.2 Watermark Extraction Process

Figure 2 shows the blind extraction process, which is the reverse of the embedding procedure. As a result, the approach does not need the original data to recover the hidden watermark. The watermarked video is divided into n-frames sequence. The corresponding watermark is then extracted from each video frame using the technique described below. First, the RGB to YCbCr color space of the selected video frame is changed. The DWT decomposition is then applied to Y component to get the LL1, LH1, HL1, and HH1 sub-bands, respectively. A 4 × 4 block decomposition is performed on watermarked LL1 sub-band. Then SVD is employed to retrieved watermark from S'_w matrix using entropy measure of watermarked singular values. As singular values provide stability and robustness, the entropy value of watermarked singular values are used for extraction of watermark. Therefore, the proposed watermark detection process is blind as the original information is not required. Watermark is extracted using the following criteria.

```
IF Sw'(1,1)+Sw'(2,2)>= Sw (1,1)+Entw*a
      W'(i, j)=1;
Else
      W'(i, j)=0;
End
```

Where, S'_w is watermarked singular value, Ent_w is an entropy value of singular values and γ is the scaling factor used to extract watermark information.

.

Simulation studies are carried out on several types of standard videos to confirm the robustness and imperceptibility of the proposed video watermarking technique. The input videos and binary watermark taken into consideration are shown in Fig. 3.

The maximum capacity for each video sequence is equal to the total number of 4 × 4 sub-blocks produced by the decomposition of the respective luminance component Y for each frame. The proposed watermarking technique has a large

Fig. 3 Snapshots of sample videos and binary watermark

capacity, as shown by the maximum size of the embedded message for a 256 × 256 frame, which is 4096 bits.

4 Experimental Results

Simulation studies are carried out on several types of standard videos to confirm the robustness and imperceptibility of the proposed video watermarking technique. The input videos and binary watermark taken into consideration are shown in Fig. 3.

The maximum capacity for each video sequence is equal to the total number of 4 × 4 sub-blocks produced by the decomposition of the respective luminance component Y for each frame. The proposed watermarking technique has a large capacity, as shown by the maximum size of the embedded message for a 256 × 256 frame, which is 4096 bits.

4.1 *Imperceptibility Analysis*

The Peak Signal to Noise Ratio (PSNR) and the Structural Similarity Index (SSIM) are used to analyze the imperceptibility requirements of watermarked video. The Peak Signal to Noise Ratio is used to evaluate the visual quality of watermarked video from a human visual perspective. This is described as follows using Eq. 1:

$$\text{PSNR} = 10 \times \left(\log \frac{X^2}{\text{MSE}}\right) \tag{1}$$

where, X is the maximum pixel value in the frame and MSE stands for mean square error.

The similarity between watermarked video frame and original video frame is measured using SSIM measurement technique. It is calculated using following Eq. 2.

$$\text{SSIM} = \frac{(2\mu_m\mu_n + a)(2\sigma_{xy} + b)}{(\mu_m^2\mu_n^2 + a)(\sigma_m^2\sigma_n^2 + b)} \tag{2}$$

where _μ_n and _${\sigma_n}^2$ denote the mean and variance of the intensities contained in the watermarked video frame, respectively, while μ_m and ${\sigma_m}^2$ define the mean and variance of the intensities contained in the original video frame, respectively. a and b are two variables used to stabilize the division, and σ_{xy} represents the covariance of original and watermarked video frames.

The average of the PSNR and SSIM values of all frames of the video is used to calculate PSNR and SSIM for each tested video. We compare the original and watermarked videos' visual similarity in order to determine the appropriate values

Fig. 4 Original (first row) and watermarked (second row) frames

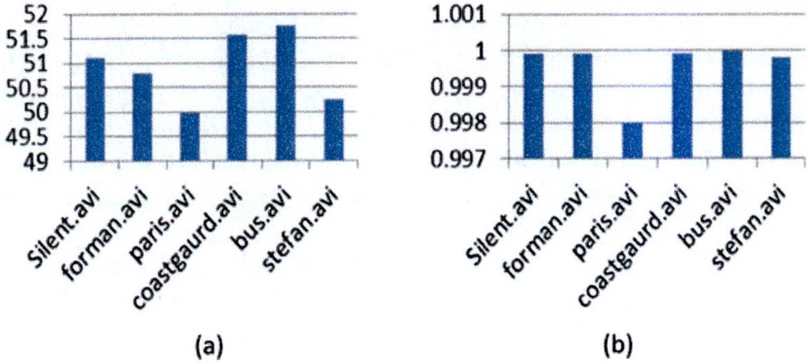

Fig. 5 **a** PSNR values obtained for different videos **b** SSIM values obtained for different videos

for (α; β). The values of α and β are chosen during experimentation to determine the Peak Signal to Noise Ratio (PSNR) for this purpose.

The frames of three host videos are shown in Fig. 4, along with the corresponding watermarked versions. The watermark is totally transparent, and the frames of watermarked and original videos cannot be visibly distinguished from one another. Figure 5a shows the obtained PSNR values, which range from 49 to 52 dB. It indicates that the videos with watermarks have good visual quality. As shown in Fig. 5b, the SSIM values are also getting closer to 1, indicating the strong similarity between the host videos and the watermarked ones. A possible option that allows for this high level of imperceptibility is to incorporate the watermark into the S component of SVD applied to the Low frequency sub bands of DWT.

4.2 *The Robustness Analysis*

The robustness of proposed scheme is analyzed using normalized correlation coefficient NCC [1, 9]. The NCC is used to calculate similarity between extracted and original binary watermark. NCC is calculated using following formula (3).

$$\mathrm{NCC} = \frac{\sum_{p=1}^{m}\sum_{q=1}^{n}\mathrm{OrgW}(i,j)\mathrm{ExtW}'(i,j)}{\sqrt{\sum_{p=1}^{m}\sum_{q=1}^{n}\mathrm{OrgW}^2(i,j)}\sqrt{\sum_{p=1}^{m}\sum_{q=1}^{n}\mathrm{ExtW}^2(i,j)}} \tag{3}$$

where, OrgW(i, j) is the original watermark of size $m \times n$. and ExtW is the extracted watermark of size $m \times n$.

Different distortions and attacks are first carried out on the watermarked videos in order to evaluate the resilience of the proposed watermarking system. Second, the NCC value is determined after the extraction procedure. The NCC values of the attacked videos are displayed in Fig. 6. shows that proposed watermarking scheme is resistant to noise attacks like Gaussian noise and salt and pepper. After applying salt and pepper, and various white Gaussian noise, the obtained NCC value is 1. Using the DWT, which is resistant to noise addition, ensures this high correlation. It is also noted that the proposed scheme is resistant to rotational attack. In fact, the watermark can be successfully removed simply adjusting the rotation degrees. The NCC values that have been obtained against this attack reach 0.9998. This resilience is made possible by the use of the SVD, which is resistant to geometrical attacks, particularly rotational attacks. Moreover, figure shows that the robustness of proposed scheme against cropping and median filter attacks. The NCC value for the cropping attack and median filter attack are 1. The NCC value for compression attack is about 0.9999. This high level of robustness is made possible by inserting the watermark in the DWT low frequency sub-bands. The robustness of proposed scheme against frames dropping attack is strengthened by the redundant embedding of a full watermark into each frame of the video. 32% of the frames were removed and NCC value observed to be 1.

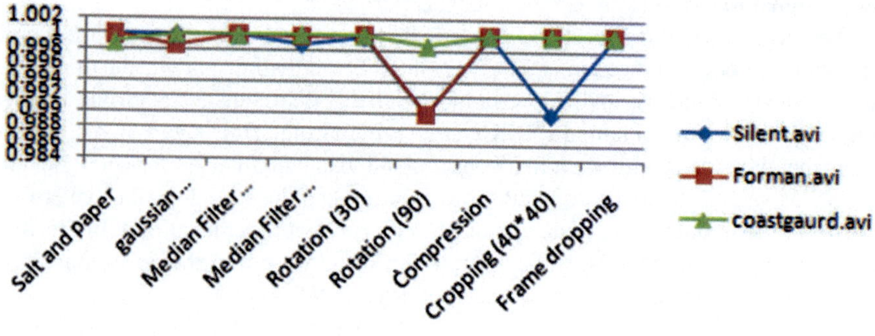

Fig. 6 Robustness of proposed scheme against several attacks

4.3 Comparison with Existing Methods

This section will contrast the effectiveness of our proposed technique with the three existing techniques [14, 17, 20]. Table 2 shows NCC values, it is observed that our algorithm and the method suggested by [17] can withstand attacks from Gaussian noise. In fact, our proposed technique successfully repels this attack with a maximum NCC value of 1. However, the approach described in [14] is not robust against Gaussian noise.

Additionally, both our proposed method and that of [17] successfully withstand the median filter attack with NCC value of 1. Furthermore, by providing an NCC value of 0.9999, our scheme is shown to offer a high level of robustness against compression compared to the existing techniques [17, 20]. With an NCC value of 1, proposed technique provides high robustness against cropping attacks than [14]. On the other hand, the robustness against frame dropping attack shows that the proposed scheme provides good result compared to [14, 17]. Table 1 shows the PSNR values of proposed scheme and existing scheme [17]. It is observed that the proposed method provides high imperceptibility compared to [17].

Table 1 Comparison of the PSNR of proposed scheme and existing [26] scheme

Video	[17]	Proposed scheme
Farman	37.39	50.90
Paris	33.0109	50
Coastguard	45.317	51.57
Stefan	37.7797	50.29
Bus	49.6773	51.82

Table 2 Comparison of the proposed scheme resilience to other video watermarking techniques

Noise	[20]	[14]	[17]	Proposed technique
Salt and paper	0.9206	0.9816	1	1
Gaussian noise(0.05)	0.41932		1	1
Median filter (3*3)			1	1
Rotation (30)	0.9076	0.9587	0.9996	0.9998
Compression	0.9523		0.9997	0.9999
Cropping (40*40)		0.9557	1	1
Frame dropping		0.9833	0.9998	1

5 Conclusion

In this paper, we proposed DWT and SVD based blind robust video watermarking technique. In order to obtain the trade-off between the imperceptibility and robustness requirements, the S component of the SVD is applied to the low frequency sub bands of the DWT with entropy of singular values. To evaluate the performance of the proposed approach, the Normalized Correlation Coefficient (NC), Peak Signal to Noise Ratio (PSNR), and Structural Similarity Index (SSIM) are computed. The proposed scheme successfully resists several attacks, including geometry, image processing, and compression, according to experimental findings. When compared to previous strategies, the one that is being offered exhibits a high level of robustness. The video quality remains similar in terms of imperceptibility. We can thus draw the conclusion that the proposed method is effectively acceptable for situations that demand more robustness than imperceptibility. Thus, future work will focus on key-frame based video authentication in the video surveillance context.

References

1. Omar G, Shawkat KG (2018) A survey on cryptography algorithms. Int J Sci Res Publ (IJSRP) 8(7):495–516
2. Vikrant M, Shubhanand SH (2016) A survey on cryptography techniques. Int J Adv Res Comput Sci Softw Eng (IJARCSSE) 6(6):469–475
3. Asikuzzaman M, Pickering MR (2018) An overview of digital video watermarking. IEEE Trans Circuits Syst Video Technol 28(9):2131–2153
4. Kim C, Yang CN, Leng L (2020) High-capacity data hiding for ABTC-EQ based compressed image. Electronics 9:644
5. Asim N, Yasir S, Nisar A, Aasia R (2015) Performance evaluation and watermark security assessment of digital watermarking techniques. Sci Int 27(2):1271–1276
6. Arti B, Ajay K (2017) Digital video watermarking techniques: a review. Int J Eng Comput Sci (IJECS) 6(5):21328–21332
7. Hedayath, BS, Gangatharan N, Tamilchelvan R (2016) A survey on video watermarking technologies based on copyright protection and authentication. Int J Comput Appl Technol Res (IJCAT) 5(5):295–303
8. Zhang D, Li Y (2010) A non-blind watermarking on 3D model in spatial domain. In: International conference on computer application and system modeling, pp 267–269
9. Preda RO, Vizireanu N (2011) New robust watermarking scheme for video copyright protection in the spatial domain. UPB Sci Bull 73(1):93–104
10. Priya P, Tanvi G, Nikita P, Ankita T (2014) Digital video watermarking using modified LSB and DCT technique. Int J Res Eng Technol 4(3):630–634
11. Jyothika A, Geetharanjin PR (2018) A robust watermarking scheme and tamper detection using integer wavelet transform. In: 2nd international conference on trends in electronics and informatics (ICOEI), Tirunelveli, India, pp 676–679
12. Kadu S, Naveen C, Satpute VR, Keskar AG (2016) Discrete wavelet transform based video watermarking technique. In: International conference on microelectronics, computing and communications (MicroCom), Durgapur, pp 1–6
13. Guangxi C, Ze C, Daoshun W, Shundong L, Yong H, Baoying Z (2019) Combined DTCWT-SVD-based video watermarking algorithm using finite state machine. In: Eleventh international conference on advanced computational intelligence, pp 179–183 (2019)

14. Narasimhulu CV (2017) A robust hybrid video watermarking algorithm using NSCT and SVD. In: IEEE international conference on power, control, signals and instrumentation engineering (ICPCSI), Chennai, pp 1495–1501
15. Jeebananda P, Prince G (2016) An efficient video watermarking approach using scene change detection. In: 1st India international conference on information processing (IICIP), Delhi, pp 1–5
16. Naved A (2016) A robust video watermarking technique using DWT, DCT, and FFT. Int J Adv Res Comput Sci Softw Eng (IJARCSSE) 6(6):490–494
17. Hammami A, Hamida AB, Amar CB (2020) A robust blind video watermarking scheme based on discrete wavelet transform and singular value decomposition. In: International joint conference on computer vision, imaging and computer graphics theory and applications (VISIGRAPP 2020)
18. Hammami A, Hamida AB , Amar CB (2021) Blind semi-fragile watermarking scheme for video authentication in video surveillance context. Multimed Tools Appl. https://doi.org/10.1007/s11042-020-09982-4,2020
19. Mostafa S, Ali A (2016) A Multiresolution video watermarking algorithm exploiting the block-based motion estimation. J Inf Secur Eng (JIS) 7(4):260–268
20. Dolley S, Manisha S (2018) Robust scene-based digital video watermarking scheme using level-3 DWT: approach, evaluation and experimentation. Radioelectr Commun Syst 61(1):1–12
21. Ponni alias Sathya S, Ramakrishnan S (2020) Non-redundant frame identification and keyframe selection in DWT-PCA domain for authentication of video. IET Image Process 14(2):366–375
22. Tuan TN, Duan DN (2015) A robust blind video watermarking in DCT domain using even-odd quantization technique. In: International conference on advanced technologies for communications (ATC), Ho Chi Minh City, pp 439–444
23. Cao Z, Wang L (2019) A secure video watermarking technique based on hyperchaotic Lorentz system. Multimed Tools Appl. 78(18):26089–26109. https://doi.org/10.1007/s11042-019-07809-5
24. Wong K, Chan C, Maung MA (2020) Lightweight authentication for MP4 format container using subtitle track. IEICE Trans Inf Syst 103.D(1):2–10
25. Maung MAP, Tew Y, Wong K (2019) Authentication of Mp4 file by perceptual hash and data hiding. Malays J Comput Sci 32(4):304–314

Predictive Analytics in Financial Services Using Explainable AI

Saurabh Suryakant Sathe and Parikshit Mahalle

Abstract Large financial institutions have a lot of products for earning money, but a major source of their income is from lending money. This income mainly depends on the ability of the loan taker to repay the loan. Every year these institutions incur huge losses due to loan defaulters. Through this research, we will try to predict the possibility of an individual defaulting on a loan using various features in our dataset. Then, will try to explain our results using the novel concept of explainable AI. This research uses a random forest classifier to predict if the individual will have difficulties in repayment of loans. The dataset is sourced from Kaggle. This research has followed a machine learning pipeline wherein we have pre-processed dataset, trained model, predicted outcomes using a model trained, and explained results using explainable ai. Random Forest and XGBoost algorithms are used. The models predicted results with accuracy as high as 97%. For the explainable part, we have used the Shapash library. In our study, we have explained the factor that contributes most to global as well as individual predictions. The results of these predictions can be used by financial institutions to enforce rules and restrictions surrounding this parameter to avoid losses. Furthermore, the explanations for the individual predictions can be used for personalized loan profile improvement recommendations.

Keywords Explainable AI · Random forest · Data · LIME · SHAPASH · XGBoost · ROC

S. S. Sathe
Department of Software Engineering, San Jose State University, San Jose, CA 95192, USA
e-mail: saurabh.sathe@sjsu.edu

P. Mahalle (✉)
Department of Artificial Intelligence and Data Science, Vishwakarma Institute of Information Technology Kondhwa-Budruk, Pune, Maharashtra 411048, India
e-mail: parikshit.mahalle@viit.ac.in

H. O. Bansal et al. (eds.), *Next Generation Systems and Networks*, Lecture Notes in Networks and Systems 641, https://doi.org/10.1007/978-981-99-0483-9_35

1 Introduction

Algorithms for machine learning are at the forefront of giving effective solutions to challenges in today's world. Business solutions, healthcare improvements, work process automation, and much more are all made possible by AI-powered technologies. A typical machine learning application life cycle includes dataset creation or importing from external resources, data preprocessing for missing values or outliers, followed by selecting a suitable use case-based model selection, then training the model, followed by model evaluation. Finally, predictions are generated using the best model. When the predictions are obtained using the model, we get the labels or predictions for our rows.

A variety of financial services in the form of insurance and loans are offered by financial organizations. Financial organizations encounter huge monetary losses due to loan defaulters. In the United States alone, financial organizations have experienced approximately 151 billion dollars in losses. While most of us have never defaulted even on a credit card bill, there are a lot of people who fail to pay loans in time. These losses ultimately affect the economy of the nation. Then this economic imbalance has adverse effects like inflation, recession, and much more. Hence, people defaulting not only has an adverse effect on the development of the country but also on our daily lives. Financial organizations generate huge amounts of data relating to the financial condition of customers, their properties, their transaction histories, and much more. This data can significantly help in the development and progress of the institution. Multiple research studies have been presented to exploit the data in developing AI-powered solutions. Applications range from churn modeling to creating marketing strategies, sentiment analysis to gauge customer success towards a product, customer segmentation using clustering, and more. One such application is to predict the financial credibility of an individual applying for a loan at a financial organization.

Multiple contributions and research studies have been presented in the same field. However, none of these studies have presented suitable explanations for the predictions that their classification or regression models have presented. The explanations can include valuable information like what factors the model considered while making the predictions or what decisions were taken by the model to reach a decision. Financial institutions will be able to make business decisions based on the results and explanations, which will have a substantial impact on the loan approval process. For example, if we know that dangerous credit applicants belong to a given age group, then higher limitations on loan candidates in that age group could be enforced.

Explainable AI (XAI) is a new technique that can assist us in understanding the findings of machine learning and AI models. Explainable AI is a library and algorithms that can help you comprehend and interpret machine learning predictions. Explainable AI provides us with increased transparency, results tracking, and an improved model. These advantages have been presented by Urja Pawar et al. and their application of XAI in the healthcare sector through their research [4]. In our research, we have implemented efficient classification models to consider the financial and economic capabilities of an individual in predicting the probability of the individual's ability to repay the loan within a given time frame. The next part of this research concentrates on explaining the predictions made by our classifiers using the XAI Shapash and LIME libraries. This can help us understand what's the factor that affects this decision the most. The explanations will be used to provide recommendations that will significantly affect the loan-sanctioning decision-making process. The results of this research can also provide personalized recommendations to the loan applicants to make their applications stronger in case of applications are declined.

This research essentially provides the following contributions:

1. Presents a generic methodology that can be applied to any problem in the financial sector domain.
2. The machine learning pipeline proposed can be used to provide similar interpretations in other sectors too.
3. Through this research, we also present what advantages can XAI provide over AI. We have also explained, how using XAI over AI can aid in business decisions and enhance user experience.

The whole research paper is organized into sections. Starting with the Introduction where we have discussed the topic of this research followed by the motivation section which discusses what significance this research has on our economy as well as daily lives, we next have a related works section where we have discussed previously done related research work, followed by the project and dataset description. The next section is the methodology or technical aspects section discussing the various subsections like architecture, data preprocessing, model training, and model evaluation sections. The next section is the experimental results section where we have discussed the results obtained from our model training and evaluation. The last 2 sections comprise of explainable AI section and the future score section. In the explainable AI section, we have tried to explain the explainable ai part and its merits and demerits. The section following is the future work section where we have discussed the future scope of our research. Finally, we have the references section where we have cited the various research studies and resources that were used in this research.

2 Motivation

While many of us have never defaulted even on a credit card bill, there are a lot of people who default on loans every year. As a result, the losses incurred by the financial institutions adversely affect the economy of the nation. Hence, such a damaged economy then leads to adverse effects like recession, inflation, and much more. Hence, our lives are heavily affected by the happenings in this financial sector.

Protecting and ensuring that the financial sector performs well overall can significantly improve our lives. Machine learning techniques are used today in many financial applications. However, I feel that the use of XAI can significantly benefit these techniques.

3 Related Work

A detailed explanation of related work is given below.

Throughout the years, multiple studies have been presented to use binary classification using financial data. These studies address various issues in the financial sector using various machine learning and data mining techniques. An 82% accuracy was obtained by Shivanna and Agrawal et al. [1] in their research study that used "deep support vector machines, boosted decision trees", and "averaged perceptron and Bayes point machines" to predict loan defaulters. Some research studies have also used various data mining techniques in financial services. The research study presented by Patel et al. [2] has techniques like "KNN, Naive Bayes, logistic regression, cat boost, and gradient boosting classifiers" to predict the probability of a candidate repaying the loan within a given time frame. When it comes to classifying and identifying the risky candidates among loan applicants, the Random Forest classification algorithm has proven to be the best. This has been proved in the research study presented by Gahlaut et al. [3]. Through this research, we also learn the advantages of ensemble machine learning models which has helped our research select models that could yield the best results. Though this research presents us with ensemble methods that achieve very high accuracies, they do lack the interpretability that our research is presenting using XAI.

The explainable AI technique is a fairly novel technique today. The applications are limited as explainable ai still suffers from the interpretability and accuracy tradeoff. Many of the complex machine learning models are still impossible to interpret. This research paper refers to 2 research studies. Pawar et al. [4] discussed the benefits of XAI in the healthcare industry, including enhanced transparency, model refinement, and result tracking. In the second one, Marcin Kapcia, Hassan Eshkiki, Jamie Duell, Xiuyi Fan, Shangming Zhou, and Benjamin Mora have developed Ex Med which is an AI tool that loads data, preprocesses the data, performs dimensionality reduction, trains model, visualizes results and presents explanations using explainable AI [5]. Both research papers concentrate on explainable AI in the healthcare sector. This research adopts a similar approach but in the finance sector. We aim to achieve transparency and interpretability just like the work discussed in these 2 research studies. We have also explored the explainable ai techniques used by Qinghao Ye, Jun Xia, and Guang Yang for their comparison study which used different classifiers and class activation maps for classifying CT scan results of covid patients [10]. These research studies involve the application of explainable ai techniques in the healthcare sector. However, no research studies have attempted to apply explainable ai in finance sector. Through our research we are presenting a generic solution that can be applied to many application scenarios.

While the random forest algorithm has proven to be the best at classifying data, various other research studies have been presented using different machine learning algorithms and different preprocessing strategies. G. Arutjothi and C. Senthamarai have used the KNN classifier to classify the loan status of loan applications in a commercial bank [6]. Various data preprocessing techniques have been presented by Mohammad Ahmad Sheikh, Amit Kumar Goel, and Tapas Kumar to clean the data before training machine learning algorithms on it [7]. A loan approval system was developed by Vishal Singh, Ayushman Yadav, Rajat Awasthi, and Guide N. Partheeban using a decision tree algorithm to approve or reject loans [8]. A probabilistic approach has been adopted in the research study by Ashlesha Vaidya using logistic regression towards predicting loan approval [9]. Though probabilistic models are easier to interpret, they lack the efficiency of ensemble techniques.

4 Proposed Work

Our research aims to achieve to two goals. First, the research aims at developing a system that can efficiently predict if the individual with the given attributes will be able to pay the loan in time or will default. Then the second goal of the research aims at developing an Explainable AI (XAI) part which can explain the results of the system developed during the first goal. Our project can suffer from the interpretability versus accuracy trade-off issue, but we will try to select and configure our models to maintain a balance between both. Hence our goal is not to develop a system that can predict the loan defaulting probability with the highest accuracy but to develop a system that can efficiently predict and explain the results as well. The explanations obtained from our research can be used by financial institutions in making better business decisions and avoiding losses. The explanations can be used to develop personalized loan profile improvement suggestions for candidates having the probability to default on a loan. This will in the larger picture help our economy and help deliver a better life for everyone.

5 Dataset

The dataset is sourced from Kaggle. The dataset has 307,511 rows and around 122 columns. The column target contains our dependent variable which we will try to predict. Other columns have information about the various financial attributes of the candidate like income, occupation, properties owned, characteristics of properties, and much more. The dataset is open for public use.

6 Methodology

This research uses the pipeline proposed by Kapcia et al. [5]. In this pipeline, we have added an additional step for the explainable AI part (Fig. 1). The outcome of this module is our results explanations. Traditionally, we have steps like data preprocessing, model training, model predictions, and model evaluation and then the best-optimized model is used to give predictions for unseen data. We have similar steps in our model. Instead of directly giving predictions, we are feeding the model generated to our explainable ai part from where results are predicted along with the explanations for the results.

a. **Data preprocessing**

The In this step we will be using various preprocessing methodologies presented by Sheikh et al. [7].

(i) Missing values

There were 2 techniques to resolve missing values issues. The first was to delete the rows with missing values. There are a lot of columns. Hence, they would also have a lot of missing values as well. Deleting rows with missing values would have ultimately led to a loss in data and variation in that data. Hence, we went with the second technique in which we assigned missing cells a value based on the type of attribute. For continuous attributes, we have assigned mean values for missing values using imputation. For categorical variables, we have assigned random values among the values available for the missing values.

(ii) Encoding

The majority of the categorical values in the dataset were attributes having binary values. For categorical attributes having binary values like gender, we have used 0

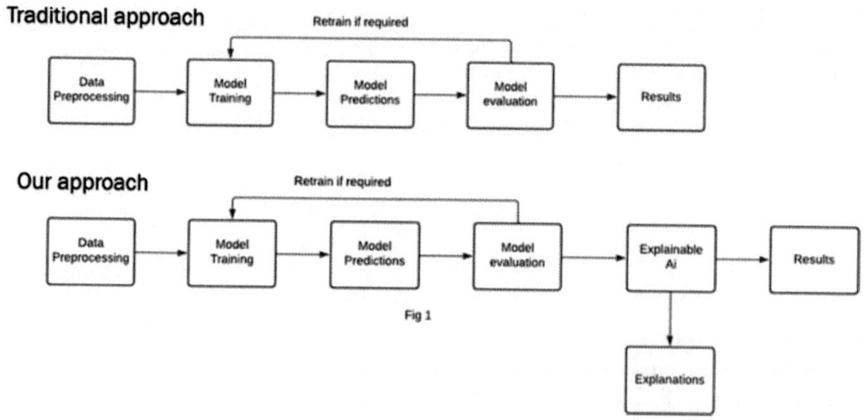

Fig. 1 Traditional approach versus our approach

and 1 to represent the different values. For other categorical variables where we have more than 2 classes of values like occupation, we have used one hot encoding.

(iii) Feature scaling

We are mainly using tree-based machine learning algorithms. Tree-based machine learning algorithms do not require feature scaling.

b. **Implementation details**

Initially, our data had a problem of class imbalance. The data set had 282,686 rows belonging to the non-defaulter class and 24,825 belonging to the defaulter class. There were 2 techniques which could solve this issue.

(i) **Down resampling**

The first technique was to resample and reduce the number of rows belonging to the majority class to match the number of rows belonging to the minority class. The problem with this approach is that it leads to the loss of valuable data, and we may lose some of the important patterns that could help our model train well. Hence, we did not adopt this approach.

(ii) **Up resampling**

The second technique was to resample the minority class to resample and duplicate the number of rows belonging to the minority class to match the number of number of rows belonging to majority class. Although we would now have duplicate data in our dataset, this would certainly not harm the model training as we are preserving the variations as well.

So, in this research, resampling has been used to solve this issue.

Ensemble Models

An ensemble machine learning model is one in which sever weak classifiers are trained and their results are combined and considered together to yield the final result. A single predictor or estimator faces challenges like high variance, low accuracy, and feature noise and bias A random forest is a decision tree-based supervised machine learning technique. A random forest has a very versatile nature. It can not only be used in classification challenges but also in regression challenges where we intend to predict a continuous value. Many decision trees come together to form a forest. This algorithm is called the random forest algorithm. The final result is anticipated by combining the results of several trees.

c. **Model Evaluation**

Our study uses 2 techniques for evaluating the model developed. The first technique is the "F1-score" which evaluates the model based on the number of current predictions by the model. The second is a graphical method known as the AUC score using the ROC curve. ROC curve helps us see how effectively our machine learning classifier is functioning.

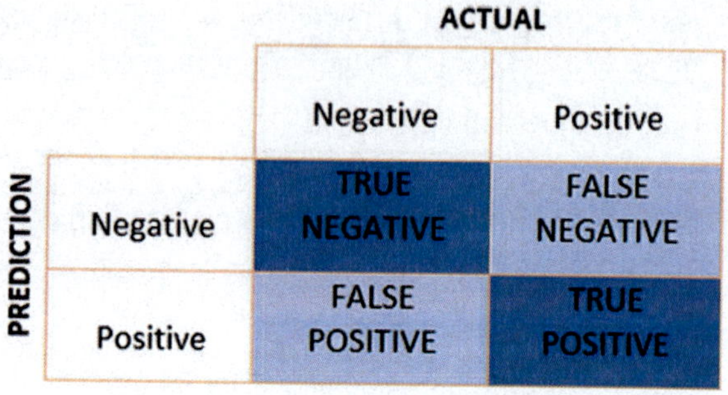

Fig. 2 Confusion matrix

Confusion Matrix

A "confusion matrix" forms the basis for evaluating the performance of any binary or multiclass classifiers. It's basically a square matrix where each dimension is n where n is equal to 2 in the case of a binary classifier or equal to the number of distinct classes in our target variable. The matrix compares the values expected in the test datasets and the value which are predicted using our trained models. This confusion matrix has a variety of evaluation metrics.

The following Fig. 2 shows a typical confusion matrix from [11].

Then using the confusion matrix, we calculated the precision and recall for the model developed.

i. Precision

"Precision" is a measure of calculating the ratio of the number of values correctly predicted positive and the number of values that are predicted positively by our model. Precision looks at how many false positives were included in the analysis. The model is 100 percent accurate if it has no false positives (FPs). The more FPs you throw in, the more inaccurate the precision becomes.

The formula for precision is:

$$\text{Precision} = (\text{number of true positives})/(\text{number of true positives} + \text{false positives})$$

ii. Recall

Recall goes in a different direction. Recall considers the number of erroneous negatives that were included in the mix rather than the number of false positives predicted by the model.

$$\text{Recall} = (\text{number of true positives}) \,/\, (\text{number of true positives} + \text{false negatives})$$

iii. F1-score

The F1 score is a combination of precision and recall. It is also known as the "harmonic mean of precision and recall". The harmonic mean is a method of determining an "average" of integers that is more suitable for ratios (such as precision and recall) than the traditional arithmetic mean. A lower total score will come from low precision or recall. As a result, it assists in balancing the two metrics. F1-score can assist balance the measure among positive and negative samples if you choose the positive class with the fewest samples. It integrates several of the other metrics into a single one, capturing numerous features at once, as seen in the first graphic in this article.

Finally, F1-score is calculated as:

$$\text{F1} - \text{score} = 2 * (\text{precision} * \text{recall}) / (\text{precision} + \text{recall})$$

Receiver operating characteristic curve (ROC Curve)

Since our classifier is a binary classifier, we can use the ROC curve to examine how well our model performed using a graphical representation. It's defined as a "curve that plots the True Positive Rates (TPR) against the False Positive Rates (FPR) at various threshold values". The "area under curve AUC" is how well our model distinguishes between classes. The greater the AUC, the better. AUC greater than 0.5 is always considered best.

7 Result and Discussion

We have trained Random Forest and XGBoost models for classification. The following table illustrates the results using a confusion matrix.

Sr. no.	Model	Accuracy (%)	F1-score
1	Random forest	99.5	0.992
2	XGBoost	80	0.75

With random forest, we have achieved a very high accuracy of 99.5% and anF1 score of 0.992. XGBoost obtained an accuracy of 80% and an F1 score of 0.75. Any accuracy beyond 80% is considered great and an F1 score above 0.7 is considered great for a classification model.

The next will be the ROC curves for random forest and XGBoost algorithms.

(1) For random forest the observer ROC curve was as follows (Fig. 3)

As visible, the random forest tree algorithm has given an AUC (area under the curve) score of 0.993. An AUC score of 0.7 and above is always considered great for classification algorithms. One reason why random forest algorithms perform best is due to their ensemble nature and ability to capture complex patterns from data with utmost ease and simplicity.

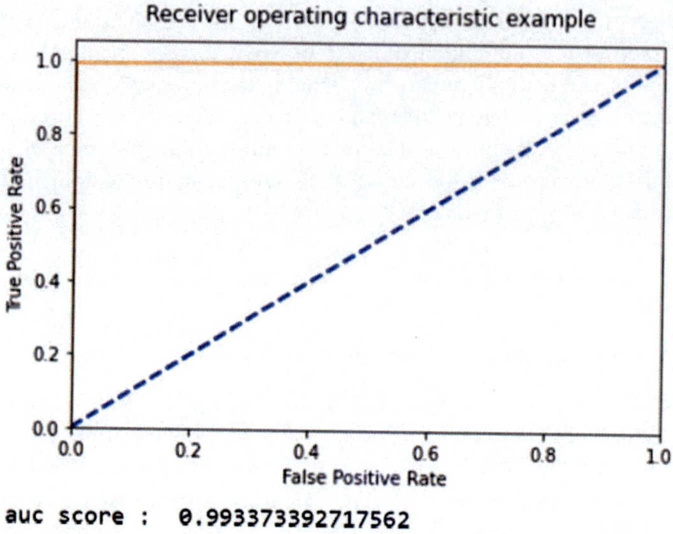

Fig. 3 ROC Curve for random forest algorithm

(2) For XGBoost the observed ROC curve was as follows (Fig. 4)

As visible, the XGBoost algorithm has given an AUC (area under the curve) score of 0.706. An AUC score of 0.7 and above is always considered great for any classification algorithm.

8 Explainable AI

1. Libraries and results

There are libraries like SHAP, LIME, SHAPASH, Eli5 and many more that are available for developing an explainable ai model. However, not all of them support non-probabilistic or tree-based machine learning algorithms. SHAPASH and LIME are the only libraries today that support developing explainable ai analysis for tree-based algorithms.

(a) LIME

Lime was one of the first interpretability methods to gain popularity. Lime is an acronym that stands for "local interpretable model agnostic explanations." This research examines Lime's technique. Lime can now explain tabular data, image, and text classifier predictions. Lime develops local surrogate models that are trained

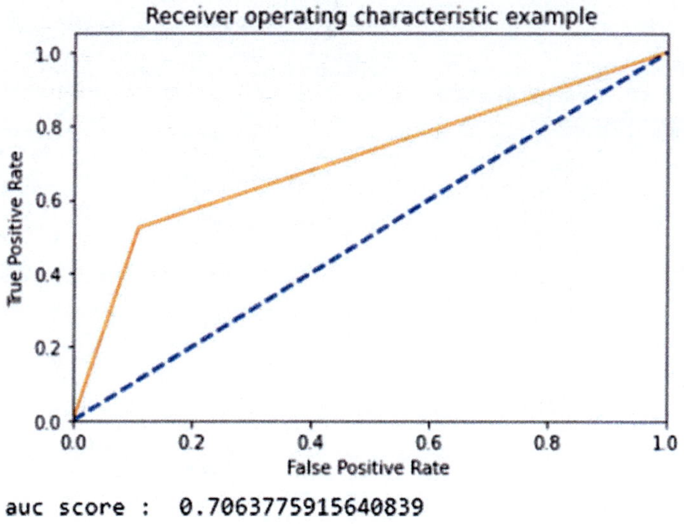

Fig. 4 ROC Curve for XGBoost algorithm

to match the ML model's predictions, resulting in a local linear approximation of the model's behaviour. While global relevance has an average effect throughout the entire data set, each variable may have a distinct impact on a local-level observation. This surrogate model can be anything from GLMs to decision trees if it tries to understand how local relevance fluctuates.

(b) SHAPASH

Compiling AI/ML results into a notebook or a web app is the best way forward for business and data scientists/analysts to communicate and engage with them. Shapash is taking the correct steps. It's a Python library developed by data scientists at the French insurance firm MAIF. This package assembles a number of SHAP/Lime explainability visuals and makes them available as a web app. The SHAP or Lime backends are used to calculate contributions. Shapash uses the several processes required to develop an ML model to make the findings intelligible. As the random forest algorithm has given us the best results, we have chosen it for developing the explanations. Following were the results obtained.

Feature importance

It is clear from the SHAPASH monitor bar chart (Fig. 5) that the feature with the highest importance is EXT_SOURCE_3. This attribute as per the column description given denotes the normalized score from external data source 3. EXT_SOURCE_2 and AMT_CREDIT which account for the normalized score from external data source 2 and the credit amount of the loan are other top attributes contributing most to the predictions made by the model. Using this inference, we can build business

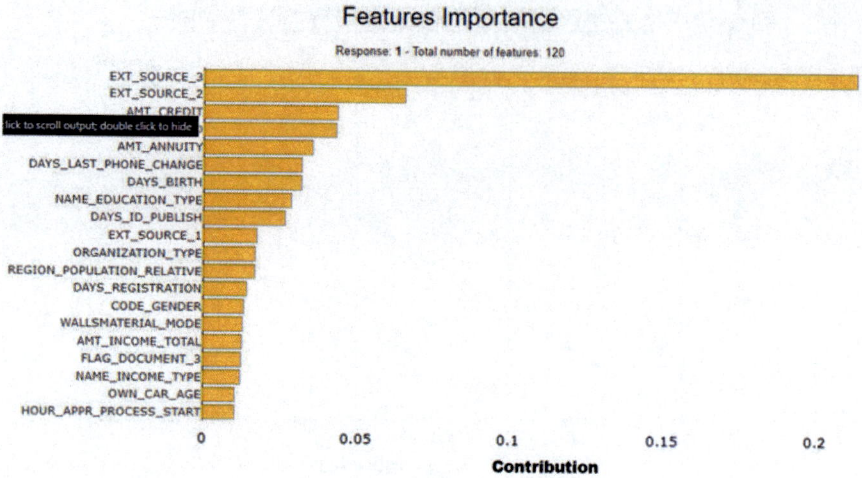

Fig. 5 Shapash Monitor Bar Chart

rules and regulations that can help improve the current business model. Stricter policies can be implemented that best judge the credibility of the individual using these attributes. These policies can also be implemented using a weighted method wherein higher weights are assigned to parameters with higher importance.

For the individual prediction explanations, we have used the lime library.

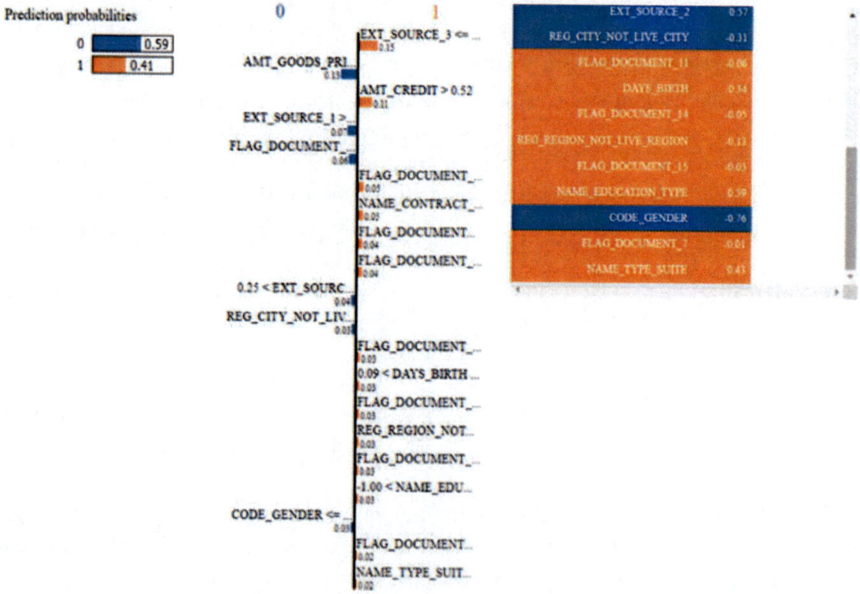

We have picked up a random sample to test the local explainability of our model. The model predicted the class as 0. 0 represents the non-defaulter class. While EXT_SOURCE_3 was one of the top contributors and was indicating class 1, all the other important attributes contributed to the class 0 probability. Hence, the class predicted was 0. This local explainability helps us understand what factors our model is considering while giving the predictions for the instance under consideration. Local explainability can be used to generate personalized recommendations for people whose loan applications are rejected by the system. To instantiate, let's say our XAI model rejected a loan application because of lower credit (amt_credit) column then suitable message can be provided to the applicant that his application was rejected due to insufficient credit. This way the applicant can improve and present a better application next time. This will also help the organization deliver better customer service and experience.

2. **Merits and Demerits**

Next, we will discuss the merits and demerits of the Xai techniques used.

(i) **Merits**

The main advantage of our model is it gives local as well as global interpretations. The global explanations will help in creating business rules and regulations while the local explanations are useful in providing personalized recommendations. The visualized results provided in form of graphs are also easy to understand for a novice user.

(ii) **Demerits**

Fitting the Shapash model to the dataset takes a lot of time. This is because our dataset has a lot of attributes, and this makes the fitting process very complicated. Explanation generation is thus a very compute-intensive process with the current model. Applying dimensionality reduction techniques and selecting the top attributes could be one option but we lose the attribute names after applying PCA. Hence tracking the reduced columns to the reduced attribute namespace would be a daunting task.

9 Future Scope and Conclusion

The model currently uses a class imbalanced dataset. More data could be collected which can certainly make the model developed in this research more accurate and powerful. The explanations generated can be used to generate rules and regulations for the organization using this model. Clustering can also be used to classify customers based on the top attribute values, and people can be classed as a low or high risk depending on the value ranges. Recommender systems could be developed based on the local explanations provided. Hence through our research, we have tried to present an efficient and easy-to-interpret solution to the black box problem discussed.

References

1. Shivanna A, Agrawal DP (2020) Prediction of defaulters using machine learning on azure ML. In: 2020 11th IEEE annual information technology, electronics and mobile communication conference (IEMCON), pp 0320–0325. https://doi.org/10.1109/IEMCON51383.2020.9284884
2. Patel B, Patil H, Hembram J, Jaswal S (2020) Loan default forecasting using data mining. In: 2020 international conference for emerging technology (INCET), pp 1–4. https://doi.org/10.1109/INCET49848.2020.9154100
3. Gahlaut A, Tushar, Singh PK (2017) Prediction analysis of risky credit using Data mining classification models. In: 2017 8th international conference on computing, communication and networking technologies (ICCCNT), pp 1–7. https://doi.org/10.1109/ICCCNT.2017.8203982
4. Pawar U, O'Shea D, Rea S, O'Reilly R (2020) Explainable AI in healthcare. In: 2020 international conference on cyber situational awareness, data analytics and assessment (CyberSA), pp 1–2. https://doi.org/10.1109/CyberSA49311.2020.9139655
5. Kapcia M, Eshkiki H, Duell J, Fan X, Zhou S, Mora B (2021) ExMed: an AI tool for experimenting explainable ai techniques on medical data analytics. In: 2021 IEEE 33rd international conference on tools with artificial intelligence (ICTAI), pp 841–845. https://doi.org/10.1109/ICTAI52525.2021.00134
6. Arutjothi G, Senthamarai C (2017) Prediction of loan status in commercial bank using machine learning classifier. In: 2017 international conference on intelligent sustainable systems (ICISS), pp 416–419https://doi.org/10.1109/ISS1.2017.8389442
7. Sheikh MA, Goel AK, Kumar T (2020) An approach for prediction of loan approval using machine learning algorithm. In: 2020 international conference on electronics and sustainable communication systems (ICESC), pp 490–494https://doi.org/10.1109/ICESC48915.2020.9155614
8. Singh V, Yadav A, Awasthi R, Partheeban GN (2021) Prediction of modernized loan approval system based on machine learning approach. In: 2021 international conference on intelligent technologies (CONIT), pp 1–4https://doi.org/10.1109/CONIT51480.2021.9498475
9. Vaidya A (2017) Predictive and probabilistic approach using logistic regression: application to prediction of loan approval. In: 2017 8th international conference on computing, communication and networking technologies (ICCCNT), pp 1–6.https://doi.org/10.1109/ICCCNT.2017.8203946
10. Ye Q, Xia J, Yang G (2021) Explainable AI for COVID-19 CT classifiers: an initial comparison study. In: 2021 IEEE 34th international symposium on computer-based medical systems (CBMS), pp 521–526. https://doi.org/10.1109/CBMS52027.2021.0010
11. Jayaswal V (2020) Performance metrics: confusion matrix, precision, recall, and F1 score. https://towardsdatascience.com/performance-metrics-confusion-matrix-precision-recall-and-f1-score-a8fe076a2262. Accessed 21 May 2022
12. Marom ND, Rokach L, Shmilovici A (2010) Using the confusion matrix for improving ensemble classifiers. In: 2010 IEEE 26-th convention of electrical and electronics engineers in Israel, pp 000555–000559. https://doi.org/10.1109/EEEI.2010.5662159

Comparative Analysis of Induction Motor Speed Control Methods

Saunik Prajapati and Jigneshkumar P. Desai

Abstract The number of industrial applications for induction motor drives is growing by the day. Various mathematical models in current control theory have explained induction motors based on the control methods used. There are several traditional inverter topologies to boost the growth of industrial applications employing induction motor driving. This study investigates MATLAB simulation for induction motor driving with different topologies such as 180° mode, Sinusoidal Pulse Width Modulation (SPWM), and Field-Oriented Control (FOC). The details of the FOC mechanism are discussed in the middle of the paper. In the end, EV dynamic modeling is described and used with the FOC method. Based on simulation investigations and results, it was concluded that FOC control is superior among all three methods used in this work. Finally, FOC-controlled induction motor has been used with the vehicle load. The results reveal that induction motor control with vehicle load is challenging.

Keywords Electrical vehicle · Induction motor · Inverter · FOC method · Control

1 Introduction

Because of their great consistency, efficiency, and closeness to ground application, induction motors are widely utilized in industry today [1]. Applications include machines, industrial mechanization, and automobiles [2, 3]. Nonetheless, the controller of induction motors is an important challenge because of their advanced mathematical models, linked torque and flux, and nonlinear behavior influenced by the working environment, such as external disturbances and parameter changes [4, 5]. Field-oriented control (FOC), commonly known as vector control, is the most widely used alternating current (AC) machine control method in high-performance trading applications [6]. Similar to a direct current (DC) machine, the FOC of an

S. Prajapati · J. P. Desai (✉)
Electrical Engineering Department, U. V. Patel College of Engineering, Ganpat University, Mehsana 384012, India
e-mail: jpd.fetr@gmail.com

H. O. Bansal et al. (eds.), *Next Generation Systems and Networks*, Lecture Notes in Networks and Systems 641, https://doi.org/10.1007/978-981-99-0483-9_36

induction motor enables torque and flux decoupling. As a basic component of FOC, the rotor flux is generally measured by a flux observer, which demands a correct motor model [7]. Any parameter change results in a flux estimation inaccuracy, which affects control performance [8]. There are several types of flux estimation models, which are often classified depending on the input signals used, such as current and voltage model observers [9, 10]. Because these flux models are sensitive to changes in motor settings, sophisticated observer methods and adaptive control algorithms may be used [1]. An adaptive sliding rotor-flux observer with an adaptive rotor resistance mechanism was described in [11]. Reference [12] defined simple and effective design requirements for induction motor rotor flux reduced-order observers, as well as a sensitivity analysis in the presence of variations in all motor parameters. Reference [4] investigated a variety of methodologies for online and offline induction machine parameter estimation, as well as parameter sensitivity analyses on full-order induction machines. To power AC loads, they can be connected to the grid and powered by DC sources such as solar panels. Because of the advantages of multilevel inverters over traditional 2- and 3-level inverters, by way of the lower price of semiconductor device power switches, an innovative cohort of inverters using an amalgamation of DC causes and switches referred to as multilevel inverters has been presented, attracting the attention of industries. Multilevel inverters generate output voltage that is more even and responds faster to sinusoidal waves when the situation has numerous levels. As a consequence, harmonics are decreased, and voltage stress is diffused owing to the usage of many switches, resulting in fewer total power losses. As a result of technical advancements in the power electronics and semiconductor sectors, adjustable speed drives (ASDs) and variable frequency drives (VFDs) are among the most efficient and trustworthy drive systems for IMs. ASDs are characterized by attractive characteristics such as consistent transient response, continuous speed control, and large energy savings [1]. ASD torque control performance is also significantly superior to DC machine drives. This is owing to amazing technical breakthroughs in digital microprocessors and Digital Signal Processors (DSP), which efficiently aid in data management.

A block schematic of an induction motor drive is shown in Fig. 1, in which the AC power supply is assumed to be the 1-phase rectifier, which converts the AC power to DC. The DC supply is then connected to the 3-phase inverter, and the inverter converts the DC supply to AC. Induction motors continue to be powered by three-phase power. In order to run an induction motor, an inverter was necessary. Induction motor characteristics may be regulated using inverter gating pulses. There are several techniques for modulating the inverter's gating pulse, including 120° and 180° conduction, sinusoidal pulse width modulation (SPWM), and space vector pulse width modulation (SVPWM).

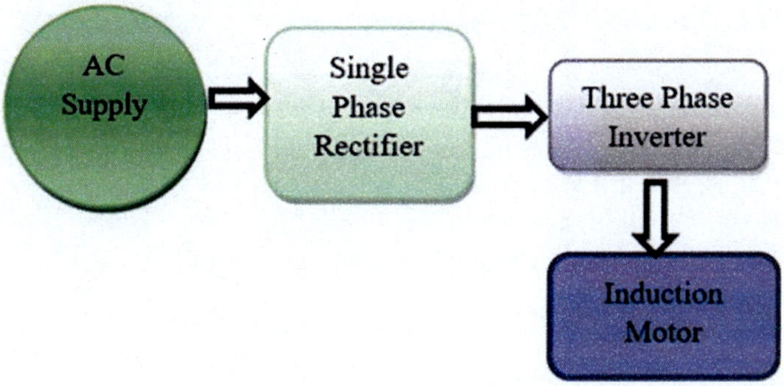

Fig. 1 Schematic block diagram of SPWM inverter-based speed control

1.1 180° Conduction Mode

The circuit shown in Fig. 2 is required for 180° conduction mode implementation. 3-phase inverters are used for VFD applications as well as high-power applications such as HVDC power transmission. A basic 3-phase inverter is made up of three single-inverter switches, each of which is connected to one of the three load terminals.

The load voltages are

$$V_{RN} = \frac{V}{3},\ V_{YN} = -2\frac{V}{3},\ V_{BN} = \frac{V}{3}$$

The line voltages are

$$V_{RY} = V_{RN} - V_{YN} = V,\ V_{YB} = V_{YN} - V_{BN} = (-\mathrm{V}),\ V_{BR} = V_{BN} - V_{RN} = 0$$

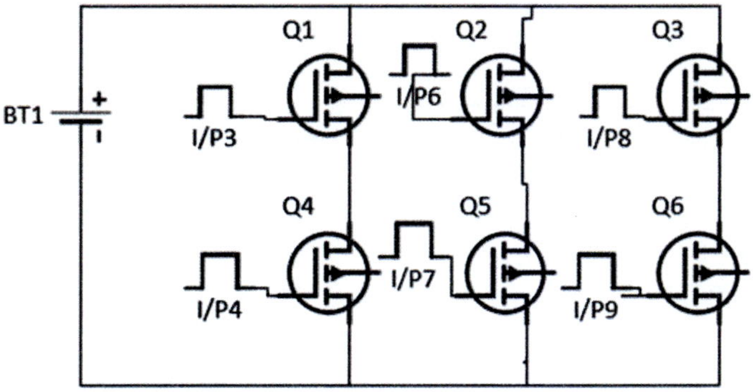

Fig. 2 Switching circuit of 180° conduction mode

1.2 Simulation of 180° Inverter

The simulation model developed for 180° conduction mode is shown in Fig. 3. The output speed response has been taken for step load change. The time domain response of speed is shown in Fig. 4. It can be seen that speed is not able to be stalled till 1 s. however, initially it achieved nearby speed in 0.15 s. Overshoot is also seen as acceptable for some suitable applications where a slight increase is not a problem.

2 Sinusoidal Pulse Width Modulation (SPWM)

Initially, carrier-based PWM systems were widely used in most applications, and they are still used today. Figure 5 depicts the circuit diagram of SPWM. For carrier-based PWM, sinusoidal PWM is the unity of the early variation signals (SPWM).

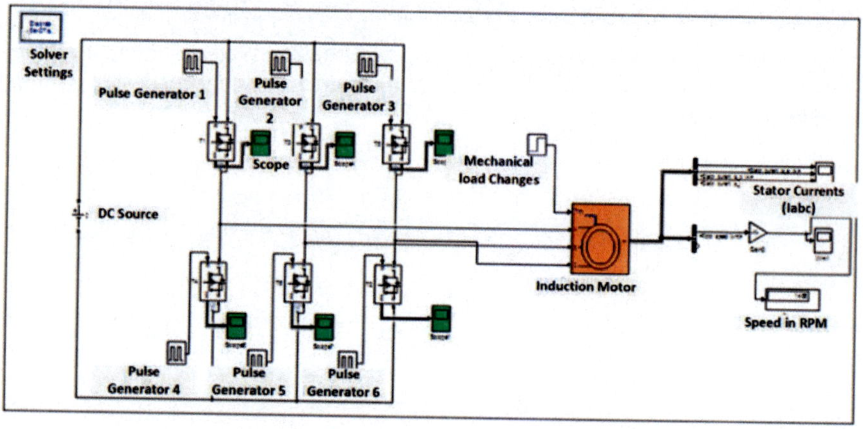

Fig. 3 MATLAB simulation circuit of 180° inverter mode

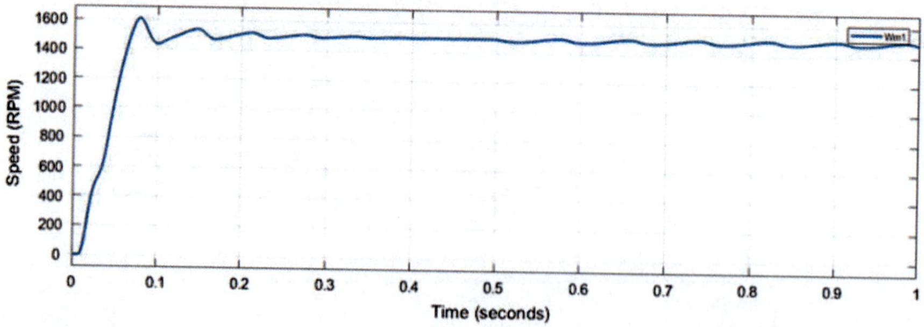

Fig. 4 Output waveform of Speed response in time domain for 180° inverter mode

In the SPWM technique, a carrier signal and a pure sinusoidal modulation signal are compared. The linkages of the circumstance waveform and the carrier waveform govern the opening and closing times of the switches. SPWM is widely used in a wide range of applications, including motor speed control, converters, and audio amplifiers. The MATLAB simulation implementation is shown using Fig. 6. Despite the fact that the SPWM technique is the easiest to understand and implement in a software system or hardware, the situation is insufficient to fully utilize the DC bus supply power available to the voltage source inverter. A triangular voltage waveform (V_r) is equivalent to three sinusoidal control voltages that remain 120 degrees out of phase in a three-phase SPWM, and the waveform's comparative levels are used to regulate the switching of devices in each phase leg of the inverter. Figure 6 represents a MATLAB simulation of an SPWM inverter in which gating signals are given by a sinusoidal pulse width modulation scheme. As compared to 180° conduction mode, in SPWM mode, the overshoot observed is quite less. However, huge speed ripples have been seen in Fig. 7; this may be owing to substantially higher switching frequencies, which cause larger strains on related switching devices and, as a result, de-rating of such devices which produce high-frequency harmonic components.

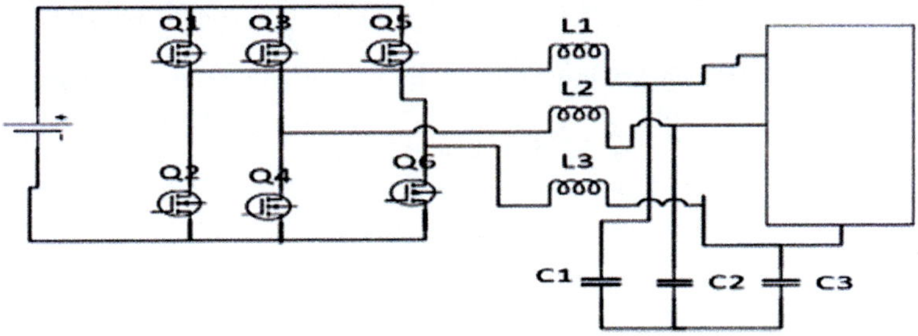

Fig. 5 SPWM Inverter circuit diagram for motor

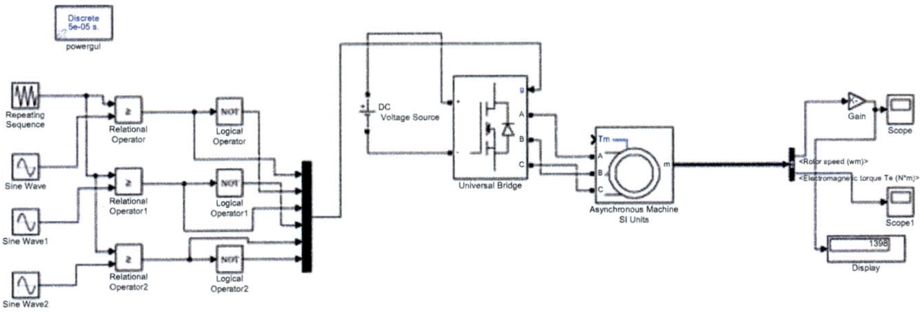

Fig. 6 MATLAB simulation circuit of SPWM Inverter control induction motor drive

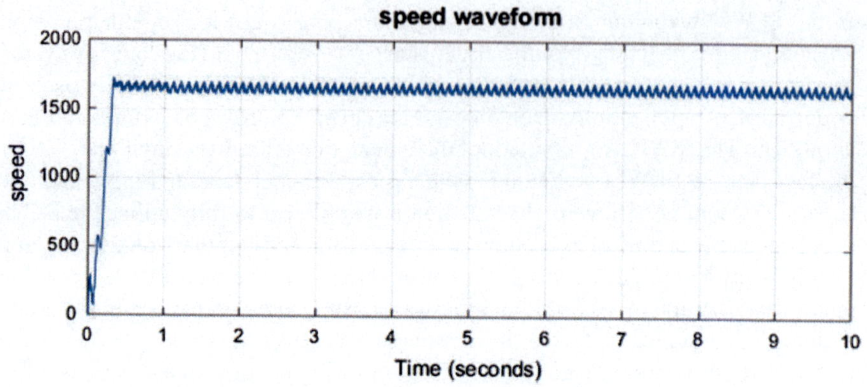

Fig. 7 Speed response of SPWM control in time domain

3 Field-Oriented Control (FOC) of Induction Motor

3.1 *Equation of Field-Oriented Control*

The vector control of a three-phase system requires solving complex equations. In this method, voltage of all three phases is transformed into a 2-phase stationary system (α–β) first, and then transformed into a revolving system that is often known as d–q coordinates. Now the equation of stator voltage (V_{abcs}) and rotor voltage (V_{abcr}) can be written as

$$\mathrm{V}_{abcs} = (\mathrm{r_s} \times \mathrm{i}_{abcs}) + \left(\frac{d\lambda_{abcs}}{dt}\right) \tag{1}$$

$$\mathrm{V}_{abcr} = (\mathrm{r_r} \times \mathrm{i}_{abcr}) + \left(\frac{d\lambda_{abcs}}{dt}\right) \tag{2}$$

where $\mathrm{r_s}$ = stator winding resistor; $\mathrm{r_r}$ = rotor winding resistor; $\frac{d\lambda_{abcs}}{dt}$ = change in flux linkage to stator $= \mathrm{L_s} \times \mathrm{i}_{abcs} + \mathrm{L_s} \times \mathrm{i}_{abcr}$; $\frac{d\lambda_{abcr}}{dt}$ = change in flux linkage to rotor $= \mathrm{L_r} \times \mathrm{i}_{abcs} + \mathrm{L_r} \times \mathrm{i}_{abcr}$.

While matrix $\mathrm{L_s}$ and $\mathrm{L_r}$ can be given as

$$\mathrm{L_s} = \begin{bmatrix} \mathrm{L_{ls}} + \mathrm{L}_{ms} & -\frac{1}{2}\mathrm{L}_{ms} & -\frac{1}{2}\mathrm{L}_{ms} \\ -\frac{1}{2}\mathrm{L}_{ms} & \mathrm{L_{ls}} + \mathrm{L}_{ms} & -\frac{1}{2}\mathrm{L}_{ms} \\ -\frac{1}{2}\mathrm{L}_{ms} & -\frac{1}{2}\mathrm{L}_{ms} & \mathrm{L_{ls}} + \mathrm{L}_{ms} \end{bmatrix}$$

$$\mathrm{L_r} = \begin{bmatrix} \mathrm{L}_{lr} + \mathrm{L}_{mr} & -\frac{1}{2}\mathrm{L}_{mr} & -\frac{1}{2}\mathrm{L}_{mr} \\ -\frac{1}{2}\mathrm{L}_{mr} & \mathrm{L}_{lr} + \mathrm{L}_{mr} & -\frac{1}{2}\mathrm{L}_{mr} \\ -\frac{1}{2}\mathrm{L}_{mr} & -\frac{1}{2}\mathrm{L}_{mr} & \mathrm{L}_{lr} + \mathrm{L}_{mr} \end{bmatrix}$$

the equation of clerk transform can be represented as

$$\begin{aligned} i_{\acute{\alpha}} &= i_a \\ i_{\beta} &= \frac{1}{\sqrt{3}} \times i_a + \frac{2}{\sqrt{3}} \times i_{\beta} \end{aligned} \tag{3}$$

The inverse of the clerk transform is mentioned in the equation

$$\begin{aligned} i_a &= i_{\acute{\alpha}} \\ i_b &= -\frac{1}{2} \times i_{\acute{\alpha}} + i_{\beta} \times \frac{\sqrt{3}}{2} \\ i_c &= -\frac{1}{2} \times i_{\acute{\alpha}} - i_{\beta} \times \frac{\sqrt{3}}{2} \end{aligned} \tag{4}$$

The equation of rotating frame transformation using park transform can be written as

$$\begin{aligned} i_{ds} &= i_a \times \cos\theta + i_b \times \sin\theta \\ i_{qs} &= -i_a * \sin\theta + i_b \times \cos\theta \end{aligned} \tag{5}$$

Now inverse park transform is given by

$$\begin{aligned} i_{\acute{\alpha}} &= i_{ds} \times \cos\theta - i_{qs} \times \sin\theta \\ i_{\beta} &= -i_{ds} \times \sin\theta + i_{qs} \times \cos\theta \end{aligned} \tag{6}$$

The equation of direct axis current is given by

$$i_{ds} = \frac{\lambda_r}{L_m} \tag{7}$$

The equation for the quadrature-axis reference current is given by

$$i_{qs} = \frac{2}{3} \times \frac{2}{P} \times \frac{L_r}{L_m} \times \frac{T_e}{\lambda_r} \tag{8}$$

The rotor flux angle is achieved by the calculation of the speed of rotation of the rotor speed (wr) and slip (ω_{sl}) which is obtained by

$$\theta_e = \int (\omega_m + \omega_{sl}) \tag{9}$$

Now slip speed is obtained by

$$\omega_{sl} = \frac{L_m}{\lambda_r} \times \frac{R_r}{L_r} \times i_{qs} \tag{10}$$

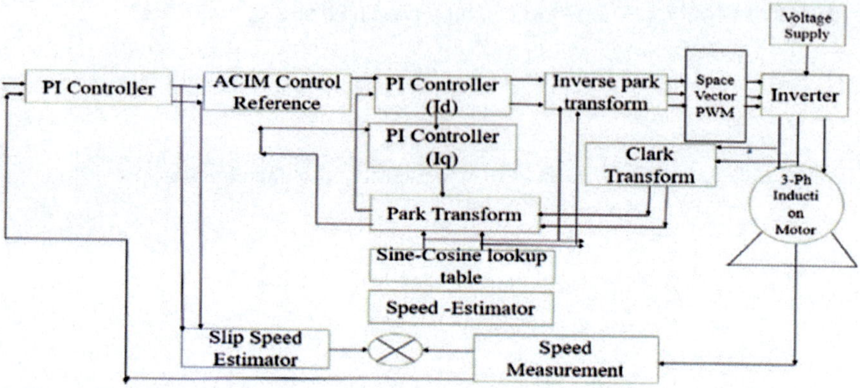

Fig. 8 Schematic block diagram of FOC for induction motor

3.2 *Block Diagram of FOC*

Figure 8 depicts a block schematic of the FOC technique for driving induction motors, which includes numerous computations such as *d*- and *q*-axis current and so on, in which ACIM control reference calculates the reference value of FOC's *d*-axis and *q*-axis currents. The block receives the torque reference value and provides mechanical speed and output corresponding to the d-axis and q-axis current as feedback. By solving the mathematical equation, this block computes the base value of the *d*- and *q*-axis currents.

3.3 *Simulation of FOC for Induction Motor*

FOC simulation in MATLAB using PI controller is shown in Fig. 9 which contains ACIM control reference, 3-phase induction motor, and subsystems of all transformations such as park, clerk, and so on. In this case, an inverter is employed to deliver power to the induction motor. Simulink's universal bridge is utilized for the inverter.

Figure 10 depicts the speed waveform of the FOC method. Because there are fewer harmonics and ripples, this waveform is quite smooth when compared to the 180° conduction and SPWM control method.

4 EV Model Load Modeling

After simulating FOC with an induction motor as load and achieving proper control as compared to the previous two methods, it is decided to implement FOC with EV load application. The modeling of vehicle dynamics includes factors such as the

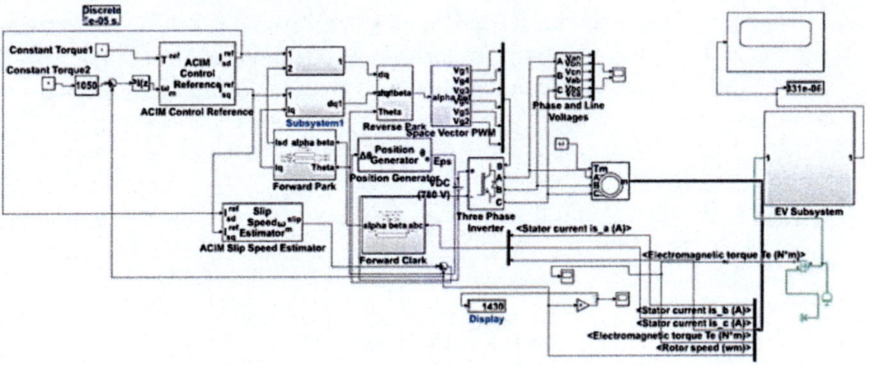

Fig. 9 MATLAB simulation of FOC for 3-phase induction motor

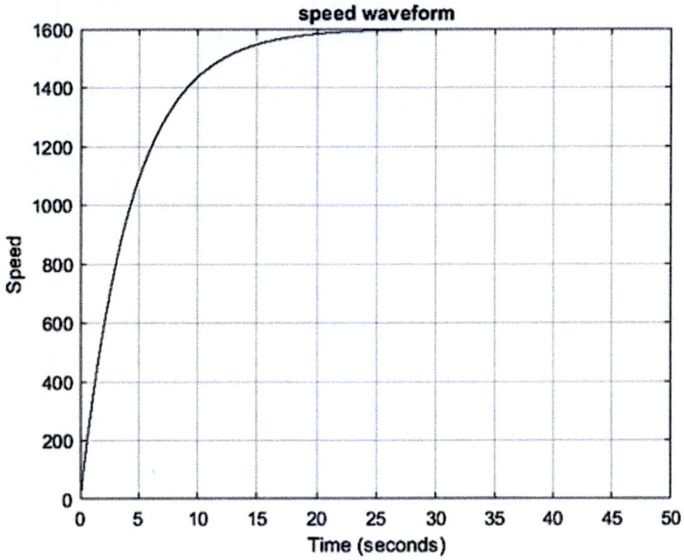

Fig. 10 Speed waveform in time domain for FOC of 3-phase induction motor

weight of the vehicle, aerodynamic drag resistance, rolling resistance, and gradient resistance. For selecting a motor rating for an electric vehicle of gross weight 170 kg is considered. The force required for driving a vehicle is calculated as

$$F_T = F_r + F_g + F_a \tag{11}$$

where F_r = Force due to rolling resistance, F_g = Force due to gradient resistance, F_a = Force due to aerodynamic drag, and F_T = Total tractive force.

The output of the motor must overcome F_T in order to move the vehicle. Rolling resistance is the resistance offered to the vehicle due to the contact of the tire with the road. The formula for calculating force due to rolling resistance is given by

$$F_r = C_{rr} \times M \times g \tag{12}$$

where C_{rr} = Coefficient of rolling resistance, M = mass in kg, and g = acceleration due to gravity = 9.81 m/s^2.

Aerodynamic drag is given by

$$F_a = 0.5 \times C_D \times A_f \times \rho \times v^2 \tag{13}$$

where C_D = Drag coefficient; A_f = Frontal area; ρ = Air density (kg/m^3); V = Velocity (m/s).

$$F_g = \pm M \times g \times \sin\theta$$

where g = Acceleration due to gravity = 9.81 m/s^2.

$$\text{Poweroutput} = \text{Force} \times \text{Velocity} \times (1000 \div 3600) \tag{14}$$

From Eqs. (11)–(14), values for vehicle dynamics are calculated as shown in Table 1.

Using Eqs. (11)–(14) and Table 1, simulation subsystem has been modeled which is shown in Fig. 11. There are two tires connected to the vehicle body. This means that this depicts a vehicle with differential mechanical drive. The model is connected with an induction motor drive and the drive is controlled with the help of the FOC method as compared to 180° conduction and SPWM; it has been found superior in our analysis.

Table 1 Vehicle dynamics parameters values

Parameter	Selected value	Parameter	Calculated value
θ (inclined angle)	2.5°	F_r	11.772 N
A_f = Frontal area	0.7	F_g	128.37 N
V = Velocity in kmph	35	F_a	19.2606 N
C_{rr}	0.4	F_T	159.402 N
g = acceleration due to gravity	9.81 m/s^2	Power	1549 W
Mass of electrical vehicle	300 kg	Motor Power	2238 W
C_D	0.5	Voltage of motor	460 V
Air density(ρ)$\left(\frac{Kg}{m}\right)$	1.1644	Current of motor	4.86 A

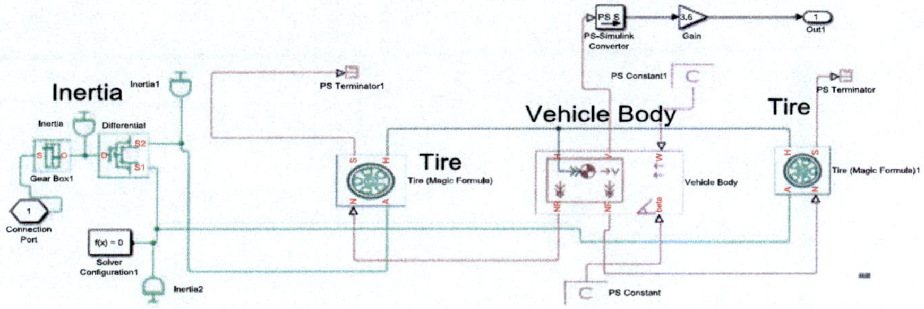

Fig. 11 EV Subsystem modeling

Table 2 Comparison of all different methods implemented

Sr. no.	180° mode	SPWM	FOC
1	72% THD	31% THD	18% THD
2	Switching loss high	Switching loss low	Switching loss very low

5 Results and Discussions

Results of Table 2 show the total harmonic distortion production in three different methods compared in this work. As FOC gives the lowest THD, the EV modeling-based load is connected with the FOC method only. Figure 12 shows the response of speed in the time domain for the EV model which shows that the machine started with a somewhat low speed and gradually achieve a higher speed. However, the time taken by the FOC control to achieve such a rated speed is quite unreasonable and required fast control.

6 Conclusion

The performance of the 180° mode inverter, SPWM inverter, and FOC technique for induction motor driving is compared in this study. When comparing percent of THD and overshoot performance, the SPWM inverter has larger switching losses. The simulation results reveal that the speed waveform of SPWM has a higher overshoot than the field-oriented control (FOC) technique. By comparing all of the aforementioned results, it is possible to infer that the field-oriented approach (FOC) is a good option for induction motor drive, with improved dynamic performance. Hence, EV load with calculated dynamics has been connected to test FOC performance. The implementation of FOC-based control with vehicular load demonstrates that speed control has ripples when vehicle loading variance and higher speed are achieved with considerable time.

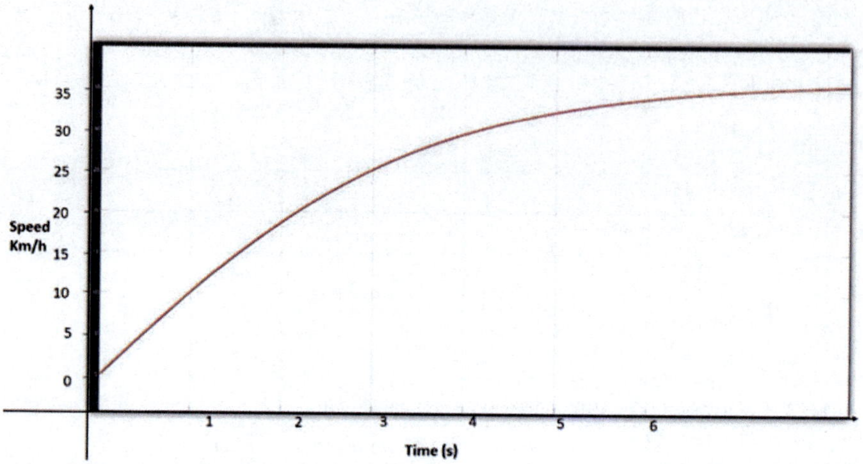

Fig. 12 Speed of EV in km/h in time domain

References

1. Varshney A, Sharma U, Singh B (2020) A grid interactive sensorless synchronous reluctance motor drive for solar powered water pump for agriculture and residential applications. In: m2020 IEEE international conference on power electronics, drives and energy systems (PEDES), pp 1–6. https://doi.org/10.1109/PEDES49360.2020.9379885
2. Singh B, Shukla S, Chandra A, Al-Haddad K (2016) Loss minimization of two stage solar powered speed sensorless vector controlled induction motor drive for water pumping. IECON 2016:1942–1947
3. Kumar R, Singh B (2020) Position sensorless BLDC motor drive for single stage PV based water pumping. IEEE
4. Murshid S, Singh B (2020) An improved SMO for position sensorless operation of PMSM driven solar water pumping system. In: N2020 IEEE international conference on power electronics, smart grid and renewable energy (PESGRE2020), pp 1–5
5. Appaso P, Gaikvad A (2016) Comparative study of 3 and 5 level inverter. IJAREEIE
6. Bhimbra PS (2012) Power electronics, 4th ed. Khanna Publishers
7. Rai R, Shukla S, Singh B (Sep–Oct 2020) Sensorless field oriented SMCC based integral sliding mode for solar PV based induction motor drive for water pumping. IEEE Trans. Ind. Appl. 56(5):5056–5064
8. Shukla S, Singh B Single-Stage PV (July–Aug 2018) Array fed speed sensorless vector control of induction motor drive for water pumping. IEEE Trans. Ind. Appl. 54(4):3575–3585
9. Noroozi MA, Moghani lS, Mili Monfared l, Givi H (2012) Sensorless starting method for Brushless DC motors using 180 degree commutation. IEEE Catlog
10. Alahmad A, Kaçar F (2021) Simulation of induction motor driving by bridge inverter at 120°, 150°, and 180° operation. In: 2021 8th international conference on electrical and electronics engineering (ICEEE), pp 121–125
11. Loss minimization of two stage solar powered speed sensorless vector controlled induction motor drive for water pumping
12. Kumar R, Singh B (2020) Position sensorless BLDC motor drive for single stage PV based water pumping. In: 2020 IEEE 5th international conference on computing communication and automation (ICCCA), pp 634–639

13. Lulhe AM, Date TN (2015) A design & MATLAB simulation of motor drive used for electric vehicle. In: 2015 International conference on control, instrumentation, communication and computational technologies (ICCICCT), pp. 739–743

A Comparative Investigation of Deep Feature Extraction Techniques for Video Summarization

Bhakti D. Kadam and Ashwini M. Deshpande

Abstract Video summarization refers to creating a temporally abridged version of video containing all the important highlights required for context understanding without the loss of original information. Segmentation and feature extraction are the two pre-processing tasks to be carried out on the input video sequence in any summarization framework. Segmentation divides the video into non-intersecting temporal segments while feature extraction process represents entire video in the form of feature vectors. This paper investigates video feature extraction using pre-trained deep neural networks, viz., GoogleNet, ResNet, and ResNeXt. These deep networks are employed to extract feature vectors from the video frames and the extracted features are used to summarize videos using summarization models. The efficacy of these deep networks for feature extraction step is then compared in terms of F1-scores of summarized videos. Our experimentation revealed that the performance of a deep network for feature extraction depends on nature of videos and summarization approaches. It is observed that GoogleNet is optimum choice for feature extraction in video summarization application.

Keywords Deep neural networks · Feature extraction · Segmentation · Video summarization · F1-score

B. D. Kadam (✉) · A. M. Deshpande
Department of Electronics & Telecommunication Engg., MKSSS's Cummins College of Engineering for Women, Pune, Maharashtra, India
e-mail: bhakti.kadam@cumminscollege.in

A. M. Deshpande
e-mail: ashwini.deshpande@cumminscollege.in

B. D. Kadam
Department of Electronics & Telecommunication Engg., SCTR's Pune Institute of Computer Technology, Pune, Maharashtra, India

H. O. Bansal et al. (eds.), *Next Generation Systems and Networks*, Lecture Notes in Networks and Systems 641, https://doi.org/10.1007/978-981-99-0483-9_37

1 Introduction

Video has become a common communication medium for sharing of information with proliferation of Internet [1]. The task of automatic video summarization refers to creating a temporally abridged version of the video containing all the important highlights required for context understanding without the loss of original informativeness. The process of generic video summarization system is shown in Fig. 1. The key components of any summarization framework are segmentation and feature extraction of input video sequence, importance score computation and selection of keyframes or keyshots. This paper focuses on deep learning-based feature extraction techniques used in video summarization.

Segmentation and feature extraction play an important role in video summarization frameworks as it helps in concise representation of given video. Video feature extraction is a process of dimensionality reduction as shown in Fig. 2 in which the entire input video sequence is represented in terms of feature vectors. The huge amount of available video datasets results in more computational resources and time in various video-processing applications. Feature extraction techniques are the methods that select and/or combine variables into features or feature vectors that effectively reduce the amount of data that must be processed maintaining the accuracy in representation of videos.

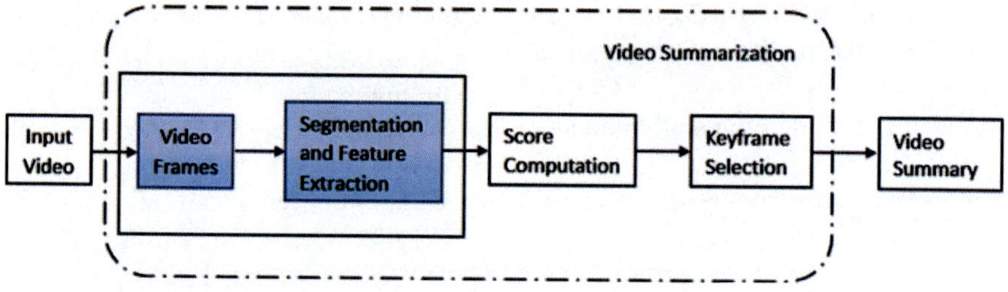

Fig. 1 A generic video summarization system

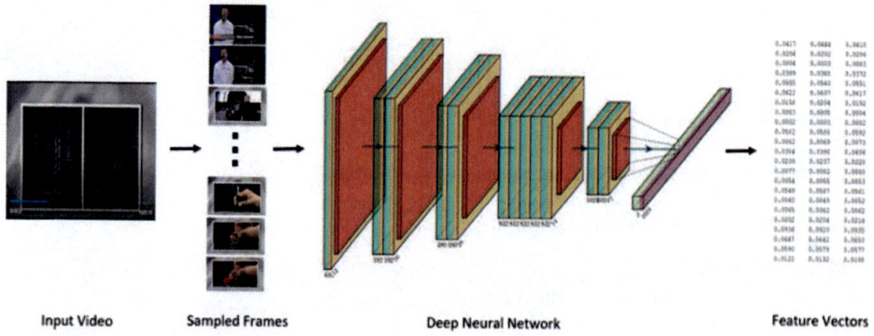

Fig. 2 Video feature extraction using deep neural network

The fundamental step to extract features from video is to split or segment the entire video sequence into frames. The features can be extracted using classical approaches exploring local hand-crafted features as well as state-of-the-art deep neural networks. The commonly considered video features are space-time, spatio-temporal, global motion, dense motion, and optical flow [1]. The motion vectors and feature vectors are also extracted as per the required video representation.

Feature extraction plays an important role in many applications including video representation and understanding due to following advantages: (i) reduction in computational requirements as it reduces the number of resources needed for processing of images or videos without losing important or relevant information, (ii) reduction in the amount of redundant data for an efficient analysis, (iii) simplification of learning processes by reducing the amount of data and the machine's efforts in modeling variable combinations (features), and (iv) improvement in the accuracy of proposed techniques or models.

Different video-processing applications require different kinds of features to be extracted from video data. Spatial features, temporal features, i.e., optical flow, textual features [2], trajectories, and content-based features are commonly considered in analysis of videos in most of the computer vision tasks. The structure of the paper is as follows: Section 2 describes some commonly used deep learning models from literature employed for the task of video feature extraction. The detailed implementation and obtained results are presented in Section 3 and conclusions are provided in Section 4.

2 Deep Learning Models for Feature Extraction

Video summarization and video feature extraction are the long-standing research areas in the field of computer vision. The video features can be extracted using conventional techniques such as histogram of gradient (HoG), space-time interest point (STIP), and scale-invariant feature transform (SIFT) which deals with hand-crafted features. With rapid growth in the area of deep learning, many deep neural networks are developed for extraction of features from videos such as convolutional neural network (CNN), RNN, LSTM, and feature fusion models. Video segmentation and feature extraction reduce the computational overheads by reducing the pre-processing of all the frames in the video.

Many deep learning-based approaches have been proposed in literature for extracting feature vectors from video data. Some of the methods include two-stream inflated 3D ConvNet (I3D) [3], 3D convolutional network (C3D) [4], two-stream networks using combinations CNNs or LSTMs [5, 6], and hybrid multi-modal networks [7, 8].

In video summarization frameworks proposed in literature, GoogleNet is the most common deep network employed for extracting the feature vectors. In addition to GoogleNet, feature vectors are extracted using ResNet152, ResNeXt and fusion of GoogleNet with ResNet and ResNeXt during the experimentation and comparative analysis is presented. The details of deep networks explored during the investigation are

- **GoogleNet**: GoogleNet (or Inception V1) was introduced by researchers at Google in 2014 [9]. It is a deep neural network designed for image classification task with significant decrease in error rate. The network is 22 layers deep having convolution, pooling, inception, and classification layers. In video summarization frameworks, GoogleNet pre-trained on ImageNet dataset [10] is the most commonly used network to extract features of input video sequence. The last two classifier layers are excluded and output is taken from dropout layer.
- **ResNet152**: ResNet was proposed in 2015 by researchers at Microsoft Research which has introduced a new architecture called Residual Network [11]. When the depth of deep neural network increases, training error increase leading to degradation of accuracy. The authors have provided solution to the problem of degrading accuracy and vanishing gradients by introducing residual blocks and skip connections. The last softmax classifier layer is excluded taking the features from the output of convolution layer.
- **ResNeXt**: ResNeXt was developed by researchers at UC San Diego and Facebook AI Research in 2015 [12]. The network has added a next cardinality dimension on top of the ResNet. The network is constructed "by repeating a building block that aggregates a set of transformations with the same topology". This leads to improvement in performance and decrease in error rate. The last softmax classifier layer is excluded taking the features from the output of convolution layer. For feature extraction, ResNeXt-50 (32×4d) is used, where 32 indicates grouped convolutions with 32 groups.
- **Fusion of GoogleNet and ResNet152**: The feature vectors are extracted using both GoogleNet and ResNet152 and combined at the output layers.
- **Fusion of GoogleNet and ResNeXt-50 (32×4d)**: The feature vectors are extracted using both GoogleNet and ResNeXt-50 (32×4d) and combined at the output layers.

3 Experimentation and Results

This section provides details of experimental settings (datasets, segmentation methods used and metrics). Then, the quantitative results are discussed comparing performance of proposed feature extraction methods using existing GoogleNet.

3.1 Datasets

The performance of proposed feature extraction techniques is evaluated on two benchmark summarization datasets, SumMe [13] and TVSum [14]. The SumMe dataset has 25 videos of various genres such as sports, holidays, and cooking. The TVSum dataset consists of 50 videos from YouTube covering 10 categories, including documentary, educational, and egocentric. Both datasets are provided with multiple user annotations in terms of user selected keyframes and shot level importance scores. As the number of videos in both the datasets are comparatively less, another two datasets, i.e., OVP [15] and YouTube [15], are used to augment the training datasets. This helps in tackling the requirement large training dataset. The OVP dataset has 50 videos and the YouTube dataset is composed of 39 videos. The key statistics of these datasets are presented in Table 1.

3.2 Segmentation Techniques

- **Uniform Sampling**: Uniform sampling is the most basic approach used in video summarization for segmentation and keyframe extraction [16]. In this segmenta-

Table 1 Key statistics of datasets for video summarization

Dataset	Number of videos	Duration	Specifications	Types of annotations	Number of annotations per video
SumMe [13]	25	1-6 min	The videos covering holidays, sports, and events in webm or mp4 format.	User selected keyframes	15
TVSum [14]	50	1–4 min	The videos are in MPEG-1 format with 30 fps (352 × 240 pixels) distributed among several genres such as documentary, educational, ephemeral, historical, lecture, and egocentric.	Shot level importance scores	20
YouTube [15]	50	1–10 min	The videos of various genres like cartoons, sports, TV shows, commercials, and home videos.	Multiple sets of keyframes	5
OVP [15]	50	1–4 min	The videos of various genres like documentary, educational, historical, etc.	Multiple sets of keyframes	5

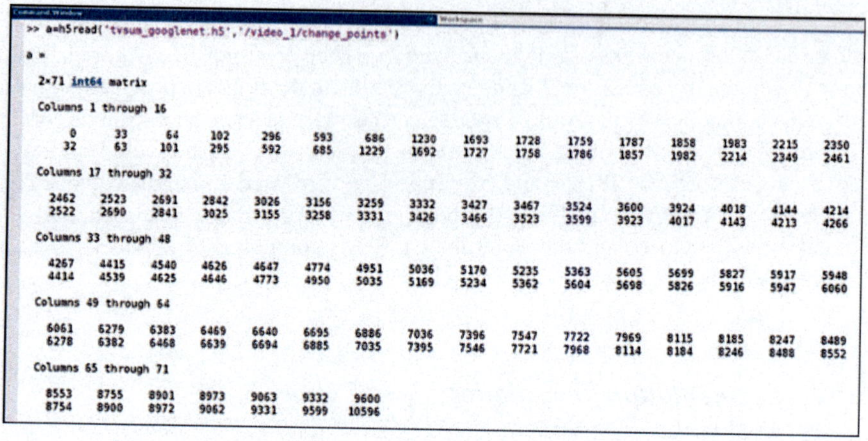

Fig. 3 Change points detected using KTS for video_1 from TVSum

tion method, every kth frame is selected where the value of k is decided as per required number of frames to be extracted from video. This is a very fundamental approach which does not consider any semantic relationship between the frames. Segmentation with sample rate of 15 is used before extracting the features, i.e., features are extracted from every 15th frame of input video with consideration that there are very few significant changes within a group of 15 frames in a video. This reduces pre-processing of all frames decreasing computational overheads.

- **Kernel Temporal Segmentation**: Kernel Temporal Segmentation (KTS) is a method to split video into a set of non-intersecting temporal segments [17]. It is a statistical framework based on change point detection. The method is more accurate when operated on high-level descriptors. When KTS is used in video summarization, the matrix of extracted feature vectors from input video sequence is considered as an input. The change points are computed and the algorithm outputs the frame numbers where significant changes are observed. The method is applied on videos after extracting the features to segment the video considering semantic relationships between the frames. The sample change point positions using KTS for video_1 from TVSum dataset are shown in Fig. 3. It can be seen from the obtained change points that the first noticeable change occurs at frame number 32, i.e., first video segment is from frame number 0–32, second segment is from frame number 33 to 63, and so on.

3.3 Evaluation Metrics

The performance of summarization frameworks is measured in terms of the F1-score [18]. The F1-score is calculated between predicted and reference summaries using

Eqs. 1, 2 and 3. The summarization datasets are provided with user annotations which are considered as reference summaries while computing F1-scores.

Consider $x_i = \{0, 1\}$ and $x_i^* = \{0, 1\}$ be the labels indicating the frames included in the final predicted and reference summaries, respectively. Let, $x_i = 1$ if the i-th frame is included in summary otherwise 0. F1-score is defined as

$$F1 - score = \frac{2 * Pre * Rec}{Pre + Rec}, \tag{1}$$

where

$$Pre = \frac{\sum_{i=1}^{N} x_i * x_i^*}{\sum_{i=1}^{N} x_i} \tag{2}$$

and

$$Rec = \frac{\sum_{i=1}^{N} x_i * x_i^*}{\sum_{i=1}^{N} x_i^*} \tag{3}$$

are the frame-level precision and recall scores and N is the total number of frames in the original video sequence.

3.4 Results

This section investigates the obtained results. Two deep learning-based video summarization models, diversity representative deep summarization network (DR-DSN) [19], and self-attention mechanism (VASNet) framework [20] are used to evaluate the performance of feature extraction methods.

The input video sequences are downsampled from 30 fps to 2 fps for feature extraction, i.e., every 15th frame is used to extract the features. This reduces temporal redundancy and computational time. In most of the video summarization frameworks, GoogleNet pre-trained on ImageNet dataset [10] is employed to extract the feature vectors. It outputs a feature vector of size $1024 \times N$ where N depends on number of video frames or sampling rate of the video sequence. In addition to GoogleNet, feature vectors are also extracted using two other deep networks and the results are compared. ResNet152 and ResNeXt-50 (32×4d) pre-trained on ImageNet dataset are used for feature extraction. A feature vector of size $2048 \times N$ is extracted at the output excluding the last classifier layer where N depends on number of video frames or sampling rate of the video sequence. From 2048 extracted features, 1024 features are considered for uniformity and fair performance comparison. Additionally, video features are also extracted by fusing GoogleNet with ResNet152 and ResNeXt leveraging benefits of both deep networks.

Tables 2 and 3 provide the comparison of F1-scores of the DR-DSN summarization framework on the two publicly available benchmark summarization datasets,

Table 2 Comparison of F1-scores of DR-DSN evaluated on TVSum

Sr. no.	Video ID	GoogleNet	ResNet152	ResNeXt	GoogleNet + ResNet152	GoogleNet + ResNeXt
1	Video_1	65.2	68.2	69.4	67.3	69.4
2	Video_12	65.4	62.8	65.4	62.6	62.6
3	Video_18	70.6	64.4	66.0	69.9	65.5
4	Video_27	54.8	41.0	37.9	43.5	42.7
5	Video_29	55.2	54.3	54.9	54.9	54.8
6	Video_42	59.5	63.8	65.0	65.8	65.8
7	Video_45	72.7	72.7	72.7	72.7	72.7
8	Video_5	71.7	64.6	58.7	64.3	71.7
9	Video_6	66.7	58.7	62.3	62.5	63.4
10	Video_7	57.5	50.5	56.1	50.0	55.5
	Average	63.9	60.1	60.8	61.4	62.4

Table 3 Comparison of F1-scores of DR-DSN evaluated on SumMe

Sr. No.	Video ID	GoogleNet	ResNet152	ResNeXt	GoogleNet + ResNet152	GoogleNet + ResNeXt
1	Video_1	60.0	57.2	59.4	57.3	59.4
2	Video_16	28.6	34.9	34.9	35.1	36.2
3	Video_22	49.2	44.4	46.0	49.9	55.5
4	Video_23	57.8	54.0	58.3	59.5	62.7
5	Video_4	49.3	39.2	28.0	38.2	40.6
	Average	49.0	45.9	45.3	48.0	50.8

TVSum, and SumMe, respectively. Table 4 on the similar lines provides the comparison of the F1-scores of VASNet framework [20].

The videos from both the dataset belong to various categories. After analyzing the obtained F1-scores, it can be said that GoogleNet performs better when the videos are having less dynamic motion, i.e., the frames are changing slowly with more spatial and temporal redundancy. For videos with rapid changes in frames having less spatial and temporal redundancy, ResNet152 or ResNeXt is good choice for extraction of feature vectors.

In addition to this, it can be noted from Table 4 that the fusion of deep networks yields comparatively better results (6% improvement in F1-scores) for VASNet summarization model. It indicates that summarization method also contributes to the efficacy to feature extraction technique.

Table 4 Comparison of F1-scores of VASNet

	GoogleNet	ResNet152	ResNeXt	GoogleNet + ResNet152	GoogleNet + ResNeXt
TVSum	61.56	61.43	61.06	65.07	65.11
SumMe	46.99	45.78	45.34	50.7	51.2

4 Conclusions

In this paper, a comparative analysis of video feature extraction using deep neural networks for an application of video summarization is carried out. Unlike existing method of using only GoogleNet for feature extraction, feature vectors are extracted using ResNet, ResNeXt, and fusion of GoogleNet with ResNet and ResNeXt. The performance comparison of these deep networks is presented in terms of F1-scores. The experimentation shows that GoogleNet performs better for one of the summarization models while fusion of networks gives good results for another summarization model. So, it can be concluded that the feature extraction technique should be selected based on nature of videos and summarization framework.

References

1. Suresha M, Kuppa S, Raghukumar D (2020) A study on deep learning spatiotemporal models and feature extraction techniques for video understanding. Int J Multimed Inf Retr 9(2):81–101
2. Mirza A, Zeshan O, Atif M, Siddiqi I (2020) Detection and recognition of cursive text from video frames. EURASIP J Image Video Process 2020(1):1–19
3. Carreira J, Zisserman A (2017) Quo vadis, action recognition? a new model and the kinetics dataset. In: proceedings of the IEEE conference on computer vision and pattern recognition, pp 6299–6308
4. Tran D, Bourdev L, Fergus R, Torresani L, Paluri M (2015) Learning spatiotemporal features with 3d convolutional networks. In: Proceedings of the IEEE international conference on computer vision, pp 4489–4497
5. Peng Y, Zhao Y, Zhang J (2019) Two-stream collaborative learning with spatial-temporal attention for video classification. IEEE Trans Circuits Syst Video Technol 29(3):773–786. https://doi.org/10.1109/TCSVT.2018.2808685
6. Zhang H, Liu D, Xiong Z (2019) Two-stream action recognition-oriented video super-resolution. In: 2019 IEEE/CVF international conference on computer vision (ICCV), pp 8798–8807. https://doi.org/10.1109/ICCV.2019.00889
7. Jiang YG, Wu Z, Tang J, Li Z, Xue X, Chang SF (2018) Modeling multimodal clues in a hybrid deep learning framework for video classification. IEEE Trans Multimed 20(11):3137–3147. https://doi.org/10.1109/TMM.2018.2823900
8. Wu Z, Wang X, Jiang YG, Ye H, Xue X (2015) Modeling spatial-temporal clues in a hybrid deep learning framework for video classification. In: Proceedings of the 23rd ACM international conference on Multimedia, pp 461–470
9. Szegedy C, Liu W, Jia Y, Sermanet P, Reed S, Anguelov D, Erhan D, Vanhoucke V, Rabinovich A (2015) Going deeper with convolutions. In: Proceedings of the IEEE conference on computer vision and pattern recognition, pp 1–9

10. Russakovsky O, Deng J, Su H, Krause J, Satheesh S, Ma S, Huang Z, Karpathy A, Khosla A, Bernstein M et al (2015) Imagenet large scale visual recognition challenge. Int J Comput Vis 115(3):211–252
11. He K, Zhang X, Ren S, Sun J (2016) Deep residual learning for image recognition. In: Proceedings of the IEEE conference on computer vision and pattern recognition, pp 770–778
12. Xie S, Girshick R, Dollár P, Tu Z, He K (2017) Aggregated residual transformations for deep neural networks. In: Proceedings of the IEEE conference on computer vision and pattern recognition, pp 1492–1500
13. Gygli M, Grabner H, Riemenschneider H, Van Gool L (2014) Creating summaries from user videos. In: European conference on computer vision, pp 505–520. Springer
14. Song Y, Vallmitjana J, Stent A, Jaimes A (2015) TVSum: summarizing web videos using titles. In: Proceedings of the IEEE conference on computer vision and pattern recognition, pp 5179–5187
15. De Avila SEF, Lopes APB, da Luz Jr, A, de Albuquerque Araújo A (2011) VSUMM: A mechanism designed to produce static video summaries and a novel evaluation method. Pattern Recognit Lett 32(1):56–68
16. Nixon M, Aguado A (2019) Feature extraction and image processing for computer vision. Academic
17. Potapov D, Douze M, Harchaoui Z, Schmid C (2014) Category-specific video summarization. In: European conference on computer vision, pp 540–555. Springer
18. Otani M, Nakashima Y, Rahtu E, Heikkila J (2019) Rethinking the evaluation of video summaries. In: Proceedings of the IEEE/CVF conference on computer vision and pattern recognition, pp 7596–7604
19. Zhou K, Qiao Y, Xiang T (2018) Deep reinforcement learning for unsupervised video summarization with diversity-representativeness reward. In: Proceedings of the AAAI conference on artificial intelligence, vol 32
20. Fajtl J, Sokeh HS, Argyriou V, Monekosso D, Remagnino P (2018) Summarizing videos with attention. In: Asian conference on computer vision, pp 39–54. Springer

Optimum Positioning of Electric Vehicle Charging Station in a Distribution System Considering Dependent Loads

Ranjita Chowdhury, Bijoy K. Mukherjee, Puneet Mishra, and Hitesh D. Mathur

Abstract Large-scale penetration plug-in electric vehicles (PEV) in a distribution system (DS) is a mandate due to its environment friendly applicability but it also increases the overall demand for the system. Extensive use of PEV demands for properly positioned electric vehicle charging station (EVCS) as the improper placement will affect the technical performance of the system in terms of voltage profile and stability, losses, to name a few. To facilitate proper EVCS positioning, this paper puts forward a lucid, apparent power loss-dependent technique applicable for voltage, frequency, and time-dependent loads. Reformulating and then considering the initial State of Charge (SOC_{ini}) of the PEV, the net optimum demand on a charging station is computed using Bayesian optimization technique. A best possible combinational set of PEVs with proportionate distance coverage is considered for demand computation at a charging station and finally EVCS positioning is validated by the results obtained by computation of two standard assessment indices.

Keywords Electric Vehicle Charging Station (EVCS) · Dependent loads · Bayes optimization · Dynamic Loss Evaluation Indicator (DLEI) · Initial State of Charge (SOC_{ini}) · Assessment indices

1 Introduction

The prevailing sources of fuel are reducing very fast. To compensate for this deficit, electric vehicles (EV) are being extensively used. Their eco-friendly nature has been another reason for their extensive application. EVs contribute toward a technically, economically, and environmentally sustainable system if positioned aptly [1, 2]. However, unregulated EV charging can lead to loss of reliability [3] and extra load

R. Chowdhury (✉)
Department of Electrical Engineering, IEM, Kolkata, India
e-mail: p20200507@pilani.bits-pilani.ac.in

B. K. Mukherjee · P. Mishra · H. D. Mathur
Department of Electrical and Electronics Engineering, BITS Pilani, Pilani, Rajasthan, India

H. O. Bansal et al. (eds.), *Next Generation Systems and Networks*, Lecture Notes in Networks and Systems 641, https://doi.org/10.1007/978-981-99-0483-9_38

on the existing system [4]. In the subsequent discussion, the different techniques put forward by the research fraternity are considered.

In order to optimally place an EVCS with a purpose of loss reduction, queuing theory was used in [5]. With the same objective of loss reduction and monitoring of voltage deviation and its profile, a voltage-dependent power flow methodology for EVCS placement was used by Kongjeen et al. [6]. With an ultimate target of loss reduction and minimization of voltage deviation, a Pareto optimality technique was proposed in [7]. In [8], the authors suggest a two-step optimization methodology using gray wolf optimization technique and genetic algorithm method with an ultimate objective of loss reduction. In order to obtain a better siting and sizing position for EVCS in a system, a dual optimization technique based on Genetic Algorithm supported Improved Particle Swarm Optimization was proposed in [9]. Various social factors like initial state of charge (SOC_{ini}), location of CS, charging start time, and its duration were taken into consideration by Chaudhari [10] to sit and size EVCS optimally. The researchers in [11] economically and technically evaluated PEV fast charging stations using optimal power flow method with no major advancements in the existing system. The waiting time of vehicles is also considered here. Placement of slow and fast EVCS in a distribution network with an objective minimization in voltage deviation is modeled in [12]. Differential evolution is applied to achieve the targeted objectives and validated using Harris Hawks optimization technique. Optimal placement of EVCS with an objective of cost minimization is proposed [13] using a two-step linear programming process. With an objective of determining the optimum location of an EVCS so as to reduce the cost of charging, power losses, and voltage deviation, the multi-agent system technique is used to estimate the charging requirement of a PEV [14]. Domínguez-Navarro et al. [15] applied Monte Carlo technique to find the optimum demand for EV in presence of wind and solar empowered resources. Non-dominated sorting genetic algorithm is applied for identifying the optimal size, location, and degree of penetration of the units. To optimally allocate EVCS and to decide on their appropriate size, an optimization model is suggested using second-order conic programming [16]. Shrivastava et al. [17] evolved photovoltaic charging method in unison with EV to ensure continuous supply of power to the vehicles during maximum loading period. Additionally, this method enhances the profitability, reliability, and distance coverage of the EV into consideration.

In order to address the above issue and to optimally position a set of EVCS considering their proper sizing, a novel Dynamic Loss Evaluation Indicator (DLEI) technique has been put forward in the current work. This method identifies ten EVCS positions in an IEEE 33 bus radial distribution system (i.e., 33% of the total number of buses) and a single and combination of two charging stations are placed at a time. Two standard assessment indices proposed in literature are considered to validate the positioning. These indices are Line Loss Reduction Index (LII) and Voltage Stability Margin Index (VSM) and are obtained in [18]. In order to optimize the power demanded at a station, Bayesian optimization technique [19] is used to decide on the possible combination of vehicles and their total power requirement.

The main contributions of this work are Reformulating SO_{Cini} considering the above-mentioned factors such as driving ratio, type of vehicles entering the station and their combinations, formulating an optimum combination of vehicles based on Bayes optimization technique and their power requirement, proposal of a novel, apparent power loss driven technique to optimally position EVCS in an IEEE 33 bus distribution system, validation of the proposed technique by application of assessment indicators.

The rest of the paper is designed as follows. Section 2 deals with the proposed technique for optimum EVCS placement and formulation of the initial state of charge while Sect. 3 deals with the results obtained by simulation and its analysis followed by the conclusion of the entire work in Sect. 4.

2 Proposed Methodology

In this section, a new, lucid apparent power empowered technique namely Dynamic loss evaluation indicator (DLEI) is proposed for an optimum EVCS positioning. Since two different load models are considered here, so two different representation of this technique is put forward. Upon confirmation of the best position for charging station placement, next the total optimum power requirement at a station is decided upon by reformulating SOC_{ini}. Finally, EVCS is designed and DLEI method is validated by the use of assessment indicators.

2.1 *Dynamic Loss Evaluation Indicator (DLEI)*

This subsection puts forward the proposed DLEI technique for two different load models namely Voltage and Frequency Dependent Load Model (VFDLM) and time-dependent Dynamic Load Model (DLM). DLEI values are computed as per the equations proposed below and the buses with highest DLEI values are considered the best position for EVCS positioning.

DLEI for Voltage and Frequency Dependent Load Model (VFDLM)

Since this indicator is based on the apparent power loss of a system, so the loss beyond any bus 'k' is computed as

$$S_{loss} = \frac{P_{pot}^2 + Q_{pot}^2}{V_k^2} Z_k \tag{1}$$

Now, DLEI is computed as per the expression given below

$$\zeta_{DLEI} = \frac{\partial S_{loss}}{\partial P_{pot}} \Delta P_{pot} + \frac{\partial S_{loss}}{\partial Q_{pot}} \Delta Q_{pot} \tag{2}$$

Now, the total potential load at a bus is dependent on both voltage and frequency and obtained as [20]

$$P_{pot} = P_k \left(\frac{V}{V_0}\right)^{K_{pv}} [1 + K_{pf}(f - f_0)] \tag{3}$$

$$Q_{pot} = Q_k \left(\frac{V}{V_0}\right)^{K_{qv}} [1 + K_{qf}(f - f_0)] \tag{4}$$

where

S_{loss} = the total apparent loss of the system; P_k, Q_k = total resultant active and reactive loads at bus 'k'; P_{pot} = total potential active load beyond bus 'k'; Q_{pot} = total potential reactive load beyond bus 'k', Z_k = total impedance of the line connected in between the buses; V_k = magnitude of voltage at the bus 'k', K_{pv}= voltage slope for active power; K_{qv}= voltage slope for reactive power; K_{pf}= frequency slope for active power and K_{qf}= frequency slope for reactive power (values are obtained from [20]).

In this case, DLEI is recalculated on the basis of (3) and (4) as

$$DLEI_{VFDLM} = \frac{2P_k\left(\frac{V}{V_0}\right)^{K_{pv}}[1 + K_{pf}(f - f_0)]Z_k}{V_k^2}\Delta P_{pot} + \frac{2Q_k\left(\frac{V}{V_0}\right)^{K_{qv}}[1 + K_{qf}(f - f_0)]Z_k}{V_k^2}\Delta Q_{pot} \tag{5}$$

DLEI for Dynamic Load Model (DLM)

The total potential loads for dynamic load modeling considering time-driven load are recomputed as

$$P_{pot}(t) = P_k + \Delta P(t) \tag{6}$$

$$Q_{pot}(t) = Q_k + \Delta Q(t) \tag{7}$$

where

$$\Delta P(t) = \Delta P_F(t) + \Delta P_V(t) \tag{8}$$

and

$$\Delta Q(t) = k_4 \Delta V(t) \tag{9}$$

where $\Delta P_F(t)$ is the time domain solution of

$$\frac{\Delta P_F(s)}{\Delta F(s)} = k_2(1 + sT_F) \tag{10}$$

and $\Delta P_V(t)$ is the time domain solution of

$$\frac{\Delta P_V(s)}{\Delta V(s)} = k_5 \frac{(s + b)(s + a)}{(s + b)^2 + \omega^2} \tag{11}$$

where s represents the Laplace transform operator, k's are the gain constants, T_F is the time constant and ω is the frequency in radian and their values are given from [20].

Therefore, the indicator for DLM is calculated as

$$DLEI_{\mathrm{DLM}} = \frac{2\mathrm{P_{pot}(t)Z_k}}{\mathrm{V_k^2}}\Delta\mathrm{P_{pot}} + \frac{2\mathrm{Q_{pot}(t)Z_k}}{\mathrm{V_k^2}}\Delta\mathrm{Q_{pot}} \tag{12}$$

Solving Eq. (12) by putting the values of P_{pot} and Q_{pot} from (6) to (11) the optimum positions of EVCS units are determined.

2.2 Computation of Initial State of Charge (SOC_{ini})

For optimum power requirement at a charging station, the state of charge of the vehicle at the moment of its arrival at a station is very vital. So, proper computation of SOC_{ini} is a mandate. In this subsection, SOC_{ini} is reformulated considering certain vital factors as mentioned previously and is given by

$$SOC_{ini} = 1 - \left[\rho B'_{cap}(y)\right] \tag{13}$$

$$B'_{cap} = yB_{cap} \tag{14}$$

where $\rho = \frac{d}{d_m}$ is coined as driving ratio with $0.1 \leq \rho \leq 1$; d = daily driving distance of the vehicle; d_m = maximum distance the vehicle can cover, i.e., its maximum mileage; $B'_{cap}(y)$ = optimum battery capacity of a vehicle in pu; y = proportion of the type of vehicles obtained as mentioned in the next section; B_{cap} = optimum battery capacity of a specific type of vehicle in pu.

2.3 Designing of Charging Station

For obtaining the optimum power rating of a charging station, four types of electric vehicles are considered, namely, bike, small car, three-wheeler, and large car and their probable combinational ratio are computed using Bayes' optimization technique. Now, the final power demand at a station by considering SOC_{ini} for the specific type of vehicles is given by

$$EV_n^{exp}(i) = \begin{cases} \left(0.9 - SOC_{ini}^{t_c}\right) for t_{k,n}^{dep} \leq t_{c+s-1} \\ \left(0.9 - SOC_{ini}^{t_c}\right) \frac{s}{\frac{\left(t_{k,n}^{dep} - t_c\right)}{\Delta t}} for t_{k,n}^{dep} > t_{c+s-1} \end{cases} \quad (15)$$

where i is the type of vehicles considered in this paper; $EV_n^{exp}(i)$ is the expected charging amount of an EV; t_c is the current time slot; $t_{k,n}^{dep}$ is the EV departure time slot; Δt is the time slot duration, s is the time slot quantity of a time horizon; c is the current time slot number.

After computing the desired demand at a station, DLEI method is applied and EVCS positioning is determined. Initially, single station is placed and thereafter two stations are simultaneously considered and their technical feasibility is monitored by assessment indices and results are demonstrated in the next section of the paper.

3 Results and Analysis

This section deals with the results obtained on the basis of DLEI method suitable for EVCS positioning and thereafter the results are validated by the assessment indices.

3.1 *Optimum Positioning of EVCS as Per DLEI Method*

As per the previous section, the appropriate position of a charging station is determined by the proposed DLEI technique. Top ten values of DLEI are considered for EVCS positioning and represented in Table 1.

For DLM, the whole time frame is divided into quarters, months, and hours with four sets, 12 sets, and 17 sets of load data, respectively. Out of ten probable slots, first stations are placed singly and then combination of two at a time. The combinations are made on the basis of the nearness of CS from the source (bus 5, 9, and 20), load is less (bus 16, 9, 5), or a combination of less load-far from the source and vice-versa. Finally, the indices are computed to validate the results obtained.

Table 1 Positioning of EVCS as per the result of DLEI

Bus No	REDG placement/Charging station No	Magnitude of active power in kW(P) at the corresponding station	Magnitude of active power in kvar (Q) at the corresponding station
16	CS1	60	20
5	CS2	60	30
7	CS3	200	100
8	CS4	200	100
9	CS5	60	20
12	CS6	60	35
13	CS7	60	35
15	CS8	60	100
18	CS9	90	40
20	CS10	90	40

3.2 Determination of Optimum Combination of Different Types of Electric Vehicles

In this section, the combinational ratios of different types are initially stated and then on the basis of optimization technique, the optimum power demand at a station is determined and tabulated in Table 2. The probable combination of bikes, small cars, large cars, and three-wheelers arriving at a CS are as follows: Bike-0.164, three-wheeler-0.196, small car-0.2, and large car-0.44. Their battery capacities are different, so their power requirements are also different. On the basis of these combinational ratios, the optimum mix of vehicles and their demand are calculated on the basis of the proportional distances traveled by them and shown below. First CS is not considered owing to the standard practice followed in their placement.

3.3 Evaluation of Assessment Indices

In this subsection, after EVCS positioning, power flow analysis is performed, two assessment indices are taken into consideration and their results are portrayed below one by one.

Line Loss Reduction Index (LII)

LII occupies a vital position in analyzing a system's technical performance. For stable functioning of a system, LII should be less than 1. Borderline case arises when LII is very close to 1. The results of LII are shown below for two types of load models and for two EVCS positioning. Figures 1 and 2 show that the LII value is on the borderline case, i.e., just slightly more than 1 but not much more than 1 for

Table 2 Optimum power requirement at different charging stations in pu values

Driving ratio (ρ)	Apparent power required for 20% of distance coverage	Apparent power required for 30% of distance coverage	Apparent power required for 50% of distance coverage	Apparent power required for 70% of distance coverage
0.2	0.01331	0.021831	0.049106	0.090032
0.3	0.034257	0.042807	0.070167	0.112917
0.4	0.06327	0.071849	0.09918	0.140163
0.5	0.0741	0.082622	0.109953	0.150965
0.6	0.04674	0.051015	0.064695	0.085215
0.7	0.116337	0.12483	0.152019	0.192831
0.8	0.127082	0.135575	0.162792	0.203576
0.9	0.140534	0.149084	0.176444	0.217484
1.0	0.144153	0.152703	0.180063	0.221103

all load models which signifies that DLEI positioning will be justified only if some additional sources of generating units which improve the index. A comparative study of the values of LII at the 2CS for the two combinational sets is illustrated in Fig. 3 which justifies single EVCS positioning.

Voltage Stability Margin Index (VSM)

For any method to be effective, the voltage stability of the system should be another point of concern. DLEI method of EVCS positioning will be justified only if the voltage stability of the system is enhanced. For a measure of voltage stability, VSM is used as a technique. Figures 4, 5 and 6 depict VSM values for two types of load models and also for two different EVCS positioning.

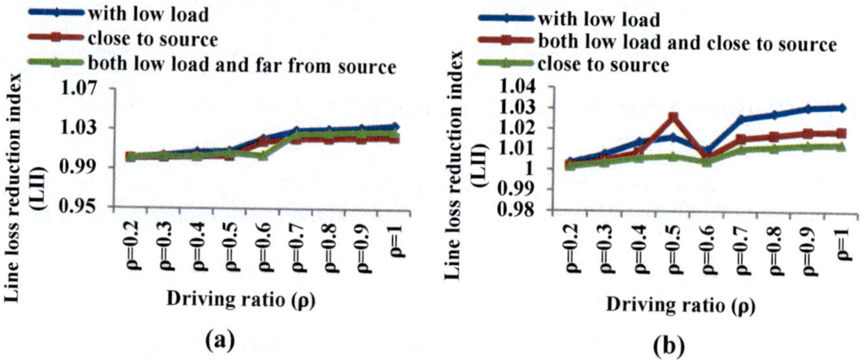

Fig. 1 Variation of LII for VFDLM with (**a**) 1 and (**b**) 2 CS for optimum EVCS positioning

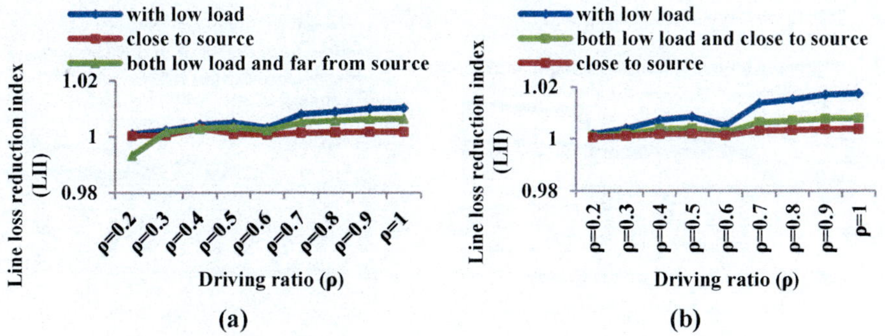

Fig. 2 Variation of LII for DLM with (**a**) 1 and (**b**) 2 CS for optimum EVCS positioning

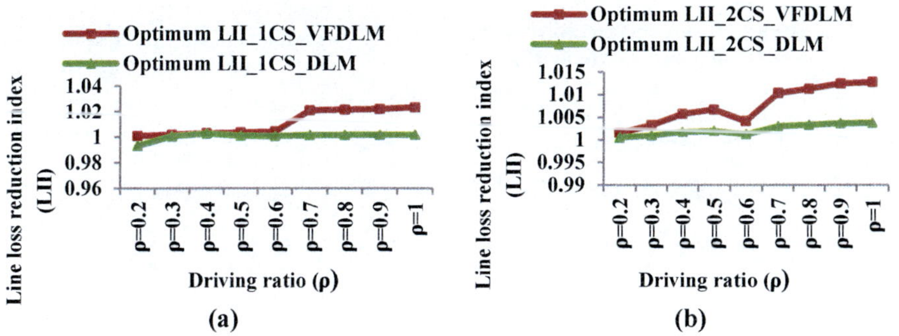

Fig. 3 Variation of LII for different load models with (**a**) 1 and (**b**) 2 CS for optimum EVCS positioning

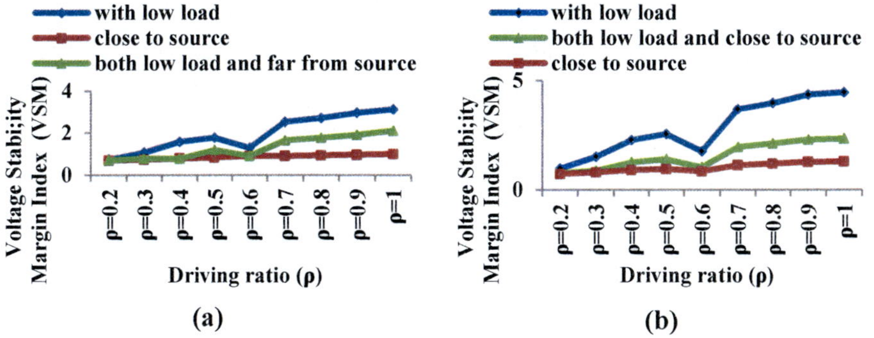

Fig. 4 Variation of VSM for VFDLM with (**a**) 1 and (**b**) 2 CS for optimum EVCS positioning

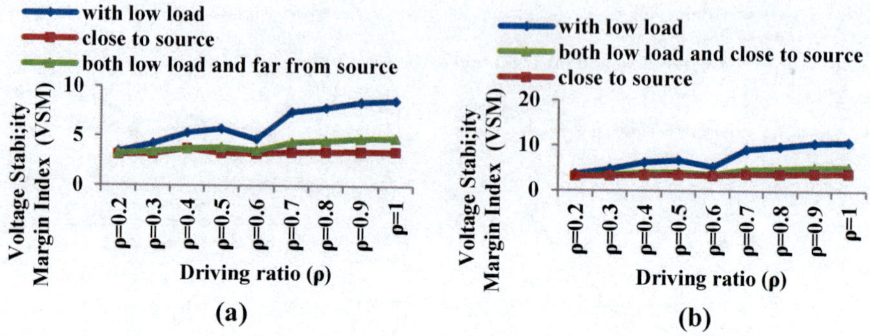

Fig. 5 Variation of VSM for DLM with (**a**) 1 and (**b**) 2 CS for optimum EVCS positioning

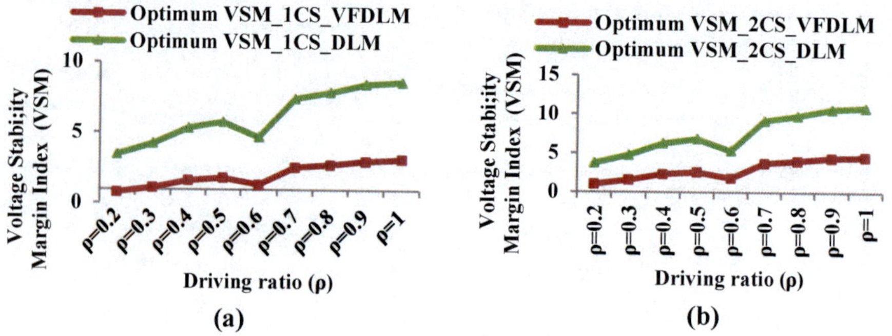

Fig. 6 Variation of VSM for different load models with (**a**) 1 and (**b**) 2 CS for optimum EVCS positioning

It is evident from all the figures above that the voltage stability of the system is maintained with EVCS placement at the proposed buses as the VSM value is positive and consistently maintained for all types of load models.

4 Conclusion

The drastic increase in demand for electric vehicles necessitates EVCS positioning at various locations in a distribution system. But improper positioning can hamper the stability of the existing system and create technical hindrances. So an efficient method is desirable to cater to the aforesaid need. This paper proposes a novel technique of EVCS positioning in an IEEE 33 bus distribution system. Initially, the demand at a charging station is computed by considering the various combinations of vehicles and their capacity. Then EVCS are positioned, both singly and in sets of two. Validation of the method is obtained by two assessment indices. It has been observed that the line loss increases marginally but the voltage stability of the system is enhanced after positioning. So, there is a need for some small, alternate source of generation in the system to improve LII and enhance the system's performance.

References

1. Un-Noor F, Padmanaban S, Popa LM, Mollah MN, Hossain E (2017) A comprehensive study of key electric vehicle (EV) components, technologies, challenges, impacts, and future direction of development. Energies 10(8): 1–82
2. Richardson DB (2013) Electric vehicles and the electric grid: a review of modeling approaches, impacts, and renewable energy integration. Renew Sustain Energy Rev 19 (C):247–254
3. Reddy GH, Goswami AK, Choudhury NBD (2018) Impact of plug-in electric vehicles and distributed generation on reliability of distribution systems. Eng Sci Technol, Elsevier 21(1):50–59
4. Chunhua L, Chau KT, Wu D, Gao S (2013) Opportunities and challenges of vehicle-to home, vehicle-to-vehicle, and vehicle-to-grid technologies. Proc IEEE 101(11):2409–2427
5. Zhu J, Li Y, Yang J, Li X, Zeng S, Chen Y (2017) Planning of electric vehicle charging station based on queuing theory. J Eng 13:1867–1871
6. Kongjeen Y, Bhumkittipich K (2018) Impact of plug-in electric vehicles integrated into power distribution system based on voltage-dependent power flow analysis. Energies 11(6):1571
7. Ul-Haq A, Cecati C, El-Saadany E (2018) Probabilistic modeling of electric vehicle charging pattern in a residential distribution network. Electr Power Syst Res 157:126–133
8. Singh J, Tiwari R (2019) Real power loss minimization of smart grid with electric vehicles using distribution feeder reconfiguration. IET Gener Transm Distrib 13(18):4249–4261
9. Awasthi A, Venkitusamy K, Padmanaban S, Selvamuthukumaran R, Blaabjerg F, Singh AK (2017) Optimal planning of electric vehicle charging station at the distribution system using hybrid optimization algorithm. Energy 133(C):70–78
10. Chaudhari K, Kumar KN, Krishnan A, Ukil A, Gooi HB (2019) Agent based aggregated behavior modeling for electric vehicle charging load. IEEE Trans Ind Inf 15(2):856–868
11. Bitencourt L, Abud TP, Dias BH, Borba BSMC, Maciel RS, Quir´os-Tort´os L (2021) Optimal location of EV charging stations in a neighborhood considering a multi-objective approach. Electr Power Syst Res 199:1–9
12. Rajesh P, Shajin FH (2021) Optimal allocation of EV charging spots and capacitors in distribution network improving voltage and power loss by Quantum-Behaved and Gaussian Mutational Dragonfly Algorithm (QGDA). Electr Power Syst Res 194:1–11
13. Faridpak B, Gharibeh HF, Farrokhifar M, Pozo D (2019) Two-step LP approach for optimal placement and operation of EV charging stations. In: IEEE PES innovative smart grid technologies (ISGT-Europe), Bucharest, Romania, September
14. Jiang C, Jing Z, Ji T, Wu Q (2018) Optimal location of PEVCSs using MAS and ER approach. IET Gener Transm Distrib 12:4377–4387
15. Domínguez-Navarro JA, Dufo-López R, Yusta-Loyo JM, Artal-Sevil JS, Bernal-Agustín JL (2019) Design of an electric vehicle fast-charging station with integration of renewable energy and storage systems. Int J Electr Power Energy Syst 105:46–58
16. Erdinç O, Taşcıkaraoğlu A, Nikolaos G, Paterakis NG, Dursun I, Sinim MC, Catalão JPS (2018) Comprehensive optimization model for sizing and siting of DG units, EV charging stations and energy storage systems. IEEE Trans Smart Grid 9:3871–3882
17. Shrivastava P, Alam MS, Asghar MSJ (2019) Design and techno-economic analysis of plug-in electric vehicle-integrated solar PV charging system for India. IET Smart Grid 2:224–232

18. Hedayati H, Nabaviniaki SA, Akbarimajd A (2008) A method for placement of DG units in distribution networks. IEEE Trans Power Deliv 23:1620–1628
19. Shahriari B, Swersky K, Wang Z, Adams RP, Freitas ND (2016) Taking the human out of the loop: a review of bayesian optimization. Proc IEEE 104:1–24
20. Casper SG, Nwankpa CO, Bradish RW (1995) Bibliography on load models for power flow and dynamic performance simulation. IEEE Trans Power Syst 10:523–553

A Study on DC Fast Charging of Electric Vehicles

C. B. Ranjeeth Sekhar, Surabhi Singh, and Hari Om Bansal

Abstract As the penetration of electric vehicles (EVs) is increasing, their efficient charging becomes very important. This paper presents the simulation of various charging algorithms for the Li-Ion Batteries of EVs and their comparison with each other. Various charging algorithms like Constant Current (CC), Constant Voltage (CV), and Constant Current-Constant Voltage (CC/CV) algorithms have been discussed along with various DC-DC charging topologies like the Buck converters and the LLC Resonant Converter have been discussed. Finally, voltage matching algorithm has been proposed, simulated, and incorporated into the CC/CV algorithm and compared to previous results.

Keywords EVs · CC · CV · CC/CV · Buck converter · Resonant LLC converter

1 Introduction

With the increase in pressure on fossil fuels and pollution by the internal combustion engines (ICE) of vehicles, electric and hybrid vehicles have become the go-to replacement option. EVs provide many advantages over traditional ICE vehicles, i.e., they are less polluting and quieter. Research on EVs can lead toward zero-carbon electric vehicles, which is one of the central pillars of a "3Z" concept (i) Zero poverty, (ii) Zero unemployment, (iii) Zero net-carbon emissions [1, 2]. One of the main concerns is the range of the EVs, which is the maximum distance that can be traveled once fully charged. Recent advancements in this field have significantly increased the range of EVs and have ranges upward of 400 km. Another concern is the time taken to charge the EVs. The aim of this paper is to find and optimize ways to reduce the time taken to charge the battery of an EV.

EV battery uses DC to charge. The battery can be charged directly from a DC source or converted from an AC source to DC using a power electronic converter (onboard charger). DC charging is preferable to AC charging [3]. DC fast chargers

C. B. Ranjeeth Sekhar (✉) · S. Singh · H. O. Bansal
Birla Institute of Technology and Science, Pilani, India
e-mail: f20200316@pilani.bits-pilani.ac.in

H. O. Bansal et al. (eds.), *Next Generation Systems and Networks*, Lecture Notes in Networks and Systems 641, https://doi.org/10.1007/978-981-99-0483-9_39

Table 1 Specification of EVs [12]

EV model	Battery capacity (kW)	Range (km)
Tata Tigor EV	26	306
Tata Nexon EV	30.2	312
Mahindra E Verito	21.2	110
MG ZS EV	44.5	461

(off-board chargers) offer faster charging than onboard chargers used in High voltage E-cars (30–80 kWh capacity). The aim of this paper is to design an off-board charger with a large power output for fast charging. Table 1 contains a list of EV models available in India along with its battery capacity and range.

2 Fast Charging Circuitry for EVs

2.1 DC Microgrid

A DC bus is used to charge the EV. The bus can be powered using the power grid (using ac/dc converters). Alternate sources such as solar can be used as sources (using dc/dc converters). Power is transferred from bus to battery pack through some control feedback mechanism. A schematic diagram of a DC microgrid [4] is depicted in Fig. 1.

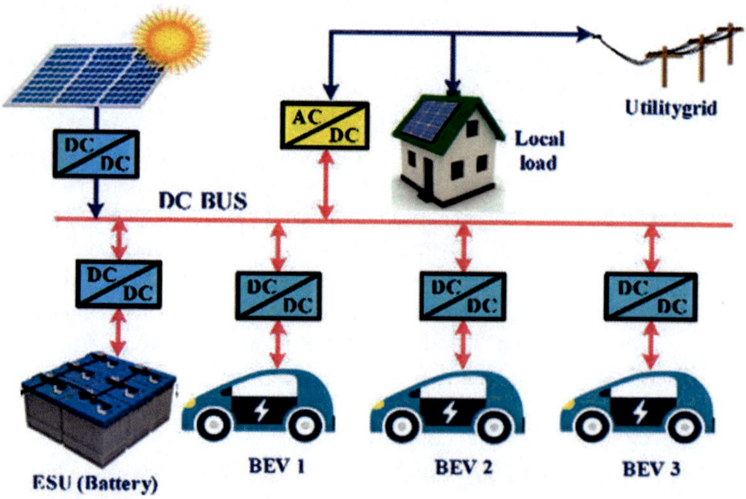

Fig. 1 Schematic diagram of a DC microgrid [5]

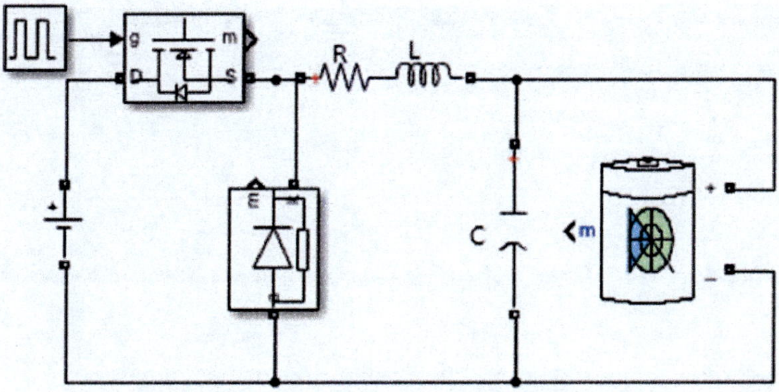

Fig. 2 Schematic circuit diagram of buck converter

2.2 *Comparison of Some Charging Circuitry*

There are various converter circuitries have been used in the charging mechanism of EVs. Some of them have been compared here. The simulation parameters used are DC bus is taken to be 750 V; Battery simulated is that of MG ZS EV (V terminal = 394 V) and Battery = 44.5 kWh Li-ion [6]

2.2.1 Buck Converter

The buck converter steps down the voltage from 750 to 394 V. The voltage transfer characteristics of the buck converter is controlled by Eq. 1. V_{out} here is 394 V and Vin is 750 V. So, D (the duty cycle chosen) is 0.52533 = 52.533%. Figure 2 shows the circuit diagram of a Buck converter.

$$D = \frac{Vout}{Vin} \tag{1}$$

Simulations were run taking parameters as given below:

$$L = 10^{-3}H;\, C = 10^{-6}F,\, R = 1\Omega;\, Switching\ frequency\ f = 5000\,\text{Hz}$$

The results of the simulation are shown in Fig. 3.

2.2.2 LLC Resonant Converter

LLC resonant converter topology uses MOSFETs controlled using Pulse Width Modulation as well as a transformer to control voltage output. The MOSFETs fed

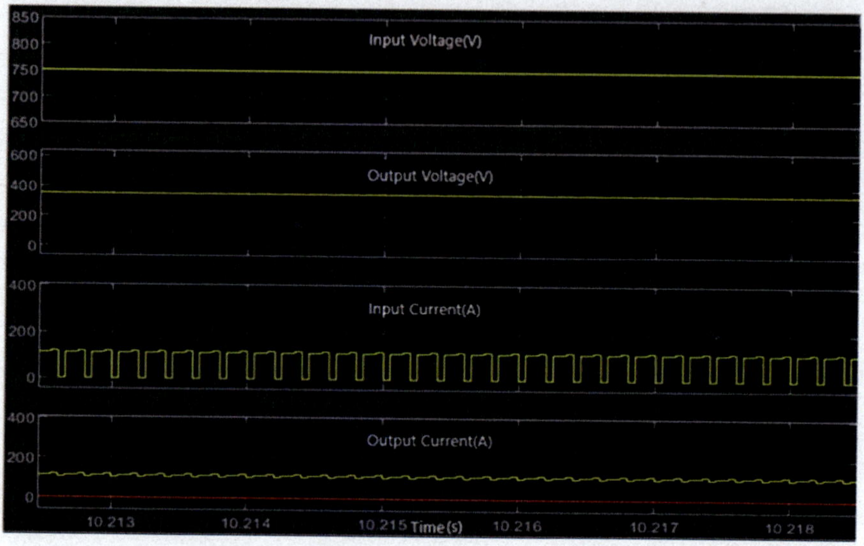

Fig. 3 Input voltage, output voltage, input current, and output current with respect to time of a buck converter circuit from top to down, respectively

by pulse waves are used to convert the dc voltage into a square wave, the output will be a square wave of amplitude 750 V and the duty cycle of the wave is controlled by the input pulse waves. The circuit diagram of LLC Resonant Converter is shown in Fig. 4.

Following parameters are used in the simulation:

$$C = 10^{-6} F, Lr, m = 10^{-3} H, Switching\ Frequency\ f = 5000\,\text{Hz}.$$

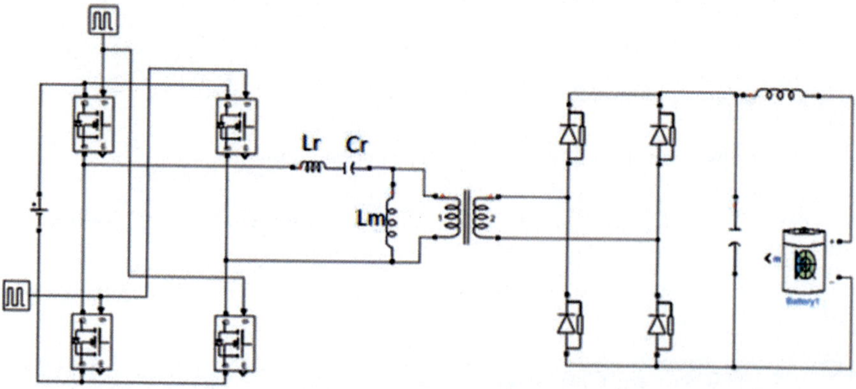

Fig. 4 Schematic circuit diagram of LLC resonant converter

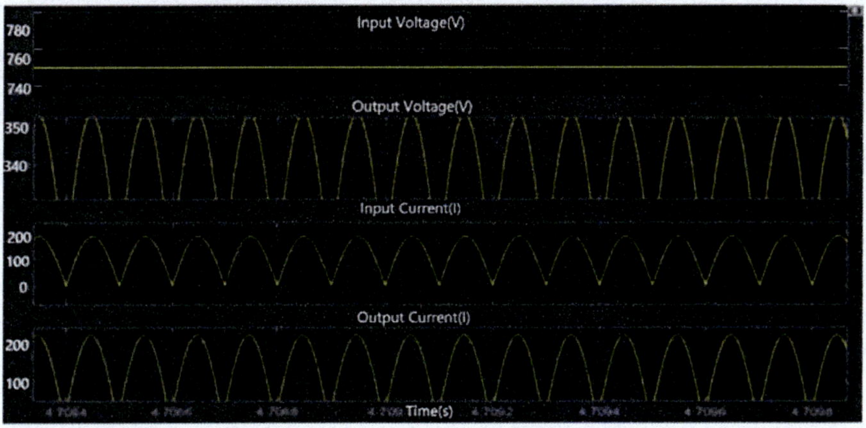

Fig. 5 Input voltage, output voltage, input current, and output current of a LLC resonant circuit from top to down, respectively

The results of the simulation are shown in Fig. 5.

Buck DC/DC converters have lesser efficiency as compared to LLC resonant circuits. In this paper, authors have chosen the LLC resonant topology and tried improving upon it by tweaking the values of its components and adding control mechanisms in order to implement algorithms.

3 LLC Resonant Circuit

3.1 Setting Parameters for LLC Resonant Circuit

Values of capacitance and inductances of the resonant tank must be configured based on input and output requirements in order to get best functioning circuit [7, 8]. Figure 4 shows the LLC Resonant Circuit with its passive components labeled

$$L_m = Magnetizing\, inductance;\;\; L_r = Resonant\, inductance$$
$$C_r = Resonant\ Capacitance$$

$$Q = \frac{W_0 L_r}{R_{ac}} = \frac{1}{W_0 R_{ac} C_r}; = \frac{L_m}{L_r};\, R_{ac} = \frac{8}{\pi^2}\frac{N1^2}{N2^2}R_{laod} \tag{2}$$

R_{load} was taken to be the internal resistance of the battery. Q value is picked depending on range of input gain (high G value and low drop off preferable). The variation of resonant tank gain with respect to frequency is shown in Fig. 6.

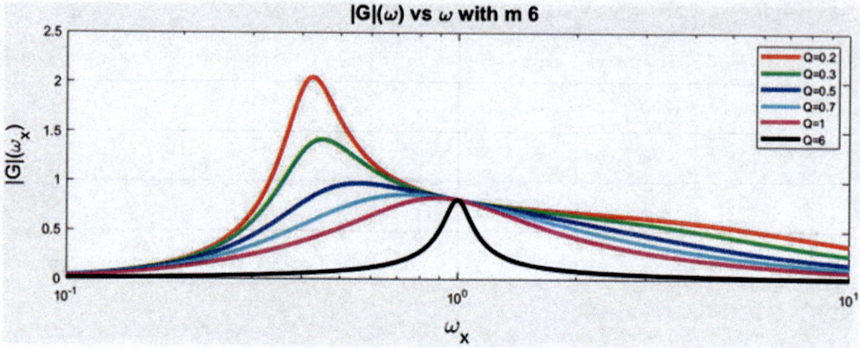

Fig. 6 Graph of variation of resonant tank gain with respect to frequency

Value of m has been chosen as 6 because it gives the most desirable graph, i.e., it can operate in the desired gain range. Gain of the resonant tank should be 1. For this range of operation, the calculated values are as follows:

$$L_r = 6.082 \times 10^{-5} C_r = 1.64 \times 10^{-5};\ L_m = 3.64 \times 10^{-5};\ \text{Efficiency} = 94.8\%.$$

These values provide higher efficiency and operating at the above values enables soft switching, zero current switching, and lower losses.

4 Comparison of Various Charging Algorithms

There are many algorithms for fast charging which are constant voltage, constant current, and constant current followed by constant voltage charging. In this section, they are explained in detail and are analyzed through simulations. These algorithms are tested on the resonant LLC circuit as discussed above.

4.1 Constant Current/Constant Voltage (CC/CV)

In this algorithm, rapid charging first happens in the beginning with constant current being supplied to the battery pack. After about 80–90% state of charge (SoC), the charging switches from constant current to constant voltage while the current gradually drops down as depicted in Fig. 7. This was implemented using PID controllers. In the beginning, a current of 50A is outputted with a voltage output of 340 V. Efficiency at the beginning is around 91%. At 25% SoC, voltage rises to 421 V while current reaches a steady state value of 33A. At 50% SoC, voltage has risen to 427 V while current remains the same. At 75% SoC, voltage rises to 430 V. This was the

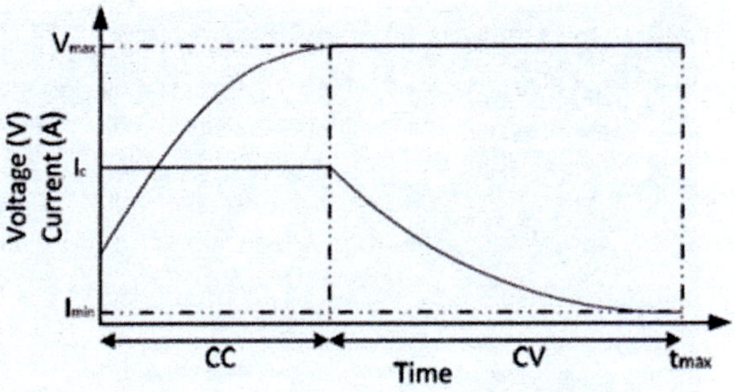

Fig. 7 Graph showing ideal CC/CV control algorithm [9]

constant current portion. The overall efficiency was calculated to be around 92%. In CV mode, the voltage remains constant while current drops off to a very small value.

4.2 Constant Current

Here, the output current is maintained at a constant value by varying input voltage or by variable resistance. This is controlled by using a PID controller. The current flow to the battery stays constant throughout charging. The current, voltage, and efficiency of this mode are identical to the previous mode (CC/CV), the only difference being the voltage does not become constant after a point.

4.3 Constant Voltage

In this mode, voltage is kept constant while charging. This method is very inefficient for charging Li-Ion battery packs because the terminal voltage of the battery keeps changing, if we keep the voltage constant, the current output becomes very low and charging does not take place. The voltage and current characteristics are shown in Figs. 8 and 9, respectively. It can be seen that the current input is very small in this mode of charging.

The CC/CV mode of charging is the preferred mode of charging as it is the most efficient and easiest to implement [10]. The drawback is that current output is low and in the CV mode, current has a steep uncontrollable drop. A new method is proposed in the following section to overcome this issue.

A brief summary of the three algorithms can be found in Table 2.

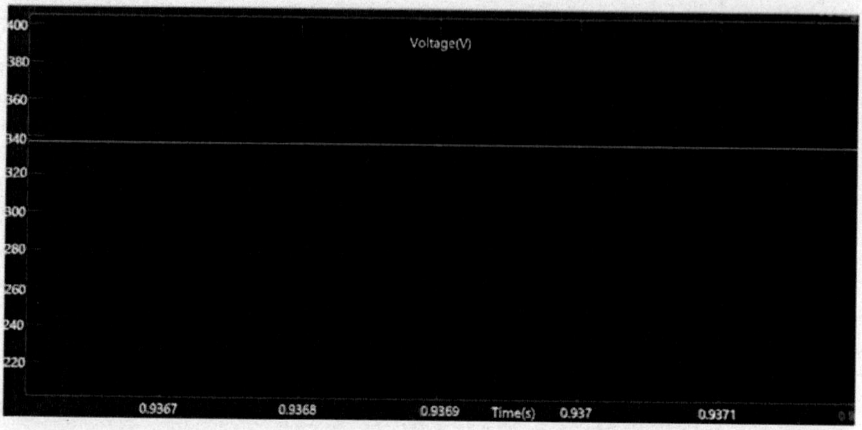

Fig. 8 Voltage characteristics of CV charging

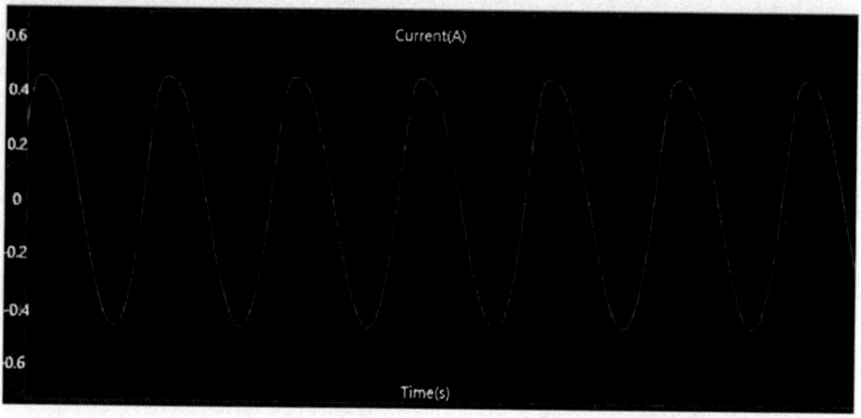

Fig. 9 Current characteristics of CV charging

Table 2 Review of various charging algorithms

Constant voltage	Constant current	Constant current/ Constant voltage
The output voltage remains constant	The output current remains constant	Initially, the current is constant while voltage increases. After a point, at around 80–90% SoC, voltage becomes constant and current drops off
The current output is very low	The current output is high	The current output is high
Battery can be safely charged	Battery cannot be safely charged due to high current close to fully charged state	Battery can be safely charged

5 Voltage Matching

The terminal voltage (V_t) of the Li-Ion battery depends on the battery SoC. The V_t of the battery increases from 337.5 V at an SoC of 0% to a voltage of 461 V at an SoC of 100%. When the voltage provided by the LLC Resonant converter at its output matches with the V_t of the battery, the current and by extension the power output to the battery is much greater than when the voltages are not matched.

To match these voltages, a best fit curve between V_t and SoC is obtained using LaGrange interpolation [11], and a variable winding transformer with the help of a controller is used to keep the voltage at the output same as that obtained from the best fit curve. The equation obtained for the best fit curve is shown in Eq. (3).

$$\begin{aligned} V = {} & 1.31 \times 10^{-16}C^{11} - 7.45 \times 10^{-14}C^{10} + 1.86 \times 10^{-11}C^{9} - 2.68 \times 10^{-9}C^{8} \\ & + 2.47 \times 10^{-7}C^{7} - 1.51 \times 10^{-5}C^{6} + 6.24 \times 10^{-4}C^{5} - 1.72 \times 10^{-2}C^{4} \\ & + 0.31C^{3} - 3.42C^{2} + 22.51C + 337.5 \end{aligned} \tag{3}$$

The algorithm was tested on LLC Resonant circuit and compared the circuit without said algorithm. The algorithm gives much better current output compared to without it. This can be incorporated into algorithms to solve the issue of low current output.

6 New Proposed Algorithm

A PI controller is used to keep the current constant in CC phase. A variable winding transformer is used to regulate the current output. If the voltage is matched, the current output is high. At the end of CC phase, the transformer ratio is gradually increased to drop off the current in the shape of a curve while keeping the voltage constant. To control the variable winding transformer, output voltage is taken as shown in Eq. (2). Input voltage is taken from the DC bus that supplies the power. The variable winding transformer is controlled by the relationship between the terminal voltage (V_t) and the SoC. The control system mainly controls current while the transformer controls the voltage.

Here, the CC phase stops after 85% SOC and CV starts. During the CC phase, current is always around 100–105A throughout the phase. Voltage undergoes a steady increase till the CV phase. Efficiency is around 80%. Figures 10 and 11 show the voltage and current characteristics, respectively. The advantages of the new algorithm can be seen in Table 3.

Result obtained: Average power output calculated = **44.6 kWh**, Efficiency = **80%**

The battery of MG ZS EV (capacity = 44.5 kW) can be charged in less than an hour.

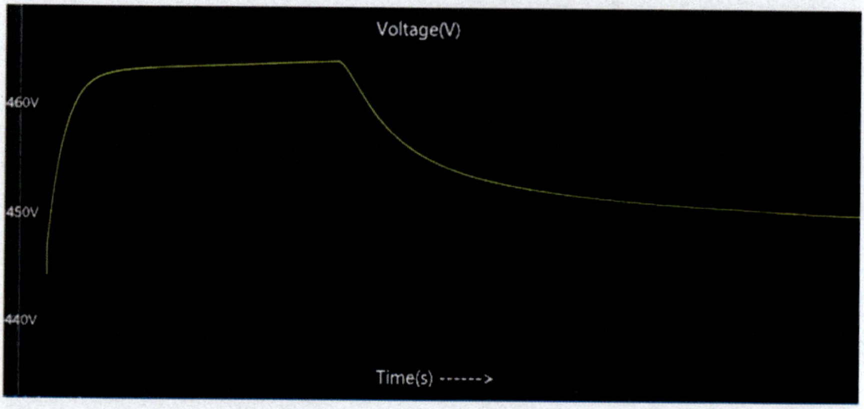

Fig. 10 Graph shows the voltage characteristics at the end of CC and the beginning of CV phase

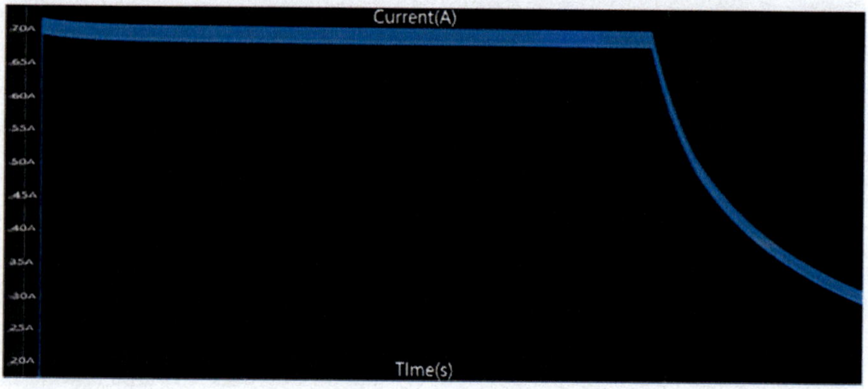

Fig. 11 Graph shows current characteristics when CC phase ends

Table 3 Advantages of the new improved algorithm

Proposed work	Previously reported works
Current output around 100A	Current output around 30A
Current drop after CC phase is like a parabola	Current drop after CC phase is a vertical line
Rate of current drop can be controlled	Rate of current drop cannot be controlled
CV phase control via transformer	CV phase control via MOSFETs
Faster charging	Slower charging

7 Conclusion

In this paper, the authors have simulated and compared the various existing charging topologies for DC-DC converters and algorithms for DC fast charging. It was found out from the simulation that the LLC resonant converter is more efficient than a buck converter. CC/CV charging method proved to be the most efficient. A voltage matching technique along with a related control system was introduced and a new charging algorithm based on it was implemented. It was compared with previous algorithms and proved to be superior in terms of current and power output. The power output was 44.6 kWh. This will fully charge the battery of an MG ZS EV (battery capacity = 44.5 kW) in an hour. This is considerably faster than traditional methods such as AC charging and older DC charging methods. All simulations were done using MATLAB.

References

1. Chakraborty S, Vu H-N, Hasan MM, Tran D-D, Baghdadi ME, Hegazy O (2019) DC-DC converter topologies for electric vehicles, plug-in hybrid electric vehicles and fast charging stations: state of the art and future trends. Energies 12(8):1569
2. Yunus M, Weber K (2017) A world of three zeros: the new economics of zero poverty zero unemployment and zero net carbon emissions. 1st edn. Public Affairs, New York
3. Lim KL, Speidel S, Bräunl T (2022) A comparative study of AC and DC public electric vehicle charging station usage in Western Australia. Renew Sustain Energy. Transit 2:100021
4. Ronanki D, Kelkar A, Williamson SS (2019) Extreme fast charging technology: prospects to enhance sustainable electric transportation. Energies 12(19):3721
5. Savio DA, Juliet VA, Chokkalingam B, Padmanaban S, Holm-Nielsen JB, Blaabjerg F (2019) Photovoltaic integrated hybrid microgrid structured electric vehicle charging station and its energy management approach. Energies 12(1):168
6. Collin R, Miao Y, Yokochi A, Enjeti P, Jouanne A (2019) Advanced electric vehicle fast-charging technologies. Energies 12:1839
7. Choi H (2007) Design consideration of half-bridge LLC resonant converter. Korean Institute of Power Electronics, Korea
8. Steigerwald RL (1988) A Comparison of Half-bridge resonant converter topologie. IEEE Trans Power Electron 3(2)
9. Amin A, Ismail K, Hapid A (2018) Implementation of a LiFePO4 battery charger for cell balancing application. J Mechatron, Electr Power, Vehic Technol 81–88
10. Thomson SJ, Thomas P, R A, Rajan E (2018) Design and prototype modelling of a CC/CV electric vehicle battery charging circuit. In: 2018 international conference on circuits and systems in digital enterprise technology (ICCSDET), pp 1–5
11. LaGrange Calculator, Function Equation Finder. https://www.dcode.fr/function-equation-finder. Accessed 11 March 2022
12. Battery Specifications, CarDekho. https://www.cardekho.com. Accessed 1 Oct 2022

Handling Uncertain Environment Using OWA Operators: An Overview

Saksham Gupta, Ankit Gupta, and Satvik Agrawal

Abstract Fuzzy sets were presented by Zadeh in 1965 as a method of describing and managing data that was not concrete, but rather fuzzy. Fuzzy logic theory gives a mathematical foundation for capturing the inconsistencies inherent in human cognitive processes such as thinking and reasoning. Yager in 1988 presented a unique aggregation approach focusing on ordered weighted averaging (OWA) operators in response to the application of fuzzy logic. It was referred to as membership aggregation cumulative operators by him. Following on from this concept, other academics have highlighted the importance of the OWA weighting vector in a wide variety of implementations such as modelling and decision-making. The objective of this study is to provide an overview of OWA operators while also demonstrating their application in various domains.

Keywords OWA · Fuzzy · Weights · Ordered weighted operators · Machine learning · Applications · Fuzzy logic · Decision making · Modelling

1 Introduction

Data that corresponds to the practical phenomenon and real-world scenarios is hardly ever precise, however, the computational and mathematical theories that utilize such data work on exact numbers. Such issues have resulted in a gap between the invention of theories and their actual application to solve practical problems. Representation of data under any logical structure is bound to yield results that place existing data points under multiple categories with varying degree of representation for each category. Data may not always conform completely to one label and may take forms that portray vastly varying characteristics. The computing approach based on this relative degree

S. Gupta (✉) · A. Gupta
Chandigarh College of Engineering and Technology, Sector 26, Chandigarh, India
e-mail: dev.sakshamgupta@gmail.com

S. Agrawal
Kalinga Institute of Industrial Technology, Patia, Bhubaneswar, Odisha, India

H. O. Bansal et al. (eds.), *Next Generation Systems and Networks*, Lecture Notes in Networks and Systems 641, https://doi.org/10.1007/978-981-99-0483-9_40

and scale of "truth" between existing instances, in contrast to the existing paradigm of Boolean representation of values (false/0 and true/1) utilized by current computation theory, can be termed as fuzzy logic.

Mapping of data that is imprecise or comes with certain levels of approximation can be handled by fuzzy sets [1]. Fuzzy sets and the theory behind handling data using these sets were introduced in 1965 [1]. Fuzzy sets represent the data with various intermediate degrees of truth between the extreme cases represented by Boolean logic and as such provide a way to accurately represent large-scale real-world data that cannot always be mapped precisely onto their Boolean counterparts. Hence, in a way, the Boolean representation is a special case of fuzzy logic where data is mapped only to the extreme ends of the degrees of truth. There is a singular membership value in the fuzzy set for each element, which means the evidence for $x \in X$ and the evidence against $x \in X$ are effectively equivalent. In the real world, however, imprecise information or knowledge is most likely to result in vagueness and hesitation. Specifically, 2 characteristic functions (degrees of belonging and degree of non-belonging) are incorporated in intuitionistic fuzzy sets (IFS) to tackle this problem [2].

Due to the complexity of data represented by fuzzy sets, it becomes necessary to condense multiple data streams into a single representation value which indicates, with sufficient accuracy, the properties of its constituent members. Such an action may seem trivial at first but with the multitude of practical usage scenarios along with the diversity of application fields, a need for operators which are robust and have a broad spectrum of information integration applications like statistical analysis, sensor fusion, artificial intelligence, computer vision, etc. is clear to comprehend. The aggregation functions are among the most important tools for blending information from a theoretical perspective, especially the averaging functions, such as the arithmetic median and mean [3].

The Ordered Weighted Averaging operators (OWA) are an excellent example of these averaging operations introduced in 1988 by Yager [4]. OWA operators represent information fusion functions that distribute weights based on the initial input parameter and further aggregate the membership and non-membership properties of these inputs in a singular operator. OWA operators give a set of data aggregation functions that can accommodate input parameters and are compatible with many currently existing operators.

A mapping F from $I^n \rightarrow I$ (where $I = [0,1]$) is called an Ordered Weighted Averaging operator of dimension n if related with F. is a weighting vector $\boldsymbol{W} = \begin{bmatrix} W_1 \\ W_2 \\ W_n \end{bmatrix}$ given that

1. $W_1 \in (0, 1)$
2. $\sum_1 W_1 = 1$

and where $F(a_1, a_2, \ldots, a_n) = W_1 b_1 + W_2 b_2 + \ldots W_n b_n$, where b_i is the ith biggest element in the group of $a_1, a_2, \ldots, a_n$.

These are a conspicuous collection of parameterized data aggregation operators that have been widely used in a variety of fields which have been summarized in Fig. 1.

Since OWA operators are a direct result of data aggregation, they have seen immense acceptance in both theoretical and practical fields in which fuzzy logic can be applied. Over the years various researchers have used the OWA operator in a large variety of applications, for instance, fuzzy logic controllers, neural networks, market research, database systems, linguistic quantified propositions, decision-making, expert systems, lossless image compression and mathematical programming. Some specific instances of OWA operators in action can be seen during the calculation of engine loads and cylinder compression in automobile industry, medical diagnosis using computer vision-based telemetry and while understanding semantic connections between sentences in natural language processing.

OWA operators may specify a diverse variety of data aggregations, including the median, maximum, mean, minimum and many others. This flexibility to provide many varying aggregations has one drawback: it is difficult to determine the corresponding weights for an OWA operator in a particular instance, which can lead to the creation of multiple techniques to set their values [5–7]. Many of these strategies are focused on converting an expert's opinion to a definite vector corresponding to a particular weight, typically limiting the inherent intricacy of the given operator to a group of OWAs that are relatively easy to specify using parameters. Other ideas derive the weights from data, although the OWA operators are still confined to parameterized variations in most of these proposals [7, 8].

The focus of this work is to provide detailed information on the overview of OWA operators along with the various applications of these operators. We have examined the current literature and backgrounds and provided details for the uses of different OWA operators under specific domains.

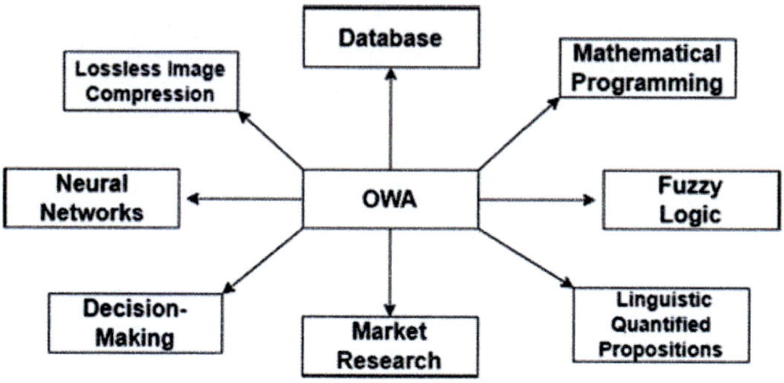

Fig. 1. Use cases of OWA

2 Background/Previous Surveys

Table 1 shows the notable surveys done on OWA operators. We have provided data on the key points from every survey and provided details on the type of survey conducted.

Table 1 Previous surveys on OWA operators

Authors	Type of survey	Key takeaways
Yager [9]	Introduced several parametrized groups of operators; S-OWA, maximum entropy, window and step	Introduced the idea of aggregate dependent weights
Mesiar et al. [10]	Surveys OWA operators based on generalizations like GOWA, IOWA, OMA, etc.	OWA generalizations are done as symmetric means or Choquet integrals using given symmetric measures or as additive functionals which are comonotonic in nature
Fullér [11]	OWA operators can realize trade-offs between objectives, by allowing a positive compensation between ratings	Short survey of OWA operators and illustrate their applicability by a real-life example
Yi et al. [12]	Based on the family of IGOWA operators like IOWQA, Quasi-IOWA	The paper surveys IGOWA operators based on weighting average and parameter
Gorzin et al. [13]	A survey on ordered weighted averaging operators and their application in recommender systems	Short review of the papers which have combined RS and OWA operator
He et al. [14]	Based on 1213 bibliographic records obtained by using topic search from Web of Science	This paper studies the publications about OWA operators between 1988 and 2015
He et al. [14]	Performed a bibliometric analysis of research work done in OWA operators from 1988–2015	Studied the publications about OWA operators based on 1213 item records present in Web of Science
Arya and Kumar [15]	Surveyed different papers that studied intuitionistic fuzzy operators	The paper surveyed other research done on intuitionistic fuzzy operators and generated six aggregation operators based on a picture fuzzy set environment

3 Weights

Selecting the weights associated with OWA operators is significant [5]. The specification of weights and their associated values play a fundamental role as changing their values may lead to vastly different resultant models which may provide distinct results for the same data, especially in the case of decision-making systems [16]. The primary strategy devised by O'Hagan is based on utilizing the maximum value of entropy for a given orness [17]. Lagrange Multipliers have also been utilized to provide analytical solutions to optimization problems in a constrained space and determine the best weights for such scenarios [18]. A method based on minimum variance has also been suggested to provide weight values that have limited variability [19]. Two significant characteristic measures to determine the weight parameters have been presented by Yager, where one is based on the orness of data aggregation and can be represented as

$$orness(W) = \frac{1}{n-1}\sum_{i=1}^{n}(n-i)w_i \tag{1}$$

and it presents the degree of an operation as provided by the data aggregation. The second one, the dispersion measure of the data aggregation, is given as

$$disp(W) = \sum_{i=1}^{n} w_i \ln w_i \tag{2}$$

and it provides the value of the degree to which W considers all data once it is aggregated.

The form and type of the weight vector determine the final aggregation applied by an operator. Multiple strategies have been presented to get the value of weights associated with an operator based on learning systems, guided using a quantifier or using exponential smoothing [4, 9, 20, 21].

Characterization of weights determination was done in 2009 by Ahn under the following approaches: 1. Programming-based, 2. Experience-based, 3. Analytic formula-based, 4. Quantifier-guided [22].

Based on maximizing/minimizing dispersion or degree of orness, various methods of calculating weights can be formulated. Hagan [17], provided one such approach where

$$Maximize - \sum_{i=1}^{n} w_i \ln w_i \tag{3}$$

subject to,

$$\frac{1}{n-1}\sum_{i=1}^{n}(n-i)w_i = \alpha, 0 \leq \alpha \leq 1 \tag{4}$$

and

$$\sum_{i=1}^{n} w_i = 1, 0 \leq w_i \leq 1, i = 1, \ldots, n \tag{5}$$

Equation (5) was transformed into a polynomial by applying Lagrange Multiplier [18]. Solving this equation resulted in an optimal weighting vector as

$$w_1[(n-1)\alpha + 1 - nw_1]^n = ((n-1)\alpha)^{n-1}[((n-1)\alpha - n)w_1 + 1] \tag{6}$$

$$w_n = \frac{((n-1)\alpha - n)w_1 + 1}{(n-1)\alpha + 1 - nw_1} \tag{7}$$

$$w_j = \sqrt[n-1]{w_1^{n-j} w_n^{j-1}} \tag{8}$$

4 Uses

Primary applications of OWA operators can be divided into two specific categories depending on their use and representation. They can either be used in mathematical systems to generate statistical models, perform data aggregation or regression and can be further utilized in decision-making models to provide efficient means for data representation [21]. On the other hand, OWA operators have also seen use in specific industry-based domains where imprecise data has been fed into fuzzy systems to generate practical and scalable solutions to problems that cannot be tackled by conventional computational theories.

4.1 Use in Mathematical Systems and Decision Models

Statistical regression utilizes fuzzy connectives where multiple operations performed on the data can be condensed into a single OWA operator which can perform the tasks of other functions such as LAD (Least Absolute Deviation), Standard Least Squares (SLS) and Maximum Likelihood Criteria (MLC) [23].

Many numerical techniques have been proposed to solve the regression-related problems [24]. There are specific cases where this aggregation using OWA has achieved better results and a case of data outliers is one of them. There is a strong correlation between the quantifiers used in fuzzy systems and the weight vectors in

OWA operators. A threshold value that creates a distinction between wanted and unwanted data can be used to shape a monotonic quantifier, and hence the metrics of threshold, i.e., entropy and attitudinal character are fundamentally related to weight vectors in an OWA operator. While these metrics can be termed as independent, it has been shown that when it comes to the quantifier threshold, these metrics can be related and can be used in conjunction for verification of the quantifier threshold [25].

MCDM (Multiple criteria decision-making) models have been proposed based on a dynamic fuzzy system utilizing OWA operators that can assist used to model and represent decision-making problems. They are also able to provide efficient solutions to problems when significant details or information is missing for the data to be analyzed [26]. Table 2 lists the different domains in which OWA operators have been used in mathematical systems.

4.2 Use in Industrial Domains

Industrial applications of OWA operators have been diversifying over the past years and have been aided in that regard by the increase in AI research and the inclusion of fuzzy systems in domains other than theoretical research.

A data querying system has been developed to work on the hemodialysis database for a local hospital situated in Taiwan which aids doctors and healthcare professionals to provide fast and accurate diagnosis in case of hemodialysis [42]. The model has been built and its parameters have been tuned based on expert opinions, but it also has the provision to include a fuzzy OWA query system that the doctors can use to update the values of weights dynamically. These fuzzy systems will then be able to change the weight vectors of each attribute, and hence the model will be able to provide synthetic recommendations to doctors. Table 3 lists the different domains in which OWA operators have been used along with their respective applications.

5 Conclusion

The ordered weighted averaging (OWA) operators are well-known information aggregation operators and have seen usage in multiple domains across industries and research. They have primarily been utilized in fuzzy systems where data is not precise and can vary within a range. The OWA operators are typically utilized to aggregate data based on weights which can be adjusted to provide the best fit for the required use case. OWA operators are not just limited to statistics and mathematics but have also proven to be immensely useful in modern applications like AI-based classifications, linguistic systems and business decisions. With the growing number of operators emerging to satisfy nearly every research domain, OWA operators have solidified their place in the research industry as some of the most robust yet flexible tools for data aggregation and representation.

Table 2 Applications of OWA operators in mathematical systems

Domains	Author	Applications
Set theory	Verbiest et al. [27]	An OWA based selection system for rough sets in fuzzy set theory
	Riza et al. [28]	Implementation of fuzzy rough set theory algorithms using OWA operators
Data aggregation	Chiclana et al. [29]	Use of OWA operators in descriptions and explanation of reciprocity in aggregation of preference relations
	Yager [30]	Use of OWA operators to implement a quantified and guided data aggregation scheme
	Yager [31]	Data aggregation tooling and mechanisms as utilized in modelling, classification, and decision-making problems
Linguistic systems	Gou et al. [32]	Definition of MULTIMOORA method and fuzzy hesitant language sets
Decision making	Herrera-Viedma et al. [33]	Consistency-based preference and relation decision modelling for group decisions
	Zhou and Chen [34]	Continuous generalization of time interval data in decision-making processes
	Yager [35]	Data aggregation and mapping over continuous interval argument in decision-making models
	Liu and Wang [36]	Multi attribute decision-making using neutrosophic prioritized and interval mapped OWA operator
Data modelling	Jin and Qian [37]	Generalization of complex data and feature simplification using OWA operators
	Merigó et al. [38]	A study of OWA operators to determine distance measures, weighted averages and general data modelling
Adaptive systems	Herrera and Lozano [39]	Adaptive algorithms classification and design using OWA operators
	Derrac et al. [40]	Implementation of nearest neighbour algorithms and their experimental analysis
Defuzzification	Torra and Good [41]	OWA operators' application in cases of defuzzification and data expansion

Table 3 Application of OWA operators in industrial domains

Domains	Authors	Applications
Automobile	Eckert et al. [43]	Gear selection for automatic transmission based on engine load, road conditions, transmission response and driving style
	Garcia-Triviño et al. [44]	Power delivery calculations and controller logic for charging stations of electric cars
	Bastian [45]	Modelling fuel injection and corresponding control maps in vehicles based on engine parameters and vehicle sensor data
Healthcare	Puente et al. [46]	Emergency room queuing system **when influx and density of patients** along with service time requires is uncertain
	Phuong and Kreinovich [47]	Diagnosis using computer-based models based on symptoms, exposure and medical history with real time patient data
Business	Imamverdiev and Derakshande [48]	Security risk assessment model to make secure business decisions based on existing data
	Vigier et al. [49]	Prediction of business failures based on market sentiment, operational negligence and employee conduct
	Bowles and Enrique Peláez [50]	Calculation of system-level failure probability
	Spott et al. [51]	Fuzzy decision-making systems based on expected results to improve performance indicators according to predetermined criteria
Natural language processing	Gupta et al. [52]	Understanding of connotation and semantic interaction between ideas represented words, linguistic notations and thought-based speech
Supply chain management	Lu et al. [53]	Supplier selection based on raw materials, quality of product, lead-time, transportation cost and transportation path
	Celikbilek et al. [54]	Distribution network creation based on cost, vehicle type and supply requirement

(continued)

Table 3 (continued)

Domains	Authors	Applications
Electronics	Singhala et al. [55]	Temperature control, target and current temperature determination in environment control systems
	Wakami et al. [56]	Washing mode, energy requirements and standby duration calculation in dishwashers and washing machines
	Hopkins [57]	Drum voltage calculation based on humidity, picture density, output quality and temperature in copy machines and printers
Aerospace	Jian-guo and Jun [58]	Altitude control systems for satellites and spacecrafts based on environmental factors
	Rawea and Urooj [59]	Autopilot mechanisms based on travel route, aircraft speed, wind velocity and flight duration
Artificial intelligence	Sharma et al. [60]	Image recognition based on CNN models for pooling, image data feature extraction and weight calculation
	Huang and Cheng [61]	Time series forecasting models providing accurate predictions based on past data utilizing data consolidation techniques

References

1. Goguen JA, Zadeh LA (1965) Fuzzy sets. Inf Contr 8:338–353. Zadeh LA (1971) Similarity relations and fuzzy orderings. Inf Sci 3:177–200. (1973) J Symbolic Logic 38:656–657
2. Atanassov KT (1986) Intuitionistic fuzzy sets. Fuzzy Sets Syst 20:87–96
3. Beliakov G, Pradera A, Calvo T (2007) Aggregation functions: a guide for practitioners. In: Studies in fuzziness and soft computing
4. Yager RR (1988) On ordered weighted averaging aggregation operators in multicriteria decisionmaking. IEEE Trans Syst, Man, Cybern 18:183–190
5. Xu Z (2005) An overview of methods for determining OWA weights. Int J Intell Syst 20:843–865
6. Kishor A, Singh AK, Sonam S, Pal NR (2020) A new family of OWA operators featuring constant orness. IEEE Trans Fuzzy Syst 28:2263–2269
7. Beliakov G (2003) How to build aggregation operators from data. Int J Intell Syst 18:903–923
8. León T, Zuccarello P, Ayala G, de Ves E, Domingo J (2007) Applying logistic regression to relevance feedback in image retrieval systems. Pattern Recogn 40:2621–2632
9. Yager RR (1993) Families of OWA operators. Fuzzy Sets Syst 59:125–148
10. Mesiar R, Stupnanova A, Yager RR (2015)Generalizations of OWA operators. IEEE Trans Fuzzy Syst 23:2154–2162
11. Fullér R (1996) OWA operators in decision making
12. Yi P, Dong Q, Li W (2021) A family of Iowa operators with reliability measurement under interval-valued group decision-making environment. Group Decis Negot 30:483–505

13. Gorzin M, Hosseinpoorpia M, Parand F-A, Madine SA (2016) A survey on ordered weighted averaging operators and their application in recommender systems. In: 2016 eighth international conference on information and knowledge technology (IKT)
14. He X, Wu Y, Yu D, Merigó JM (2017) Exploring the ordered weighted averaging operator knowledge domain: a bibliometric analysis. Int J Intell Syst 32:1151–1166
15. Arya V, Kumar S (2020) A new picture fuzzy information measure based on Shannon entropy with applications in opinion polls using extended Vikor–TODIM approach. Comput Appl Math 39
16. Fuller R (2007) On obtaining OWA operator weights: a sort survey of recent developments. 2007 In: IEEE International conference on computational cybernetics
17. O'Hagan M (1988) Aggregating template or rule antecedents in real-time expert systems with fuzzy set logic. In: Twenty-Second Asilomar conference on signals, systems and computers
18. Fullér R, Majlender P (2001) An analytic approach for obtaining maximal entropy OWA operator weights. Fuzzy Sets Syst 124:53–57
19. Fullér R, Majlender P (2003) On obtaining minimal variability OWA operator weights. Fuzzy Sets Syst 136:203–215
20. Beliakov G, James S (2011) Induced ordered weighted averaging operators. In: Recent developments in the ordered weighted averaging operators: theory and practice, pp 29–47
21. Filev D, Yager RR (1998) On the issue of obtaining OWA operator weights. Fuzzy Sets Syst 94:157–169
22. Ahn BS, Park KS (2008) Comparing methods for multiattribute decision making with ordinal weights. Comput Oper Res 35:1660–1670
23. Csiszar O (2021) Ordered weighted averaging operators: a short review. IEEE Syst, Man, Cybern Mag 7:4–12
24. Yager RR, Beliakov G (2010) OWA operators in regression problems. IEEE Trans Fuzzy Syst 18:106–113
25. Yager RR (2009) On the dispersion measure of OWA operators. Inf Sci 179:3908–3919
26. Chang J-R, Ho T-H, Cheng C-H, Chen A-P (2005) Dynamic fuzzy OWA model for group multiple criteria decision making. Soft Comput 10:543–554
27. Verbiest N, Cornelis C, Herrera F (2013) Owa-FRPS: a prototype selection method based on ordered weighted average fuzzy rough set theory. In: Lecture notes in computer science, pp 180–190
28. Riza LS, Janusz A, Bergmeir C, Cornelis C, Herrera F, Ślezak D, Benítez JM (2014) Implementing algorithms of rough set theory and fuzzy rough set theory in the R package "Roughsets." Inf Sci 287:68–89
29. Chiclana F (2003) A note on the reciprocity in the aggregation of fuzzy preference relations using OWA operators. Fuzzy Sets Syst 137:71–83
30. Yager RR (1998) Quantifier guided aggregation using OWA operators. Int J Intell Syst 11:49–73
31. Yager RR (1992) Applications and extensions of OWA aggregations. Int J Man-Mach Stud 37:103–122
32. Gou X, Liao H, Xu Z, Herrera F (2017) Double hierarchy hesitant fuzzy linguistic term set and Multimoora method: a case of study to evaluate the implementation status of haze controlling measures. Inf Fusion 38:22–34
33. Herrera-Viedma E, Chiclana F, Herrera F, Alonso S (2007) Group decision-making model with incomplete fuzzy preference relations based on additive consistency. IEEE Trans Syst Man Cybern, Part B (Cybernetics) 37:176–189
34. Zhou L-G, Chen H-Y (2011) Continuous generalized OWA operator and its application to decision making. Fuzzy Sets Syst 168:18–34
35. Yager RR (2004) Owa aggregation over a continuous interval argument with applications to decision making.: IEEE Trans Syst Man Cybern, Part B (Cybernetics) 34:1952–1963
36. Liu P, Wang Y (2015) Interval neutrosophic prioritized OWA operator and its application to multiple attribute decision making. J Syst Sci Complex 29:681–697
37. Jin LS, Qian G (2016) Owa generation function and some adjustment methods for OWA operators with application. IEEE Trans Fuzzy Syst 24:168–178

38. Merigó JM, Palacios-Marqués D, Soto-Acosta P (2017) Distance measures, weighted averages, OWA operators and Bonferroni means. Appl Soft Comput 50:356–366
39. Herrera F, Lozano M (2003) Fuzzy adaptive genetic algorithms: design, taxonomy, and future directions. Soft Comput—A Fusion of Foundations, Methodologies and Applications 7:545–562
40. Derrac J, García S, Herrera F (2014) Fuzzy nearest neighbor algorithms: taxonomy, experimental analysis and prospects. Inf Sci 260:98–119
41. Torra V, Godo L (2002) Continuous wowa operators with application to defuzzification. Aggreg Oper 159–176
42. Wang J-W, Chang J-R, Cheng C-H (2005) Flexible fuzzy owa querying method for hemodialysis database. Soft Comput 10:1031–1042
43. Eckert JJ, Santiciolli FM, Yamashita RY, Corrêa FC, Silva LCA, Dedini FG (2019) Fuzzy gear shifting control optimisation to improve vehicle performance, fuel consumption and engine emissions. IET Contr Theory Appl 13:2658–2669
44. Garcia-Trivino P, Fernandez-Ramirez LM, Torreglosa JP, Jurado F (2016) Fuzzy logic control for an electric vehicles fast charging station. In: 2016 international symposium on power electronics, electrical drives, automation and motion (SPEEDAM)
45. Bastian A. Modeling fuel injection control maps using fuzzy logic. In: Proceedings of 1994. IEEE 3rd international fuzzy systems conference
46. Puente J, Gomez A, Parreno J, de Fuente D (2003) Applying a fuzzy logic methodology to waiting list management at a hospital emergency unit: a case study. Int J Healthc Technol Manag 5:432
47. Phuong NH, Kreinovich V (2001) Fuzzy logic and its applications in medicine. Int J Med Inf 62:165–173
48. Imamverdiev YN, Derakshande SA (2011) Fuzzy OWA model for information security risk management. Autom Contr Comput Sci 45:20–28
49. Vigier HP, Scherger V, Terceño A (2017) An application of OWA operators in fuzzy business diagnosis. Appl Soft Comput 54:440–448
50. Bowles JB, Peláez CE (1995) Fuzzy logic prioritization of failures in a system failure mode, effects and criticality analysis. Reliab Eng Syst Saf 50:203–213
51. Spott M, Sommerfeld T, Dorne R (2008) Using fuzzy techniques in business performance management. In: NAFIPS 2008—2008 annual meeting of the North American fuzzy information processing society
52. Gupta C, Jain A, Joshi N (2018) Fuzzy logic in natural language processing—a closer view. Procedia Comput Sci 132:1375–1384
53. Lu K, Liao H, Kazimieras Zavadskas E (2021) An overview of fuzzy techniques in supply chain management: Bibliometrics, methodologies, applications and future directions. Technol Econ Dev Econ 27:402–458
54. Celikbilek C, Erenay B, Suer GA (2015) A fuzzy approach for a supply chain network design problem. In: Annual production and operations management society conference
55. P S, D N S, B P (2014) Temperature control using fuzzy logic. Int J Instr Contr Syst 4:1–10
56. Wakami N, Araki S, Nomura H. Recent applications of fuzzy logic to home appliances. In: Proceedings of IECON '93—19th annual conference of IEEE industrial electronics
57. Hopkins M (1995) Three-input, three-output fuzzy logic print quality controller for an electrophotographic printer
58. Guo J-G, Zhou J (2008) Altitude control system of autonomous airship based on fuzzy logic. In: 2008 2nd international symposium on systems and control in aerospace and astronautics
59. Rawea A, Urooj S (2015) Design of fuzzy logic controller to drive autopilot altitude in landing phase. In: Advances in intelligent systems and computing, pp 111–117
60. Sharma T, Singh V, Sudhakaran S, Verma NK (2019) Fuzzy based pooling in convolutional neural network for image classification. In: 2019 IEEE international conference on fuzzy systems (FUZZ-IEEE)
61. Huang S-F, Cheng C-H (2008) Forecasting the air quality using OWA based time series model. In: 2008 international conference on machine learning and cybernetics

Fuzzy Logic Application for Early Landslide Detection

Nimish Rastogi, Harshita Piyush, Siddhant Singh, Shrishti Jaiswal, and Monika Malik

Abstract A landslide detection System is an Arduino-based system that aims to detect landslides in different areas prone to them. The phenomena of landslide being accelerated through several things, which also include heavy rainfall. Heavy rainfall increases moisture content, which causes movement in the soil and hence inclination changes. With the illustration of landslides through IoT, the concept of fuzzy logic has also been implemented and the vulnerability of landslides is measured. In the hardware model, an output of 'ALERT' is notified on the mobile phone through the Wi-Fi module ESP8266 and also the readings of different sensors are shown on the LCD attached to the model.

Keywords Landslide · Early warning system · Sensor · Fuzzy logic

1 Introduction

You must have heard about the massive destruction caused to human lives and the environment due to the geographical distress in certain areas; majorly hilly areas, plateaus, and mountain ranges. A landslide is the rapid sliding of a large mass of bedrock that occurs due to the reason that when rocks and debris are dislodged, they are pulled by gravity down the slope. To save human lives and property, we have thought of making a detection system that alerts nearby authorities about the danger which can occur in the near future. Our LANDSLIDE EARLY WARNING DETECTION.

SYSTEM warns us about the danger by checking certain parameters like rainfall, soil shifting, and slope of the desired area. We have also used MATLAB to simulate the conditions using FUZZY LOGIC.

N. Rastogi (✉) · H. Piyush · S. Singh · S. Jaiswal · M. Malik
JSS Academy of Technical Education, Noida, Uttar Pradesh, India
e-mail: nimishrastogi72@gmail.com

M. Malik
e-mail: monikamalik@jssaten.ac.in

H. O. Bansal et al. (eds.), *Next Generation Systems and Networks*, Lecture Notes in Networks and Systems 641, https://doi.org/10.1007/978-981-99-0483-9_41

We have used Wi-Fi-based WSN (Wireless Sensor Network) technology in our prototype to communicate alerts/warnings to the desired authorities. WSN technology is very useful as it quickly captures, processes, and responds to rapid changes in data and transmits it to the receiver side.

Since Arduino only returns the value in 0 and 1, but in the case of ambiguity, Arduino fails to provide outcomes, then fuzzy logic comes into play. It gives us values between 0 and 1, and hence it minimizes the chance of false alarms.

1.1 Landslide

The displacement of tectonic plates causes two natural disasters namely landslides and earthquakes. Various types of landslides include earthflow, debris flow, debris, avalanche, etc. The various parameters that affect landslides are excess melting of snow, excess rainfall, earthquake, volcano eruption, and changes in the levels of groundwater.

Some types of landslides are:

- Mudslide is one of the types of landslides which usually occurs due to heavy rainfall, flooding, and ice melting. This muddy flow interrupts anything which comes across its path.
- Glaciers can also have a landslide and there is also not much of a difference from a cliff-top, instead the ice melts and is subject to heavy rainfall which makes it unstable.

2 Related Work

A great amount of study has been already conducted in the area of landslide detection. Through the study, in paper [1], we inferred that the presence of water and internal soil erosion plays a very crucial role in landslides which can be monitored by studying the plasticity index of soil. As in [2], we got an idea of how the prototype should work, and what should be the numerical values of the various parameter ranges, which we used in our study also. From [3], we identified the drawbacks which come when we use a GSM module, so to avoid this we used a Wi-Fi module instead of a GSM module.

3 Methodology

By prototype methodology, developers develop a system through which theoretical solutions are implemented. As in [5], for designing the Early Warning System prototype, the sensors that are being used such as rain sensor, soil shifting sensor, tilt sensor with the microcontroller-based Wi-Fi module from [4], etc., in the model need to follow some basic steps, which are as follows:

- Planning a landslide prototype mechanism with the available sensors.
- To form construction with soil similar to the structure of termite land.
- Creating a hardware model of the Landslide detection system.
- Using the Wi-Fi module in the hardware model.
- The raw data is used to support the decision-making.
- The data is then fed to database and processed using fuzzy logic to support the responses of landslides.
- The data is then processed from database and then it shows the level of threat of landslide.
- Simulating the whole prototype in MATLAB using fuzzy logic design.
- Combining both hardware and software with communication.

As in [6], there are many sensors that we have used in this project, namely rain sensor, accelerometer sensor, Wi-Fi module, soil shifting sensor, LED display, buzzer, etc. In the hardware implementation, we have made a prototype that senses all the external stimuli, and based on that, alerts the authorities about the danger.

The trapezoidal MF is selected for all the variables so that it takes the highest span for better computation efficiency. The trapezoidal MF is defined by four variables a, b, c, and d. Spans b and c define the element's highest membership function. Also, if the element has a value between (a, b) or (c, d), then the membership function will have a membership value between 0 and 1. The trapezoidal function is defined in Eq. (1) (Table 1).

$$\mu(x : a, b, c, d) = \begin{cases} 0 & x \leq a \\ \frac{x-a}{b-a} & a \leq x \leq b \\ 1 & b \leq x \leq c \\ \frac{d-x}{d-c} & c \leq x \leq d \\ 0 & d \leq x \end{cases} \quad (1)$$

3.1 Fuzzy Logic Implementation

As in [7], the four processes on which fuzzy logic works and are used to evaluate FIS values are:

1. **Fuzzification**: First, we put the input values, which represent our parameters: soil shifting, rainfall, and slope, into the FIS system model. Depending on the above three parameters, we get the membership functions, which basically is the point of intersection of parametric values along with the degree of membership functions.

Table 1 Value of linguistic variables

S. No	Variable	Criteria	
1	Soil shifting	0–2	Less
		2–4	Moderate
		4–10	More
2	Rainfall	0–58	Low
		58–68	Adequate
		68–90	High
3	Slope	30°–50°	Low
		51°–60°	Medium
		61°–90°	High

2. **Rule Evaluation**: After completing the fuzzification step, the IF-THEN rules use the membership values to evaluate new fuzzy output sets. The IF–THEN rules of FIS system that we have made, have multiple number of inputs. Now the fuzzy AND operator, which is simply used to choose the minimum of the three membership values, is used to obtain a single number.
3. **Aggregation**: The aggregation is basically the process of combining all the outputs we get by applying all the rules (27 rules in our FIS model). OR fuzzy logic operator is used after combining all our rules. The OR operator is basically used to select the highest of our values after the rule evaluating step to get the new aggregate fuzzy set, which we will use in the next step.
4. **Defuzzification**: The final step is the defuzzification process, in which we obtain our final output value. Mamdani technique is taken into use to calculate the values, and the Centroid defuzzification method is used to obtain the output values which are used to form a cluster. Therefore, the Centre of Area (COA) will be calculated and is then used in this method (Fig. 1)

Rainfall

Rainfall has a major role in landslides. For the rainfall parameter, we have considered three trapezium mathematical functions (mf). According to their ranges, i.e., when rainfall is between 0–58 cm, it is Low, while for 58–68 cm of rainfall, it is Medium, and 68–90 cm of rainfall, it is High (Fig. 2).

Slope

The slope of a landmass is very helpful for the detection of a landslide. For the slope, the trapezium mathematical functions are taken which represents the different range and their probability, i.e., when the slope ranges from 0–50, the probability is Low, while for 51–60, it is Adequate, and for 60–90, the probability is High (Fig. 3).

Soil Shifting

Soil shifting depends not only on the other two parameters, but also plays an important role individually. For the soil shifting, the trapezium mathematical function

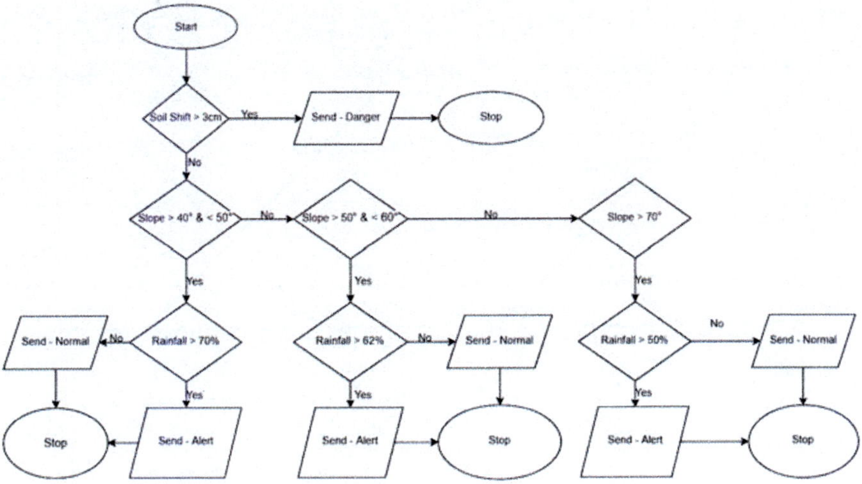

Fig. 1 Flow chart of program

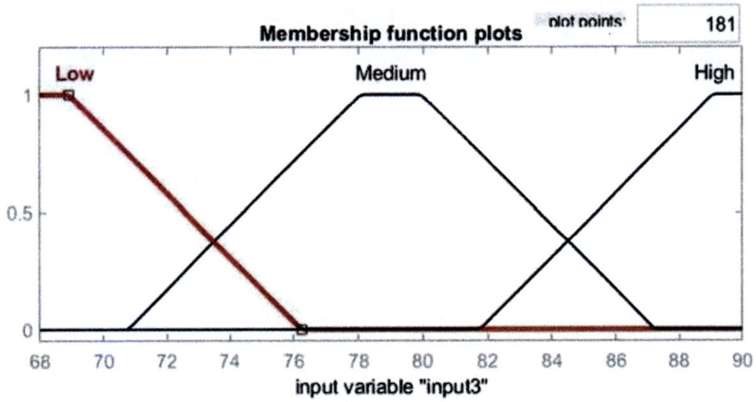

Fig. 2 Rainfall graph

takes three values i.e., Low, Moderate, and More ranging from 0–2, 2–4, and 4–10 respectively.

Depending upon the three defined ranges and their results, many more combinations are formed which help us to get the final probability for landslide as these three parameters depend on each other. The probabilities obtained from the combination of these three inputs from three parameters, results in better detection of landslide.

Guidelines and references for determining the evaluation rules after the fuzzification step are displayed in Table 2. These evaluation rules were developed on the basis of the rules approach and correlation calculation method to find the priorities of the variable. Thus, after the process, 27 evaluation rules were developed.

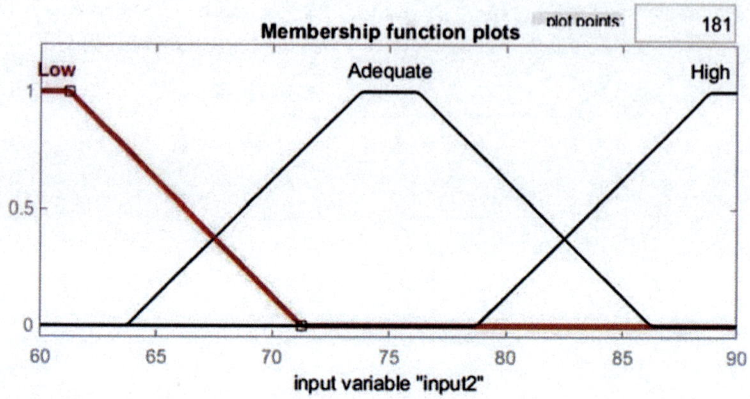

Fig. 3 Slope

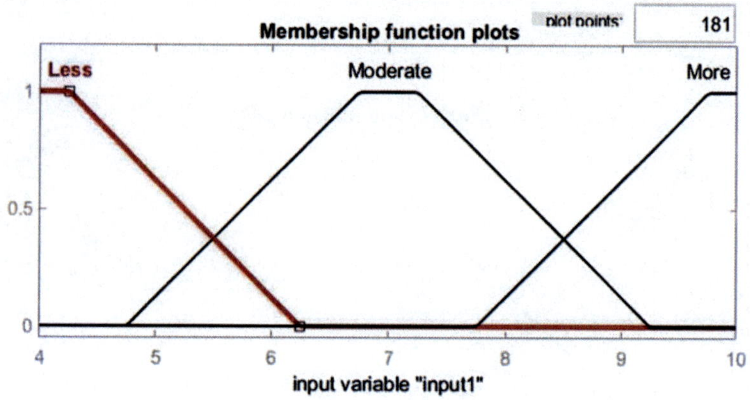

Fig. 4 Slope shifting

Fuzzy memberships can be obtained by the use of defuzzification method. The fuzzy membership level is found from each of the variables—rainfall, slope, and soil shifting. The 3 variables are obtained as the output level. We have used 9 different classifications of output levels, they are very weak, weak, little weak, little medium, medium, high medium, little strong, strong, and very strong (Fig. 5).

4 Results

Through the rule representation graph, we can find the center of the area of the system and through that, we can check the vulnerability of landslides. The result obtained is given a particular linguistic variable. By these, we can also plot the surface view

Table 2 Evaluation rule

S. No	Soil shifting	Slope	Rainfall	Vulnerability of landslide
1	Less	Low	Low	Very Weak
2	Less	Low	Medium	Weak
3	Less	Low	High	Little Weak
4	Less	Adequate	Low	Weak
5	Less	Adequate	Medium	Little Weak
6	Less	Adequate	High	Little Medium
7	Less	High	Low	Weak
8	Less	High	Medium	Little Medium
9	Less	High	High	Medium
10	Moderate	Low	Low	Little Weak
11	Moderate	Low	Medium	Little Medium
12	Moderate	Low	High	High Medium
13	Moderate	Adequate	Low	Little Medium
14	Moderate	Adequate	Medium	Medium
15	Moderate	Adequate	High	High Medium
16	Moderate	High	Low	Medium
17	Moderate	High	Medium	High Medium
18	Moderate	High	High	Strong
19	More	Low	Low	Medium
20	More	Low	Medium	High Medium
21	More	Low	High	Strong
22	More	Adequate	Low	High Medium
23	More	Adequate	Medium	Strong
24	More	Adequate	High	Very Strong
25	More	High	Low	Very Strong
26	More	High	Medium	Very Strong
27	More	High	High	Very Strong

graph, which basically uses two parameters at a time and gives the output. The output graph has more of the red and yellow shade where the conditions meet and at that condition, the landslide is more probable to happen.

5 Conclusion

In this paper, we are concluding that according to the landslide early warning system, landslides can occur when the ground slope is 40°and rain sensor reaches about 70%,

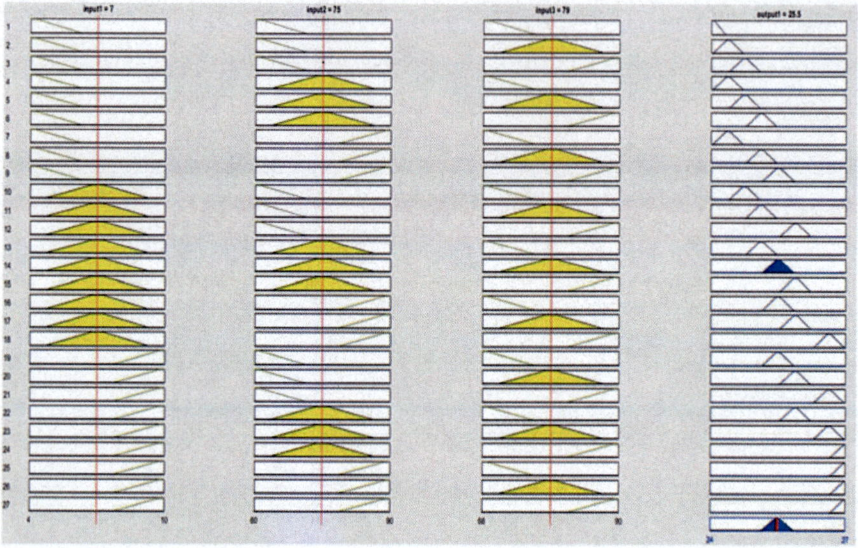

Fig. 5 Rule representation

or when the slope is between 50° and 60° and the rain sensor reaches approximately 68%; meanwhile, if the slope is above 60°, and the rain sensor reaches 58%. Then the prototype also shows that if the soil shift is more than 3, then the area is dangerous. Also, the warning alert message will be sent every minute repeatedly before the landslide actually occurs.

Here we have presented an approach for the formation of clusters in Wireless Sensor Networks that uses Fuzzy Logic to enhance the lifetime of the network. We have also analyzed and evaluated the performance of our protocol through the simulations and have compared its performance (Fig. 6).

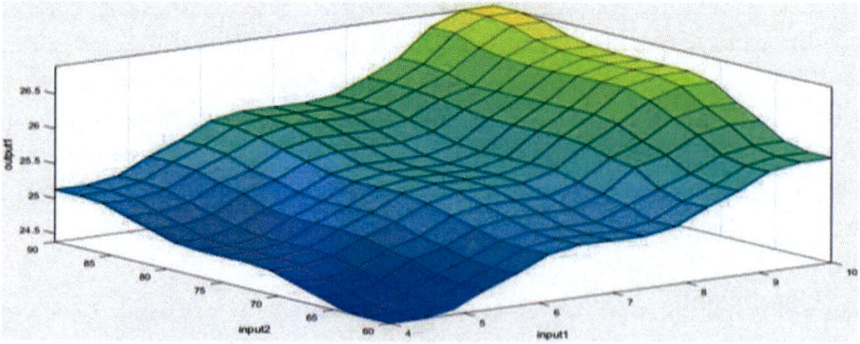

Fig. 6 Surface view

References

1. Abdul Khader S, Anitha K. Landslide warning system
2. Landslides early warning system with GSM modem based on microcontroller using rain, soil shift and accelerometer sensors. Int J Geomate 19(71):137–144
3. Bhoomi S (2017) Landslide early warning system. Int J Sci Res Public 7(7)
4. Wireless sensor network-based land slide detection and early warning system. Zorigmelong: A Tech J 3(1) (2016)
5. Landslide warning system using Arduino. Int Res J Eng Technol (IRJET) 8(4) (2021)
6. Sensor based landslide early warning system—SLEWS development of a geoservice infrastructure as basis for early warning systems for landslides byintegration of real-time sensors
7. Wardhana D, Sofwan A, Setiawan I. Fuzzy logic method design for landslide vulnerability made

An Investigation of Right Angle Isosceles Triangular Microstrip Patch Antenna (RIT-MPA) for Polarization Diversity

Murali Krishna Bonthu, Kankana Mazumdar, and Ashish Kumar Sharma

Abstract An asymmetrical feed-based Right Angle Isosceles Triangular Patch Antenna (RIT-MPA) design is proposed for polarization diversity. Here the RIT-MPA is investigated for linear and circular polarization radiation through reshaping of the two truncated corners of the triangular patch by using two PIN diodes. Here the proposed RIT-MPA design with two PIN diodes exhibits linear polarization in ON configuration and circular polarization in OFF configuration for the resonant frequency of 1.43 GHz with an adequate gain of 0.72 dB and 0.84 dB, respectively. The antenna proposed here shows potential use for S-band-based satellite communication.

Keywords Microstrip patch · Polarization · Reconfigurable antenna · PIN diode

1 Introduction

Recent wireless communication systems such as 5G, satellite and radar applications require modern RF front-end systems with multiple functionalities with less space and structure complexity. Therefore, Reconfigurable Antennas (RAs) are developed to enable diverse communication capabilities for various wireless systems requirements according to adaptively changing surrounding conditions. These RA can modify their geometrical structures to change their resonant frequency, radiation, and polarization characteristics [1, 2]. Microstrip Patch Antennas (MPAs) are usually designed for linear polarization. However, circularly polarized antennas gained much attention recently due to their limited transmission losses and reduced multipath reflections compared to linearly polarized antennas for relative movement between transmitting and receiving antennas [3, 4].

The circular polarization in microstrip patch antennas is widely accomplished through two orthogonally feed-based transmission lines, which ultimately increases

M. K. Bonthu · K. Mazumdar · A. K. Sharma (✉)
Department of Electronics and Telecommunications, Veer SurendraSai University Technolgy, Burla, Odisha 768018, India
e-mail: aksharma_etc@vssut.ac.in

H. O. Bansal et al. (eds.), *Next Generation Systems and Networks*, Lecture Notes in Networks and Systems 641, https://doi.org/10.1007/978-981-99-0483-9_42

the undesired radiations through feed point and excitation of higher order modes [5]. Therefore, the dual-feed-based MPAs affect their axial ratio performance. Hence, the single feed-based circularly polarized MPAs have gotten much attention in recent years. In the single feed technique, the required orthogonal modes with 90° phase differences and equal amplitude are achieved through truncated patch corners [6]. The corner truncating technique with a single feed is widely used to achieve circular polarization in real-time applications. Sung et al. [7] designed the equilateral triangular microstrip patch for circular polarization through the single feed-based technique.

The circular polarization was mainly accomplished through three triangular slots in the ground plane, which exhibit various polarization characteristics based on truncated corners. This type of design introduces higher mode multiband resonance with limited polarization diversity [8]. Hence, in this investigation, a single feed-based Right angle isosceles triangular patch antenna (RIT-MPA) has been designed for polarization diversity by using only two truncated corners with two PIN diodes with less space and structure complexity.

2 Design and Configuration of RIT-MPA

The proposed design of single feed RIT-MPA is shown in Fig. 1. Side length (D) of RIT-MPA is 50 mm and imprinted on a substrate (FR4 epoxy) of height (h = 3.2 mm) with relative permittivity ε_r of 4.4. Here the single microstrip line-based feeding technique is used by break-jointure where the break is indicated by G. Further, a transformer is used as a corresponding device between the feed and transmission line. To get good impedance matching, we can adjust the size of the feed length (Lg) and width (Wg). A standard MPA is basically represented by linear polarization. To get a circularly polarized antenna, diverging factors needs to be uneven. Designs that present uneven diverging factors typically use truncated corner and stubs-based perturbations. Here the uniformity of the proposed RIT-MPA shape is perturbed through corner cut, which makes it feasible to excite degraded orthogonal modes for circular polarization. Here three corners of the right angle isosceles triangle with a side length of l were truncated to produce the perturbation for circular polarization operation. Here the resonant frequency of the proposed RIT-MPA is obtained through the cavity model for perfect magnetic walls [9] as per the given equation.

$$f_r = \frac{c}{2D\sqrt{\varepsilon_r}}(\sqrt{m^2 + n^2}) \tag{1}$$

where D is the side length of RIT-MPA, m and n are the integer for dominant mode (TM10), ε_r is the relative permittivity of the substrate, and c is the light velocity. In the proposed design of RIT-MPA, maximum RF power is coupled at point P at the right diagonal face, whereas a small amount of RF power is coupled at coupling point Q. Therefore, the circular polarization of the RIT-MPA design can be obtained through the radiation along both axes PP1 and QQ1 on coupling points x and M,

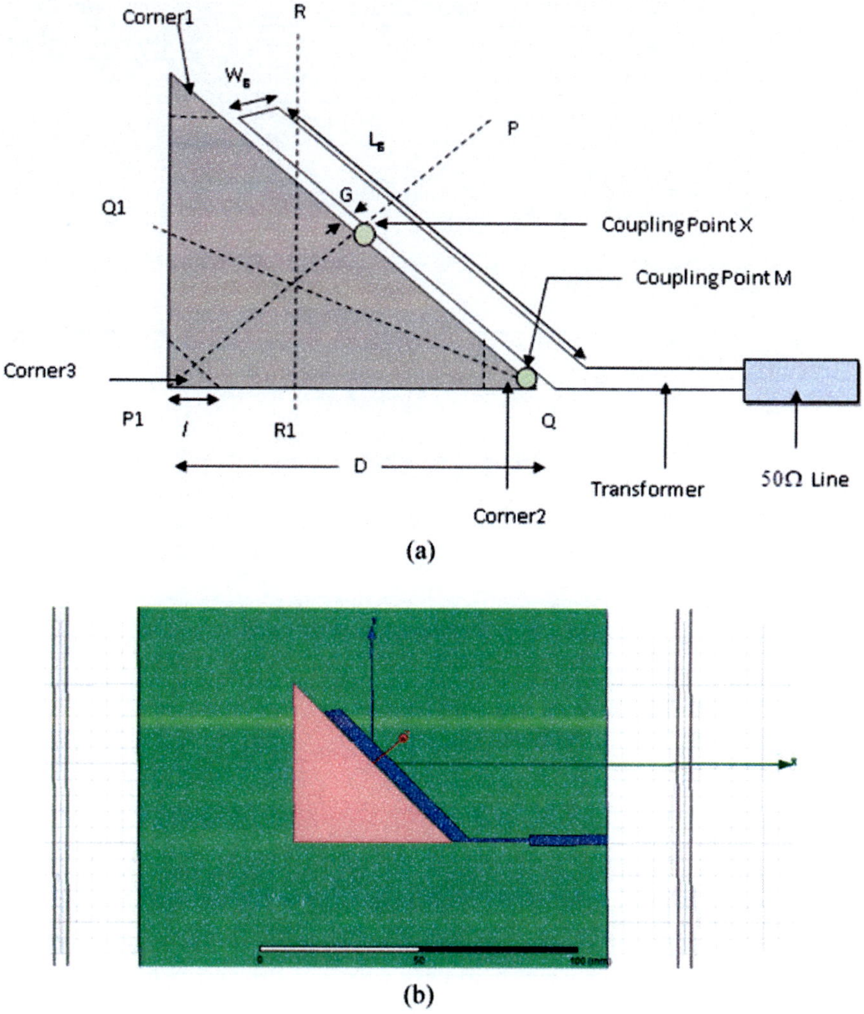

Fig. 1 **a** A schematic view, **b** Ansys HFSS model of Right Angle Isosceles Triangular Patch Antenna (RIT-MPA)

respectively. As axis RR1 does not exhibit a coupling point, circular polarization is not achieved even though the two other faces alongside this axis are perturbed. The simulation analysis of the RIT-MPA is done through the Ansys HFSS for the frequency range of 1 GHz to 2.5 GHz. Here it is seen that the conventional (without corner cut) RIT-MPA resonates at 1.40 GHz frequency as shown in Fig. 2. Here it is seen that this conventional RIT-MPA shows an axial ratio of >> 2 dB and gain of 0.41 dB at the resonant frequency of 1.40 GHz with linearly polarized radiation as shown in Figs. 3, 4, and 5, respectively (Table 1).

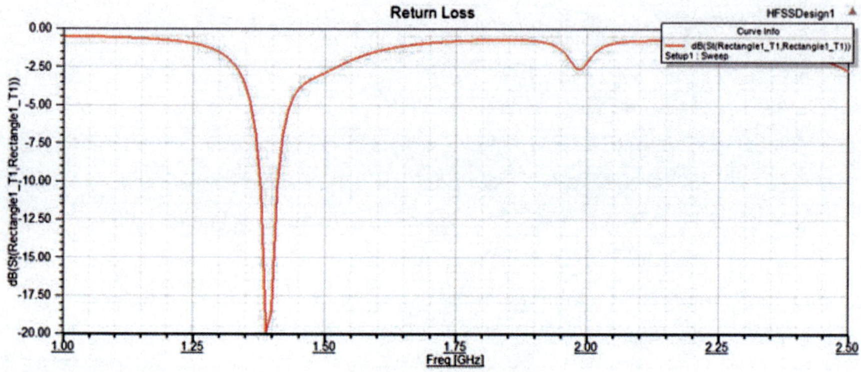

Fig. 2 Simulated return loss plot for conventional RIT-MPA

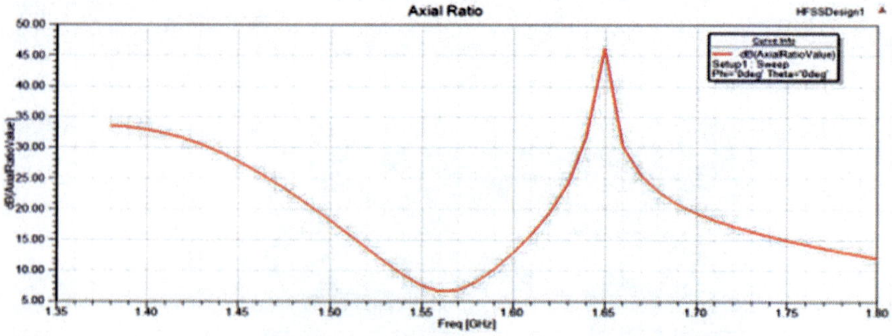

Fig. 3 Simulated axial ratio plot for conventional RIT-MPA

Table 2 shows the performance of different antenna configurations. Here it is observed that conventional RIT-MPA (without a corner cut) shows linear polarization at the resonant frequency of 1.40 GHz with 0.41 dB gain. When all three or some of the corners of the patch are removed, the polarization diversity (LP to CP) is exhibited by the patch. For all four antenna configurations with truncated corners, the gain is improved with a minimal resonant frequency shift. Hence, by using the truncated corners, a diverse polarization can be achieved without changing the antenna resonance, radiation pattern, and size [10, 11].

Here, various combinations of all three corner cuts have been analyzed through the Ansys HFSS simulation in order to analyze the polarization diversity without changing the resonant frequency as shown in Table 2. Here terms 'O' depict the available corners and 'X' depicts the corner cuts for corresponding corners in the RIT-MPA. Here it is observed that antenna configuration 4 shows circular polarization due to the absence of corners 2 and 3. Here it is observed that coupling point P shows the maximum power transmission to the patch, whereas corner cuts 2 and 3 show the

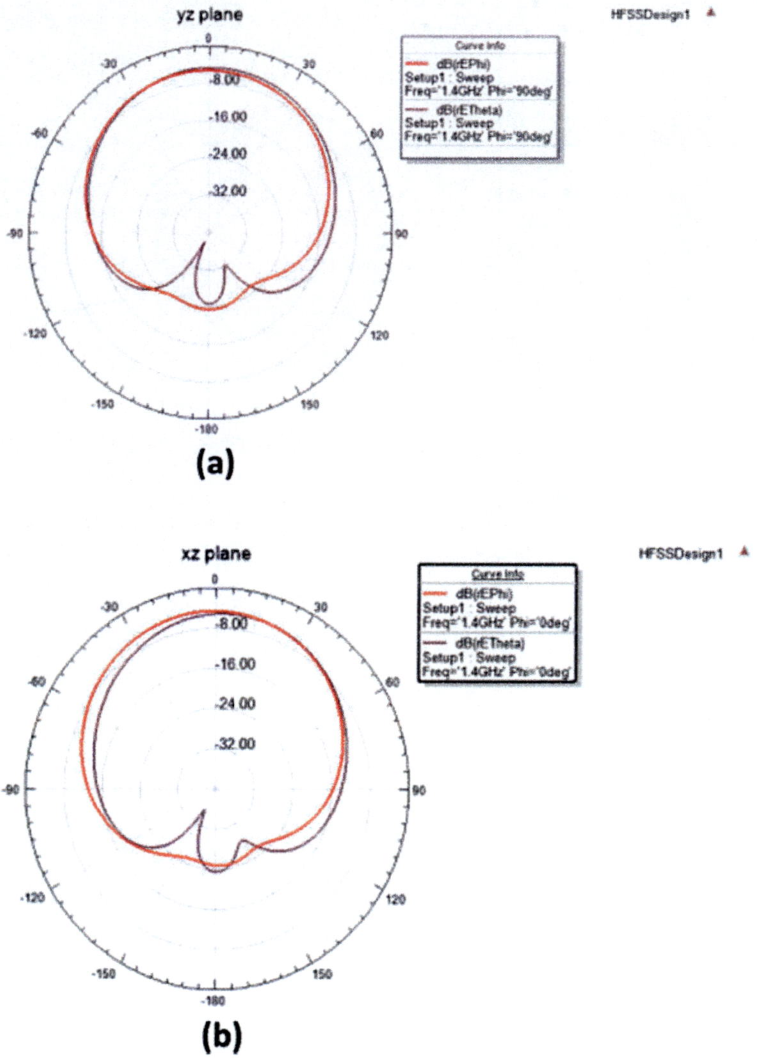

Fig. 4 Simulated y–z and x–z plane patterns for conventional RIT-MPA

perturbed field distribution on axis PP1 and QQ1, which ultimately exhibit circular polarization [11].

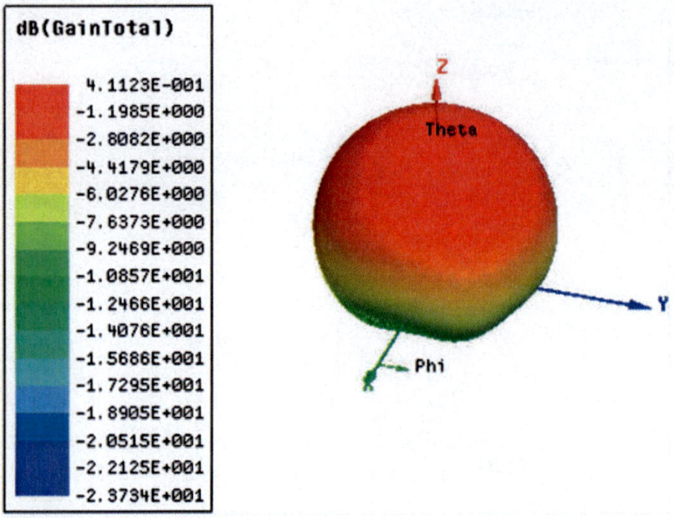

Fig. 5 Simulated 3D gain plot for conventional RIT-MPA

Table 1 Right angle isosceles triangular patch Antenna (RIT-MPA) design parameters

Parameter name	Value
Patch length (D)	50 mm
Substrate length and width	100 mm
Substrate height (h)	3.2 mm
Substrate dielectric constant (ε_r)	4.4
Feed width (W_g)	6 mm
Feed length (L_g)	39.7 mm
Corner cut (l)	4 mm
Transformer length and width	29.5 mm, 1 mm
Break-jointure (G)	0.5 mm

3 Configuration of RIT-MPA with PIN Diode for Polarization Diversity

This section represents a new configuration of RIT-MPA with two PIN diodes for polarization diversity. The proposed RIT-MPA with truncated corner patches produces diverse polarization characteristics without changing the structure of the feed as shown in Fig. 6. All the dimensions of this configuration are the same as Fig. 1. Here the two microsemi MPP4203 PIN diodes are placed into the gap G1 and G2 of 1 mm for corners 3 and 2, respectively. The equivalent electrical parametric values of the PIN diode for OFF and ON configuration are given in Table 3. These PIN diodes are simulated as passive RLC elements using Ansys HFSS [3].

Table 2 The performance of different antenna configurations of RIT-MPA

Configuration	Corner1	Corner2	Corner3	Geometry	Pol	f_r (GHz)	AR (dB)	Gain (dB)
Antenna without corner cut	O	O	O	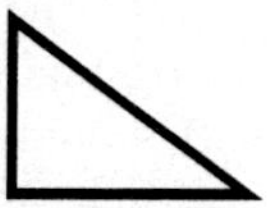	LP	1.40		0.72
C 1	X	X	O	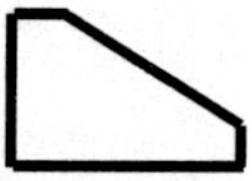	LP	1.44		0.71
C 2	O	O	X	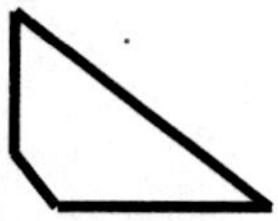	LP	1.39		0.4
C 3	X	O	O	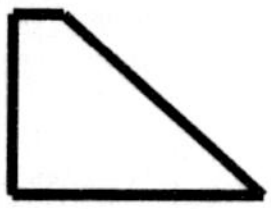	LP	1.42		0.52
C 4	O	X	X	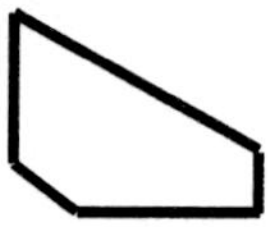	CP	1.43	2	0.84

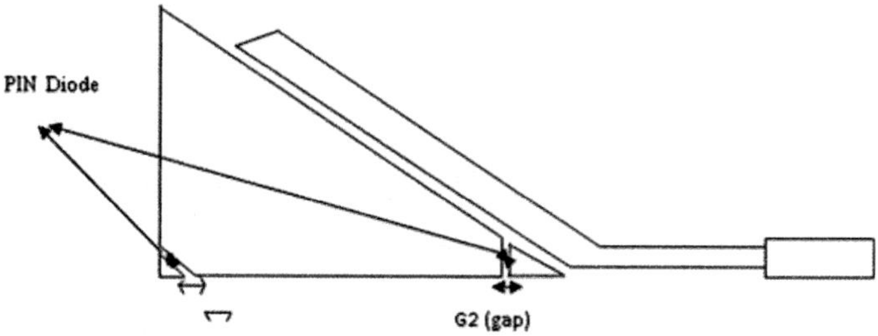

Fig. 6 Proposed design of RIT-MPA with PIN diodes for polarization diversity

Table 3 The performance of RIT-MPA for different switching configurations

PIN diode configuration	Parameters			
	R_S (Ω)	R_P (kΩ)	L_S (nH)	C_T (pF)
ON	3.5	–	0.45	–
OFF	–	3.5	0.45	0.08

4 Results and Analysis

Here it is noticed that the polarization diversity can be obtained by switching the PIN diodes in ON and OFF configuration. The proposed RIT-MPA with two PIN diodes shows the linear to circular polarization diversity for the S-band with a resonant frequency of 1.43 GHz. Figure 7 shows the return loss graphs for both switching configurations. Here it is observed that the proposed RIT-MPA show an adequate bandwidth of 45 MHz at the resonant frequency of 1.43 GHz for both switching configuration. Further, the axial ratios for both configurations are shown in Fig. 8. During the first switching configuration when both pin diodes are OFF, it exhibits the axial ratio of 2 dB for circular polarization, while in ON configuration, the proposed RIT-MPA shows the linear polarization. Further, it is also observed that narrowband RIT-MPA offers antenna miniaturization and pre-filtering, which mainly reduces the interference signal at the receiver front end by avoiding the undesired signal at the same frequency. Figures 9, 10, 11 and 12 show the radiation pattern for both switching configurations for both x–z and y–z planes along with the 3D gain plots. Here it is observed that during ON and OFF configuration, the proposed design shows a gain of 0.72 dB and 0.84 dB, respectively. The proposed RIT-MPA show the improved gain for linear polarization from 0.41 dB to 0.72 dB compared to the conventional RIT-MPA due to the perturbation at both corner when both PIN diodes are in ON configurations. All the given return loss, axial ratio, and their corresponding radiation pattern plots are adequate for S-band-based satellite communication [8].

5 Conclusion

In this study, the design of single feed-based RIT-MPA with truncated corners is investigated for linear to circular polarization diversity. Here two corners of the triangular patch are truncated to achieve diverse polarization without changing the resonant frequency and size of the antenna. Further, the polarization diversity is obtained with truncated RIT-MPA by using two PIN diodes. Here the proposed PIN diode-based RIT-MPA design exhibits linear polarization in ON configuration and circular polarization in OFF configuration at a resonant frequency of 1.43 GHz. These RIT-MPA designs show an adequate gain of 0.72 dB and 0.84 dB for linear and circular polarization, respectively. Furthermore, the proposed RIT-MPA antenna shows the pre-filtering at the receiver front end through the single resonance for all switching configurations due to their narrow bandwidth. Therefore, the proposed RIT-MPA show the potential application in S-band-based satellite communication application.

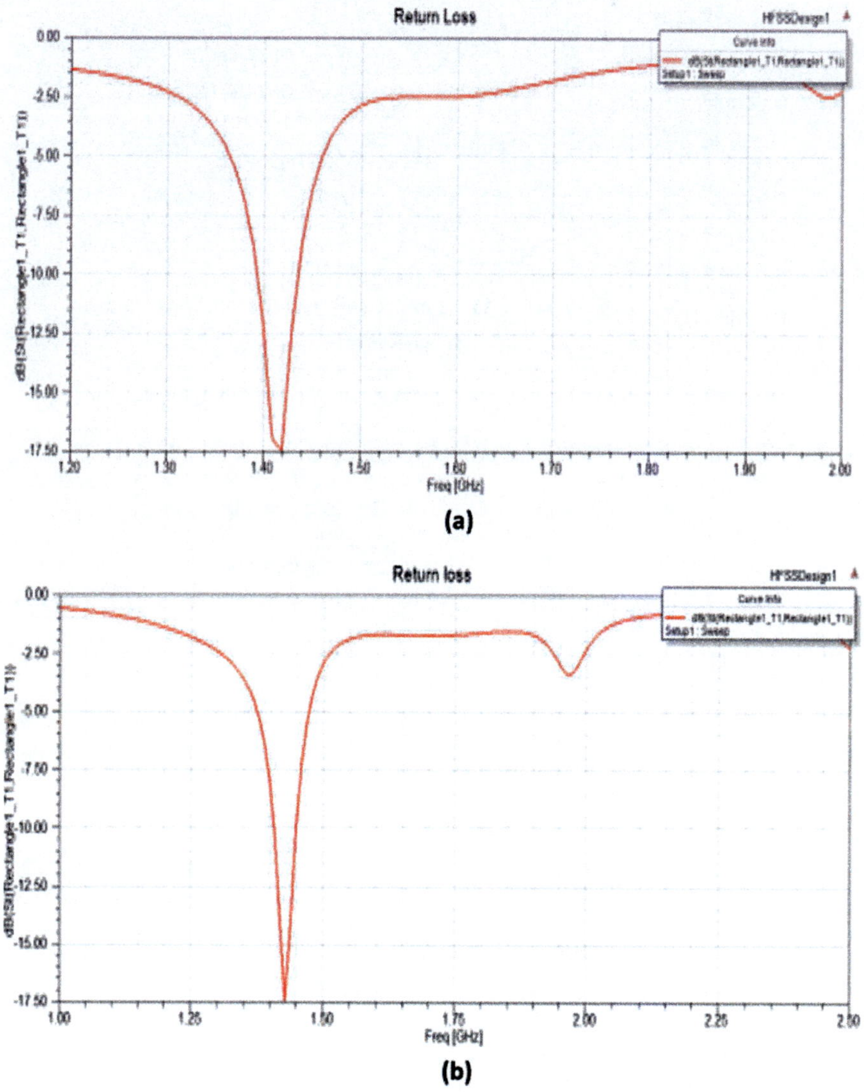

Fig. 7 Simulated return loss graph when PIN diodes in (**a**) OFF and (**b**) ON configuration

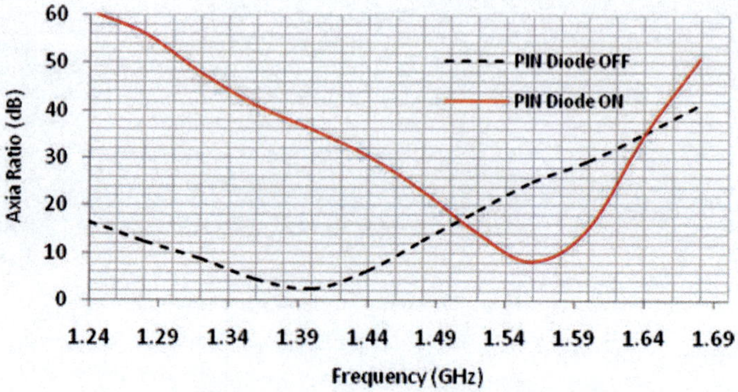

Fig. 8 Simulated axial ratio graph when PIN diodes are in OFF and ON configuration

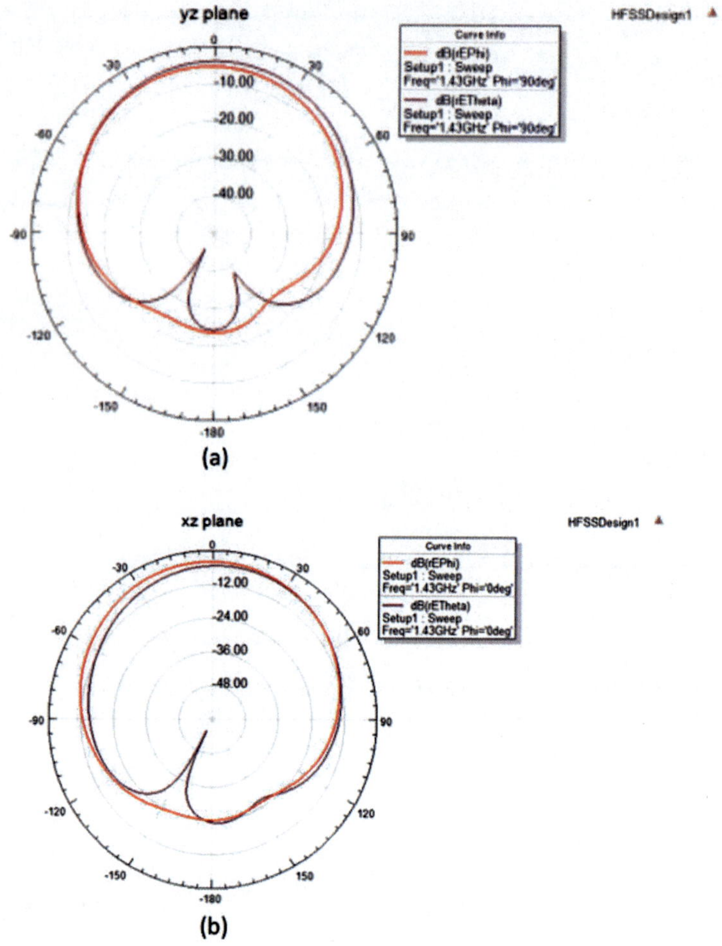

Fig. 9 Simulated (**a**) y–z and (**b**) x–z plane plot when PIN diode are in OFF configuration

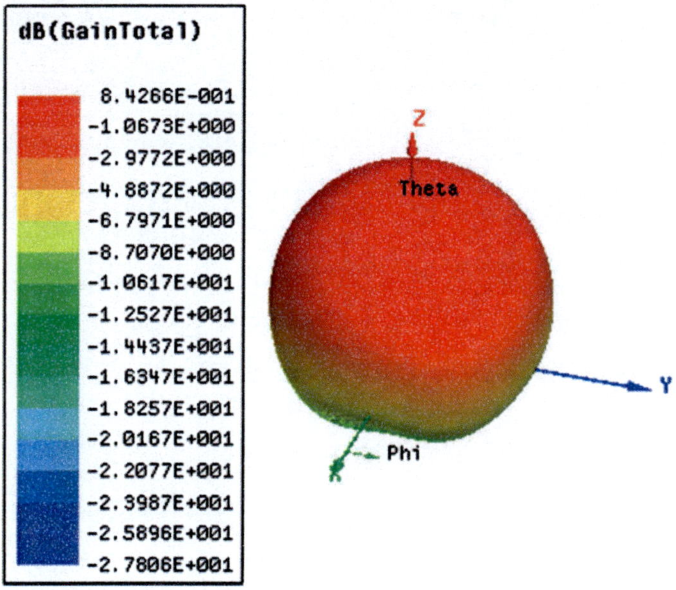

Fig. 10 Simulated 3D gain plot when PIN diodes are in OFF configuration

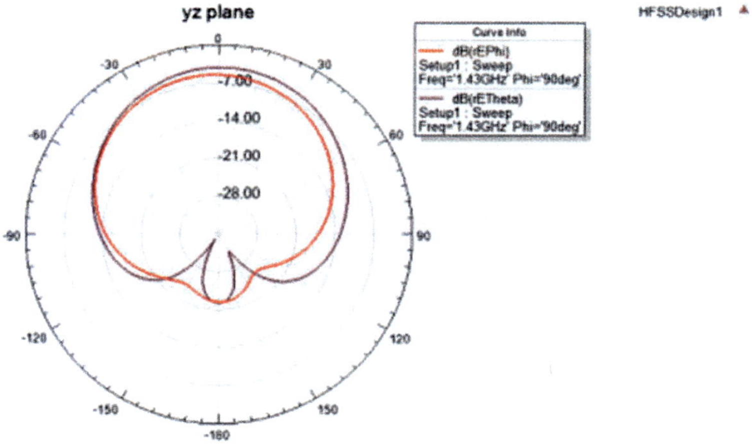

(a)

Fig. 11 Simulated (a) y–z and (b) x–z plane plot when PIN diodes are ON configuration

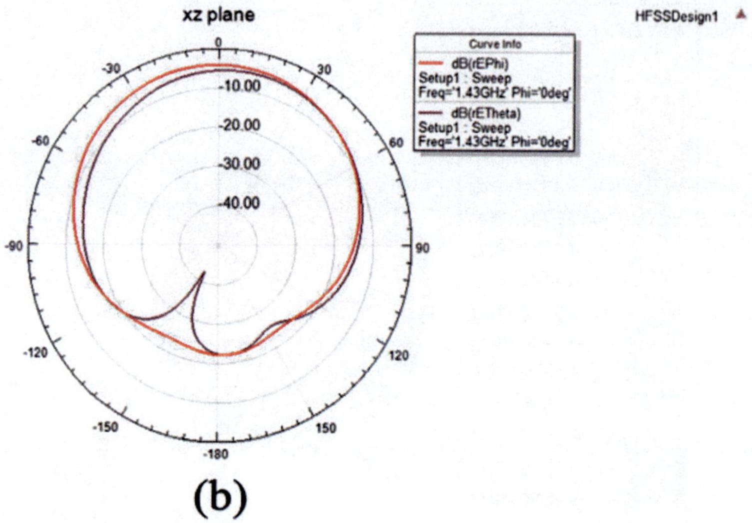

Fig. 11 (continued)

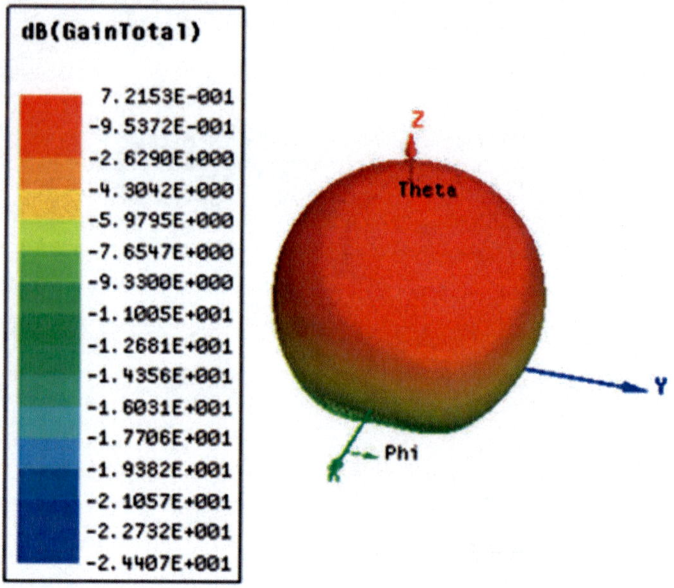

Fig. 12 Simulated 3D gain plot when PIN diodes are in ON configuration

Acknowledgments The authors would like to thank Microwave Lab at the Department of ETC Engineering, VSSUT Burla, for Ansys HFSS simulations work.

References

1. Johnson RC, Jasik H (1984) Antenna engineering handbook. McGraw-Hill Book Company, New York
2. Bhartia RP, Bahl I, Ittipiboon A (2000) Microstrip antenna design handbook (Artech House Antennas and Propagation Library) Artech House, London
3. Sahu NK, Sharma AK (2017) An investigation of pattern and frequency reconfigurable microstrip slot antenna using PIN diodes. In: Progress in electromagnetics research symposium-spring (PIERS). IEEE, St. Petersburg Russia, pp 971–976
4. Sharma AK, Gupta N (2015) Pattern reconfigurable antenna using non-uniform serpentine flexure based RF-MEMS switches. In: Progress in electromagnetics research symposium-spring (PIERS), PIERS Proceedings, Prague, Czech Republic, pp 2840–2843
5. Wong KL, Pan SC (1997) Compact triangular microstrip antenna. Electron Lett 6(33):433–434
6. Helszajn J, James DS (1978) Planar triangular resonators with magnetic walls. IEEE Trans Microw Theory Tech 2(26):95–100
7. Sung Y (2009) Investigation into the polarization of asymmetrical-feed triangular microstrip antennas and its application to reconfigurable antennas. IEEE Trans Antennas Propag 58(4):1039–1046
8. Tang CL, Lu JH, Wong KL (1998) Circularly polarised equilateral-triangular microstrip antenna with truncated tip. Electron Lett 34(13):1277–1277
9. Khraisat YS, Olaimat MM (2012) Comparison between rectangular and triangular patch antennas array. In 2012 19th international conference on telecommunications (ICT). IEEE, April, pp 1–5
10. Mazumdar K, Sharma AK (2019) A study of 30°-30°-120° triangular microstrip patch miniaturization using shorting pin. In: 2019 international conference on wireless communications signal processing and networking (WiSPNET). IEEE, March, pp 10–12
11. Sung YJ, Jang TU, Kim Y-S (2004) Reconfigurable microstrip patch antenna for switchable polarization. IEEE Microw Wireless Compon Lett 14(11):534–536

Author Index

H. O. Bansal et al. (eds.), *Next Generation Systems and Networks*, Lecture Notes in Networks and Systems 641, https://doi.org/10.1007/978-981-99-0483-9